Testing and Diagnosis of VLSI and ULSI

NATO ASI Series

Advanced Science Institutes Series

A Series presenting the results of activities sponsored by the NATO Science Committee, which aims at the dissemination of advanced scientific and technological knowledge, with a view to strengthening links between scientific communities.

The Series is published by an international board of publishers in conjunction with the NATO Scientific Affairs Division

A Life Sciences
B Physics

Plenum Publishing Corporation
London and New York

C Mathematical
   and Physical Sciences
D Behavioural and Social Sciences
E Applied Sciences

Kluwer Academic Publishers
Dordrecht, Boston and London

F Computer and Systems Sciences
G Ecological Sciences
H Cell Biology

Springer-Verlag
Berlin, Heidelberg, New York, London,
Paris and Tokyo

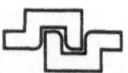

Testing and Diagnosis of VLSI and ULSI

edited by

Fabrizio Lombardi

Department of Electrical and Computer Engineering,
University of Colorado at Boulder, Boulder, Colorado, U.S.A.
Department of Computer Science, Texas A&M University,
College Station, Texas, U.S.A.

and

Mariagiovanna Sami

Dipartimento di Elettronica,
Politecnico di Milano, Milano, Italy

Kluwer Academic Publishers

Dordrecht / Boston / London

Published in cooperation with NATO Scientific Affairs Division

Proceedings of the NATO Advanced Study Institute on
Testing and Diagnosis of VLSI and ULSI
Como, Italy
June 22– July 3, 1987

Library of Congress Cataloging in Publication Data

NATO Advanced Study Institute on "Testing and Diagnosis of VLSI and
    ULSI" (1987 : Como, Italy)
    Testing and diagnosis of VLSI and ULSI / edited by Fabrizio
Lombardi and Mariagiovanna Sami.
        p.   cm. -- (NATO ASI series. Series E, Applied sciences ; no.
151)
    "Published in cooperation with NATO Scientific Affairs Division."
    "Proceedings of the NATO Advanced Study Institute on "Testing and
Diagnosis of VLSI and ULSI," Como, Italy, June 22-July 3, 1987."
    Includes index.
    ISBN-13: 978-94-010-7134-5
    1. Integrated circuits--Very large scale integration--Testing-
-Congresses.   I. Lombardi, Fabrizio, 1955-   . II. Sami,
Mariagiovanna.   III. North Atlantic Treaty Organization.   Scientific
Affairs Division.   IV. Title. V. Series.
TK7874.N345 1987
621.381'73--dc19
                                                    88-26623
                                                    CIP

ISBN-13: 978-94-010-7134-5   e-ISBN-13: 978-94-009-1417-9
DOI: 10.1007/978-94-009-1417-9

Published by Kluwer Academic Publishers,
P.O. Box 17, 3300 AA Dordrecht, The Netherlands.

Kluwer Academic Publishers incorporates the publishing programmes of
D. Reidel, Martinus Nijhoff, Dr W. Junk, and MTP Press.

Sold and distributed in the U.S.A. and Canada
by Kluwer Academic Publishers,
101 Philip Drive, Norwell, MA 02061, U.S.A.

In all other countries, sold and distributed
by Kluwer Academic Publishers Group,
P.O. Box 322, 3300 AH Dordrecht, The Netherlands.

TABLE OF CONTENTS

PREFACE

This volume contains a collection of papers presented at the NATO Advanced Study Institute on "Testing and Diagnosis of VLSI and ULSI" held at Villa Olmo, Como (Italy) June 22 - July 3, 1987.

High Density technologies such as Very-Large Scale Integration (VLSI), Wafer Scale Integration (WSI) and the not-so-far promises of Ultra-Large Scale Integration (ULSI), have exasperated the problems associated with the testing and diagnosis of these devices and systems. Traditional techniques are fast becoming obsolete due to unique requirements such as limited controllability and observability, increasing execution complexity for test vector generation and high cost of fault simulation, to mention just a few.

New approaches are imperative to achieve the highly sought goal of the "three months" turn around cycle time for a state-of- the-art computer chip. The importance of testing and diagnostic processes is of primary importance if costs must be kept at acceptable levels.

The objective of this NATO-ASI was to present, analyze and discuss the various facets of testing and diagnosis with respect to both theory and practice.

The contents of this volume reflect the diversity of approaches currently available to reduce test and diagnosis time. These approaches are described in a concise, yet clear way by renowned experts of the field. Their contributions are aimed at a wide readership: the uninitiated researcher will find the tutorial chapters very rewarding. The expert will be introduced to advanced techniques in a very comprehensive manner.

An integral part of the NATO-ASI was also to provide an insight into future research directions in the proposed field. This has been incorporated in this volume by a series of articles which describe in more detail particular research topics as innovative extensions of more established techniques.

The topics addressed in this volume aim at different levels of application: from an initial characterization such as fault models at a physical level, the reader is exposed to a diverse spectrum of techniques ranging from probabilistic to spectral methods, from gate level to system level analysis. The overall organization reflects a structured engineering approach which bridges many disciplines ranging from basic electrical engineering, to computer science and mathematics.

Technological issues such as CMOS, electron-beam methods, memory manufacturing and array testing and reconfiguration, are also addressed to provide the reader with a fully comprehensive understanding of the problems involved with special devices. These features make this volume an attractive reference book for a variety of graduate courses including VLSI, testing and fault tolerant computing.

We hope that the reader will enjoy this volume as much as we enjoyed organizing the NATO-ASI. Our last thank you is to all the lecturers and the participants who have made the Advanced Study Institute such a remarkable success.

The articles in this volume have been judged and accepted on their scientific quality, and language corrections may at times have been sacrificed in order to allow quick dissemination of knowledge to prevail.

Fabrizio Lombardi
Mariagiovanna Sami
Directors and Editors

TRENDS IN DESIGN FOR TESTABILITY

T. W. Williams

IBM Corporation
General Technology Division
Boulder, CO 80302
USA

INTRODUCTION

From the onset of the first designs of logic networks, testing has
been an important issue. However, testing has increased in importance over the past ten years, since the cost per logic gate as a
percentage, has been growing with increasing density. This paper
will discuss the different Design for Testability techniques which
are popular today. The Ad Hoc approaches, consist of In-Circuit
Testing, along with Functional Testing, supplemented by Signature
Analysis. In the Structural approaches, the Scan Design techniques
will be discussed. In the area of Self-Test, the techniques known
as BILBO and Exhaustive Testing will be discussed. Finally, the
trends of these techniques will be discussed.

FAULT MODELS

In 1959, a technique was proposed by Eldred in a paper entitled
<u>Test</u> <u>Routines</u> based upon symbolic logic statements, which essen-
tially was the inception of the Stuck-At-Fault model.

Basically, the model is one where the inputs and the outputs of the
logic gate are assumed fixed to a value, either Logic 0 or Logic 1,
irrespective of what value is applied or should be coming out of
the network. For example, the AND gate shown in Figure 1 with a
Stuck-At-Fault on the A input implies that that A input is per-
ceived by the AND gate as a 1, irrespective of what value was
applied to its input. Furthermore, if the output of the gate was
stuck, or fixed, to a given value, its output would be that value,
irrespective of what value was applied to the inputs of the gate.

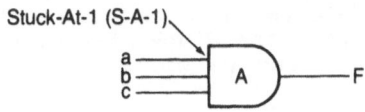

Figure 1 Example of Stuck-At-Fault.

Figure 2a shows a good machine with the input pattern of 011. The
output for the good machine and the AND of those three values is 0.
With the faulty machine shown in Figure 2b, the first input being
Stuck-At-1 with the pattern 011, the A input is thought of as a 1
by the AND gate, and therefore, the three variables are AND'ed
together A, B and C and the output is equal to 1 for the faulty

1

F. Lombardi and M. Sami (eds.), Testing and Diagnosis of VLSI and ULSI, 1–31.
© 1988 by Kluwer Academic Publishers.

machine. This would be a test pattern for the Stuck-At-1 fault on the A input.

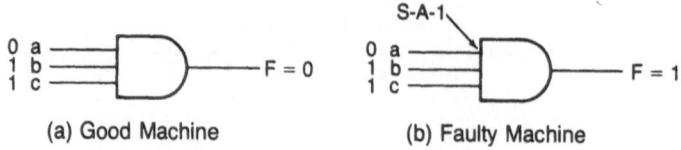

(a) Good Machine   (b) Faulty Machine

Figure 2 Test for a Stuck-At-Fault.

The basic attribute of a test is that you apply a pattern to the network, and you get a different response between the good machine and the faulty machine.

The single Stuck-At-fault assumption comes into play here, in that only one fault is assumed at a time. This is to keep the test generation problem as simple as possible, and historically, it has worked well. The job of test generation is one of putting together input patterns to a combinational logic network so that each Stuck-At-Fault taken one at a time is tested.

The job for the fault simulator, once the test generator has delivered a set of patterns, is to determine which faults are tested by this set of test patterns. The fault simulator assumes one fault at a time in general and applies the patterns, determines whether that fault is tested or not, and puts out a report of the number of tested faults versus the number of faults that are untested by that given set of test patterns. If a fault is untested by a given set of test patterns, it does not mean it's untestable; it just means that this set of test patterns was not able to detect it.

Is there some way to determine, or at least give a good feeling, of what kind of test coverage is required in order to ship a product of a certain quality level? An approach is described here which gives a derivation of an equation that relates yield, Stuck-At-Fault coverage and defect level. [23] Stuck-At-Fault coverage has been used as a figure of merit for many years for determining how good a set of test patterns are. However, the Stuck-At-Fault model does not take into account shorts, sequential faults or multiple defects in a network. As an approximation, though, let's assume that the Stuck-At-Faults are, in fact, independent of one another, and we will ignore for the moment these other fault models.

Let P_n be the probability of a Stuck-At-Fault occurring when n, such faults are on a chip. The probability of having a good chip, then, is the yield, Y, and it can be calculated to be:

$$Y = (1-P_n)^n \tag{1}$$

Let A represent the case in which no Stuck-At-Faults are on the chip. Let B represent the case in which we know for certain that no Stuck-At-Faults are in m of the sites tested. The results of the test patterns, testing m out of n of the total faults with no defects observed, is then

$$P(B) = (1-P_n)^m \tag{2}$$

To determine the probability of A (a good chip), given that the test of m of the faults has been performed successfully, we use the conditional probability

$$P(A|B) = \frac{P(A \cap B)}{P(B)} \qquad (3)$$

The probability of A B is the probability of no Stuck-At-Faults and no faults in the m sites tested by this set of test patterns. That is

$$P(A\ B) = (1-P_n)^n \qquad (4)$$

The probability that a defective chip is shipped is one minus the probability that the chips sent are good. If DL is the defect level, then the defect level is equal to
DL = 1 - P(A|B)

$$= 1 - \frac{P(A \cap B)}{P(B)}$$

$$= 1 - \frac{(1-P_n)^n}{(1-P_n)m}$$

$$= 1 - (1-P_n)^{n-m} \qquad (5)$$

Substituting the value of Y from Equation 1 into 5 gives the following results:

$$DL = 1-Y^{(n-m)/n} = 1-Y^{1-(m/n)} \qquad (6)$$

If T represents the fault coverage for a given test that tests only m of the n Stuck-At-Faults, then the defect level is

$$DL = 1-Y^{(1-T)} \qquad (7)$$

This equation is shown for different yields in Figure 3. Let's take a short look at this figure. If the yield is .25 and we do absolutely no testing at all, we will ship, by definition of yield, 25% good product and 75% bad product. We'll save the cost of a tester, but we'll be shipping a lot of bad chips.

If we want to improve the defect level, drop it down to a very low level, what is obviously needed from the curve are very, very high fault coverages in the upper 90's if we want fault coverages down in the low 1% or less. The curve that has been derived is basically one that makes many assumptions, however, it does track fairly well with actual data. This section has given a relationship between the fault model, yield and defect level, which is intrinsic to the basic testing problem, i.e., how much testing is enough? The difficulties of trying to generate a test for complex networks have also been alluded to.

It should be pointed out that there is another fault model which is not modelable by a single Stuck-At-Fault in MOS integrated circuits. This is known as the CMOS Sequential fault or CMOS Open fault. In essence, when this fault occurs, it does not act like

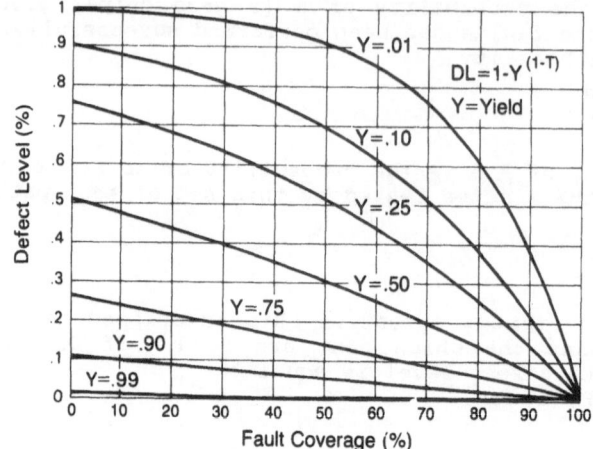

Figure 3 Defect Level as a Function of Yield and Fault Coverage.

any of the Stuck-At-Faults. It requires a sequence of patterns in order to test it, even if this fault is in a simple AND block implementation. Today, it is controversial as to whether one should go through and generate tests for this fault directly, or whether one should let the Stuck-At-Fault coverage try to do as good a job as possible, or design the devices such that the fault does not exist. That is, the network will fail in a way which will cause a regular combinational Stuck-At-Fault to occur as well and therefore, be tested by the Stuck-At-Fault test.

AD HOC TECHNIQUES:

The Ad-Hoc techniques assume that the designer at the board level does not have the ability to affect the chip designs which will be used on that board, and probably, in the case of microprocessors, won't even know the actual gate level implementation. This implies that generating Stuck-At-Fault tests and trying to evaluate them is totally out of the question.

With In-Circuit Testing, a probe is used on the underside of the board to make contact with all the inputs and outputs, the primary inputs and primary outputs of a module soldered on the other side of the board. Other modules will be connected to these modules, and this particular module will drive others. With the In-Circuit approach, current is either forced or drawn from a given node to obtain a value. This can be a very power intensive situation, however, it is done for a very short period of time on the order of milliseconds or less, if possible. The idea is to apply a short burst of patterns through the given module which will test to make sure that there are no shorts between the pins; that there is some identification of the proper function on the module on the other side of the board; and that the inputs and the outputs are neither Stuck-At-1 or Stuck-At-0.

The complete test of the modules usually cannot be performed by virtue of the fact that it would take too long, greater than some milliseconds, which would cause the junction temperatures of devices to overheat and therefore, affect their reliability. This technique has become very popular. However, with the advent of

surface mounted components, the ability to probe is becoming more
and more difficult. This also makes probing very difficult on
boards which have components on both sides. The problem is further
exacerbated by the reliability of the probe contact onto the board,
all causing difficulties at the test environment. However, if one
can obtain controllability and observability of the module I/O's,
then the testing function for a subset of the patterns can be
achieved.

If the In-Current test of a module on a board can be achieved
without overdriving, then the complete module test can be
performed. This is becoming a more popular approach since there
are many technical problems associated with the overdriving technique. Once the In-Circuit test is completed usually a functional
test is performed on the board as close to machine speeds as
possible. These functional tests can be run without having to know
the gate level description of the networks used (i.e. microprocessors, controllers, etc.). It is helpful in functional testing to
observe some nodes internal to the board (i.e. not driven to
primary outputs). To help compact the test data, a technique known
as Signature Analysis is employed.

In essence, Signature Analysis is the compaction of functional test
data in a Linear Feedback Shift Register [21]. An example of one
is shown in Figure 4.

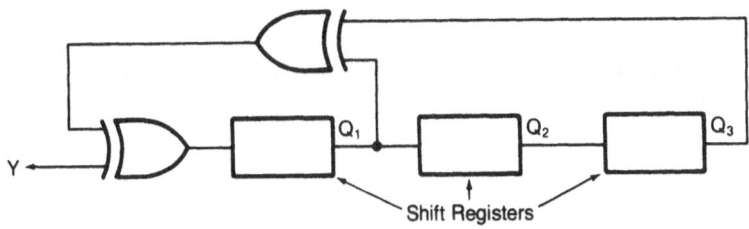

Figure 4 Signature Analysis.

The three block shift registers are basically straightforward shift
registers, shifting the values of Q1, Q2 from left to right, except
Q3 is exclusive OR'ed with Q1, and that, in turn, is exclusive
OR'ed with the output Y from some other source and stored in the Q1
position. We'll discuss the other source in just a moment.

This Linear Feedback Shift Register, LFSR, is called linear because
the feedback elements are all exclusive OR's, and it's a shift
register, hence its name. If Y is held to a 0, the shift register
will, in this case, go through a count of 7. You can see that if
the 000 value was stored into the Linear Feedback Shift Register,
it would stay at 000 all the time. If any other count is obtained,
for example, 001, and the Linear Feedback Shift Register would
sequence through seven patterns, the last of which would be 010,
and then return to 001. This Linear Feedback Shift Register is
maximal in length since it counts 2^n-1, where n is the number of
shift registers, in this case 3; therefore, the count is a count of
7. If the values coming in from Y are synchronized with the
shifting of the shift register, they don't change when samples are
made on the input side of the shift register, a change in the
pattern of the count will occur.

In essence, this compaction technique takes many input values from Y and in this case, compacts them into three bits, the Signature. The problem associated with this technique is that there may be some sequences of patterns other than the one we're looking for, i.e., the good machine response Y_1 that will also end up at the exact same value as the good machine value. If this happens, we now have an aliasing problem. It is felt, although it has not been conclusively proven, that the aliasing problem, the probability of making mistakes, is approximately $1/2^k$ where k is the number of shift register positions. Thus, if 16 bits of shift register were used, then it would be $1/2^{16}$. The aliasing problem will be covered later in this chapter.

In order to do Signature Analysis on a board, it is required that the board be initializable, that is, one should be able to set every latch element to a given value; that global loops on the board level be broken, since the diagnostic resolution of this testing method would be hampered. For instance, if you checked an output of a module which was bad (checked all its inputs), then found an input that was bad, traced this back to another module, etc. The object being to find a module with all good inputs and a bad output, i.e., the defect source. If you had a global loop, you'd wind up at the first point from which you started. This mean that every module in that loop would then be suspect bad. If the global loop was broken, what would then happen is that a module would be identified with a bad output, all of whose inputs are good.

In order to obtain these Signatures, an initialization pattern of the board plus the initialization of the Signature Analysis tool (which is not part of the actual board under test itself), a fixed number of sequences would have to be gone through with a golden ("perfect") board for those nodes of interest. Certainly, on a microprocessor board, data buses and address buses are prime candidates to begin collecting signatures.

To clarify what this signature is, it is the last value left in the Linear Feedback Shift Register after a predetermined number of samples taken of Y with a fix initialization of board and the Signature Analysis register. This predetermined number always has to be the same for a given test. The identification of a bad sequence of patterns versus a good sequence of patterns is one that would leave a different signature in the Signature Analysis register from the good machine signature.

The use of In-Circuit tests followed by functional tests is very common especially where logic networks, such as µ processors, where the logic network is not known. The use of In-Circuit testing of modules without overdriving is becoming more popular since it elevates many of the complications associated with overdriving. Signature Analysis continues to be used at board level functional testing.

STRUCTURED DESIGN TECHNIQUES:

With the increasing complexity of networks in the late '60's, a need for a solution to the testing problem was necessary. It was well-known that controllability and observability were the keys to assist testing. Furthermore, it was known that if one could control and observe the latch variables within a sequential network, then the test generation problem could be reduced to one of just testing the combinational logic. It was also apparent that as densities increased, the sensitivity to races became an issue with respect to the test patterns, i.e., if the test pattern had a race in it, then there was a chance that sometimes it would work on a tester and sometimes it wouldn't; or if the product was all one-way, all fast, for example, it would never work on a tester, resulting in what is called a zero yield situation, which would be very difficult to analyze.

As a result, a number of approaches came forth, some publicly, some privately, and the end result was that a wealth of designs fall into the category of structured design approaches, which became known in the mid to late '70's. These techniques became known as "Scan" techniques since they employed shift registers to scan data into and out of the system latches [1].

Let's begin by taking a look at Figure 5, the basic structure of the Huffman model. This model is one of a sequential machine that separates the memory elements from the combinational logic. The combinational logic is fed by primary inputs and the memory elements are on the output side of the combinational logic network. If we can control and observe the memory elements via some straightforward mechanism, then test patterns for the combination logic network can come directly from a test generation program that generates tests for combinational logic rather than the more difficult case of sequential logic. This will reduce the complexity of the task and make very difficult problems into solvable problems.

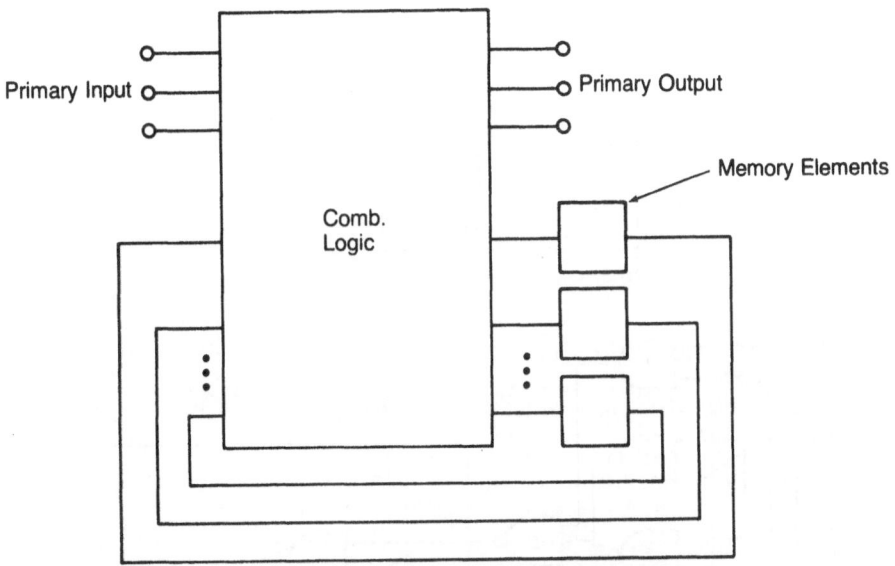

Figure 5 Basic Structure of the Huffman Model.

One approach for obtaining the controllability and observability of the memory elements is to cut all the feedback loops, observe these on the tester, then reconnect them in a higher packaging level. However, I/O's are the most critical thing in LSI logic, and therefore, this would not have been a reasonable solution. Furthermore, this is not a reasonable solution for VLSI since the same constraint holds.

Another approach would be to treat the memory elements as an array and address them directly. This would give controllability and observability to the memory elements and solve the testing problem. One of the techniques Random Access Scan does will in fact use this concept to obtain the controllability and observability.

An approach which is somewhat different than treating the memory elements as an array is to treat the memory elements as a shift register. Thus, being able to shift values into the memory elements satisfies the controllability constraint, and the ability to shift out satisfies the observability constraint required for the memory element.

A number of companies have employed this technique. NEC, Nippon Electronic Company, Ltd. uses this technique called Scan Test; Sperry Corporation (now UNISYS) uses one called Scan Set; and IBM uses one called LSSD, just to mention a few [6-11]. In order to do test generation for a large network such as this one, it would be prudent to couple not only these design techniques with a design automation system which would automatically supply the test patterns for the network in order to make the overall process an efficient one.

Figure 6 shows the symbolic representation of a shift register latch, (SRL). Figure 6a is the symbolic representation; Figure 6b is the implementation in AND invert gates.

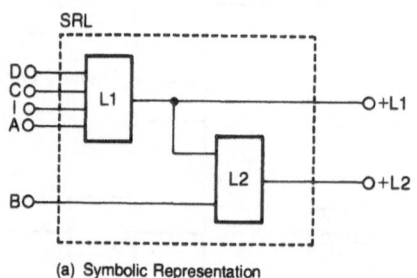

(a) Symbolic Representation

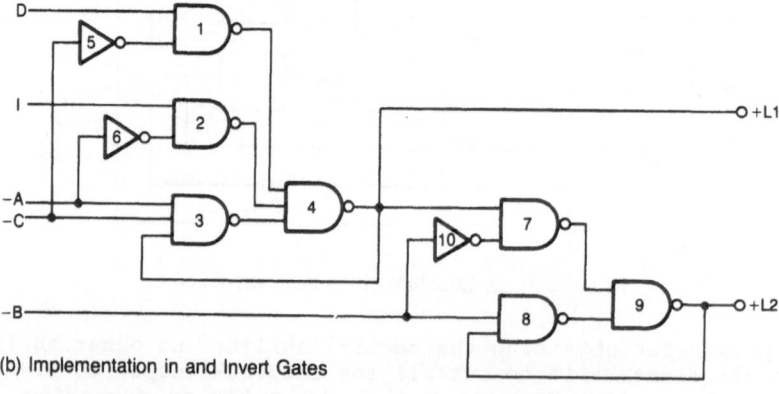

(b) Implementation in and Invert Gates

Figure 6 Polarity — Hold SRL Shift Register Latch.

On the lefthand side are the inputs D, C, I, A and B to the symbolic representation. D is the data input to the polarity hold latch; C is the system clock; I is the scan input, in other words the input for the shift register; A is the A clock; B is the clock for the L2 latch, the second latch.

Looking at Figure 6a, you can see that there are two blocks inside the symbolic representation of an SRL, an L1 latch and an L2 latch. Every SRL contains two latches, this will become clearer.

The most important design structure is the double-latch design which is shown in Figure 7. Primary inputs, P, come into the combinational network plus the previous state, Y, to make up the new values, X_1 through X_n, which are the inputs to the SRL's. The clocks are C_1, which is the system clock for all the L1 latches; the A shift clock, which essentially latches data into the L1 latches on the heavy, dotted-line input; the C_2 or B clock, which is the same clock input or same net, and it essentially clocks the L2 latch. Therefore, whatever is in L1 gets loaded into L2, and then that output is ready to be fed back around to the combinational network. Note, that in every loop there are two latches clocked by different clocks. The clocks, C_1 and C_2, are fairly close together; when C_1 goes "off" and C_2 goes "on"; that is because there is no logic between the L1 latch and the L2 latch. As long as they are not overlapping, one can have a race-free design. If the clocks are overlapped when they arrive at L1 and L2, there is then the exposure of having a race which could be critical, or winding up in the wrong state in the latches.

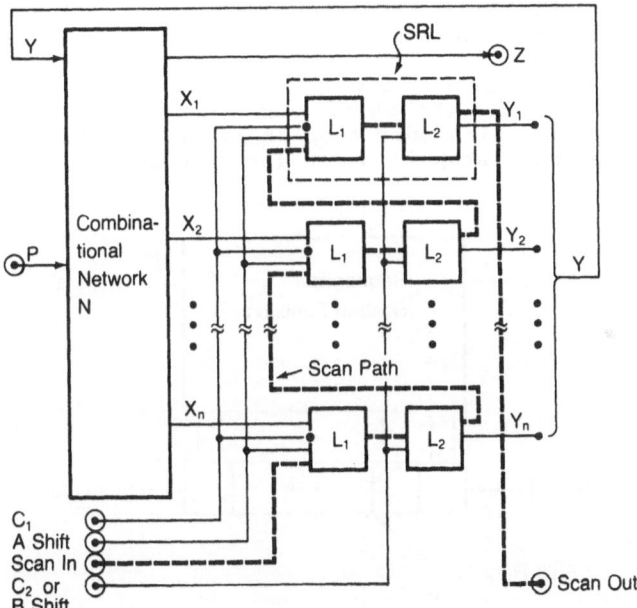

Figure 7 LSSD Double-latch Design.

One thing to note about the double-latch design is that both the L1 and L2 latches participate in the system function. Thus the basic testing structure for this network is as follows. The values necessary to be fed to the combinational network, Y, are loaded into the system latches by the scan mechanism using the A, B clocks with the operate valve on the Scan input. Next the Primary inputs, P, are applied their required valves. Next the primary outputs are measured, Z, then the "C" clock is turned on and off to measure the outputs of the combinational network; This is followed by a "B" clock. Now the shift register is ready to be shifted out such that the values stored in the latches can be seen at the Scan Out primary output. This constitutes the text process for one pattern for the combinational network. This process is repeated over and over until the test is complete for the combinational logic; Complete is measured against the Stuck-At-Fault coverage.

LSSD is just one of many techniques that constitutes the Scan Path technique. Nippon Electronics Corporation, Ltd. has another approach called Scan Path. This technique is very similar to LSSD in that they both use the shift register function to establish and observe the state of the latches. However, with the Scan Path approach, a single clock drives the latches so that this clock is a master/slave clock. Furthermore, there is gating employed on a Scan Path basis so that each scan path can be addressed, and only one scan path is scanned out at any one time.

On the other hand, UNISYS Computers uses the Scan Path approach where needed, and also has latches outside of the system function to be able to control and observe nets within the system function. This technique is called Scan Set, see Figure 8. However, with these latches external to the system function, observations can be made without disturbing system functions. That is, these latches can be scanned while the system is doing productive work. With this function, the shift register can sample points within the sequential system function and then scan them out without ever disturbing the operation of the system function. This type of activity cannot be performed with the Scan Path technique or the LSSD technique. Since the shift registers are part of the system latches, system function must stop in order for shifting to occur. System function must stop in order for shipping to occur with both LSSD and Scan Path.

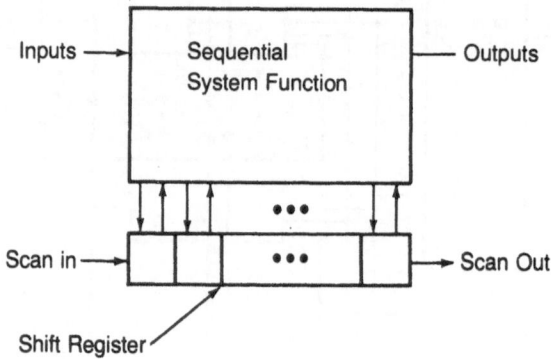

Figure 8 Scan Set Funciton.

A somewhat different technique with the same objectives, that is, controllability and observability of the system latches, is the Random Access Scan approach used by Fujitsu. This approach treats each one of the latch elements as a bit in memory. Each bit in the memory has its own unique address, and it has a port which can load data into the latch so that the contents of the latch can be observed. The general structure is shown in Figure 9a, and the latch implementation is shown in Figure 9b. For the whole RAM, there is only one scan-in and one scan-out.

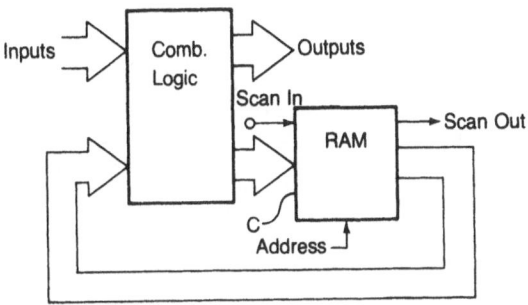

(a) General Structure

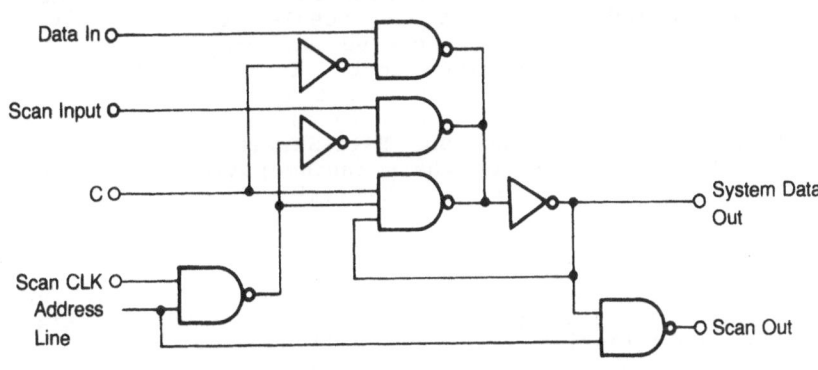

(b) Latch Structure

Figure 9 Random Access Scan Structure.

With normal system operation, the scan clock is "off", which is 0, and the system clock, when it goes "on" and "off", will load the data into the polarity hold latch and will make it available for system use, i.e., it´s just a normal latch.

However, when one wants to preload this latch to a given value, the system clock is turned "off"; the address for this latch is raised; the address line for this latch is raised; and the scan clock is turned "on" and "off" with the appropriate data on the scan input. This scan input feeds all the latches in the array. However, since only one latch is activated at a time, only one latch receives the scan clock and that value is loaded into the latch.

Similarly, if one wanted to observe a particular latch, the address line for that particular latch would be activated by an appropriate address mechanism which is part of the structure, and the result is the contents of the latch is observable on the scan out, and only this latch will control the single scan out pin. Thus, if the objective is to test the combinational logic, then pre-setting all the latches to a given value is straightforward; turning the scan clock "off", hitting the system clock, capturing tests in the latches and going through the address lines and observing the contents of the latches, which may have changed, will then give results of the tests.

All of these approaches of Structured Design techniques have become more and more popular in vertically integrated companies. They are very slow at best in coming. The Application Specific Integrated Circuits, ASIC, which are done at gate array foundries, do not have, in general, good support for scan designs. For example, the shift register books are far too large, and there is no coupled automated support for test generation. Scan has been employed in partial ways by some µ processor designers to help test portions of the network which traditionally have had problems. It will take very large chips ≥ 50K gates which cannot successfully be tested with functional patterns before Scan techniques become more popular.

SELF-TEST:

Self-Testing is the ability of a network to apply patterns to its own subnetwork and to compact the results so that they can be observed after the test is completed and compared with the good machine response for that given test. This could be a situation where many, many functional patterns are applied to the network and the results of the tests are stored in one of the registers within the network. The Microprocessor Self-Stimulated Test is of this type. Another type would be where random patterns are applied to the network, in particular, to the combinational portion of the network, and the results of each test pattern compressed. This is another form of Self-Testing.

In this section, we will look at two different forms of Self-Testing: one Insitu and one Exsitu Self-Testing structures. The Insitu Self-Testing structure will use system registers to generate and compact test data; The Exsitu structure will use registers external to the system function to generate and compact test data. Figure 10 shows the relationship of these Self-Test structures.

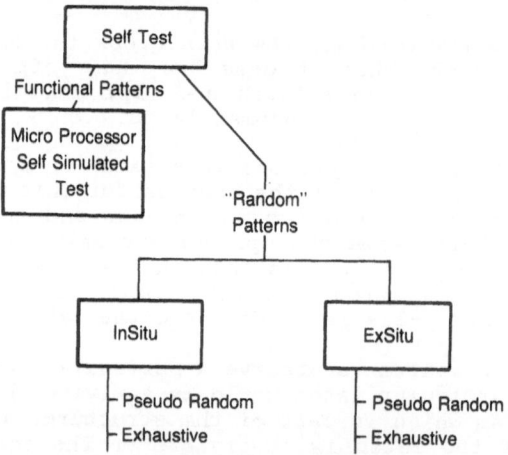

Figure 10 Self Test Relationships.

Under both the Insitu and Exsitu are two categories: one is pseudorandom, which is a collection of random patterns (pseudo because there is no pure source of random patterns); and the second is exhaustive, a situation where all 2^n patterns will be applied to a network with n inputs. There will be some refinement on the size of n, and that will be discussed in a later section. Also in this section, we will discuss Linear Feedback Shift Registers and how they work along with aliasing analysis.

With a board containing a microprocessor and other support logic, functional tests can be written to test the microprocessor and to test other chips on the board, for example; memories, interface chips, controller chips, etc. Of course, if all of this gets integrated into a single chip, then that chip can also use this same kind of test sequence.

The sequence consists of applying functional patterns to different portions of the network and accumulating the results so that at the end of a test, some registers can be observed in the microprocessor and can be compared to the good machine response. These functional patterns are usually executed at machine speed and obviously are helpful in performing AC testing, that is, testing at speeds close to the speeds in which the product will be used. It is usually sufficient to supply power to the board and clock signals, with a minimum of tester board interaction.

Since logic models for the processors and other chips are probably not known, then the only way to obtain a good machine output is from a high-level model, which is sometimes difficult to write accurately; or it can be done right at the tester, this is called learn mode, where the tester looks at the results of a "golden" board. This is the response to be expected for good boards.

This approach to testing has been around for a number of years and is a way of life, particularly, when one does not have logic models for the devices on the boards. Unfortunately, it creates an interactive environment. Since you don't know the actual logic model of the devices, one cannot be certain that the test is an adequate test, then the test may have to be improved. Therefore, the test has occurred at subsequent testing levels, for example, system or field use. That is why it is alluded to as an "interactive approach." As mentioned previously, this form of Self-Testing was in use long before the other forms of tests that we will now go into.

Before the other Self-Testing techniques can be discussed, a quick review of the attributes of Linear Feedback Shift Registers will be discussed. The structure in Figure 11 is an example of a Signature Analysis Register [21]. The network is supplied with a pattern. The output of the network, in this situation, is a single output which goes into the first exclusive OR. The output of this exclusive OR then feeds a shift register of length k. The second and the last bits of the shift register are exclusive OR'ed and are fed back around to the input side of the shift register.

In general, a shift register is initialized to a given value, and patterns are applied to the network. Let us assume, for a moment, that the network is combinational so that one does not have to deal with the initialization of the network.

Patterns are applied to the network synchronously with the shifting of the shift register. This means that the first pattern is

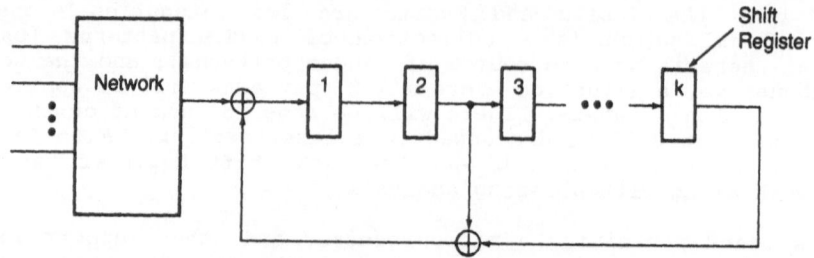

Figure 11 Signature Analysis Register.

applied to the network after a certain time has passed such that the results are available in the output of the network and available at the input of the first shift register, so that the shift register shifts one bit to the right. As a result, a new value is in the shift register, and it is ready to accept its next value from the output of the network. This is obtained then by having the second pattern applied to the network, and so on.

This kind of network has been proposed to be an integral part of a sequential logic design such that it can be used to both generate and compact results of a test. A bound on aliasing in the Signature Analysis environment, and also, some results of aliasing from a program which models the Markov process associated with the Signature Analysis Register will be shown.

Figure 12 shows the general form of a k bit Linear Feedback Shift Register.

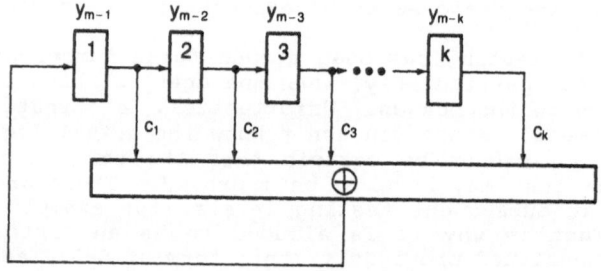

Figure 12 General form of a Linear Feedback Shift Register.

It is linear because all the elements in the Feedback Shift Register are exclusive OR's which are linear elements over the field GF2.

The Y_m bit of the LSFR can be represented as follows:

$$y_m = \sum_{i=1}^{k} c_i \cdot y_{m-i} \qquad (8)$$

The characteristic polynomial for the LSFR is shown below.

$$1 + \sum_{i=1}^{k} c_i \cdot x^i \qquad (9)$$

The last element of the LSFR is fed back, and that makes it a kth-order polynomial, which can be seen from the characteristic polynomial. The derivation of this polynomial can be found in [1, 20].

The operation of this Linear Feedback Shift Register can be described as well in the matrix notation, as shown below. This would be a vector y_{m-1} to y_{m-k} as the present state of the shift register times the matrix, the first row of which are the feedback tabs, c_1 through c_k. The shift register function which could be a diagonal row of ones. The product of the matrix and the state vector will then be the next state of the LSFR. This can be represented in a simple form as shown below.

$$
\begin{bmatrix}
y_m \\
y_{m-1} \\
y_{m-2} \\
\vdots \\
y_{m-k+1}
\end{bmatrix}
=
\begin{bmatrix}
c_1 & c_2 & c_3 & \cdots & c_k \\
1 & 0 & 0 & & 0 \\
0 & 1 & 0 & & 0 \\
\vdots & & \vdots & & \vdots \\
0 & 0 & 0 & \cdots 1 & 0
\end{bmatrix}
\begin{bmatrix}
y_{m-1} \\
y_{m-2} \\
y_{m-3} \\
\vdots \\
y_{m-k}
\end{bmatrix}
$$

$$\underline{Y}_m = \underline{C} \cdot \underline{Y}_{m-1} \tag{10}$$

If the initial state vector for the shift register is $\underline{Y}_{-1}$, then the recursion formula can be used to find the life value of the LSFR, which is shown in the following equations:

$$
\begin{aligned}
\underline{Y}_0 &= \underline{C} \cdot \underline{Y}_{-1} \\
\underline{Y}_1 &= \underline{C} \cdot \underline{Y}_0 = \underline{C}^2 \cdot \underline{Y}_{-1}
\end{aligned}
\tag{11}
$$

in general then

$$\underline{Y}_i = \underline{C}^{i+1} \cdot \underline{Y}_{-1}$$

Figure 13 shows a simple Signature Analysis Register. If we select D_i to equal the incoming bits in the Signature Analysis register, then we can write a recursion formula again to show the operation of the LSFR. The general equation is shown below for this operation.

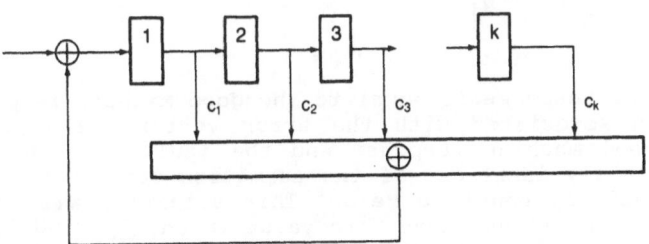

Figure 13 Simple Signature Analysis Register.

$$
\underline{D}_i =
\begin{bmatrix}
d_i \\
0 \\
0 \\
0
\end{bmatrix}
\qquad
\underline{Y}_0 = \underline{C} \cdot \underline{Y}_{-1} + \underline{D}_0
$$

$$\underline{Y}_1 = \underline{C} \cdot \underline{Y}_0 + \underline{D}_1 = \underline{C}^2 \cdot \underline{Y}_{-1} + \underline{C} \cdot \underline{D}_0 + \underline{D}_1$$

$$\underline{Y}_2 = \underline{C} \cdot \underline{Y}_1 + \underline{D}_2 = \underline{C}^3 \cdot \underline{Y}_{-1} + \underline{C}^2 \cdot \underline{D}_0 + \underline{C} \cdot \underline{D}_1 + \underline{D}_2$$

in general then

$$\underline{Y}_i = \underline{C}^{i+1} \cdot \underline{Y}_{-1} + \sum_{j=0}^{i} \underline{C}^{i-j} \cdot \underline{D}_n \tag{12}$$

Now let us consider the operation if a vector $\underline{D}_{ie}$, $i=0,1,\ldots,n$ is inputted into the Signature Analysis Register.

This vector is the sum of the good machine response vector $\underline{D}_i$ plus the error vector $\underline{E}_n$. The error vector has a 1 whenever there is a difference between the good machine and the faulty machine response. (All addition is modulo 2).

$$\underline{D}_{je} = \underline{D}_j + \underline{E}_j$$

Given this form of error vector, we can write the response for the error vector which will be $\underline{Y}_{ie}$

$$\underline{Y}_{ie} = \underline{C}^{i+1} \cdot \underline{Y}_{-1} + \sum_{j=0}^{i} \underline{C}^{i-j} \cdot \underline{D}_{je}$$

$$= \underline{C}^{i+1} \cdot \underline{Y}_{-1} + \sum_{j=0}^{i} \underline{C}^{i-j} \cdot (\underline{D}_j + \underline{E}_j)$$

$$= (\underline{C}^{i+1} \cdot \underline{Y}_{-1} + \sum_{j=0}^{i} \underline{C}^{i-j} \cdot \underline{D}_j)$$

$$+ \sum_{j=1}^{i} \underline{C}^{i-j} \cdot \underline{E}_j$$

$$= \underline{Y}_i + \sum_{j=0}^{i} \underline{C}^{i-j} \cdot \underline{E}_j \tag{13}$$

Thus, the error response is equal to the good machine response plus the summation associated with the error vector, as shown above. Hence, the good machine response and the faulty machine response will be absolutely identical if the summation containing the error vector is precisely equal to zero. This situation will be called aliasing, if there is one none zero value in any $\underline{E}_i$, $i=0,1,\ldots$,n.

In order to study aliasing, we can limit our observations to just considering a Signature Analysis Register which is initialized to zero with the error vector being the only source--in other words, assuming that the good machine is always zero for the moment. This is clear from the following equation:

Let

$$\underline{Y}^{*}_{ie} = \underline{Y}_{ie} + \underline{Y}_{i}$$

Then

$$\underline{Y}^{*}_{ie} = \underline{Y}_{i} + \sum_{j=0}^{i} \underline{C}^{i-j} \cdot \underline{E}_{j} + \underline{Y}_{i}$$

$$= \sum_{j=0}^{i} \underline{C}^{i-j} \cdot \underline{E}_{j}$$

$$\underline{Y}^{*}_{ie} = \underline{C}^{i-1} \cdot \begin{bmatrix} 0 \\ 0 \\ : \\ 0 \end{bmatrix} + \sum_{j=0}^{i} \underline{C}^{i-j} \cdot \underline{E}_{j} \qquad (14)$$

Of course, from a similar kind of analysis, one can get a super position theorem, as is given below.

Let the input of the Signature Analysis Register be

$$D_{i} = D_{i1} + D_{2i} \cdot \quad i=1,2,\ldots,n$$

Then

$$\underline{Y}_{i} = (\underline{C}^{i+1} \cdot Y + \sum_{i=0}^{i-n} \underline{C}^{i-j} \cdot \underline{D}_{j1}) + \sum_{i=0}^{i-n} C^{i-j} \cdot D_{2j}$$

Bounds on Aliasing Errors

Given the error vector sequence shown below,

$$\underline{E}_{0} = \begin{bmatrix} 1 \\ 0 \\ 0 \\ : \\ 0 \end{bmatrix} \quad \ldots \quad \underline{E}_{i} = \begin{bmatrix} c_{i} \\ 0 \\ 0 \\ : \\ 0 \end{bmatrix} \qquad (15)$$

with i=1,2,......,k. If the error vectors are applied in the following order: first, $\underline{E}_{0}$, then $\underline{E}_{1}$, $\underline{E}_{2}$ and $\underline{E}_{k}$, this will give the shortest sequence beginning with an error and winding up back in the all zero state--in other words, in the situation where aliasing will occur. There is no other vector shorter which will do this same function. One could write this error sequence which results in alising in terms of x for the shift value or delay value as follows:

$$c_{k} \cdot x^{k} + c_{k-1} \cdot x^{k-1} + \ldots + c \cdot x + 1 \qquad (16)$$

Clearly, all linear combinations of this error vector also result in aliasing. This follows from the algebraic structure of the field. Furthermore, this vector is unique; that is, if there is a vector which causes aliasing, it must be a linear combination of error vectors which are shifted in time. This is done by multiplying by X^i to obtain a shift of i in time. Hence, the error vector can be represented as follows:

$$\sum_{i=1}^{\infty} \alpha_i \cdot (c_k \cdot x^k + c_{k-1} \cdot x^{k-1} + \ldots + c \cdot x + 1) \tag{17}$$

where α_i ε GF2 (Note: Not all alpha equal zero. This corresponds to the error-free vector.) Thus in a sequence of length L, where L is greater than k, all aliasing vectors can be represented by the following summation:

$$\sum_{i=1}^{L-k} \alpha_i \cdot (c_k \cdot x^k + c_{k-1} \cdot x^{k-1} + \ldots + c \cdot x + 1) \cdot x^{i-1} \tag{18}$$

It is quite easy to count the number of error vectors that exist. That is, of the total 2^L vectors of length L, and from this equation, it is clear that there are

$$2^{L-k} - 1 \tag{19}$$

vectors which will cause aliasing. The -1 is for the situation when all of the coefficients in the linear combination are zero. This is equal to having a good machine response and having us wind up in the all zero state, and we certainly do not want to count that as an alias sequence.

With the very strong assumption of equal errors--that is, one error sequence is not anymore likely than the other error sequence, one can obtain the following equation for the aliasing probability.

$$(2^{L-k} - 1) / 2^L = 1/2^k - 1/2^L \qquad L \geq k$$
$$= 1/2^k \cdot (1 - 1/2^{L-k}) \tag{20}$$

If L is much greater than k, then the limit is clearly 1 divided by 2^k. These results were the ones obtained by Frohwerk, however, as Smith [22] correctly pointed out, why not just look at the last k bits out of the total stream of L or for that matter any k bits on the stream of L! Also implicit in this equal likelihood of errors is the assumption that the probability of a 1 in each bit position of the error vector is precisely 0.5, and looking at any k bits would give exactly the same aliasing error as we would obtain by using the Signature Analysis Register, given the assumption's equal likelihood. The assumption is clearly too strong. There is no proof that networks do, in fact, obey the equal likelihood function. It would be nice if someone could prove this, however, nothing has come close to that to date. In fact, counter examples do exist, so clearly, no one could prove it.

Thus, let us assume that the probability of an error in each bit of the error vector is p. In order to go any further, the following theorem needs to be proved:

Theorem: Assume that the Signature Analysis Register is in state

$\underline{Y}_1$, and given any other state $\underline{Y}_f$, there exists a unique sequence of exactly length k which will take the Signature Analysis Register from State $\underline{Y}_1$ to $\underline{Y}_f$.

Proof: Let $\underline{Y}_1 = (y_k, y_{k-1}, y_{k-2}, \cdots , y_1)$

$$\underline{Y}_f = (y_{kf}, y_{k-1f}, y_{k-2f}, \cdots , y_{1f})$$

By Construction.
The sequence we need is $S_c = <S_{1c}, \cdots , S_{kc}>$ where S_{1c} is the first error bit fed to SA Register to get us from $\underline{Y}_1$ to $\underline{Y}_f$, followed by S_{2c}, S_{3c}, etc.

We know that

$$y_m = \sum_{i=1}^{m} c_i \cdot y_{m-i}$$

hence S_j, j=1,2,...,k is given by

$$S_{1c} = (\sum_{i=1}^{k} c_i \cdot y_{k-i+1}) + y_{1f}$$

$$S_{2c} = (\sum_{i=1}^{k-1} c_{i+1} \cdot y_{k-i}) + c_1 \cdot S_{1c} + y_{2f}$$

in general

$$S_{jc} = (\sum_{i=1}^{k-j+1} c_{i+j-1} \cdot y_{k-i-j+2}) + (\sum_{i=1}^{j-1} c_i \cdot S_{j-ic}) + y_{kf} \qquad (21)$$

QED.

Consider now an error vector of length L coming into the Signature Analysis Register, which is of length k, where L is greater than k. Assume that the error was detected somewhere in the sequence and before L-k. At this point the LFSR may be in any state--it may be in the all zero state; it may be in any of the other states that would indicate that an error has occurred. No matter what state the Signature Analysis Register ends up in, it would take exactly k more inputs which is unique to get it to the all zero state. Given that the probability of error in the error vector for each bit is p, the probability of ending up in the all zero state P_0 is bounded by the following equation, depending on whether p is less than or equal to a half or greater than or equal to a half.

$$(1-p)^k \leq P_0 \leq p^k \qquad 1 \leq p \leq 0.5 \qquad (22)$$

$$p^k \leq P_0 \leq (1-p)^k \qquad 0 \leq p \leq 0.5 \qquad (23)$$

This gives a bound on the aliasing error as a function of p. Clearly, if p is equal to 0.5 the bound is exactly that obtained for the equal distribution of their equal likelihood assumption made before. In this situation, since the bound is a single point, it is in fact exactly the results of Smith and Frohwerk. For any value of p close to 0.5, the bound is fairly strong, as can be seen from Figure 14, which is a graph for Signature Analysis Registers of several different lengths.

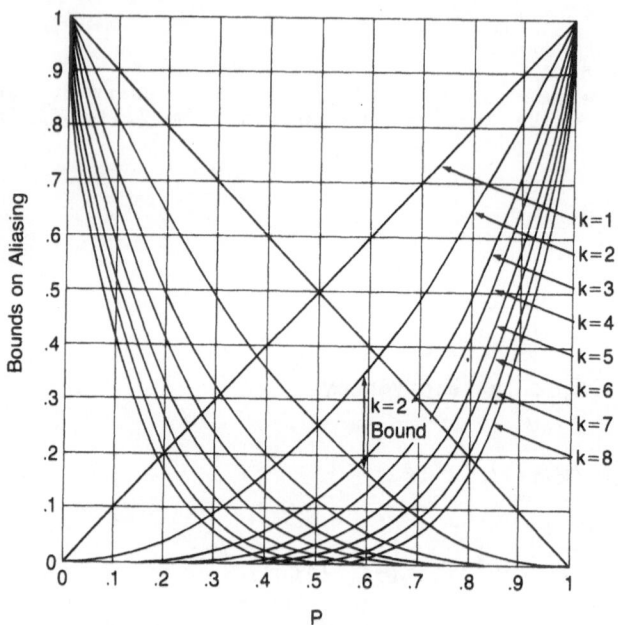

Figure 14 Aliasing probability for different register lengths and fault probabilities p.

It is clear that for faults which are very unlikely to be tested, the bound is poor. That is, if the probability of detecting the fault with a given random pattern is going to be very small or if the probability of detecting a fault is very high, this bound is not a good bound. However, in other situations, it is very good.

To analyze the dynamic properties of the Signature Analysis Register, one can use the mathematical structures represented by these Linear Feedback Shift Registers, LFSRs. For example, Figure 15 and 16 show a graph represention of the Markov process for a primative and nonprimative polynomials. The analysis of equation 12 can be done by using z-Transforms [24]. Using this technique one can show for all values of p, primitive polynomials are better than or equal to nonprimitive polynomials with respect to aliasing. Two points should be noted from the analysis of the Markov process. First, all polynomials have a pole at z=1, whether or not they are primitive and independent of the value of p. This unique pole gives rise to the final value for the aliasing error of $1/2^k$.

The second point is that at p=1/2, the poles for both primitive and nonprimitive are at the origin (except for the single pole for both polynomials at z=1). After k+1 patterns, where k is the length of the signature analysis register, the probability of aliasing is $p=1/2^k$. This explains why other analysis [21, 22] shows no differ-

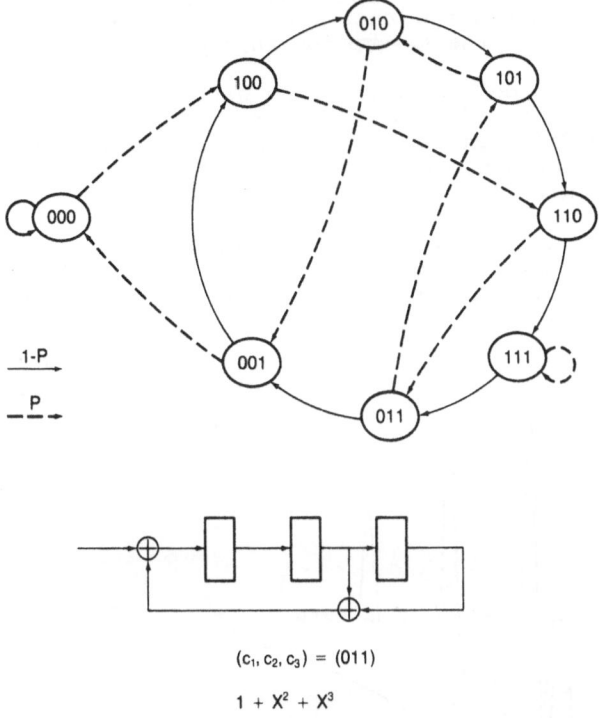

$(c_1, c_2, c_3) = (011)$

$1 + X^2 + X^3$

Figure 15 Markov process for a register with primitive feedback function.

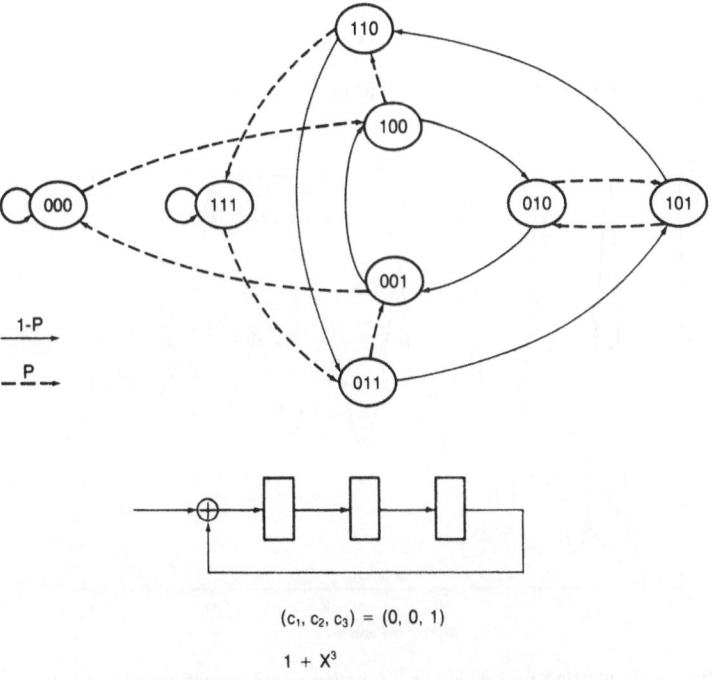

$(c_1, c_2, c_3) = (0, 0, 1)$

$1 + X^3$

Figure 16 Markov Transition diagram for the register $x^3 + 1$.

ence between primitive and nonprimitive polynomials. That analysis is based on the assumption that the error patterns are uniformly distributed, that is, the assumption that p=1/2.

Now the issue is to determine how much better primitive polynomials are. The results of simulating the Markov process for different values of p and different types of p and different types of polynomials show that responses are much better for primitive polynomials (Figures 17 through 20).

Note that an easy-to-test fault can result in aliasing problems that are just as large as or larger than those for a difficult-to-test fault. In our example, easy-to-test faults are p=.95 and hard-to-test faults are p=.05.

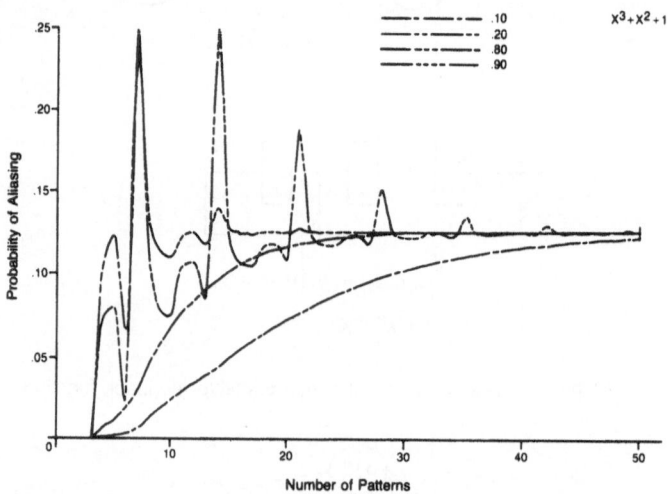

Figure 17 Aliasing probability as a function of test length for $x^3 + x^2 + 1$.

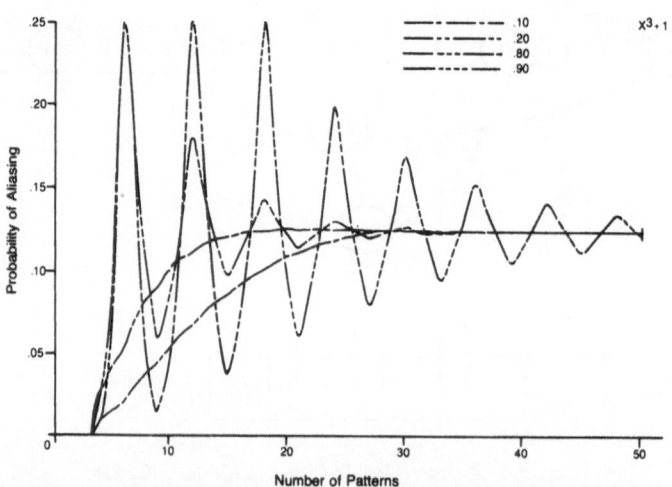

Figure 18 Aliasing probability as a function of test length for $x^3 + 1$.

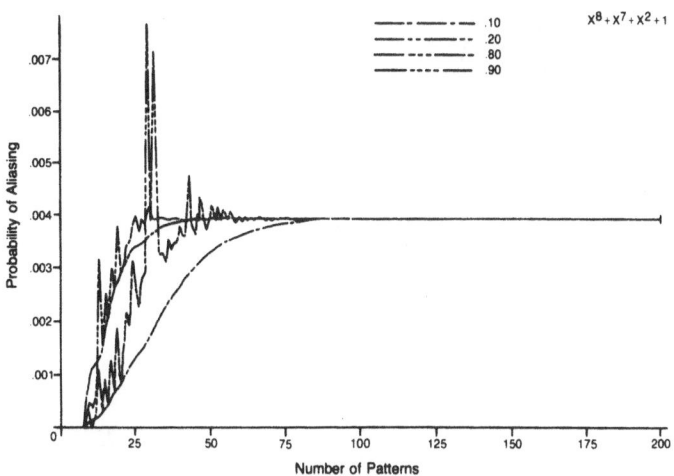

Figure 19 Aliasing probability as a function of test length for $x^8 + x^7 + x^2 + 1$.

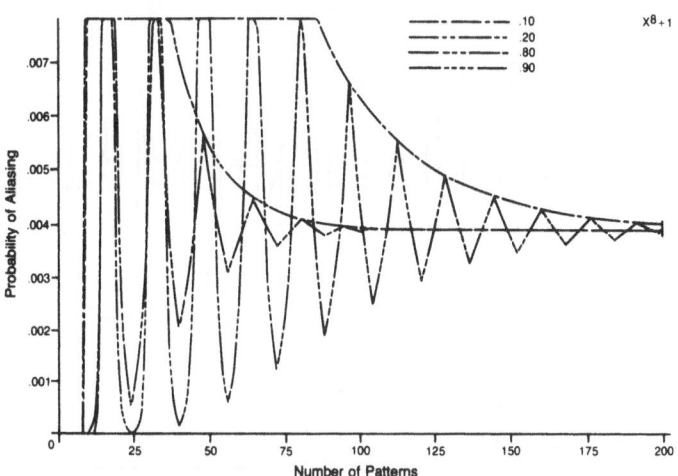

Figure 20 Aliasing probability as a function of test length for $x^8 + 1$.

Many simulations have been run, and all the results indicate that primitive polynomials are better than nonprimitive polynomials. Further research shows that a reasonable bound for the aliasing error is

$$1/2^k + (2^k-1)(1-2p)^{n(1-1/(2^k-1))}$$

This rather complex looking equation is really not so hard to calculate and gives a lot of insight into how aliasing errors relate to the number of patterns, n, and to the length of the signature analysis register, k.

Figure 21 gives two aliasing error plots for the second-degree polynomial for p=.1 and p=.9. Also, in the same plot, the bound is given. Figure 22a shows the bound for an eight-degree polynomial that is nonprimitive for p=.05--clearly not a bound for this case.

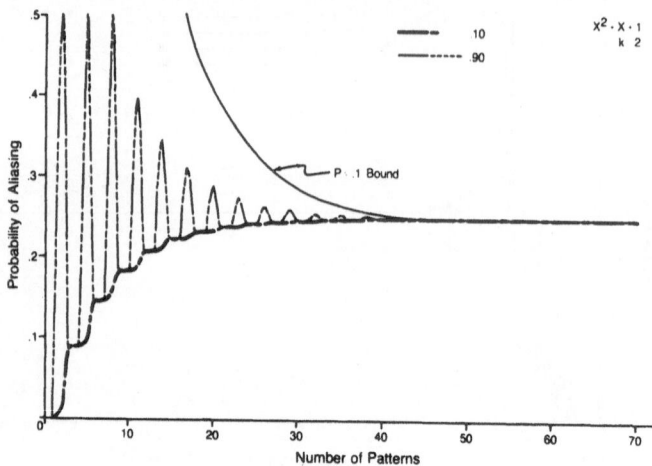

Figure 21 Aliasing error for a primitive polynomial $x^2 + x + 1$ for $p = .1$ and $p = .9$ with the bound for $p = .1$.

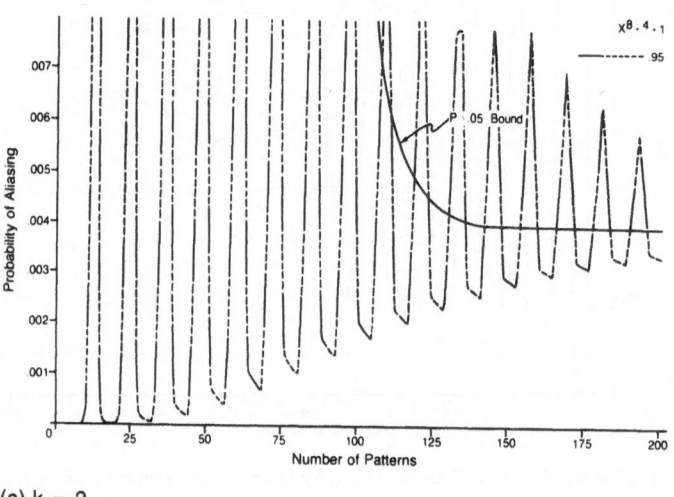

(a) k = 2

Figure 22b, however, shows the same bound with the aliasing response for an eight-degree primitive polynomial. The response is well within the bound. Thus, we can design a signature analysis register for a design if the value of p, k, and n are known or can be estimated.

INSITU SELF-TESTING STRUCTURES:

The Insitu Structures are ones that use the system latches as part of the Linear Feedback Shift Registers. It was not shown in the previous section, but a characteristic of the output of a Linear Feedback Shift Register of maximal length is a random property. Thus, they can be used as a pseudorandom source of patterns. These random patterns will be adequate to test a combinational logic structure, given enough of them.

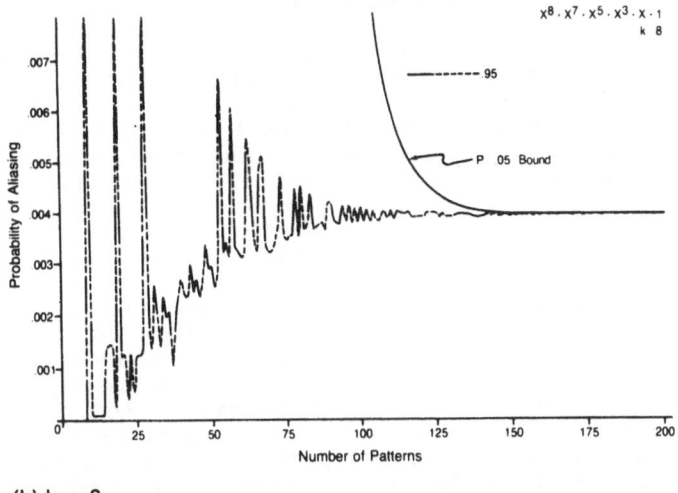

(b) k = 8

Figure 22 Aliasing error for two polynomials with the bound for p = .05: Non-primitive
polynomial $x^8 + x^4 + 1$ (a) and primitive polynomial $x^8 + x^7 + x^5 + x^3 + x + 1$ (b).

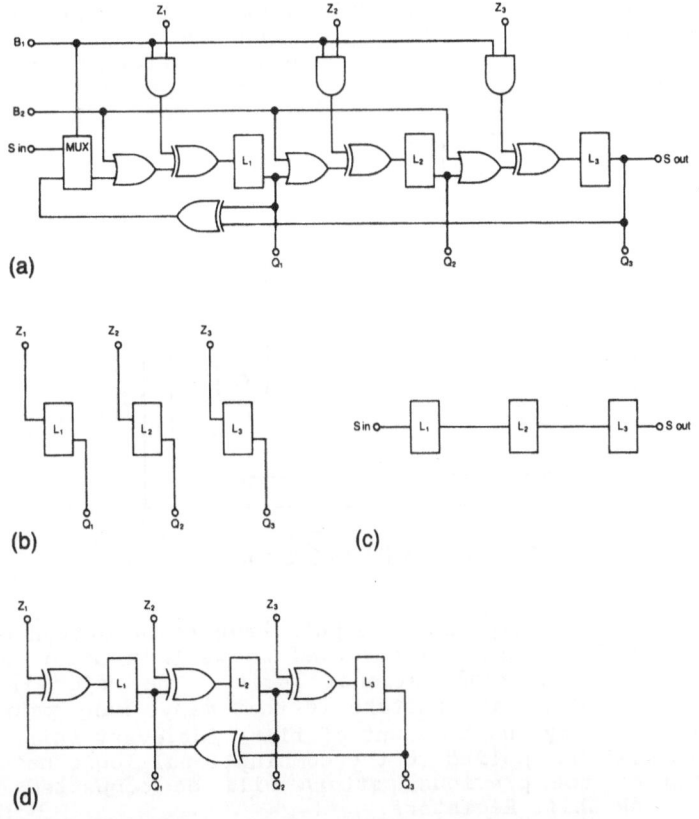

Figure 23 BILBO and Its Different Modes: (a) General Form of BILBO Register,
(b) $B_1 B_2 = 11$, System Operation Mode, (c) $B_1 B_2 = 00$, Linear Shift — Register Mode,
(d) $B_1 B_2 = 10$, Signature Analysis Register with Multiple Inputs (Z_1, Z_2, Z_3).

Figure 23 is a figure of the Built-In Logic Block Observation Technique BILBO structure, in particular, the BILBO register [13]. One can think of it as Z_1, Z_2 and Z_3 coming from a combinational logic network, and Q_1, Q_2 and Q_3 going onto the next combinational network. There are three modes for this register. The mode shown in Figure 23 (b) is a normal register function. The mode shown in (c) is a shift register function, that is, the function used for scanning data into and out of the system latches; and (d) is the Multi-input Signature Analysis structure, where the data bits are exclusive OR'ed to the shifting function, and the output of the LFSR forms a maximal length Linear Feedback Shift Register.

This register would be inserted into the position which would be occupied by the LSSD SRL's as shown in Figure 24. When testing this combinational logic, an initial value would be loaded into the BILBO register, then the mode would be switched to B1/B2 equal 10, which would be the Multiple-input Signature Analysis Register. Many shift clocks would be given, however, a fixed number would be used. At the end of this fixed number, the value stored in the BILBO register will be the signature for the good machine from which all others will be compared. If there was an error in the combinational logic and one of the random patterns coming out of the BILBO network in fact tested it, the output of the combinational network would be different than the good machine. It is then hoped that the compaction of the test results in the Linear Feedback Shift Register would not be masked out or aliased, so that after the fixed number of shift patterns have occurred, the value stored in the BILBO shift register would be different than the good machine value.

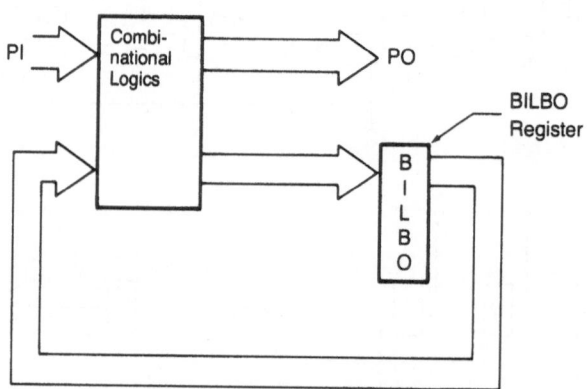

Figure 24 BILBO Structure.

The primary inputs and primary outputs have to be driven by Linear Feedback Shift Registers and observed as well in order to give a complete test of the combinational logic. The major reason for going to this kind of a structure is that many, many patterns can be applied in a very shoft amount of time; for every shift clock, a new pattern will be applied to the combinational logic network, and the results of the previous pattern will be compacted into the Linear Feedback Shift Registers.

It was originally hoped that the test patterns which come for free from the Linear Feedback Shift Register would be good enough for testing the combinational logic network. Unfortunately, the analysis of whether these patterns is good enough is as complex as doing test generation itself, and thus, the real savings of this technique comes in manufacturing with shortened test application time and the field. Some testers could apply the shift clock patterns much faster than you could apply other kinds of patterns. The product can make use of the Self-Testing function in the field environment. The latter is taking on more and more importance with the emphasis on quality.

Clearly, there are some networks which are going to have difficulty being tested by this structure. For example, networks like PLA's [1]. PLA's have an attribute of having very high fan-in to logic gates. For instance, you could have an AND block with 20 inputs. The probability of testing the third input Stuck-At-1 would be $1/2^{20}$. If you're only applying 1000 patterns, the probability that it would be tested would be 1 in 1000, and there is a very low probability that it will, in fact, be tested. Therefore, in order to have this network tested, either deterministic test patterns need to be applied, or the network has to be modified in some way.

The BILBO technique also helps with the test data volume. In Scan Path techniques or Scan Set, or LSSD, a considerable amount of test data volume is involved with the shifting in and out of test patterns. With the scan technique or the others, if P patterns are to be applied with a maximal shift register length L, the time to load these patterns is proportional to PL. The scanning of patterns in and out is not done at machine speed since measurements must be taken at the scanned primary outputs. If the tester runs slower when it's making measurements and driving the scan line, then a speed advantage can be obtained. Assume that the speed advantage ratio is K, and K is typically 100-1000 to 1. If P pseudonym patterns are to be applied to a BILBO-type network, then after the BILBO registers are loaded, the patterns can be applied in time which is proportional to P/K. Thus, there is a speed difference of K times L for an LSSD pattern to be applied, versus a pseudorandom pattern to be applied. With L equal to 100 and K equal to 1000, K times L would imply 10^5 difference in speed per pattern. Clearly, more patterns need to be applied with a BILBO structure, however, it is felt that there is not five orders of magnitude of difference.

If the tester has the ability to apply patterns at the same rate as system speed using some sort of buffer, then that would make K equal to 1, and the speed advantage is only 100 to 1. Therefore, if there are two orders of magnitude between the random patterns and the deterministic patterns, then the test time would be exactly the same. However, the test data volume for a BILBO approach is definitely smaller, since all that is needed is to have an initial feed, a number of clock pulses, and the final signature for the BILBO register (with diagnostics).

Of course, this function does not come for free. The overhead for BILBO is LSSD or Scan Path overhead, plus at least one exclusive OR per stage of shift register.

Another Insitu approach is built-in Verification Testing or Exhaustive Testing [14, 15, 16]. This is a technique which applies all possible patterns, i.e., all 2^n patterns to the combinational ogic network. If all possible patterns are applied and the fault mechanism does not change the network into a sequential network, then the test patterns will be good for "any" fault model. In essence, what is happening is that every single point in the Karnaugh map is being inspected. The hope is that every output pin of a combinational logic network is not a function of all the input pins. If this is the case, then a subset of 2^n patterns can be applied for each output pin.

The major difference between the two techniques, BILBO and Verification Testing, is that with the BILBO technique, a tool has to be put in place to determine the proper size of the number of random patterns. Without getting into the details of the complexity of run time, this could be a fault simulation program, or it could be a statistical analysis program, or a probability calculating program. However, some sort of program has to be used in order to tell whether the network is testable with random patterns. This tool would also show redundant logic that's untestable and other logic which is difficult to test with random patterns, the latter two categories being difficult to distinguish from one another; clearly, the identification of redundant logic is equivalent to a test generation problem. In other words, it is difficult.

With Verification Testing, since all 2^n patterns are going to be applied, all faults which are not redundant will get tested with the two constraints that were mentioned earlier (no sequential faults and no dependency increase). Therefore, no tools are recommended with this approach. However, the design verification or the synthesis tools must guarantee no redundancies exist since there is no check for them!

Thus, both the BILBO technique and the Verification Testing technique are Insitu Self-Testing structures since the system latches would be used to drive the combinational logic with test patterns and compact the results.

EXSITU SELF-TESTING

The third form of Self-Testing is the Exsitu structures. These structures generate pseudorandom patterns, usually with Linear Feedback Shift Registers, and compact the results with these same type of registers. However, the generators and compactors are not part of the system function, and hence, the name Exsitu. The basic structure is shown in Figure 25.

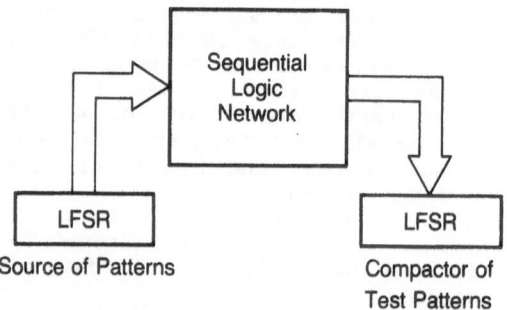

Figure 25 ExSitu Self-Testing Structure.

An example of such a structure is the STUMPS approach [17] shown in Figure 26. In essence, this technique drives the scan paths of

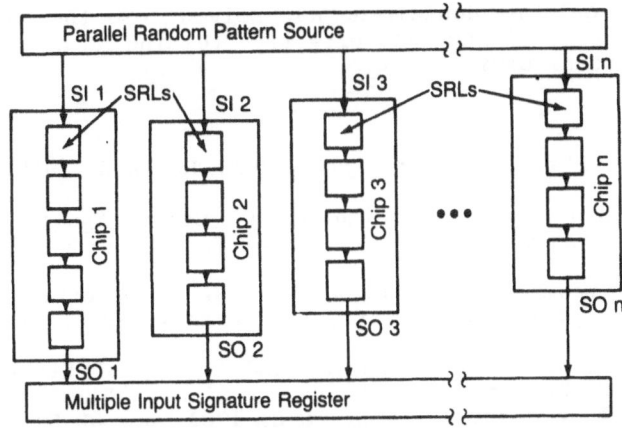

Figure 26 ExSitu STUMPS Approach.

LSSD chips with pseudorandom data from a Linear Feedback Shift Register at the top of the structure called parallel random pattern source. The output of the shift registers are compacted into Multiple-input Signature Analysis registers at the bottom which is an LFSR with multiple inputs exclusive OR´ed into the Linear Feedback Shift Register. The method of testing with this structure is as follows.

The LSSD shift registers on each of the chips are scanned into with data from the parallel random pattern source. once all the shift registers are loaded which can be done in parallel, a system clock is hit so that the results of the test are stored in some of the SRL´s on the chips. This data is then off-loaded into the Multiple-input Signature Analysis Register which compacts the test results. This technique is lower in overhead than the BILBO approach, however, it takes a longer amount of time to test the structure since the shifting occurs for every pattern. This technique acquires a factor of L times longer to apply the same number of patterns to the combinational network, where L is the length of the longest scan path. Hence, the speed advantage is K as defined in the prior section. If the tester is capable of loading and unloading patterns at machine speed, the socket time for this test would be longer since more pseudorandom patterns would be needed over deterministic patterns. Another approach of the Exsitu Self-Testing structure is the On Chip Maintenance System proposed by CDC [18].

In summary, the microprocessor self-stimulated is a very popular technique and some form of this technique is used by virtually all board manufacturers. Portions of BILBO networks have been used by a number of companies, but not complete BILBO primary because of cost overhead. Verification testing has not been used because of the overhead in cells and time. Some small portions have been used to drive complete testing of some sub-networks. These techniques are all becoming more and more popular and may eclipse the scan designs.

CONCLUSION

There are many useful and interesting techniques to help the testing problem. However, time has shown that logic testing problems can only be contained, not solved. So the quest will go on to find more efficient and effective testing methods in the hope of deducing total device cost.

REFERENCES

(1) T. W. Williams, VLSI Testing, North Holland, 1986.

(2) H. Y. Chang, E. G. Manning and G. Metze, Fault Diagnosis of Digital Systems, New York: Wiley-Interscience, 1970.

(3) A. D. Friedman and P. R. Menon, "Fault Detection in Digital Circuits," Englewood Cliffs, NJ: Prentice-Hall, 1971.

(4) T. W. Williams, "The History and Theory of Stuck-At-Faults" Proc. 6th European Conference on Circuit Theory and Design, Stuttgart, W. Germany, Sept. 6-8, 1983, pp. 80-81.

(5) T. W. Williams and K. P. Parker, "Design for Testability--A Survey," IEEE Proceedings of the IEEE, Vol. 71, No. 1, January, 1983, pp. 98-112.

(6) M. J. Y. Williams and J. B. Angell, "Enhancing Testability of Large Scale Integrated Circuits Via Test Points and Additional Logic," IEEE Trans. Comput., Vol. C-22, pp. 46-60, January, 1973.

(7) E. B. Eichelberger and T. W. Williams. "A Logic Design Structure for LSI Testing," Proc. 14th Design Automation Conf., IEEE pub. 77 H1216-1C, pp. 462-468, June, 1977.

(8) S. Funatsu, N. Wakatsuki, and T. Arima, "Test Generation Systems in Japan," Proc. 12th Design Automation Symp., pp. 114-122, June, 1975.

(9) H. Ando, "Testing VLSI With Random Access Scan," in Dig. Papers Compcon 80, IEEE Pub. 80CH1491-OC, pp. 50-52.

(10) J. H. Stuart, "Future Testing of Large LSI Circuit Cards," in Dig. Papers 1977, Semi-Conductor Test symp., IEEE Pub. 77CH1261-7C, pp. 6-17, October, 1977.

(11) A. Yamada, N. Wakatsuki, T. Fukui, and S. Funatsu, "Automatic System Level Test Generation and Fault Location for Large Digital System," Proc. 15th Design Automation Conf., IEEE, Pub. 78CH1363-1C, pp. 347-352.

(12) T. W. Williams and E. B. Eichelberger, "Random Patterns Within a Structured Sequential Logic Design," in Dig. Papers, 1977 Semi-Conductor Test Symp, IEEE Pub. 77CH1261-7C, pp. 19-27, October, 1977.

(13) B. Koenemann, J. Mucha, and G. Zwiehoff, "Built-In Logic Block Observation Techniques," in Dig. Papers, 1979, Test Conf., IEEE Pub. 79CH109-9C, pp. 37-41, October, 1979.

(14) E. J. McCluskey and S. Bozorgui-Nesbat, "Design for Autonomous Test," IEEE Trans. on Computers, Vol. C-30, No. 11, pp. 866-875, November, 1981.

(15) E. J. McCluskey, "Verification Testing," in Dig. Papers, 1982

International Test Conf., IEEE Pub. 82CH1808-5, pp. 183-190, October, 1982.

(16) D. T. Tang and C. L. Chen, "Interactive Exhaustive Pattern Generation for Logic Testing," IBM Journal of Res. and Dev. Vol. 28, No. 2, pp. 212-219, March, 1984.

(17) P. H. Bardell and W. H. McAnney, "Self-Testing of a Multi-Chip Logic Module," in Dig. Papers 1982 International Test Conf., IEEE Pub. 82CH1808-5, pp. 200-204, October, 1982.

(18) D. R. Resnick, "Testability and Maintainability With a New 6K Gate Array," VLSI Design, March/April, 1983.

(19) W. W. Peterson and E. J. Weldon, Jr., "Error-Correcting Codes," Cambridge, MA, the MIT Press, Second Edition, 1972.

(20) S. W. Golomb, "Shift Register Sequences," San Francisco, CA, Holden-Day, Inc., 1967.

(21) R. A. Frohwerk, "Signature Analysis: a New Digital Field Services Method," Hewlett-Packard Journal, May 1977, pp. 2-8.

(22) J. E. Smith, "Measures of the Effectiveness of Fault Signature Analysis," IEEE T. C., June 1980, pp. 510-514.

(23) T. W. Williams and N. C. Brown, "Defect Level as a Function of Fault Coverage," IEEE, T. C. Vol. C-30, No. 12, December, 1981, pp. 987-988.

(24) T. W. Williams, W. Daehn, M. Gruetzner, and C. W. Starke, "Comparison of Aliasing Errors for Primative and Non-Primitive Polynomials," Proc. International Test Conference, Sept. 1986, Washington, D.C., pp. 282-288.

STATISTICAL TESTING

Vishwani D. Agrawal
AT&T Bell Laboratories
Murray Hill, NJ 07974-2070

This paper is a tutorial on probabilistic and statistical techniques used in VLSI testing. Specific problems discussed include test data analysis, testability analysis, test generation, and fault coverage evaluation. Methods applied to solve these problems can be classified as statistical data analysis, probabilistic analysis, and Monte Carlo techniques. Since many of these solutions may be unconventional for practicing engineers, the paper emphasizes concepts and illustrates the advantage of such techniques in handling computationally complex problems.

1. INTRODUCTION

It is a normal practice to use abstraction for solving the problems of synthesis and analysis. In VLSI design, simple examples of abstraction are Boolean gates and state machines. Such models not only reduce the level of complexity but also allow the use of computer-aids. For very large circuits, however, even the computer aids become too expensive. There is, in fact, a two-fold explosion of complexity. First, the program must process a large number of circuit components (e.g., gates, interconnections, etc.). Second, the input space (e.g., the number of possible input stimuli) grows with the circuit size. Applications of probability and statistics can significantly reduce the complexity of VLSI design problems. This paper is an introductory tutorial on such methods applicable to testing.

Statistical methods can be classified into three categories:

1. **Statistical Inference** - These involve decision making on the basis of nonexhaustive set of observations. For example, simulation of a randomly selected subset of faults can be used to estimate the fault coverage of tests. In contrast, a deterministic method will require simulation of all faults. Methods of this type are discussed in Sections 2 and 5.

2. **Probabilistic Analysis** - In general, a deterministic method must compute the circuit response separately for each input signal (vector or sequence of vectors). A probabilistic method, on the other hand will analyze the entire circuit only once to compute signal probabilities. Testability analysis given in Section 3 is a method of this type.

3. **Monte Carlo Method** - This involves searching for a solution among randomly

33

F. Lombardi and M. Sami (eds.), Testing and Diagnosis of VLSI and ULSI, 33–47.

generated data. An example of a Monte Carlo method is test generation using random vectors as described in Section 4. Since rapid generation and analysis of large amount of data are required, use of computers is necessary for these methods. In general, whenever analysis is much simpler than synthesis, many synthesis problems can be solved by Monte Carlo procedures.

2. FAULT COVERAGE AND QUALITY

The quality of tested product must be, in some way, related to the completeness of tests. The completeness of tests is measured in terms of their *fault coverage*. Fault coverage is the fraction (or percentage) of single stuck faults that are detectable by tests. Obviously, a circuit can have multiple faults. It can also have faults that may not simply map onto single stuck faults. We must, therefore, establish how the quality of the tested product is related to the coverage of single stuck faults.

We will use *reject ratio* as a measure of the quality of the tested product. Reject ratio is defined as the fraction of faulty items among those tested as good. For perfect testing the reject ratio should be zero. In the following, we will establish a relationship between reject ratio of the tested VLSI devices and the fault coverage of tests.

Suppose f is the fault coverage of tests. That is, the tests are capable of detecting a fraction f of all stuck faults. Thus f can be regarded as the probability of detecting a randomly selected fault. If a VLSI chip has exactly n faults then the probability of its passing the tests is $(1-f)^n$. The probability of any chip passing the tests can be computed as

$$y(f) = \sum_{n=0}^{N} (1-f)^n p(n)$$

where N is the total number of possible faults over which the fault coverage is determined and n is assumed to be a random integer with a probability density function $p(n)$. In the above equation, $y(f)$ is the yield of tested chips which includes a fraction $y(1)=p(0)$ of really good chips and the remainder consisting of faulty chips tested good due to imperfect tests.

The causes of faults in a VLSI chip are the imperfections in materials and processing. Chip processing is carried out in a number of steps like application of photo-resist, mask alignment, exposure, etching, etc. Each step processes the entire chip area simultaneously. If something goes wrong, a large area may become defective. For the purpose of yield prediction, therefore, the number of random defects can be modeled by a clustered distribution known as the *negative binomial distribution* [1]. Thus, the probability of finding exactly x physical defects on a chip of area A are given by

$$\frac{(x+a-1)!}{x!\,(a-1)!}(Ab)^x(1+Ab)^{-x-a}$$

This distribution is characterized by two parameters, a and b. The parameter a is responsible for *clustering*. As a gets closer to zero, the spatial distribution becomes more clustered. The defect density, i.e., the average number of defects per unit area, is given by the product ab.

In general, a physical defect can be assumed to produce several logic (e.g., stuck type) faults. For the purpose of this analysis we will assume that the number of faults per defect is a Poisson random variable with mean c. According to a known result on the sum of Poisson variables [2] then, in the presence of x defects, the probability of a chip having exactly n faults will be

$$\frac{(cx)^n}{n!} e^{-cx}.$$

This expression gives the probability of n faults on a chip, conditional to the given fact that the chip has x physical defects. To obtain $p(n)$, the probability of n faults on any chip, we simply add the weighted conditional probabilities. Thus

$$p(n) = \sum_{x=0}^{\infty} \frac{(cx)^n}{n!} e^{-cx} \cdot \frac{(x+a-1)!}{x!\,(a-1)!} (Ab)^x (1+Ab)^{-x-a}$$

After substituting the above expression in the formula for $y(f)$ we use the method of probability generating functions to derive the following closed form yield equation [3]:

$$y(f) = [1 + Ab(1-e^{-cf})]^{-a}$$

Notice that A is the area of the chip and the yield equation is characterized by three parameters, a, b, and c.

During wafer testing, by recording the variation of yield as a function of the fault coverage and fitting the expression for $y(f)$ in the observed data, we can determine the parameters, a, Ab, and c.

The procedure outlined above has several applications.

2.1 Characterization of Process Yield

VLSI chips are fabricated through a complex process. Many things in this process can (and do) go wrong. To name a few, materials (silicon wafers, chemicals, gases) may have impurities, time and temperature controls may malfunction, or mask alignment may be imperfect, not counting human errors. Due to short term (lot-to-lot or day-to-day) variation of the causes of defects, the process yield may vary. Test data allows monitoring of the process yield.

The quantity $y(1)$, as obtained from the last equation, is the process yield. Process parameters, a, b, and c (area A of the chip is known), can be numerically computed to best fit $y(f)$ to the actual wafer test data. A typical result is shown in Fig. 1. For the wafer lot used, the values of parameters were found to be as expected. Variations from the expected values will signal problems in processing.

In practice, wafer processing is continuously monitored through various fabrication steps. Any diagnostic information derived from the functional testing will then provide an overall assessment of the process. Variations of the parameters in the yield equation point to changes in defect patterns. For example, an increase in a means greater clustering of defects. An increase in c would indicate large defective area or simply a large number of small defects produced by a common mechanism. Through experience, the variation in a

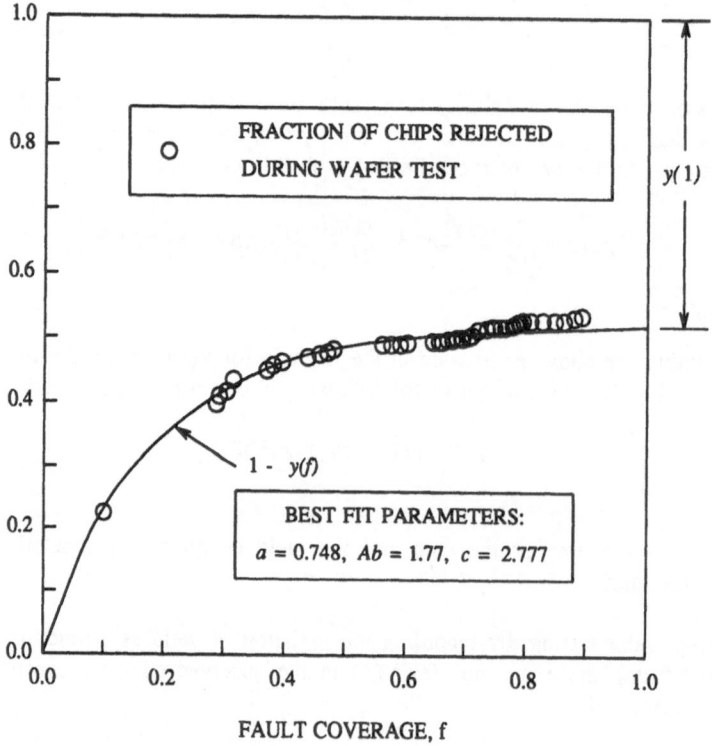

Fig. 1 Analysis of VLSI wafer test data.

yield parameter can be related to one or more specific errors in the processing.

2.2 Reject Ratio and Fault Coverage Requirement

For very large chips, the complexity of test generation usually prohibits 100% fault coverage. In practice, designers feel satisfied with 95% coverage. Critical schedules often allow time barely to make it to the 90% mark on very large chips. When chips are tested with tests having less than 100% fault coverage, some faulty chips can be expected to pass the tests. The reject ratio $r(f)$, i.e., the fraction of faulty chips in the tested lot, can be readily found as follows:

$$r(f) = \frac{y(f)-y(1)}{y(f)}$$

For the chip analyzed in Fig. 1, the fault coverage of tests was about 90%. Using the estimated values of parameters, $a=0.748$, $Ab=1.77$, and $c=2.777$, we obtain, $r(0.9)=0.01$. If 1% reject ratio was not sufficient for the required quality level, then for any reject ratio r, the following formula, easily derived from the above expressions, can be used:

$$f = -\frac{1}{c} \ln \left[1 + \frac{1-(1-r)^{1/a}[1+Ab(1-e^{-c})]}{Ab} \right]$$

For a 0.1% reject ratio (i.e., one defective part per one thousand), this gives a 99% fault coverage. Of course, an even lower reject ratio will require almost 100% fault coverage.

The above analysis leads to some interesting conclusions. Suppose we reduce the minimum feature size on the chip. The new technology will allow us to pack more circuitry on the same silicon area. As a result of the increased circuit density, we would expect more faults due to a defect covering an identical area as before. This should increase the value of the parameter c which represents the average number of faults per defect. Assuming a three-fold increase in c, if other parameters were to remain the same, the above formula gives $f=92.5$ for a reject ratio $r=10^{-4}$. This means one defective part per 10,000 (a defect level most manufacturers would like to attain or exceed) and that too with just 92.5% fault coverage. This illustrates that with higher level of integration, the testing problems may be different, not necessarily worse.

An alternative formulation of this problem has been given by Williams and Brown [4].

3. TESTABILITY ANALYSIS

Testability analysis is a topological analysis of circuit. The results give useful information on testability in the form of node *controllabilities* and *observabilities*, two quantities that play crucial roles in test generation. One nice feature of testability analysis is that it is independent of any specific set of stimuli (or vectors). Such a requirement can be satisfied by considering statistical inputs and working with signal probabilities in place of the actual signal values.

Earlier attempts at non-probabilistic forms of testability analyses produced confusing results that were inaccurate and difficult to interpret [5]. Much of the recent work is, however, based on a probabilistic foundation.

In the probabilistic definition, 1-controllability of a line in the circuit is the probability of that line assuming a value 1. Similarly, the 0-controllability is the probability of a 0 on that line. In general, the primary input controllabilities represent the input vectors in a statistical sense. A common assumption often made is to let all input vectors be equiprobable. Thus all primary input controllabilities become equal to 0.5. Also, 0 and 1 observabilities are defined as the probabilities of observing a 0 or 1 state on a line at any primary output. The observabilities for 0 and 1 states are separately defined because they are related to detection of stuck-at-1 and stuck-at-0 faults, respectively, and in general, they can have different values.

Using a simple example, we will illustrate the computations. Consider the circuit in Fig. 2(A). The binary values on the lines show the result of exhaustive simulation. Based on the result of exhaustive simulation, the signal controllabilities can be easily computed by simply counting 0s and 1s on every line. One-controllabilities are shown in Fig. 2(A). Now suppose we wish to obtain the controllabilities without simulation. We first compute the 1-controllability of line e as 1/4 by multiplying the 1-controllabilities (both 1/2) of lines a and b. This is because both a and b should be 1 for e to be 1. The 1-controllability of line f is one minus the 0-controllability of b. It is, therefore, 1/2. A 1 output from the OR gate requires that both its inputs not be simultaneously 0. Since the 0-controllabilities of these inputs are 3/4 and 1/2, a simple-minded computation will give the 1-controllability of g as 1−3/8=5/8. This is, of course, wrong since exhaustive simulation gives 3/4. The reason for error is the reconvergent fanout in this circuit that makes the signals e and f correlated. Their probabilities, therefore, can not be multiplied without introducing error in the result.

An accurate computation of controllabilities in the presence of reconvergent fanouts requires a more careful analysis. In this simple example, lines e and f can be made independent by setting the line b to a fixed value. When b is set to 1, simple computation as described above will produce the 1-controllability of g as 1/2. This is shown in Fig. 2(B). Similarly, setting b to 0 in Fig. 2(C), the output 1-controllability is obtained as 1. These two controllabilities are conditional probabilities and can be combined after weighting with appropriate controllabilities of b to obtain the correct 1-controllability of g as: $1/2 \times 1/2 + 1 \times 1/2 = 3/4$.

The numerical procedure illustrated in this example is known as PREDICT (Probabilistic Estimation of Digital Circuit Testability). PREDICT [6], [7] includes special procedures for computing observabilities. Controllabilities can also be computed exactly through a symbolic analysis [8]. Approximate methods for computing probabilistic testability are the cutting algorithm [9] and COP [10].

When originally developed, testability analysis was supposed to help a designer in assessing the circuit testability. The results of analysis were assumed to quantify the effort of test generation. However, it soon became apparent that testability analysis could not predict test generation effort which depended upon designers skill and experience or on heuristics employed in a test generation program. More realistic applications of testability analysis are in identifying difficult to test portions of a circuit and predicting (approximately) the size of test set [11].

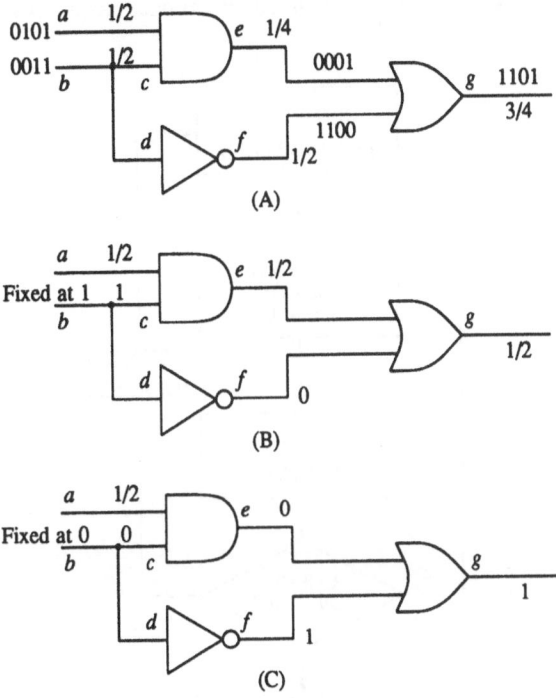

Fig. 2 Computation of probabilistic controllability.

4. TEST GENERATION

A test for a digital circuit is an input vector (or a sequence of vectors) that manifests a potentially faulty condition of the circuit at a primary output. According to this definition any vector (or sequence) that produces an output signal could qualify as a test. However, the problem of test generation is to find a set of tests that will cover a specified number of potential (normally modeled as stuck type) faults. While the complexity of deterministic test generation methods depends on circuit size and complex (e.g., reconvergent fanout) interconnections, the random method only depends on the number of circuit inputs.

In the random method the tests are generated through a random process. Once the tests have been generated, there is nothing random about them. Some knowledge of the operation of the circuit under test can be useful particularly for sequential circuits. The method, if used properly, can provide good results at a very low cost. Figure 3 gives a somewhat generalized flow of a random test generator. An automatic or interactive program can be written to produce vectors.

A fault simulator is necessary for effective random test generation. Setting or changing the probabilities of 0 and 1 bits at the circuit inputs, in general, requires some knowledge of the function. In the simplest case, applicable mostly to combinational circuits, the input probabilities can be set independent of the function. For sequential circuits it is more important to consider the function in developing the random test generator.

This is because, the test generator is required to produce sequences in order to obtain

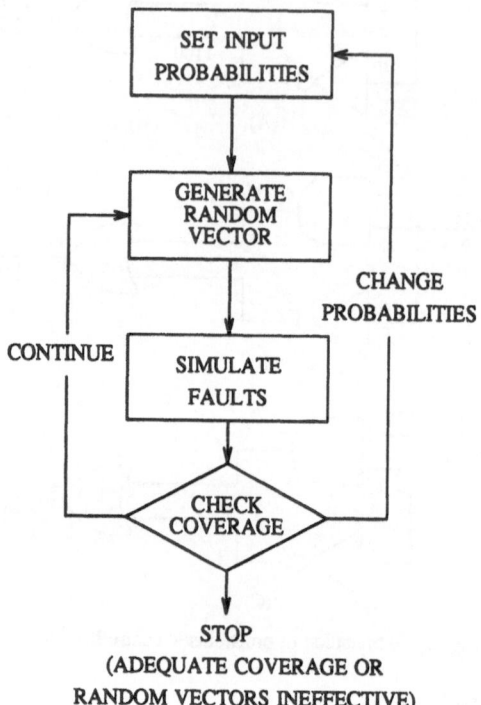

Fig. 3 Random test generation.

observable outputs. We will give a few simple examples to illustrate the method.

Consider the four-bit ALU (Texas Instruments Type SN74181) that performs a variety of functions. It has 14 inputs: four function select inputs, eight data inputs, one carry input, and one mode (arithmetic or logic) select input. A random test generator, completely ignoring the function, is given below:

```
Begin  Do i = 1,1000
         Vector[1,14] = random(0.5)
         Simulate faults
         If [any fault detected]
             Remove detected faults from consideration
             Store Vector as a test
             Stop If [fault list empty]
         End If
     End
```

This procedure can be implemented in any programming language. In the function "random(0.5)", the argument specifies the probability of 1s. The program will produce 1s and 0s with equal probability (i.e., 0.5) for each bit in the 14-bit vector. This is the simplest form of random test generator; no information about the function of the circuit is used. Suppose a designer thinks that the arithmetic functions of the ALU, selected by setting the mode select input M to 0, should be exercised more often than the logic functions. This can be done in the random test generation program simply by adding the statement "M = random(0.4)" and modifying "Vector[1,13]" to exclude the signal M. Now 60% of the generated vectors will perform arithmetic operation on random data.

Notice that the procedure requires the use of a fault simulator. Also, it rejects any vector that does not enhance the fault coverage. In actual runs, this program produced from 30 to 35 test vectors after generating approximately 200 random vectors. The tests covered all detectable faults in the ALU [12].

Of course, the random test generation is not always so effective. For combinational circuits, it has been found that the number of levels of logic and gate fanins determine the success of random test generation. For circuits with deep logic and high fanins, augmenting the random method with a deterministic method like D-algorithm is an effective strategy [13], [14]. Random test generation programs for sequential circuits require greater care [15]. In general, following points should be considered:

• Clock signals should not be random. They should be set according to functional specifications.

• Control signals, like "reset", should become active with low probability to allow valid outputs from the circuit.

• Other data input signals can change randomly but follow specified restrictions. For example, certain data signals may not change on a certain edge of clock signal to avoid race conditions.

• A vector can not be rejected just because it did not enhance the cumulative fault coverage. This is because every vector carries the circuit to a new state. A simple random vector generation scheme may thus produce too many vectors. Alternatively, the test generator can save the states of all flip-flops on the vector which detected some faults. Now, if a prespecified number of subsequent vectors do not detect any new fault, then they can be rejected and the states restored to the saved values.

As an example, a test generator program written by the author using the above guidelines for a 16-bit multiplier (having a pipeline register) produced a test sequence of 2,000 vectors with 98% fault coverage. Of the remaining faults, 1.5% were logically redundant.

As another example, a test generator for a microprocessor could randomly pick an instruction and generate a random data pattern. By repeating this sequence a specified number of times it will produce a test program which, when assembled, will test the microprocessor by randomly exercising its logic.

The basic attractiveness of random test generation lies in its simplicity and inexpensiveness. However, the produced sequences could be very long. Several approaches have been suggested for improving the efficiency of random test generation [16]-[18]. To date, real application of the random test generation technique has only been reported for combinational circuits [19] and for microprocessors [20].

5. COVERAGE EVALUATION

Fault coverage of test vectors is normally evaluated by fault simulators. Fault simulation of VLSI circuits, that have a large number of gates and a large number of test vectors, is one of the most expensive steps in the design process. We will discuss two statistical techniques that result in considerable economy.

5.1 Statistical Sampling

Only a fraction of faults are simulated in a sampling technique in order to estimate the fault coverage. In a method which has a direct analogy with the opinion polls, a randomly chosen sample of some fixed number of faults is simulated and the fraction detected by the test set is used as an estimate for the fault coverage. The particular appeal of sampling techniques lies in the fact that the confidence range of the estimate depends only on the sample size and not on the population size. Many people find this result counterintuitive. However, it is the basis for the success of Monte Carlo methods [21] in solving problems involving many dimensions (e.g., n-dimensional quadrature). Also, the estimate range narrows closer to 100% fault coverage. For example, an estimate of 95% fault coverage can be accurate within 2% for a sample size of 1,000 while a 50% estimate has a 5% tolerance [22].

Accuracy in a sampling experiment demands large sample size. A sample size of 1,000 to 2,000 faults is generally considered necessary. Simple analysis gives the following estimate (with 99% confidence) of fault coverage:

$$C = c \pm \frac{4.5\sqrt{1+Nc(1-c)}}{N}$$

Where

$c = $ *fraction of covered faults in the sample*

$N = $ *number of sampled faults* $(N \geq 1,000)$

As an example consider a sample of 2,000 faults randomly picked from the fault list. After fault simulation with the given vectors if 1,900 sample faults are found detectable, then the above formula gives the range of coverage as

$$C = 0.95 \pm .022$$

Thus the fault coverage of the simulated vectors over the fault list is estimated between 92.8% and 97.2%.

Considering the fact that a typical VLSI circuit can have 10,000, or even 100,000, faults, simulation of just 2,000 faults will result in significant saving of computing resources. One disadvantage of the sampling approach is that it does not provide a complete list of undetected faults. Designers still consider the method useful. For instance, in the above example, the sample fault simulation will give 100 undetected faults that the designer could examine to find the portions of the circuit with low detectability. On the other hand, simulation of all faults in a 100,000 fault circuit with a similar coverage would have produced a list of undetected faults with nearly 5,000 entries, a number too large to examine conveniently!

5.2 STAFAN

Another statistical technique, known as STAFAN [23] completely avoids fault simulation. Using nodal activity generated during good circuit simulation, it determines controllability (the degree to which the test vectors exercise circuit nodes) and observability (the likelihood of faults propagating to the output). Taking both factors into consideration, detection probabilities of individual faults and the fault coverage can be determined. The main overhead the analysis adds to good-circuit simulation is the collection of nodal activity statistics. As a result, the computation time grows only linearly with the number of circuit nodes.

To understand the principle of STAFAN, consider the simple circuit of Fig. 4. Suppose a fault-free simulation is performed and the signal values are as shown by the bit-streams in the figure. Statistical controllabilities are determined simply by counting the 1 and 0 bits. For example, on line a there are three 1s among five bits. Therefore, the 1-controllability $C1(a)$ is 0.6. Also, the 0-controllability $C0$ is 0.4. Similarly, controllabilities of all lines can be found directly from true-value (i.e., fault-free) simulation. For input lines, we also compute *sensitization*, i.e., the fraction of time the value of line is propagated to the gate output. Noticing that the value of line a is propagates to the output

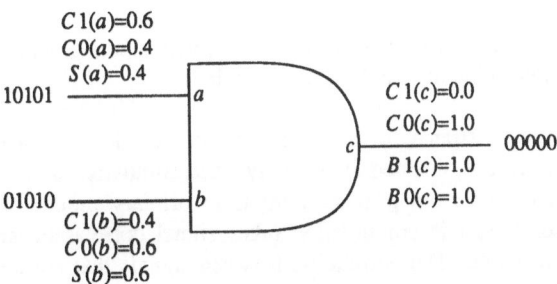

Fig. 4 An example of statistical fault analysis.

of the AND gate when line b is 1, we get the sensitization of line a as $S(a) = 0.4$.

Next, STAFAN requires computation of observabilities. One- and zero- observabilities of a line are the probabilities of observing a 1 or 0 value of that line at a primary output. In our example, let us assume that the output of the AND gate is a primary output. Obviously, the values of line c are then observable, i.e., $B1(c) = 1.0$ and $B0(c) = 1.0$. Analytical formulas allow computation of input observabilities. For example, the 1-observability of line a is the conditional probability (as defined in Section 3) of line b being 1 given that a is 1. For an AND gate, the joint probability of both inputs being 1 is the 1-controllability of its output since the output can be 1 only when both inputs are simultaneously 1. The required conditional probability is then computed by dividing the joint probability of both inputs being 1 by the probability of a being 1. Thus

$$B1(a) = \frac{C1(c)}{C1(a)} = 0.0$$

As explained earlier, the sensitization probability $S(a)$ of line a is the probability of $a=1,b=1$ or $a=0,b=1$ inputs. From this if we subtract the probability of $a=1,b=1$ input then we are left with $S(a)-C1(c)$ as the probability of $a=0,b=1$. Now, the 0-observability $B0(a)$, i.e., the conditional probability of observing a, given $a=0$, is obtained by dividing by the probability of $a=0$:

$$B0(a) = \frac{S(a) - C1(c)}{C0(a)} = 1.0$$

The detection of stuck-at-1 fault on line a requires controlling this line to 0 and then observing its value at primary output. A similar argument applies to the stuck-at-0 fault. The detection probabilities of both faults are found as

$$Prob\,(line\ a\ stuck-at-1) = C0(a).B0(a) = 0.4$$

and

$$Prob\,(line\ a\ stuck-at-0) = C1(a).B1(a) = 0.0$$

It can be verified that only 40% of the applied patterns actually detect the s-a-1 fault on line c and that none detected the s-a-0 fault.

What we have illustrated for a single gate, can be done for a large circuit. Controllabilities of all lines are obtained statistically from simulation data. Observabilities are computed in a single pass from primary outputs to primary inputs. Through the use of sensitization probabilities, this computation takes signal correlation among the inputs of the same gate into account. The correlation between signals that are farther apart is, however, neglected. The results still provide useful information about undetected faults. Fault coverage estimate can also be obtained with an accuracy of within 2 or 3 percent (Fig. 5).

The main advantage of STAFAN is that its space (memory) and time (CPU time) requirements are similar to true-value simulation rather than fault simulation. Both requirements for STAFAN and true-value simulation are linear in circuit size.

Another technique, known as parallel pattern evaluation and single fault propagation (PPSFP), has been used to statistically compute fault detection probabilities in combinational logic [24].

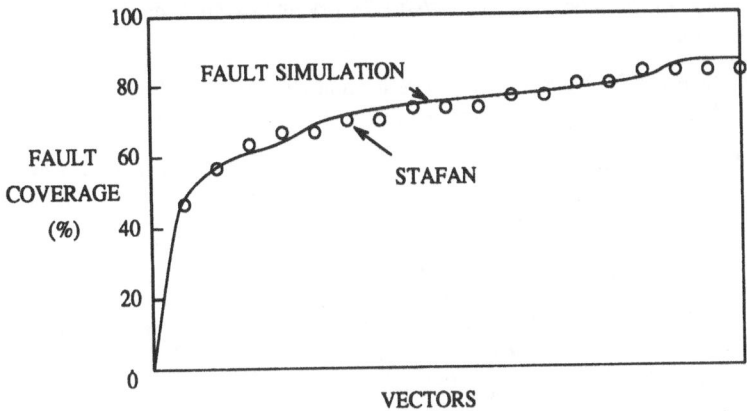

Fig. 5 Fault coverage estimate by statistical fault analysis (STAFAN).

6. CONCLUSION

Of the techniques described in this paper, statistical fault sampling has become a standard method in many design organizations. Statistical fault Analysis (STAFAN), also known as probabilistic fault grading (PFG), has been implemented by several companies and is also commercially available as a CAD tool. There is new interest in probabilistic testability analysis due to its applicability to built in self test (BIST) designs. A popular method of designing self-testing circuits is to use a linear feedback shift register to generate pseudorandom patterns for testing. Since the number of such patterns is usually large, fault simulation could be expensive. However, pseudorandom patterns statistically resemble random patterns and can be easily evaluated by testability analysis. Specific applications include finding coverage of random patterns and locating very low probability faults, often called *random pattern resistant*. Statistical analysis of test data is quite common, although the specific form of this analysis may vary. The use of such data for estimating product quality and setting the fault coverage requirement is recent and is receiving greater interest since higher fault coverage becomes increasingly difficult for higher levels of integration.

Two problems need further research. Probabilistic definition of testability allows an accurate analysis which has so far been developed for combinational logic only. Extension to sequential circuits will greatly enhance the applicability of this technique. Similarly, efficient Monte Carlo techniques of generating tests for sequential circuits are needed.

REFERENCES

[1] A. Rogers, *Statistical Analysis of Spatial Dispersions*, Pion Limited, London, England, 1974.

[2]   W. Feller, *An Introduction to Probability Theory and Its Applications*, Vol. I, Wiley, New York, 1968.

[3]   S. C. Seth and V. D. Agrawal, "Characterizing the LSI Yield Equation from Wafer Test Data,"*IEEE Trans. Computer-Aided Design*, Vol. CAD-3, pp. 123-126, April 1984.

[4]   T. W. Williams and N. C. Brown, "Defect Level as a Function of Fault Coverage,"*IEEE Trans. Computers*, Vol. C-30, pp. 987-988, December 1981.

[5]   V. D. Agrawal and M. R. Mercer, "Testability Measures -- What Do They Tell Us?," *Proc. Int. Test Conf.*, November 1982, pp. 391-396.

[6]   S. C. Seth, L. Pan, and V. D. Agrawal, "PREDICT - Probabilistic Estimation of Digital Circuit Testability," *Fault Tolerant Computing Symposium (FTCS-15) Digest of Papers*, Ann Arbor, MI, June 1985, pp. 220-225.

[7]   S. C. Seth, B. B. Bhattacharya, and V. D. Agrawal, "An Exact Analysis for Efficient Computation of Random-Pattern Testability in Combinational Circuits," *Fault Tolerant Computing Symposium (FTCS-16) Digest of Papers*, Tokyo, Japan, July 1986, pp. 318-323.

[8]   K. P. Parker and E. J. McCluskey, "Probabilistic Treatment of General Combinational Circuits,"*IEEE Trans. on Computers*, Vol. C-24, pp. 668-670, June 1975.

[9]   J. Savir, G. S. Ditlow, and P. H. Bardell, "Random Pattern Testability,"*IEEE Trans. Computers*, Vol. C-33, pp. 79-90, January 1984.

[10] F. Brglez, "On Testability Analysis of Combinational Networks," *Proc. Int. Symp. Circ. and Sys. (ISCAS-84)*, May 1984, pp. 221-225.

[11] D. M. Singer, "Testability Analysis of MOS VLSI Circuits,"*Proc. International Test Conference*, pp. 690-696, October 1984.

[12] V. D. Agrawal, "When to Use Random Testing,"*IEEE Trans. Computers*, Vol. C-27, pp. 1054-1055, November 1978. Also comments and author's reply in IEEE Trans. Comput., Vol. C-28, pp. 580-581, August 1979.

[13] V. D. Agrawal and P. Agrawal, "An Automatic Test Generation System for Illiac IV Logic Boards,"*IEEE Trans. Computers*, Vol. C-21, pp. 1015-1017, September 1972.

[14] S. Funatsu, N. Wakatsuki, and T. Arima, "Test Generation Systems in Japan," *Proc. 12th Des. Auto. Conf.*, 1975, pp. 114-122.

[15] D. M. Schuler, E. G. Ulrich, T. E. Baker, and S. P. Bryant, "Random Test Generation using Concurrent Fault Simulation," *Proc. 12th Des. Auto. Conf.*, 1975, pp. 261-267.

[16] P. Agrawal and V. D. Agrawal, "On Monte Carlo Testing of Logic Tree Networks,"*IEEE Trans. Computers*, Vol. C-15, pp. 664-667, June 1976.

[17] V. D. Agrawal, "An Information Theoretic Approach to Digital Fault Testing,"*IEEE trans. Computers*, Vol. C-30, pp. 582-587, August 1981.

[18] K. P. Parker, "Adaptive Random Test Generation,"*J. Des. Auto. and Fault-Tolerant Computing*, Vol. I, pp. 62-83, October 1976.

[19] M. Abramovici, J. J. Kulikowski, P. R. Menon, and D. T. Miller, "SMART and FAST: Test Generation for VLSI Scan-Design Circuits,"*IEEE Design & Test of Computers*, Vol. 3, pp. 43-54, August 1986.

[20] X. Fedi and R. David, "Experimental Results from Random Testing of Microprocessors," *Fault-Tolerant Computing Symp. (FTCS-14) Digest of Papers*, Kissimmee, FL, June 1984, pp. 225-230.

[21] Y. A. Shreider, *The Monte Carlo Method*, International Series of Monographs in Pure and Applied Mathematics, Vol. 87, Pergamon Press, New York, 1966.

[22] V. D. Agrawal, "Sampling Techniques for Determining Fault Coverage in LSI Circuits,"*Journal of Digital Systems*, Vol. V, pp. 189-202, 1981.

[23] S. K. Jain and V. D. Agrawal, "Statistical Fault Analysis,"*IEEE Design & Test of Computers*, Vol. 2, pp. 38-44, February 1985.

[24] J. A. Waicukauski, E. B. Eichelberger, D. O. Forlenza, E. Lindbloom, and T. McCarthy, "A Statistical Calculation of Fault Detection Probabilities by Fast Fault Simulation," *Proc. Int. Test Conf.*, Philadelphia, PA, November 1985, pp. 779-784.

FAULT MODELS

Samiha Mourad and Edward J. McCluskey

CENTER FOR RELIABLE COMPUTING
Computer Systems Laboratory
Departments of Electrical Engineering and Computer Science
Stanford University, Stanford, CA 94305-4055, USA

1. INTRODUCTION

A key requirement for obtaining reliable electronic systems is the ability to control the effects of failures in such systems. Failures in integrated circuits can be characterized according to their duration: temporary or permanent. Techniques for handling failures in digital systems use fault models to represent the logical effects of the failures. Fault models have the advantage of being a more tractable representation than physical failure modes, but risk the omission of vital effects on system operations. For example, the most common fault model assumes single-stuck faults even though it is clear that this model does not accurately represent all actual physical failures. The rationale for continuing to use the single-stuck fault model is that it has been satisfactory in the past. The arguments against the single stuck model are that it is not adequate for present VLSI or ULSI technologies.

Fault control involves first detecting the presence of a fault and then acting to eliminate its effects. There are two distinct environments in which fault detection is important: production testing and concurrent error checking. The critical issues for production testing are the cost of the test, the number and types of undetected failures, and whether faulty parts are repaired or discarded. Concurrent checking schemes differ in their cost, effect on system reliability, class of failures detected, and repair strategy. Fault models are used in developing methods for concurrent checking as well as for production test. The choice of a fault model has a dominant effect on the characteristics of the resulting technique.

This paper presents various fault models (for temporary as well as permanent failures) for VLSI and ULSI parts. It also evaluates the effectiveness of various testing and checking schemes for systems constructed of such parts. In Section 2, the relationship between faults, failure modes and failure mechanisms is established. In Section 3, processes for screening failures are described. Sections 4 through 6 examine different fault models used to represent permanent failures, the stuck-at model, MOS faults and bridging faults. In Section 7 the effectiveness of test sets generated for the single stuck-at fault model to detect multiple and bridging faults are evaluated. Delay faults are the topic of Section 8. Topics related to temporary faults will be covered in Sections 9 to 11.

2. FAULTS, FAILURE MODES, AND MECHANISMS

The underlying cause of digital system malfunctions is always some type of failure. These failures can be due to incorrect design, defects in the material or in the production, physical interference from the environment, or human error. The following classifications of failures are generally accepted in the testing community [78]:

Permanent Failures are usually caused by breaks due to mechanical rupture or some wearout phenomenon. They do not occur as frequently as *temporary failures* which are either transient or intermittent. *Transient failures* are induced by some external perturbation such as power supply fluctuations or radiation. *Intermittent failures* are usually due to some degradation of the component parameters. Permanent failures are sometimes referred to as *hard failures* and temporary failures as *soft failures*.

F. Lombardi and M. Sami (eds.), Testing and Diagnosis of VLSI and ULSI, 49–68.

The most common failure mechanisms are listed in Table 1. These mechanisms manifest themselves on the circuit level as failure modes. The most common *failure modes* are open or short interconnections, or parameter degradation. Failure mechanisms are largely dependent on the technology and even the layout of the circuit.

Table 1. Failure Mechanisms.
Basic chemical or physical failure causes

Surface and bulk effect
    Passivation pits or cracks
    Gate oxide breakdown
        Pinholes or thin spots in oxide
        Electrical overstress
    Surface Potential instability
Metalization and metal semiconductor
    Open metal at oxide steps
    Wire bonding failure
    Intermetalic compound formation
    Electromigration
Package related
    Mass transport of metal atoms
    Momentum exchange with electrons

Passivation pits and cracks are usually spotted during visual inspection They affect the yield rather the reliability of the product. Localized effects such as pinholes and thin spots in oxide may be partly eliminated by electrical screening.

These different modes cause the component to fail at different stages of its lifetime. The failure rate over the lifetime is given by the well known bathtub curve shown in Fig. 1.

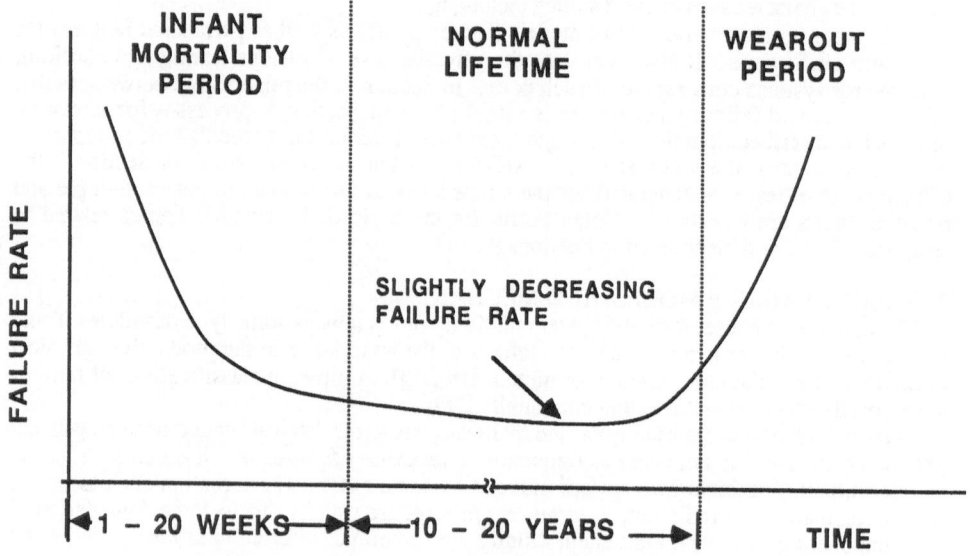

Figure 1. Failure Rate versus Lifetime.

Failures that escape visual and optical scanning cause many chips to fail within one to twenty weeks of their operation (infant mortality). At the end of this period, the failure rate tends to stabilize for 10 to 20 years (normal lifetime period). Eventually, due to excessive use of the components, there is an exponential increase in failure rate (wearout period).

The different failure modes are manifested on the logical level as incorrect signal values. A *fault model* is the representation of the effect of a failure by means of the change that is produced in the system signal [78]. The usefulness of a fault model is determined by its accuracy in representing the effect of the failure as well as its tractability as a design tool. Table 2 lists the most common fault categories. An example is given in Fig. 2, taken from [4], to illustrate the interpretation of physical defects as failure modes and their manifestation on the circuit and gate level. An attempt to map physical defects to the circuit and logical level is reported in [72]. The severity of the different physical defects is measured by their manifestation on the logical level. Results of applying of this method to one circuit, a full-adder, show that of a total of 743 simulated physical defects, 93 resulted in circuit level faults.

Table 2. Most Commonly Used Fault Models.

Fault Model	Description
Single Stuck-at (SSA) Faults	One line has a fixed value 0 or 1
Multiple Stuck-at Faults	Two or more lines have fixed values, not necessarily the same value.
Bridging Faults	Two or more lines that are normally independent, become connected when faulty.
Combinational Faults	Any change in the function realized by a combinational circuit can be caused by the fault, but the circuit remains combinational.
Stuck-Open (SOP) Faults	A failure in a pull-up or pull-down transistor in a CMOS logic device causes it to behave like a memory element under certain input conditions.
Stuck-on (SON) Faults	A transistor is always conducting.
Delay faults	A fault caused by incorrect delays in one or more paths in the circuit.
Intermittent Faults	Caused by internal parameter degradation. Incorrect signal values occur for some but not all states of the circuit. Degradation is progressive until permanent failure occurs.
Transient Faults	Incorrect signal values caused by coupled disturbance. Coupling can be via power bus capacitive or inductive coupling. Includes both internal and external sources as well as particle impact.

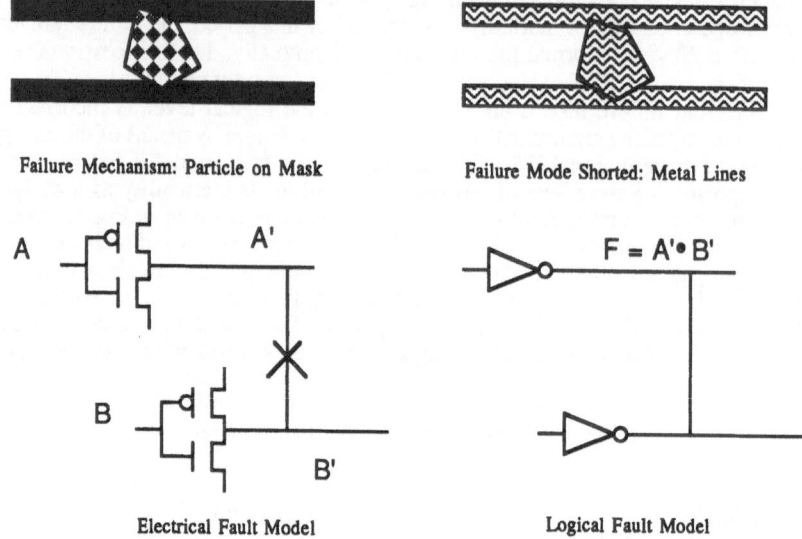

Figure 2. Interpretation of failure modes on the electrical and logical model.

The single stuck-at fault is the most widely used fault model. It is the basis of many test pattern generators and fault simulators (software programs). With the advent of MOS technology, it has become evident that other fault models are needed to represent more accurately the failure modes in this technology. Also, although temporary failures occur more often than hard failures, they have not been given as much attention. Before describing the different fault models, the processes used to screen failures in digital components will be presented.

3. FAILURE SCREENING

The first principle in developing reliable systems is to avoid building them with faulty components [82]. The issue then is how to screen the defective components. There are several steps followed in screening failed components [82]. Here we will group them in four major stages. The first stage consists of an internal visual (optical) test. The purpose of such a test is to detect some of those defects listed in Table 1. The next stage consists of parametric tests. *DC parametric testing* insures that the device will operate within the specified voltage and current levels. *AC parametric testing* is a general name for timing measurement tests which measure speed and propagation delay. Parametric tests are technology dependent [105]. For example, ECL technology is current driven and voltage requirements are different than for high input impedance TTL. The third and fourth stages are functional and stress testing. The latter is performed in order to eliminate marginal components (infant mortality) using stresses such as, vibration, temperature and humidity. The logic aspects of a design are typically checked by applying a set of input signals to assert that the circuit is performing its proper function. This is sometimes called *functional testing*. It has been proven that such a test does not necessarily guarantee that the circuit is fault free [78], [79], [80], [45]. Hence a new term *Boolean testing* was coined by [79]. A Boolean test is defined as a test in which only logical signal values are considered. It usually includes the functional test and, in addition, patterns to guarantee that the circuit is fault free. After this test has been performed, the defect level DL of the components is determined as function of the yield and the test fault coverage. The function, $DL = 1 - Y^M$ [115] was simplified by [83] to $DL = M \ln (1/Y)$, for values of $DL \leq 1000$ DPM (Defects Per Million). Here M is the *fault uncoverage*, that is, the ratio of undetected faults to the total number of faults in the circuit, and Y is the yield, the fraction of die on a wafer that are free of defects. The relation is linear in M and logarithmic in Y. Thus, it is more effective to decrease M in order to reduce the defect level. The single most important issue in eliminating defective components is the development of Boolean test sets that guarantee a high fault coverage. Such test sets are very largely dependent on the fault model used, as will be shown in the remainder of this paper.

4. THE STUCK-AT FAULT MODEL

Early techniques of testing digital circuits were mostly concerned with functional design verification which could not necessarily guarantee that the circuit was fault free. The switch to a testing method that takes into consideration the structure of the circuit was suggested in a paper presented by R. Eldred at the August 1958 meeting of the ACM [33]. The opening sentence of this paper is:

"In order for the successful operation of a test routine to guarantee that a computing system has no faulty components, the test conditions imposed by the routine should be devised at the level of the components themselves, rather than at the level of programmed orders."

In order to confirm or deny the presence of a failure in an n–input circuit representing a logical function, one can apply all 2^n input combinations and check if the output response is correct or faulty. Exhaustive testing has been used in VLSI testing by two major microprocessors manufacturers, Motorola [64] and Intel [40]. When the number of inputs is very large, more than 20 primary inputs, exhaustive testing is not practical [78]. To make testing more manageable, it is then important to characterize the defects in the circuit as logical or electrical values on the nodes connecting the different components of the circuit. That is, to represent the failure mode by a logical value. The stuck-at fault model is one such representation.

4.1 Single Stuck-at Fault.

The single stuck-at fault model assumes that a node in the logical diagram of a circuit is always fixed at 0 or at 1. In the former case, it is considered stuck-at-0 (sa0) and the latter, stuck-at-1 (sa1). This fault model can successfully represent some of the failure modes. For example, If the input 'A' of the XOR gate shown in Fig. 3 is connected to V_{GND} it is stuck at 0. A short in resistance R_2 can also be modeled by the input 'A' stuck at 0.

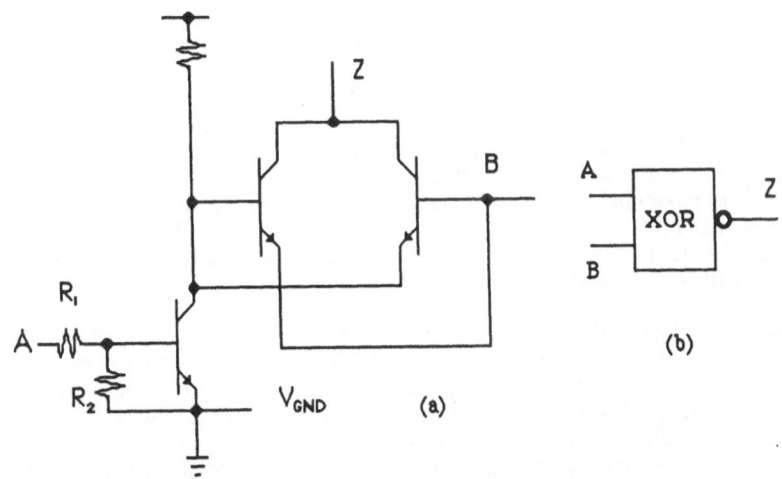

Figure 3. Bipolar Circuit: (a) Circuit level, (b) Logical level.

In the bridge nMOS circuit shown in Fig. 4 [85], the output function, Z, is given by $(A + C)(B + D)$. If C is shorted to V_{dd} (held high), this will be equivalent to C sa1. The same effect is obtained if transistor C is always on (stuck on). On the other hand, if the line ab is

54

open, the circuit is reduced to AB + CD and the failure mode cannot be modeled by the stuck-at model. In a simulation study on MOS technology circuits, [102] found that the stuck-at model can represent one of several causes: extra metal, diffusion of poly, missing diffusion or missing poly.

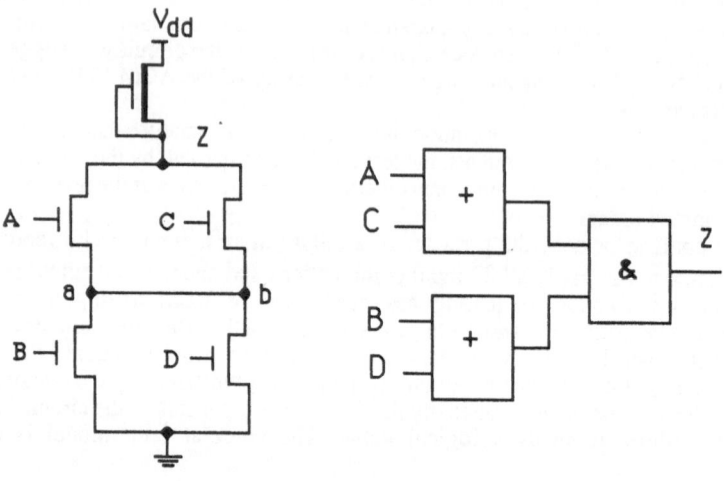

Figure 4. nMOS Circuit.

The number of SSA faults is 2m, where m is the number of signal lines in the circuit. By grouping faults according to their effect on the circuit, the number of faults to be considered during test set generation can be reduced. Two faults are *equivalent* if they have the same detecting test sets [74]. The SSA faults can thus be grouped in *equivalence classes* and it is thus sufficient to consider one fault from each class in generating test sets. The number of faults can be further reduced using *fault dominance*. A fault f_1 is said to dominate another fault f_2 if the test set of f_1 includes the test set of f_2. Consequently, test patterns that detect f_2 also detect f_1. Fanout stem stuck faults dominate fanout branch faults unless the fanout branches reconverge with odd inversion parity [82]. Using fault equivalence and fault dominance, the fault set can be collapsed to a minimal number.

The SSA fault model has proven to be successful in detecting permanent failures. It is the widely used model in software packages developed for digital testing. The attractiveness of the SSA fault model is its simplicity. There are different strategies for developing test sets for SSA faults. These will be discussed in Section 7.

4.2   Multiple Stuck-at Faults

With increased device density and decreased geometry, the likelihood that more than one SSA fault can occur simultaneously is greater [43]. In his study, Goldstein translated the probability distributions of physical defect size and location into probabilities of occurrence of single and multiple faults. The results suggest that up to 6 simultaneous multiple faults should be tested to guarantee that a circuit is fault free. The major problem involved with the generation of test sets for multiple faults is the large number of possible faults. The numbers of single, double and triple SA faults are listed in Table 3. In general, the number of r simultaneous faults in an m–node circuit is $2^r C(m,r)$, where $C(m,r)$ is the number of all combinations of r out of m things.

55

Table 3. Numbers of Possible Stuck-At Faults.

Number of nodes	Number of Faults		
	Single	Double	Triple
10	20	180	960
100	200	19800	1.3x10^6
1000	2000	1.998x10^6	1.3x10^9

In order to reduce the number of multiple stuck-at faults, fault collapsing is used. Also, SSA test sets are often used to detect multiple faults as will be discussed in Section 7.

5. FAULT MODELS FOR MOS DEVICES

The SSA fault model cannot represent all failure modes in MOS technology [9], [39], [108], [111]. The main reason for this is that MOS combinational circuits do not remain combinational under all faulty conditions.

The are several failure modes in MOS technology circuits [9], [39]: 1) transistors shorts and opens, 2) shorts between gate and drain or gate and source, 3) opens and shorts on interconnecting lines, and 4) open on gate, drain, or source contacts. The nMOS implementations of an inverter shown in Fig. 5 will be used to illustrate the interpretation of the failures modes on the logical level. The numbers indicated on the circuits refer to a certain failure mode, e.g., site 1 is open on the interconnecting line, site 2 is a short between gate emitter, and site 3 is gate contact open, etc.

The first and second categories of the failure modes can be represented by a stuck-at fault. If the transistor is stuck on, the circuit behaves as if input A is sa1. On the other hand, if the gate is shorted to the source (site 2), the fault is equivalent to A sa0. Opens on the interconnecting lines may be represented by a stuck-at model or may cause the circuit to store its previous logical value, depending on the site of the open in the circuit. In Fig. 5, an open at sites 1 and 5 is equivalent to a sa0 fault on A, while an open at site 6 will cause the output to depend on the input signal. For A = 1 the output is pulled down to 0, and for A = 0 the transistor is not conducting and it retains its previous value. Detection of the open at site 6 requires the application of the pattern pair (0,1)

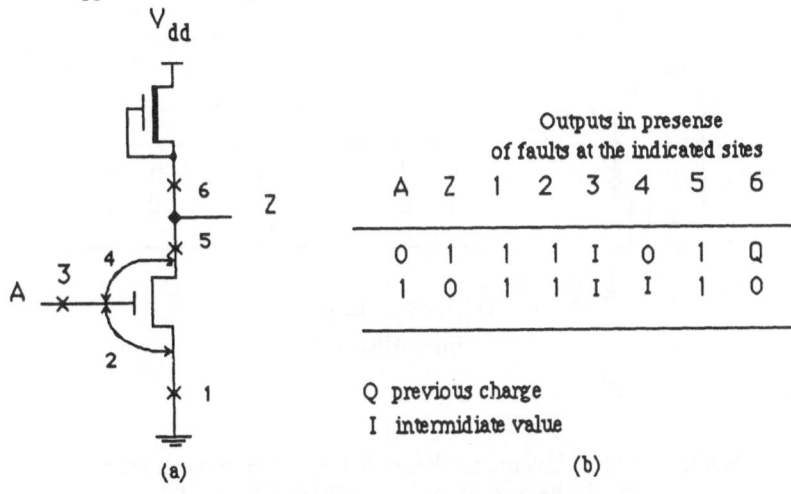

Figure 5. nMOS Inverter, a) the sites of the failure modes, b) Circuit responses.

For CMOS circuits, failures in pull-up or pull-down transistors can cause the transistor to function as a memory element. Here again, a pair of test patterns is needed to detect the fault. This type of failure is called *Stuck-Open* (SOP). It has been investigated extensively by [20], [26], [34], [61], [111], and [118]. In references [61] and [118] it has been demonstrated that Stuck-Open faults can be avoided by proper design practices. They describe a layout method for gate-arrays chips in which any SOP fault is forced to appear as a SA fault. Other design for testability techniques have been developed for CMOS circuits [19], [66], [67], [70], [92]. For the inverter in Fig. 6, opens at the indicated sites (1 through 6) are represented by stuck-open faults. The other important fault model in CMOS is the *Stuck-On* type (SON) in which a transistor is always on. Such a fault may cause a short between V_{DD} and ground. The output will then be charged or discharged to a faulty value.

Test generation algorithms for MOS technology have been reported [23], [25], [26], [42], [58]. Jain [58] suggested a modification of the D-Algorithm to represent the failures by the stuck-at model. Chiang et. al. [25], [26] have developed a graph and an approach to test pattern generation for stuck-open faults and short in complex gates. Test generation for multiple SOP and SON faults has been attempted by El-Ziq[35] for some special functions. For CMOS combinational circuits, a 3-pattern test scheme is used to detect both SOP and SON faults [68]. The advantage of this technique is that the patterns can be generated by a gate level SSA fault automatic test pattern generator.

Simulators based on gate level fault models are not suitable to represent all faulty conditions of MOS circuits. Special simulators were developed. Here, a transistor is represented as a switch [16, 17] or as a *connector-switch-attenuator* (CSA) multi-component [49], [50]. In the CSA model each ideal transistor is represented as a switch and a connector. The switch is open or closed, depending on the value of the connector. In order to allow for intermediate values between open and closed, an attenuator is added. The electrical representation of the model is by a voltage source and a finite series of non-zero resistors R_j. Using Ohm's law, the current is given by $I_j = V_j/R_j$ and represent the faulty values other than logical 0 or 1. Each transistor is then represented by the pair (V_j, I_j). Hayes [51] presents a general class of CSA and its underlying algebraic structure (pseudo Boolean). A survey of switch level algorithms used for logic and fault simulators is given in [18]. It includes an extensive bibliography on switch level representation and application in test pattern generation and simulation.

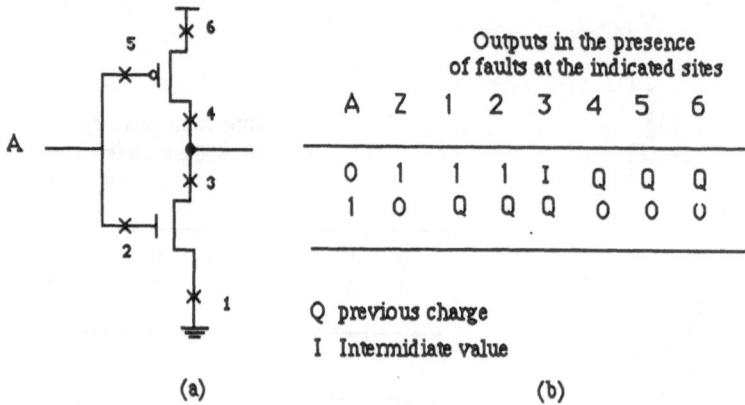

A	Z	1	2	3	4	5	6
0	1	1	1	I	Q	Q	Q
1	0	Q	Q	Q	0	0	0

Outputs in the presence of faults at the indicated sites

Q previous charge
I Intermidiate value

(a)   (b)

Figure 6. CMOS Inverter a) Sites of failure modes represented by
Stuck-Open faults, b) Responses of the circuit.

6. BRIDGING FAULTS

Bridging faults occur when two or more lines in a circuit are shorted together and create wired logic [84]. This wired logic depends on the types of gates driving the bridged lines. When the fault involve r lines with $r \geq 2$, the fault is said to be of *multiplicity r;* otherwise, it is a simple bridging fault. Multiple bridging faults are more likely to occur at the primary inputs of the chip. Bridging faults are becoming more important because of high device density and decreasing geometry. They are classified as feedback (fb) and non-feedback faults (nfb) [84]. Fig. 7 illustrates the different types of bridging faults. The total number of simple bridging faults for a circuit of m nodes is C(m,2). It is very unlikely for some pairs of the signal lines to be shorted. Thus the number of possible faults is smaller and depends largely on the circuit layout. There are several causes that produce bridging faults. Among these are 1) manufacturing defects, 2) packaging, and 3) intercomponent (chips or boards) connections.

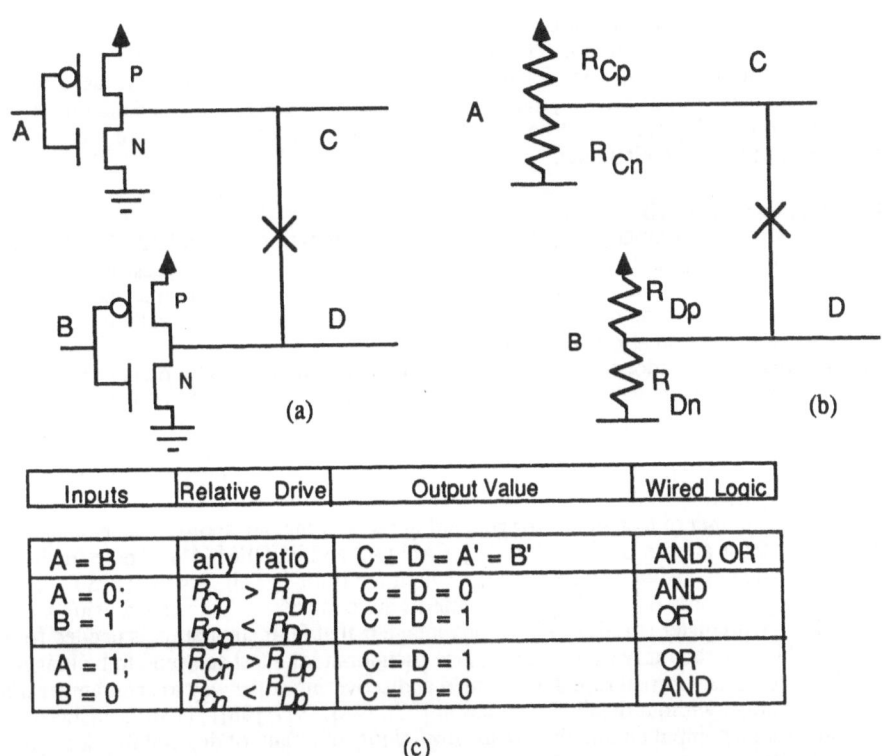

Inputs	Relative Drive	Output Value	Wired Logic
A = B	any ratio	C = D = A' = B'	AND, OR
A = 0; B = 1	$R_{Cp} > R_{Dn}$ $R_{Cp} < R_{Dn}$	C = D = 0 C = D = 1	AND OR
A = 1; B = 0	$R_{Cn} > R_{Dp}$ $R_{Cn} < R_{Dp}$	C = D = 1 C = D = 0	OR AND

(c)

Figure 7. The voting model for bridging faults: (a) Bridged signal lines, (b) Equivalent electrical circuit, (c) Possible output of the faulty circuit.

Bridging faults in TTL and nMOS circuits can be modeled by wired logic. However, in CMOS, which has become the dominant VLSI technology, the simple SSA model is not adequate. The lack of an effective logic model for bridging faults severely complicated the test pattern generation for CMOS circuits, and has resulted in the practice of monitoring the power supply current [3], [118]. Chiang [26] states that the relative strengths of the pull-up and pull-down of the driving gates determines the voltage of the shorts and that voltage may lie between the defined logic voltages. A logic model proposed by Timoc [108] uses "asymmetrical wired logic" where one signal line retains its value and the other assumes the logical function of both shorted lines. Simulation studies on CMOS circuits carried out by [3] and [36] showed that the bridged lines remain at logic 0 or 1 depending on the driving gates and the input on these gates. The *probabilistic model* [36] indicates that in the absence of information about the driving gates, a wired-AND model is most appropriate. The *voting model* [3] is illustrated in Fig. 7. Here, for the same driving gates, the logical value of the bridged lines can be represented by wired-OR or wired-AND, depending on the input signals on these driving gates and their pull-up and pull-down values as shown in Fig. 7(c).

Feedback bridging faults transform a combinational circuit into a sequential circuit, and increase the number of states in a sequential circuit. The increased number of states causes an increase in the test length [1], [53], [70], [84], and many others studied the detection of bridging faults using SSA fault test sets . Their results together with a simulation experiment [86] are discussed in the next section.

7. EFFECTIVENESS OF SINGLE-STUCK FAULT TEST SETS

Because the stuck-at fault model has been used for a long time, many effort and resources were invested in developing tests for it. It is only natural that these tests are used to detect other fault models. In this section, the effectiveness of SSA fault model in detecting multiple SA faults and bridging faults will be reported. First, we will consider different types of test generation for the SSA fault model.

7.1. Single Stuck-at Faults

McCluskey [80] compares several strategies for generating test patterns for SSA faults. The most common approach to testing is design verification. However, passing this test does not guarantee that the circuit is fault free. A fortiori this is also true for toggle and pin test sets. These approaches to testing do not assume a fault model. Even when a model is assumed, the method of test generation affects the fault coverage.

Exhaustive and pseudoexhaustive testing [75], [76] guarantee 100% detection of all SSA faults. However, exhaustive testing is not realistic when the number of inputs exceeds 20 and pseudoexhaustive testing requires efficient segmentation of the circuit. The optimal circuit segmentation for pseudoexhaustive testing has been shown to be NP-complete [6]. Techniques to partition a circuit were developed [90], [95], and [103] in which simulated annealing was used.

There is a number of techniques for generating test sets that are shorter than exhaustive test sets. Among these are the D–algorithm [96], FAN [38] and PODEM [41]. These methods are computationally expensive, and they are not suited for Built-In Self-Testing (BIST). For this type of testing, random or pseudorandom patterns are computed when needed using a Linear-Feedback-Shift-Register (LFSR). The disadvantage is that fault simulation is needed for fault grading. This is particularly troublesome since pseudorandom test sets tend to be longer than algorithmic test sets. Current research is aimed at discovering some way to accurately estimate the fault coverage without detailed simulation [27], [28], [81], [98], [113]. McCluskey [81] shows that for an n–input circuit, the *escape probability* of a fault of detectability k is given by $Q_k = e^{-kL/N}$, where $N = 2^n$, L is the length of the test set, and the *detectability* is the number of test patterns that detect the fault. The theoretical results were confirmed with simulation studies on multiplexers of different complexities and implementation [69].

7.2. Multiple Stuck-at Faults

Several studies have been conducted to examine the detection of multiple stuck-at faults using single stuck-at test patterns. Specific circuits have been identified for which a complete SSA fault test set is guaranteed to detect all multiple stuck-at faults. One such class is single output, two level circuits with primary input fanout allowed. However, these circuits are actually a special class of circuits known as restricted connected sets [99], [100]. Although the multiple fault testability of restricted connected sets is a useful property, few practical circuits correspond to the required structure.

It is not always practical to use exhaustive testing to detect multiple faults for the same reasons mentioned in previous sections. However, pseudoexhaustive and pseudorandom test sets, as well as algorithmic test sets generated for SSA faults, can be applied. Algorithms based only on the circuit structure and not the specific test set have been developed for computing a lower bound on the multiple fault coverage of single fault test sets [2]. If the guaranteed coverage is sufficient, then no fault simulation or other circuit analysis is required for the evaluation of the multiple fault coverage of specific single fault test sets. Here, the main conclusion of the analysis that the multiple fault coverage of a single stuck-at fault test set is generally close to 100% for circuits without reconvergent fanout. But the bound on the coverage decreases rapidly as the amount of reconvergent internal fanout increases. Agarwal [2] shows that the inclusion of one internal reconvergent fanout reduces the bound on the fault coverage by approximately 5%. The inclusion of a second and a third fanout reduces the bound by 3% in each case. The algorithm shows only lower bounds.

In order to assess the multiple fault coverage by SSA fault test sets, simulation studies were performed on the three types of circuits: the 74LS181 4-bit ALU [55], [56], [57], n-input balanced parity trees ($2 \leq n \leq 32$) [87], and a two-rail code checker [88]. The first benchmark circuit was selected because it is one of the most complex combinational circuits. The second circuit is a C-testable circuit that requires no more than 4 test patterns for 100% detection of all SSA faults. The last circuit uses only codeword test patterns.

For the ALU, ten SSA fault test sets were used. They included algorithmic as well as pseudorandom and pseudoexhaustive test sets. Better than 99.9% coverage was obtained for double stuck-at faults. Almost all the undetected faults involved reconverging paths through exclusive OR gates. The fault coverage determined by simulation is much higher than would be anticipated based on the results of Schertz and Agarwal. Careful analysis shows that the simulation results and the assumptions of the theoretical results suggest that much of the difference is due to test set size and multiple primary outputs. Many of the undetectable faults were *self masking*. That is, two faults which are observable at the same primary output are detected by the same test patterns. The parity tree study yielded a much lower fault coverage, 85.33%. This can be explained by the highly reconverging paths nature of the circuit that caused fault masking. For the 2-rail checker circuit, a 4 codeword test set that detects all SSA faults yields a double fault coverage of 84.5%. This coverage increased to 96% when an exhaustive codeword test was applied.

7.3. Bridging Faults

SSA fault test sets have been proposed for the detection of bridging faults by [1], [12], [37], [84] and [116]. They yield 100% fault detection for some special circuits. Mei [84] starts with an existing SSA fault test set, reorders it and augments it to detect all bridging faults. [Abromivici 83] alters the SSA test set to meet constraints imposed by bridging faults. For VLSI, these methods are not efficient. In order to gain more insight in bridging fault testing, Millman [86] carried out several simulations on the ALU 181, balanced parity trees and multiplexers. Both wired-AND and wired-OR models were used. The results of these experiments (bridging fault coverage) showed that 1) there was no significant difference between the two types of wired logic, 2) the length of the test set model seems not to be a significant factor, and 3) the patterns in a pseudoexhaustive test set can be rearranged in such a way as to increase the coverage to 100%.

8. DELAY FAULTS

It is possible for a circuit to be structurally correct but to have signal paths with delays that exceed the bounds required for correct operation. *Delay testing* of a circuit determines if it contains signal paths that are too slow or too fast in propagating input transitions. *Delay faults* may not be provoked if the operating frequency is low.

Some test structures can be used to provide a measure of device switching speed. The use of test structures becomes more difficult as device density increases. Other approaches to ensure correct timing include external scan [117] and I/O scan path [78].

60

Delay faults model the effect of physical defects on circuit response time. These faults may be made more complex by taking into account the direction of the signal transition as well as the condition of other gate inputs. Physical causes for delay faults are discussed in [54] and [60].

Delay testing consists of applying a pair of input vectors at the desired operational speed and observing the outputs for early or late transition. The application of a vector pair to sensitize and propagate the fault is illustrated in Fig. 8. The delay in the inverter is propagated to the output by the pair of vectors (011, 001). The inputs a and c sensitize a path for any transition on b to appear on the output.

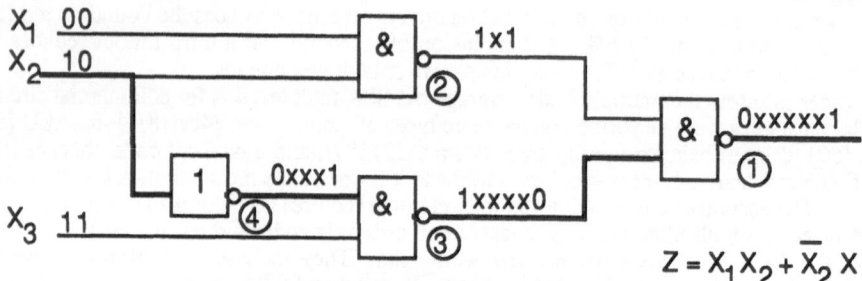

Figure 8. Sensitization and detection of a delay fault.

Test generation techniques for delay faults are proposed in [15], [54], [65], [71], and [101]. The test coverage is improved when a self-test method is used [10]. In addition to algorithmic delay test generation, pseudorandom testing [112] is also used. The latter is more suitable for BIST environment. Simulation issues in delay testing are discussed in [60] and [114].

9. TEMPORARY FAILURES

Although hard failures result in a need to change a component or repair it causing a long mean-time-to-repair, soft or temporary failures are more frequent [7], [104]. Temporary failures are much harder to track because when a component, chip or board, is tested, it is not usually possible to reproduce the fault. They are not as thoroughly studied as permanent failures. Temporary failures are encountered in different digital components — RAMs, microprocessors, etc.

There are two major types of temporary failures: transient and intermittent (recurring). The first type is usually due to some temporary external condition, while the second is due to varying hardware states — parameter degradation or improper timing. Table 4 lists the different types of temporary failures and some of their main causes.

Table 4. Temporary Failures.

Types	Causes
Transient	Power supply disturbances
	Electromagnetic interferences
	Charged particles
	Atmospheric discharges
	Electrostatic discharges
Intermittent	Parameter degradation
Timing	
Metal related (open/short)	

10. TRANSIENT FAULTS

A transient fault occurs when a logic signal has its value temporarily altered by noise signals and the resulting signal may be interpreted incorrectly by the rest of the circuit [78]. Such a fault is difficult to diagnose and correct. It is thus important to minimize the noise in the circuit and increase the noise immunity of the circuit. Transient failure may be caused by fluctuations in the power supply, metastability or cosmic radiation.

10.1. Power Supply Disturbances

Power supply disturbances are known to cause errors in the operation of digital systems. Allen [5] and Chesney [24] characterized the susceptibility of circuits to power supply disturbances by measuring the change in the outputs of gates whose inputs are kept at constant signals. These experiments related the disturbances to the noise immunity of the circuits.

Later experimentation was carried out by Côrtes [29], [30] on circuits implemented in different technologies: CMOS Gate Array, CMOS Breadboard, and LSTTL breadboard. The circuit used in the experiment was proposed by [75]. Here, the experiment was carried out under a more realistic assumption, logic signal changing with time. It was found that the susceptibility of the circuits to power supply voltage disturbances is related to the operating frequency. Errors are more likely to occur as the operating frequency increases. The results show that propagation delay variation is the dominant effect and that noise immunity plays a smaller role in error occurrence. The experimental results were confirmed with simulation on the Daisy Megalogician. It was then concluded that failures due to power supply disturbances can be modeled by delay faults. The dependency of disturbance (ΔV_{dd}) on clock frequency for the three types of technologies are shown in Fig. 9 [29]. Here the nominal supply voltage, V_{dd}, is 5 volts.

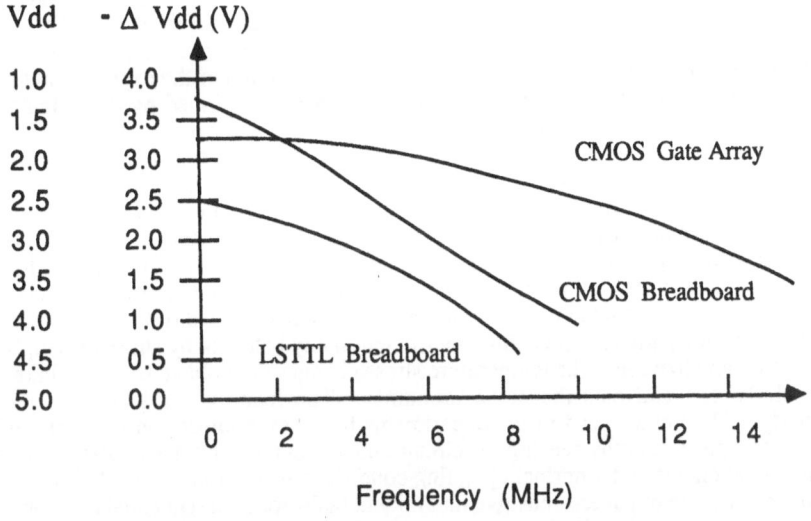

Figure 9. Tolerance of disturbances versus clock frequency.

10.2 Metastability

Another form of transient faults is caused by metastability in latches and flip flops. Metastability occurs when a latch is given only enough energy to switch its state halfway to another stable state, and when the latch enters a semi-stable, or metastable state which exists somewhere between the two stable states. The latch will remain in this metastable state for some indeterminate amount of time, but will eventually leave it for one of the stable states. No method eliminating all metastability is known [78]. It is important, however, to determine with a certain degree of certainty the probability of metastability occurrences. For this, metastability

62

sensors have been developed [21], [36], [106], and [107]. The mean time between metastability (MTBM) can be predicted in terms of latch parameters [22].

10.3. Radiation Induced Faults

Studies on the effect of radiation have examined alpha particles as well as cosmic rays disturbances in static and dynamic RAMs [11], [73], [110], [119]. Radiations cause ionization that may alter the content of the RAM cells. This is known as a *single-event* fault. Both bipolar [91] and MOS [62] technologies are susceptible to such disturbances. There is a general consensus that radiation hardening and proper packaging of integrated circuits are sufficient to decrease the occurrence of radiation induced transient faults [32].

11. INTERMITTENT FAULTS

Intermittent failures are recognized to be an important cause of field failures in computer systems. Very little is known about the failure mechanisms because spontaneous intermittents are difficult to observe and control. However, artificially induced intermittent failures can be easily produced, controlled and observed.

Several papers [14], [59], [63], [97], and [109] have addressed the problem of testing for intermittent failures. The intermittent fault models presented in these papers assume signal-independent faults. This assumption has turned out to be inappropriate after the experimental evidence of pattern-sensitive intermittent failures.

Pattern Sensitivity was first encountered in memory testing [46], [47], [93]. An interesting discussion on the causes of pattern sensitivity is presented in [94]. Hackmeister [44] reports instruction sensitivity in microprocessor chips, in a first reference to pattern sensitivity in non-memory circuits. Using standard microprocessor characterization techniques, as described in [52], Hackmeister produced *shmoo plots* (supply voltage versus speed) that have different shapes for different instruction streams. Other authors referred to Hackmeister results as pattern sensitivity due to "charge-leakage possibilities" [13] or "presence of moderately large RAMs on chip" [48]. Despite the lack of a detailed description of the experimental procedure in [44], Côrtes [31] believes that these failures are better described by delay faults than by pattern sensitivity. It is reasonable to conjecture that different instruction streams exercise different portions of the chip and failures are caused by delay faults due to supply voltage reductions as described in [30].

An attempt to collect data on intermittent failures on Sperry-Univac computers is reported by O'Neil [89]. Hard failures (field data) that are believed to have appeared earlier as intermittent failures are analyzed. Two classes of failure mechanisms are listed as responsible for this type of failure: 1) metal-related open and short circuits; 2) marginal operation of violations of operating margins. The latter class of failures has been investigated further by Côrtes [31]. In this study, catalog parts are forced into intermittent faulty behavior by stressing supply voltage, temperature and loading. The temperature stress changes the voltage transfer characteristics. The voltage stress affects the noise immunity. The loading stress reduces the driving capability. All stresses used in the experiments have some impact on the logic interfacing between two gates, thereby causing the circuit under stress to exhibit a similar behavior to that of a marginal circuit under normal operating conditions as described in [83]. The experiments reveal the existence of pattern-sensitive intermittent faults for both sequential and combinational circuits. The pattern sensitivity is particularly strong for light intermittent faults, i.e., low probability of activity. A *stress-strength* model was developed to explain the experimental results [31]. The two gates in Fig. 10 form a driver-receiver pair. The low and high values of the voltage at the input of the receiver can be related to *strength*, and the low and high values of the voltage at the driver's output as *stress*.

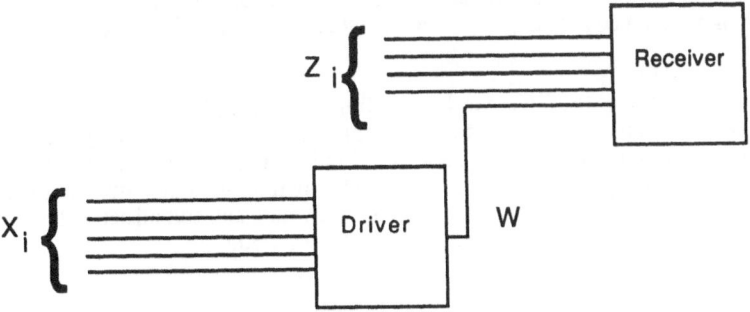

Figure 10. The Stress-Strength model, the driver-receiver pair.

A failure occurs when stress is larger than strength. The stress-strength relation is shown in Fig. 11. Côrtes further conjectures that his results can be extended to LSTTL and HCMOS catalog parts.

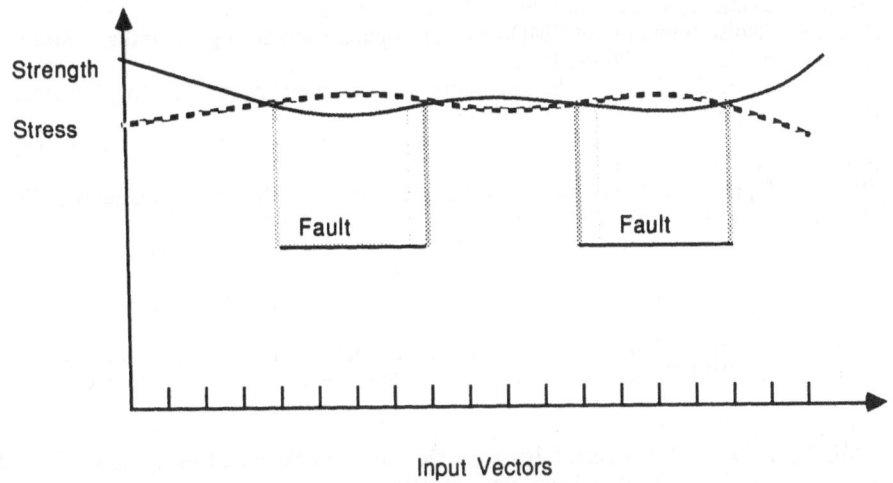

Figure 11. Stress/Strength model of intermittent faults.

12. SUMMARY

This paper presented a survey of fault models and some of their implications on the development of test pattern generators and simulators. First, the relationship between physical defects, failure mechanisms, failure modes and fault models was established. Then the screening of failures was outlined — visual, parametric, functional and stress testing. It was indicated that the functional test is not sufficient to guarantee that the circuit is fault free.

In presenting fault models, we started with the single stuck-at fault. Then we indicated the need to depart from this model for two main reasons. On one hand, the increased chip density and circuit complexity have increased the probability of occurrence of multiple and bridging faults. On the other hand, the use of MOS technology VLSI requires the development of a fault model on the switch level rather than the gate level.

Two categories of temporary faults were also presented — transient and intermittent faults. These types of faults need more attention since they occur more frequently than hard faults.

The use of the fault models presented in special structured circuits such as PLAs and RAMs was intentionally not presented for the sake of brevity rather than lack of importance.

13. ACKNOWLEDGMENTS

The authors wish to thank Prof. Nanya, I. Shperling, S. Makar, L-T Wang, and J. Udell for their valuable comments and suggestions. Thanks are also due to Daisy and AIDA corporations for providing the workstations and the simulators on which some of the results reported here were obtained. This paper was supported in part by the Innovative Science and Technology Office of the Strategic Defense Initiative Organization and administered through the Office of Naval Research under Contract No. N00014-85-K0-0600.

14. REFERENCES

1. Abromivici, M., and P. Menon, "A Practical Approach to Fault Simulation and Test Generation for Bridging Faults, "*Proc., Int'l Test Conf.*, Philadelphia, PA, Oct. 1983, pp. 138-142.
2. Agarwal, V., and A.S.f. Fung, "Multiple Fault Testing of Large Circuits by Single Fault Test Sets, "*IEEE Trans. Comput.*, Vol. C-30, No. 11, Nov. 1981, pp. 855-865.
3. Acken, J.M., "Testing for Bridging Faults (Shorts) In CMOS Circuits," *20th Design Automation Conference Proceedings*, pp. 717-718, 1983.
4. Acken, J.M., "Testing for Bridging Faults (Shorts) In CMOS Circuits," *Reliability and estability Seminar*, Center for Reliable Computing, Stanford, CA, Spring 1987.
5. Allen, A., "Noise Immunity of CMOS Versus Popular Bipolar Logic Families," *Motorola Applications Notes*, AN-708A, 1983.
6. Archambeau, E.C., "Network Segmentation for Pseudoexhaustive Testing," Center for Reliable Computing, TR 85-10, July 1985.
7. Ball, M.O., and F. Hardie, "Effects and Detection of Intermittent Failures in Digital Systems," IBM 67-825-2137, 1967.
8. Banerjee, F., and J.A. Abraham, "Fault Characterization of VLSI MOS Circuits," *Proc., IEEE Int'l Conf. Circuits and Computers*, NY, Sep. 1982, pp. 564-568.
9. Banerjee, F., "A Model for Simulation Physical Failures in MOS VLSI Circuits," *Tech. Rept. CSG-13*, Coordinated Sci. Lab., University of Illinois, Urbana, IL., January 1983.
10. Barzilai, Z., and B. Rosen, "Comparison of AC Self-Testing Procedures," *Proc., Int'l Test Conf.*, Philadelphia, PA, Oct. 1983, pp.89-94.
11. Berger, E.R., et. al., "Single Event Upset in Microelectronics: Third Cosmic Ray Upset Experiment," IBM Tech.. Directions, Federal Systems Division, Vol. 11, No. 1, 1985, pp. 33-40.
12. Bhattacharya,B.B., B. Gupta, S. Sarkar, and A.K. Choudhury, " Testable Design of RMC Networks with Universal Tests for Detecting Stuck-at and Bridging Faults," IEE Proc., Vol. 132, No. 3, May 1985, pp. 155-161.
13. Bennetts, R.G., "Techniques for testing Microprocessor Boards" Proc., *IEEE Trans. Comput.*, Vol. 128, No. 7, Oct. 1981, pp. 473-491.
14. Breuer, M.A., "Testing for Intermittent Faults in Digital Circuits," *IEEE Trans. Comput.*, Vol. C-22, No. 3, March 1973, pp. 241-246.
15. Breuer, M.A., "The Effects of Races, Delays, and Delay Faults on Test Generation," *IEEE Trans. Comput.*, Vol. C-23, No. 10, Oct. 1974, pp. 1078-1092.
16. Bryant, R.E., "MOSSIM: A Switch Level Model and Simulation of MOS LSI," *Proc., 18th Design Automation Conf.*, Nashville, TN, June 1981, pp. 786-790.
17. Bryant, R.E., "A Switch Level Model and Simulation of MOS Digital Systems," *IEEE Trans. Comput.*, Vol. C-33, No. 2, Feb. 1984, pp. 160-177.
18. Bryant, R., "A Survey of Switch Level Algorithms" *IEEE Design & Test*, Aug. 1987, pp.42-50.
19. Brzozowski, J.A., "Testability of CMOS Cells," tech. rpt. CS-85-31, Dept. of CS, University of Waterloo, Ont., Canada, Sep. 1985.
20. Chandramouli, R., "On Testing Stuck-Open Faults," *Proc., 13th Ann. Int'l Symp. Fault-Tolerant Computing Systems*, Milan, Italy , June 1983, pp. 258-265.

21. Chaney, T.J., S.M. Ornstein, and W.M. Littlefield, "Beware the Synchronizer," *COMPCON '72*, San Francisco, CA., Sep. 1972, pp. 317-319.
22. Chaney, T.J., and F.U. Rosenberger, "Characterization and scaling of MOS flip-flop Performance in Synchronizer Applications," *Proc. CALTECH Conf. on VLSI*, 1979 pp. 357-374.
23. Chen, H.H., R.G. Mathews, and J.A. Newkirk, "Test Generation for MOS Circuits," *Proc., Int'l Test Conf.*, Philadelphia, PA, Oct. 1984, pp. 70-79.
24. Chesney, T., and R. Funk, "Noise Immunity of COS/NOS B-series Integrated Circuits," *RCA Appl. Notes*, ICAN-6587, 1983.
25. Chiang, K., and Z.G. Vranesic, "Test Generation For MOS Complex Gates Networks," *Proc., 12th Ann. Int'l Symp. Fault-Tolerant Computing Systems*, Santa Monica, CA, June 1982, pp. 149-157.
26. Chiang, K., and Z.G. Vranesic, "On Fault Detection in CMOS Logic Networks," *Proc., 20th Design Automation Conference*, pp. 50-56, 1983.
27. Chin, C.K., and E.J. McCluskey, "Test Length for Pseudorandom Testing," *Proc., Int'l Test Conf.*, Philadelphia, PA,Nov. 1985, pp. 94-99.
28. Chin, C.K., and E.J. McCluskey, "Test Length for Pseudorandom Testing," *IEEE Trans. Comput.*, Vol. C-36, No. 2, Feb. 1987, pp. 252-256.
29. Côrtes, M.L., E.J. McCluskey, K.D. Wagner, and D.J. Lu, "Modeling Power Supply Disturbances in Digital Circuits," *IEEE Int'l Solid-State Circuits*, Anaheim, CA, Feb. 1986, pp. 164-165.
30. Côrtes, M.L., E.J. McCluskey, K.D. Wagner, and D.J. Lu, "Properties of Transient Errors due to Power Supply Disturbances," *IEEE Int'l Symposium of Circuits and Systems*, San Jose, CA, May 1986, pp. 1046-1049.
31. Côrtes, M.L., and E.J. McCluskey, "An Experiment on Intermittent-Failures Mechanisms" *Proc., Int'l Test Conference*, Washington D.C., Sept. 1986, pp. 435-442.
32. Diehl, S.E., et. al., "Error Analysis and prevention of Cosmic Ion-induced soft Errors in Static CMOS RAMs," *IEEE Trans. on Nuclear Science*, Vol. NS-30, No. 6, Dec. 1982, pp. 2032-2039.
33. Eldred, R.D., "Test Routines Based on Symbolic Logical Statements," *J.ACM*, Vol. 6, No. 1, 1959, pp. 33-36.
34. El-Ziq, Y.M., and R.J. Cloutier, "Functional-Level Test Generation for Stuck-Open Faults in CMOS VLSI," *Proc., Int'l Test Conf.*, Philadelphia, PA, Oct. 1981, pp. 536-546.
35. El-Ziq, Y.M., and S.Y.H. Su, "Fault Diagnosis of MOS Combinational Networks," *IEEE Trans. Comput.*, C-31, No. 2, February 1982, pp. 129-139.
36. Freeman, G.G., "Development of Logic Level CMOS Bridging Fault Models," Center for Reliable Computing, TR No. 86-10, Stanford University, CA, 1986.
37. Friedman, A., "Diagnosis of Short-Circuit Faults in Combinational Circuits," *IEEE Trans. on Computers*, Vol. C-23, NO. 7, July 1974, pp. 746-752.
38. Fujiwara, H., and T. Shimono, "On the Acceleration of Test Pattern Generation Algorithms," *IEEE Trans. on Computers*, Vol. C-32, No. 12, December 1983, pp. 1137-1144.
39. Galiay, J., Y. Crouzet, and M. Vergniault, "Physical Versus Logical Fault Models in MOS LSI Circuits: Impact on Their Testability," *IEEE Trans. Comput.*, Vol. C-29, No. 6, June 1980, pp. 527-531.
40. Gelsinger, P., "Built-in Self-test of the 80368," Proc. of Int'l Conf. on Computer Design, Rye, NY, Oct. 1986.
41. Goel, P., "Test Generation Cost Analysis and Projections,"*17th Annual Design Automation Conf.*, Minneapolis, MN, June 1980, pp. 77-84.
42. Goel, P. , "An Implicit Enumeration Algorithm to Generate Tests For Combinational Circuits," *IEEE Trans. Comput.*, Vol. C-30, No. 6, March 1981, pp. 215-222.
43. Goldestein, L.H., "A Probabilistic Analysis of Multiple Faults in LSI Circuits," IEEE Computer Society Repository R-77, Long Beach, CA.
44. Hackmeister, D., and A.C.L. Chiang, "Microprocessor Test Technique Reveals Instruction Pattern-Sensitivity," *Computer Design*, Dec. 1975, pp. 81-85.

45. Hnatek, E.R., and B.K. Wilson, "Practical Consideration in Testing Semicustom ICs," Semicustom Design Guide, VLSI, Summer 1986, p.140-151.

46. Hayes, J.P., "Detection of Pattern-Sensitive Faults in Random-Access Memories," *IEEE Trans. Comput.*, Vol. C-24, No. 2, Feb. 1975, pp. 713-719.

47. Hayes, J.P., "Testing Memories for Single-Cell Pattern-Sensitive Faults," *IEEE Trans. Comput.*, Vol. C-29, No. 3, Mar. 1980 pp. 713-719.

48. Hayes, J.P., and E.J.McCluskey, "Testability Considerations in Microprocessor-based Design," Mar. 1980, pp. 17-28.

49. Hayes, J.P., "A Unified Switching Theory with Applications to VLSI Design," *Proc., IEEE*, Vol. 70, Oct. 1982, pp. 1140-1151.

50. Hayes, J.P., "Fault Modeling for Digital MOS Integrated Circuits," *IEEE Trans. on CAD*, Vol. CAD-3, July 1984, pp. 200-207.

51. Hayes, J.P., "Pseudo-Boolean Logic Circuits," *IEEE Trans. Comput.*, Vol. C-35, No. 7, July 1986, pp. 602-612.

52. Healy, J.T., *Automatic Testing and Evaluation of Digital Integrated Circuits*, Reston Publishing Co., Inc., Prentice-Hall, 1981.

53. Hsieh, E.P., "Checking Experiments for Sequential Machines," *IEEE Trans. Comput.*, Vol. C-20, No. 10, pp. 1152-1167, Oct. 1971.

54. Hsieh, E.P., R.Rasmussen, L.Vidunas, and W. Davis, "Delay Test Generation," *Proc. 14th Design Automation Conf.*, June 1977, pp. 486-491.

55. Hughes, J.L.A., and E.J. McCluskey, "An Analysis of the Multiple Fault Detection Capabilities of Single Stuck-at Fault Test Sets," *Proc. of Int'l Test Conf.*, Philadelphia, PA, Oct. 1984, pp. 52-58.

56. Hughes, J.L.A., S. Mourad, and E.J. McCluskey, "An Experimental Study Comparing 74LS181 Test Sets," *Digest of COMPCON Spring 85*, San Francisco, CA, Feb.1985, pp. 384-387.

57. Hughes, J.L.A., , and E.J. McCluskey, "Multiple Stuck-at Fault Coverage of Single Stuck-at Fault Test Sets," *Proc., Int'l Test Conf.*, Washington, DC, Oct. 1986, pp. 368-374.

58. Jain, S.K., and V.D. Agrawal, "Test Generation for MOS Circuits Using D-Algorithm," Proc. 20th Design Automation Conference, June 1983, pp. 129-137.

59. Kamal, S., and C.V. Page, "Intermittent Faults: a Model and a Detection Procedure," *IEEE Trans. Comput.*, Vol. C-23, No. 7, July 1974, pp. 713-719.

60. Koeppe, S., "Modeling and Simulation of Delay Faults in CMOS Logic Circuits," *Proc., Int'l Test Conf.*, Washington, DC, Oct. 1986, pp. 530-536.

61. Koeppe, S., "Optimum Layout to Avoid CMOS Stuck-Open Faults," *Proc. 24th ACM /IEEE Design Automation Conf.*, Miami. FL. June 1987, pp. 829-835.

62. Kolasinski, W.A. "Single Event Upset Vulnerability of Selected 4k and 16k CMOS Static RAMs," *EEE Trans. on Nuclear Science*, Vol. NS-29, No. 6, Dec. 1982, pp. 2044-2048.

63. Koren, I., and Z. Kohavi, "Diagnosis of Intermittent Faults in Combinational Networks, "*IEEE Trans. Comput.*, Vol. C-26, No. 11, Nov. 1977, pp. 1154-1157.

64. Kuban, J.R., and J.E. Salick, "Testing Approaches in the MC68020," VLSI Design, Nov. 1984, pp. 22-30.

65. Liaw, C.S., S,V. Su, and Y. Malaiya, "Test Generation for Delay Faults using Single-Stuck Faults ," *Proc., Int'l Test Conf.*, Philadelphia, PA, Oct. 1980, pp. 281-289.

66. Liu, D.L., and E.J. McCluskey, "Design of CMOS VLSI Circuits for Testability," *Proc. IEEE Custom Integrated Circuit Conf.*, 1986, pp. 421-424.

67. Liu, D.L., and E.J. McCluskey, "A VLSI CMOS Circuit Design Technique to Aid Test Generation ," *Proc. IEEE Conf. on Computers and Communications*, 1987, pp.116-120.

68. Liu, D.L., and E.J. McCluskey, "Designing CMOS Circuits for Switch-Level Testability," *IEEE Design & Test*, Aug. 1987, pp.42-50.

69. Makar, S.R., and E.J. McCluskey, "On Testing Multiplexers," Technical Report (in preparation), Center for Reliable Computing, Stanford, CA, 1987.

70. Malaiya, Y.K., and, S. Yang, "The Coverage for Random Testing," *Proc., Int'l Test Conference*, Philadelphia, PA, Oct. 1984, pp.237-245.

71. Malaiya, Y.K., and R. Narayanaswamy, "Modeling and Testing for Timing Faults in Synchronous Sequential Circuits," *IEEE Design & Test*, Nov. 1984, pp.62-74.
72. Maly, W., F.J. Furguson, and J.P. Shen, "Systemic Characterization of Physical Defects for Fault Analysis of MOS IC Cells," *Proc., Int'l Test Conference*, Philadelphia, PA, Oct. 1984, pp. 237-245.
73. May, T.C., and M.H. Wood, "Alpha-Particle-Induced Soft Errors in Dynamic Memories," *IEEE Trans. on Electron Devices*, Vol. ED-26, No. 1, Jan. 1979, pp. 2-7.
74. McCluskey,E.J., and F.W. Clegg, "Fault Equivalence in Combinational Logic Networks," *IEEE Trans. Comput.*, Vol. C-20, No. 11, Nov. 1977, pp. 1286-1293.
75. McCluskey, E.J., and J.F. Wakerly, "A Circuit for Detecting and Analyzing Temporary Failures," *Digest of Spring COMPCON*, Feb. 1981, pp. 317-321.
76. McCluskey, E.J., and S. Bozorgui-Nesbat, "Design for Autonomous Test," *IEEE Trans. Comput.*, Vol. C-30, No.11, Nov. 1984, pp. 866-875.
77. McCluskey, E.J., "Verification Testing — A Pseudoexhaustive Test Technique," *IEEE Trans. Comput.*, Vol. C-33, No. 6, June 1984, pp. 541-546.
78. McCluskey, E.J., *Logic Design Principles: With Emphasis on Testable Semicustom Circuits*, Prentice-Hall, Englewood Cliffs, NJ, 1986.
79. McCluskey, E.J., "Comparing Causes of IC Failures," *Technical Workshop: New Directions for IC Testing*, Victoria Canada, Mar. 18-19, 1986.
80. McCluskey, E.J., "A Comparison of Test Pattern Generation Techniques," *Proc., 10th Int'l Fault-Tolerant and Systems Diagnosis Symposium*, Brno, Czechoslovakia, June 1986, pp. 11-20.
81. McCluskey, E.J., S. Makar, S. Mourad, and K.D. Wagner, "Probability Models for pseudorandom Test Sequences," *Proc., Int'l Test Conf.*, Washington, DC, Sept. 1987.
82. McCluskey, E.J., EE 488, Testing Aspects of Computers, Class notes, Spring 1987.
83. McCluskey, E.J., and S. Mourad, "Comparing Causes of IC Failures," New Directions for IC Testing, Ed. Smith, to appear in 1987.
84. Mei, K., "Bridging and Stuck-at Faults," *IEEE Trans. Comput.*, Vol. C-23, No. 7, July 1974, pp. 720-727.
85. Miczo, A., *Digital Logic Testing and Simulation*, Harper and Row, 1985
86. Millman, S., and E.J. McCluskey, "Detecting Bridging Faults with Stuck-at Test Sets," Technical Report (in preparation), Center for Reliable Computing, July 1987.
87. Mourad, S., J.L.A. Hughes, and E.J. McCluskey, "Effectiveness of Single Fault Tests to Detect Multiple faults in Parity Trees," *Comput. Math. Applic.*, Vol. 13, No. 5/6, 1987, pp. 455-459.
88. Nanya, T, S. Mourad, and E.J. McCluskey, " Multiple Fault Testing of Totally Self Checking Checkers," in preparation.
89. O'Neil, E.J., and J.R. Halverson, "Study of Intermittent Field Hardware Failure Data in Digital Electronics," *NASA Contractor Report 159269*, June 1980.
90. Patshnik, O., "Circuit Segmentation for Pseudo-Exhaustive Testing," Center for Reliable Computing, TR 83-14, Stanford, CA, 1983.
91. Price, W.E., et. al., "Single Even Upset Sensitivity of Low Power Schottky Devices," *IEEE Trans. on Nuclear Science*, Vol. NS-29, No. 6, Dec. 1982, pp. 2064-2066.
92. Reddy, S.M., and M.K. Reddy, and K/G/Luhl, "On Testable Design for CMOS Logic Circuits," *Proc.., Int'l Test Conf.*, Philadelphia, PA, Sept. 1983, pp. 435-445.
93. Reese-Brown, J., "Pattern Sensitivity in Semiconductor Memories," *Proc., IEEE Semiconductor Test Conference*, 1972, pp. 33-46.
94. Rinerson, D.D., and A. Tuszynski, "Identification of Causes of Pattern Sensitivity," *IEEE 1977 Semiconductor Test Symposium*, pp. 166-170.
95. Roberts, M.W. and P.K. Lala, "An Algorithm for the Partitioning of Logic Circuits,"*IEEE Proc.*, Vol. 131, No.4, pp. 1231–1240.
96. Roth, J.P., W.G. Bouricious, and P.R. Schneider, "Programmed Algorithms to Compute Tests to Detect and and Distinguish Between Failures in Logic Circuits," *IEEE Trans. Elect. Comput.*, Vol. EC-16, No. 10, Oct. 1967, pp. 567-580.

97. Savir, J., "Testing for Single Intermittent Failures in Combinational Circuits by Maximizing they Probability of Fault Detection," *IEEE Trans. Comput.*, Vol. C-29, No. 5, May 1980, pp. 410-416.

98. Savir, J., and P. Bardell, "On Random Patterns Test Length," *Proc., Int'l Test Conf.*, Philadelphia, PA, Oct. 1983, pp. 95-106.

99. Schertz, D.R., and G.Metze, "On the Design of Multiple Fault Diagnosable Networks," *IEEE Trans. Comput.*, Vol. C-20, No. 11, Nov. 1971, pp. 1361-1364.

100. Schertz, D.R., and G.Metze, "A New Representation for Faults in Combinational Digital Circuits," *IEEE Trans. Comput.*, Vol. C-21, No. 8, Aug. 1972, pp. 858-866.

101. Shedletskey, J., "Delay Testing LSI Logic," *Proc., 8th Int'l Conf. on Fault Tolerant Computing*, June 1978, pp. 410-416.

102. Shen, J.P., W. Maly, and F.J. Ferguson, "Inductive Fault Analysis of MOS Intergrated Circuits," *IEEE Design and Test of Computers*, Dec. 1985, pp. 13-26.

103. Shperling, I., and E.J. McCluskey, "Segmentation for Pseudoexhaustive Testing Via Simulated Annealing," *Proc., Int'l Test Conf.*, Washington, DC, Oct. 1987.

104. Siewiorek, D.P. et. al., "A Case Study of C.mmp, Cm* and C.vmp," Proc. of the IEEE 66, Oct. 1978, pp. 1178-1220.

105. Stevens, A.K., *Introduction to Component Testing: Application Electronics*, Addison–Wesley, NY 1986.

106. Stoll, P.A., "How to Avoid Synchronization Problems," *VLSI Design*, Nov. 1982, pp. 56-59.

107. Stucki, M.J., and J.R. Cox Jr., "synchronization Strategies," *Proc. CALTECH Conf. on VLSI*, Jan. 1979 pp. 375-393.

108. Timoc, C., et. al., "Logical Models of Physical Failures," *Proc., Int'l Test Conference*, Philadelphia, PA, Oct. 1983, pp. 546-552..

109. Varshney, P.K., "On Analytical Modeling of Intermittent Faults in Digital Systems," *IEEE Trans. Comput.*, Vol. C-28, No. 10, Oct. 1979, pp. 786-791.

110. Voldman, S., and L. Patrick, "Alpha Particle Induced Single Events in Bipolar Static ECL Cells," *IEEE Trans. on Nuclear Science*, Vol. NS-30, No. 6, Dec. 1984, pp. 1197-1200.

111. Wadsack, R.L., "Fault Modeling and Logic Simulation of CMOS and MOS Integrated Circuits," *Bell System Technical Journal*, Vol. 57, No. 5, May-June 1978, pp. 1449-1474.

112. Wagner, K.D., "The Error Latency of Delay Faults in Combinational and Sequential Circuits," *Proc. of the Int'l Test Conference*, Philadelphia, PA, Oct. 1985, pp. 334-341.

113. Wagner, K.D., C.K. Chin, and E.J. McCluskey, "Pseudorandom Testing," *IEEE Trans. on Comput.*, vol. C-36, No. 3, March 1987, pp. 332-342.

114. Waicukauski, J., and E. Lindbloom, "Transition Fault Simulation by Parallel Patterns Single Fault Propagation," *Proc., Int'l Test Conf.*, Washington, DC, Oct. 1986, pp. 542-549.

115. Williams, T.W., *IEEE Design and Test of Computers*, Vol. 2, No. 2, Feb. 1985, pp. 59-63.

116. Xu, S., and S. Su, "Testing Feedback Bridging Faults Among Internal, Input and output lines by Two Patterns," *IEEE Intl' Conf. on Circuits and Computers*, New York, NY, Sep. 1982, pp. 214-217.

117. Zasio, J.J., "Shifting Away from Probes for Wafer Test," *Digest of Spring COMPCON*, March 1981, pp. 395-398.

118. Zasio, J.J., "Non Stuck Fault Testing of CMOS VLSI," *Digest of Spring COMPCON*, March 1985, pp. 388-391.

119. Zoutemdyk, J.A., "Modeling of Single-Even Upset in Bipolar Integrated Circuits," *IEEE Trans. on Nuclear Science*, Vol. NS-30, No. 6, Dec. 1983, pp. 4540-4545.

FAULT DETECTION AND DESIGN FOR TESTABILITY OF CMOS LOGIC CIRCUITS*

SUDHAKAR M. REDDY AND SANDIP KUNDU

University of Iowa, Iowa City, Iowa   52242, U.S.A.

1.   INTRODUCTION

Advances in integrated circuit technologies have made complementary MOS (CMOS) the preferred MOS technology for digital logic circuits. Cost effective design and fabrication of reliable CMOS VLSI chips require understanding of various CMOS technologies, logic families, failure modes, fault detection methods and design for testability methods. In this paper we will review some of the basic methods and issues related to the design and fault detection of CMOS logic circuits.

2.   CMOS DIGITAL LOGIC CIRCUITS

In this section we briefly review design of digital logic circuits using CMOS gates. Typically a CMOS gate is constructed by a properly connected group of p-channel and n-channel enhancement FETs placed between V_{DD} and ground as shown in Figure 1. Assuming positive logic, a p-channel FET (or pFET for short) conducts when its gate terminal is logic 0 and an n-channel FET (nFET for short) conducts when its gate terminal is logic 1. It is convenient to model FETs as switches that close the path between their drain and source terminals when they are conducting. In this paper we use the symbols for FETs as shown in Figure 1.

For a given combination of input values, a CMOS gate can be in one of the following states.
- (a)   the output node is connected to V_{DD} via one or more paths through conducting pFETs and no path from output node to V_{SS} through conducting nFETs exists,
- (b)   the output node is connected to V_{SS} via one or more paths through conducting nFETs and no path from output node to V_{DD} through conducting pFETs exists,
- (c)   the output node is not connected to either V_{DD} or V_{SS} through conducting FETs, or
- (d)   the output node is connected to both V_{DD} and V_{SS} through conducting FETs.

In case (a) the gate output will be 1, in case (b) the gate output will be 0. In case (c) the output will be at high impedance with its logic state equal to its state in the immediate past. In case (d) the gate output is often marked as indeterminate, since it depends on the ratios of resistances of paths to V_{DD} and V_{SS} from output node. Only rarely are fault-free gates designed to have state (d) described above. In the event of the existence of such states in fault-free operation, designers are warned to avoid circuit inputs that could create these states.

---
*This work has been supported in part by SDIO/IST Contract No. N00014-87-K-0419 managed by U.S. Office of Naval Research.

F. Lombardi and M. Sami (eds.), Testing and Diagnosis of VLSI and ULSI, 69–91.
© 1988 by Kluwer Academic Publishers.

Next we introduce several classes of CMOS gates. A **static CMOS gate** to realize a Boolean function F is constructed by building the pFET (nFET) network of the CMOS gate (cf. Figure 1) to correspond to F(F). This correspondence is established by constructing the pFET (nFET) network such that for every input combination for which F is 1(0) there is at least one path from V_{DD} (V_{SS}) to output node through conducting pFETs (nFETs) and for every input combination for which F is 0(1) there is no conducting path from V_{DD} (V_{SS}) to the output node. Construction of the pFET and nFET networks of a CMOS gate is simplified by making the correspondence between the logical AND (OR) operation and series (parallel) connection of FETs or networks of FETs corresponding to the logic expressions involved in the AND (OR) operation. As an example consider the function F = (ab + c(d + e))f and its complement F = f + (a + b)(c + de). The static CMOS gate realizing F is shown in Figure 2. Note that the variables driving the gate inputs of pFETs are the complements of the variables appearing in the boolean expression for F. This is done to compensate for the fact that a pFET conducts when its gate is at logic 0.

Another family of CMOS gates that have found wide acceptance are the so-called **domino gates**. These are constructed from a dynamic gate and a static CMOS inverter as shown in Figure 3. The output of the dynamic gate is precharged to 1 when $\phi = 0$ and the gate output during $\phi = 1$ (called the evaluation phase) remains at 1 if the inputs applied to the gate do not create a path from the output node to V_{SS}, otherwise it will go to zero. The static inverter is used to buffer the output of the dynamic gate to reliably drive other gates from the output of a dynamic gate. The domino CMOS gates to realize function F are obtained by designing the nMOS network (cf. Figure 3) corresponding to function F. As an example in Figure 4 the CMOS domino gate realization of function F = (ab + c(d + e))f is shown.

In MOS technologies MOS transistors are often used as **transmission gates**. This application of FETs is illustrated in Figure 5. For these gates when control input CN is 1 output y is equal to input x otherwise the output is at high impedance. The transmission gate of Figure 5(c) is preferred since it transmits both logic 0 and logic 1 without reduction in voltage level. Transmission gates are used in deriving busses, logic gates and flip-flops. Two examples are shown in Figures 6 and 7.

As explained earlier, CMOS gates to realize a function F can be constructed from logic expressions for F and F. Of particular interest to us are static CMOS gates constructed from two-level sums of products (**s of p**) and/or product of sums (**p of s**) expressions for F and F. Since there are two types of two-level expressions for F and also for F there are four different types of CMOS static gate realizations possible for each function. These were called **SP-PS**, **SP-SP**, **PS-SP** and **PS-PS** realizations [23] and are illustrated in Figures 8-11. The definitions of these different realizations become clear by noting the correspondence between the type of logic expressions used to realize the pFET network and the nFET network of the static gate. For example SP-PS realization is obtained by using s of p expression for F to realize the pFET network and a p of s expression for F to realize the nFET network.

In this paper, we call AND, OR, NAND, NOR and NOT gates **primitive gates** and gates realizing functions other than these will be called **complex gates**. Figures 8-11 are examples of complex gates.

3. FAULT MODELS AND TESTS

Abstract or logic models called **fault models** represent failures that effect functional behavior of logic circuits. The failures in logic circuits are often modeled by the **line stuck-at** fault model. This model assumes that the effect of failures that modify functional behavior of logic circuits are modeled by faults that permanently force the value of appropriate inputs or outputs of gates to logic 1 or logic 0. In MOS technology the most probable failures are opens and shorts. These failures can be modeled by gate inputs **stuck-at-zero** or **stuck-at-one**, **FETs stuck-open** or **stuck-on**, gate outputs stuck-at-zero or stuck-at-one and **shorts** between circuit nodes. In this paper we consider these faults with the exception of shorts between circuit nodes. To formally define the set of faults, consider the two-input static NAND gate shown in Figure 12. The faults to be considered are the input/output nodes stuck-at-0 or stuck-at-1, the pFETs 1 and 2 stuck-open or stuck-on and nFETs 1 and 2 stuck-open or stuck-on. By a FET stuck-open (stuck-on) we mean that the fault causes the FET to be permanently off or non-conducting (on or conducting). In general we assume that the inputs and outputs of a gate may be stuck-at-0 or 1 and a FET is stuck-open or stuck-on. In this paper we consider single faults only. That is, we assume that only one of the faults cited above is present in the circuit under test. Since the faults being considered include some non-classical (i.e. other than line stuck-at) faults, tests to detect the modeled faults are to be defined also. Tests to detect gate inputs and outputs stuck-at-0 or 1 can be derived using classical methods, however tests to detect FET stuck-open and stuck-on faults need to be addressed further. Consider pFET 1 stuck-open in the two-input NAND gate of Figure 12. To detect this fault, an input pattern which attempts to turn pFET 1 on and turn pFET 2 off must be applied. That is, the input A=0 and B=1 is necessary. With this input applied and the cited fault present, the output will be at high impedance with its logic state determined by its previous state. Hence to insure detection of pFET 2 stuck-open, the output of the gate must be initialized to zero (which can be achieved by applying input 11) and then the test input 01 must be applied. This, of course, means that a sequence of two patterns, $\langle 11,01 \rangle$ must be applied to propagate the effect of a FET stuck-open fault to the output of the faulty gate. Such tests have been called two-pattern tests [12]. In general the tests to detect FET stuck-open faults in static CMOS gates must satisfy the following conditions:

To detect a pFET (nFET) stuck-open fault in a static CMOS gate G a two pattern test $\langle T_1,T_2 \rangle$ is necessary and sufficient, where T_1 initializes the output of G to logic 0(1) and T_2 is such that (i) it sets the output of gate G to 1(0) if it is fault-free and sets it to high impedance state if the fault being tested is present, and (ii) T_2 also sensitizes a path from the output of G to an observable output.

The input T_1 of the two-pattern test $\langle T_1,T_2 \rangle$ will be referred to as the **initializing input** and the second input T_2 as the **test input**.

Even though in general it is necessary to apply test sequences with at least two test patterns to detect FET stuck-open faults in static CMOS gates it is possible to argue that for certain stuck-open faults it may only be necessary to apply single pattern tests. For example if nFET 1 or nFET 2 in NAND gate of Figure 12 is stuck-open then it is never possible to set the output of the gate to zero. Hence if the inputs under test mode are such that they set the output of this gate to 1 occasionally then the output will remain at 1 (due to charge retention) and hence the cited nFET stuck-open faults become equivalent to output stuck-at-1 and hence can be detected by single pattern tests. In general FET stuck-open faults that disconnect all paths to V_{SS} (V_{DD}) may be considered as equivalent to output stuck-at-1 (stuck-at-0) if the faulty gate output can be assumed to be occasionally set to 1(0) during test. Incidentally the same argument is true for stuck-open faults in the depletion mode load transistor of an nMOS gate [22].

Tests for stuck-open faults in a domino CMOS gate can be derived similarly. Applying an input to a circuit consisting of domino CMOS gates requires going through pre-charge state with $\phi = 0$ and evaluation phase with $\phi = 1$. However because of the pre-charging phase, a nFET stuck-open fault can be detected by a single test applied during the time $\phi = 1$. The single pFET stuck-open fault in the domino gate requires a two-pattern test, with the first input setting the faulty gate output to zero during evaluate phase and the second input propagating the effect of the fault through the circuit under test in the next evaluation phase. When logic using transmission gates is used in a circuit, tests can be specified in a manner analogous to the ones for static and domino gates, but may have to be specifically tailored to the logic gate under consideration.

Next we consider FET stuck-on faults. Let pFET 1 in Figure 12 be stuck-on. A test that could detect this fault must attempt to turn pFET1 off and pFET 2 off. This implies that the test should set A=B=1, which would create a path in the faulty gate, from V_{DD} to V_{SS} through pFET 1 and nFETs 1 and 2. The logic state of the faulty gate's output under this test input will depend on the relative values of the resistances of the paths to V_{DD} and V_{SS} from the output node. Often this value may fall in the ambiguous region between logic 0 and 1. For this reason detection of stuck-on faults in static CMOS logic may require monitoring steady state supply current. Notice that under ideal conditions static CMOS gate circuits should draw zero steady state current. However with a stuck-on fault present and under an input that activates the stuck-on fault the steady state current would be non-zero. Even under non-ideal conditions the steady state currents in faulty and fault-free circuits are expected to be sufficiently different to allow fault detection through current monitoring. Another approach that has been suggested, to detect stuck-on faults is through delay testing [6]. FET stuck-on faults in the dynamic gate part of the domino CMOS gate, other than nFET 1 and pFET 1 (cf. Figure 3) stuck-on, are detectable by the normal test procedure of applying test input(s) and observing the logic value of an appropriate observable output. pFET 1 (cf. Figure 3) stuck-on fault may require observing steady state supply current and the nFET1 stuck-on fault may not be detectable. An nFET in the nFET network (cf. Figure 3) stuck-on is detectable by a test that would attempt to cut off all paths from output of the dynamic gate to V_{SS} in the fault-free gate and would create a path from output to V_{SS} in the faulty gate as well as sensitize a path from the output of the gate to an observable output [17].

In the next section we consider derivation of tests for the modeled faults.

4. DERIVATION OF TESTS

In this section we consider derivation of tests to detect single line stuck-at and FET stuck-open and stuck-on faults. We restrict our discussion to circuits constructed of primitive gates. First we consider derivation of tests in combinational logic circuits.

4.1 Derivation of Tests for Combinational Logic Circuits

Consider the modeled faults in a static CMOS two-input NAND gate. As indicated earlier two-pattern tests for FET stuck-open faults contain an initializing input and a test input. Assuming proper initializing input, the outputs of fault-free and faulty gates for all modeled faults in the two-input NAND gate of Figure 12 are shown in Figure 13. From Figure 13 it can be readily seen that the tests for all modeled faults are included in the tests for the classical line stuck-at faults. The major difference is that, to detect stuck-open faults one must properly initialize the output of the gate being tested and for detecting FET stuck-on faults one may need to monitor the supply current. The initializing inputs needed can be clearly obtained from appropriate test inputs for other faults or can be derived using only the line justification step of the standard test pattern generating procedures.

The observations made above, regarding tests for modeled faults in the two-input NAND gate are true in general for combinational logic circuits constructed from static CMOS primitive gates. For example consider derivation of tests for all modeled faults in the circuit of Figure 14. Four tests shown in Figure 14 are sufficient to detect all single line stuck-at faults in the circuit. To detect the stuck-open faults one needs to augment these four tests such that proper initializing vectors are included. One such sequence of seven tests is shown in Figure 14. As can be expected, to detect the augmented set of faults the test lengths may have to be increased.

If the combinational logic circuit contains domino gates, then the faults in the nFET network of the domino gates (cf. Figure 3) are detected by the tests for the line stuck-at faults which also detect the pFET1 stuck-on fault. The pFET 1 stuck-open fault requires an initializing input as indicated earlier, while nFET 1 stuck-on fault may not be detectable as indicated earlier.

If the circuit under test contains complex gates then a method to derive a "test equivalent" circuit that contains primitive gates and a single new logic primitive called "B block" is given in [13]. Using this equivalent circuit one can derive tests for the circuit under test. A simpler method which avoids the use of B blocks and applicable to a class of static CMOS circuits is discussed in [24].

4.2 Tests for Sequential Circuits

Deriving tests for an arbitrary sequential circuit is known to be difficult. For this reason, often sequential circuits are designed to be "easily testable." One commonly used method to design easily testable sequential circuits is the so-called scan design [9]. In this design method, the memory elements or flip-flops of the sequential machine are strung together to form a shift register. To apply a test, the secondary variables (or the contents of the flip-flops) are set, to correspond to the test, by scanning the test vector into the scan register while the results of the previous test are scanned out.

To apply two-pattern (or multi-pattern) tests in scan designs, care must be taken not to invalidate the initialization of the circuit nodes, set by the first part of the two-pattern (or multi-pattern) test. Three approaches can be taken to insure proper application of two-pattern tests. One is to design tests such that T2, the test input in a two-pattern test ⟨T1, T2⟩, is equal to T1 shifted by one position. The second approach is to insure that T2 is different from T1 in a specified manner. (For example T2 differs from T1 in one position [5]). The other approach is to introduce an extra latch to hold T1 while T2 is scanned in a manner suggested in [7]. A way to implement the last approach is illustrated in Figure 15. While T2 is being scanned-in, the circuit is in test mode with M=1 and hence, the output of the transmission gate is at high impedance, holding the previous state corresponding to T1. Thus the transmission gate is used as a dynamic latch. In the normal mode M=0 and the transmission gate is closed, thus connecting the flip-flops to the functional unit. This design has a couple of disadvantages. One is that the dynamic latch and the inverter buffering its output are in the normal operational path, thus introducing extra path delay. The second problem is that if the scan path is long, then the dynamic latch may have to be replaced with a static latch increasing circuit complexity. Better designs can be obtained by converting the latches in the scan register to dynamic latches and using static latches to hold T1 while T2 is being shifted in. These are described next.

Sequential circiuts are designed using single latch or double latch designs. A typical latch used in these designs is shown in Figure 7. In double latch designs, often two non-overlapping clocks CL1 and CL2 are used. A design of the latches to incorporate a third latch effectively to hold T1 of a two-pattern test in double latch and single latch designs are shown in Figures 16 and 17, respectively. These designs are inspired by a scan flip-flop design given in [1].

First we describe the design in Figure 16. The operation and the derivation of control signals A through F (cf. Figure 16), for double latch designs, is explained next. It is assumed that to facilitate scan design, three global signals, CL1, CL2 (non-overlapping clocks for dual latch design) and M (the mode control input to allow scan/normal mode operation) are available. SI, DI, SO and DO are scan data in, functional data in, scan data out and functional data out signals, respectively. Note that, compared to the normal master-slave flip-flop, three additional gates (G1, G3 and G4) are used in the scan flip-flop shown. The chain of gates G_1, G_2, G_3 and G_4 constitute a master-slave dynamic flip-flop of the scan register. Actually, the master of this scan flip-flop is a static latch consisting of G_2, G_5 and G_6, which also operates as the master latch for the functional data path. This can be verified by the interested reader after purusing the definitions of the control signals given below.

$$B = M.CL1, \quad C = \overline{CL1}, \quad D = M.CL2$$
$$A = \overline{M}.CL1, \quad E = \overline{M}.CL2, \quad F = \overline{CL2}$$

With M = 1, tests are scanned in and with M = 0 normal operations are achieved. Mode control input is switched to normal operation after the last bit of the test input is shifted into master latch consisting of G2, G5 and G6 and M is switched to scan mode after the functional outputs are

captured into this master latch. The advantages of the proposed design
are that it does not introduce additional delay into normal operational
logic paths and uses a minimum number of global signals.

The scan flip-flop to enable applications of two-pattern tests in
single latch designs is shown in Figure 17. Gates G1, G2, G3, G4 and G5
are the extra logic added to the original single latch. If CL is the
clock used in the original single latch design, two extra global signals
M (mode input) and SCL (Scan Clock) are needed to operate the scan flip-
flop of Figure 17. It can be noted that by properly defining control
signals B and D, the chain of gates G1, G2, G3 and G4 constitute a
dynamic master-slave flip-flop. This is accomplished by operating clocks
CL and SCL as non-overlapping two phase clocks and defining the control
signals shown in Figure 17 to be as given below.

$$B = \underline{M}.SCL, \ D = \underline{M}.CL$$
$$A = \overline{M}.CL, \ E = \overline{M}.SCL, \ F = (M+\overline{SCL})(M+\overline{CL})$$

As in the case of two-latch design, the scan flip-flop design proposed
here does not add logic delay to the operational path of single latch
designs and uses minimum number of global signals. The main difference
between the designs of Figures 16 and 17 is that both the master and
slave latches of the scan-path flip-flop in Figure 17 are dynamic where
as only the slave latch of the scan-path flip-flop in Figure 16 is
dynamic. Complexity of both designs can be reduced by using single FET
transmission gates in place of two FET transmission gates. This may
however reduce reliability of operation.

4.3 Fault Simulation

An important part of the process of generating tests for logic
circuits is fault simulation. Fault simulation is done to determine the
faults detected by a given sequence of test inputs. As indicated earlier
only stuck-open faults require two-pattern tests and in logic circuits
constructed from primitive gates since the tests for FET stuck-open
faults are also tests for gate inputs stuck-at-1 or 0 the only
modification required for fault simulation of the extended set of faults
is that the outputs of the gates in the previous time instant be
retained. As an example consider the circuit shown in Figure 18. The
current input being simulated is 0111 and the previous state of the gate
output nodes are shown in parentheses. Each lead is numbered. Let
ps(ns) indicate the pFET (nFET) driven by lead s. Also let ps/flt,
ns/flt and s/flt indicate the pFET ps or nFET ns or circuit lead 's'
faulted at flt. We use 0(1) for 'flt' to indicate line stuck-at-0(1)
faults and ON(OFF) to indicate FET stuck-on (stuck-open) fault. Even
though any of the known fault simulation techniques can be used, we use
deductive simulation to illustrate how fault simulation with the extended
fault model can be achieved by retaining the previous state of the gate
output. In deductive simulation the faults detected are deduced by
propagating lists of faults from inputs to outputs. These are shown in
Figure 18, where L_s is the list of faults detected if lead 's' was
observed. It should be pointed out that when a FET stuck-on fault is
indicated as detected we mean that the test creates a condition for a
conducting path between V_{DD} and V_{SS} and the actual detection may require
monitoring steady state supply current.

5. PROBLEMS IN TESTING CMOS LOGIC CIRCUITS

Detection of faults in CMOS logic circuits poses several difficulties that were not encountered in testing other logic families. One of these dificulties has already been mentioned. This is the problem of detecting FET stuck-on faults. Two other problems of some importance will be discussed next. First one is that of potential invalidation of tests for FET stuck-open faults and the second is the problem of undetectability of certain FET stuck-open faults in CMOS latches and flip-flops.

5.1 Test Invalidation and Robust Tests

When the test input T_2 of a two-pattern test $\langle T_1, T_2 \rangle$ is applied to detect a stuck-open fault, in say gate G, and if the fault being tested for is present in G then the output node of G will be in the high impedance state. That is, the fault indication is stored on the capacitance of the output node. This stored value may be changed or invalidated by either transients caused by circuit delays or by charge sharing with other nodes internal or external to gate G. Both types of invalidations have been demonstrated through simulation [23]. Here we want to discuss further the invalidation of tests for FET stuck-open faults due to circuit delays.

Consider the circuit of Figure 19. Suppose, we wish to test the pFET, of the output NAND gate G, connected to line 1. If the initializing vector is given to be 0110 which initalizes the output to zero, we have to find a test input. The test input should set line 1 at logic 0 and lines 2 and 3 should be at logic 1. It is seen that a test vector is 1100. When the circuit inputs change from the initializing input 0110 to test input 1100, input A may change earlier than input C thus creating a transient input of 1110 before the circuit inputs settle down to 1100. Thus line 3 may assume logic value zero for a short period before returning to logic 1. This transient in the logic value of line 3 may charge the output node of G to 1, thus destroying its initialized value. Furthermore this newly stored value of logic 1 will be retained when the circuit inputs settle at 1100. But this is the output of the fault-free circuit under input 1100 and hence the PFET stuck-open fault is not detected. Thus tests derived under the assumption of static conditions may be invalidated by transients caused by circuit delays. This leads to the concept of robust tests. If a robust test is used, timing skews in input changes or arbitrary circuit delays cannot invalidate the test. In the example above if the initializing input was chosen to be 0100 then $\langle 0100, 1100 \rangle$ constitutes a robust test.

In the example given above, the initializing vector was specified and the test vector was to be found. Other times a test vector may be specified and initializing vector is to be found or an attempt to simultaneously find both the initializing and the test vector can be made. In Figure 20 another example circuit is shown where the objective is to determine an initializing input for testing the pFET connected to line 1 when the test input is specified to be 111. As it turns out no initializing vectors exists for this test input to test the fault robustly. If however, the test input was chosen to be 110, any one of 010, 100 or 000 may be used as the initializing input vector. The last two examples were used to emphasize the point that in a general situation, to detect a stuck-open fault there may be a choice of several test inputs as well as several initializing inputs. Thus an initializing input may be picked and an attempt may be made to match it with a test input for robust testing or vice versa or we might select both the inputs simultaneously for robust testing.

Next we give formal definition of robust tests.

Definition: A two-pattern test $\langle T_1, T_2 \rangle$ is said to be a **robust test** for fault s if it detects s even in the presence of arbitrary circuit delays and timing skews in input changes when the circuit inputs are changed from T_1 to T_2.

It is known that all two pattern tests for nFET (pFET) stuck-open faults in NAND (NOR) gates are robust [20,23]. It is also known that faults in domino CMOS gate circuits are robustly testable [17,18]. However not all faults in arbitrary static CMOS gate circuits are robustly testable [20]. Necessary and sufficient conditions on robust tests as well as procedures to derive them when they exist are known and are given in [24]. Another known feature is that for some faults robust testing may be possible with test sequences of length >2 [5,23] while no two-pattern robust tests exist for these faults. The necessary and sufficient conditions on tests with greater than two patterns to be robust are given in [23].

As an example of derivation of robust tests we next consider a class of popular circuits - parity trees.

5.2 Robust Tests for Parity Trees

Parity or EOR gate trees are widely used in encoding and decoding of digital information such as in logic circuits for implementing error-correcting or detecting codes. As sub-networks they appear in implementations of logic functions based on Reed-Muller Canonic (RMC) forms [20], in data compression circuits [22] and residue arithmetic circuits, etc. Transistor stuck-open faults in linear trees were first considered in [22]. These authors demonstrated the need for stuck-open fault testing in both nMOS and CMOS technologies and suggested constant length test sequence solutions for non-robust testing of stuck-open faults and line stuck-at faults in parity trees. Non-robust tests may be invalidated due to circuit delays as demonstrated earlier. We have determined that if the test invalidation problem in non-robust testing is redressed by introducing robust tests, the test length becomes a linear function of the depth of the circuit as opposed to constant number of tests derived in previous studies by neglecting circuit transients. A parity tree of depth 3 is shown in Figure 21. When realized by EOR gates shown in Figure 22 this tree can be "robustly" tested by applying the inputs given in the following test table in the sequential order in which they appear.

78

1 A	2 B	3 A	4 C	5 A	6 D	7 A	8 E
0	0	0	0	0	0	0	0
1	0	1	0	1	0	1	0
0	0	0	0	0	0	0	0
0	1	0	0	0	1	0	0
0	1	0	1	0	1	0	1
0	0	0	1	0	1	0	1
0	0	0	1	0	0	0	1
0	0	0	0	0	0	0	1
0	0	0	0	0	0	0	0

Note that the columns corresponding to the bit positions of the test
vectors are labeled by an integer and a letter. The integer label
corresponds to the label of the primary input of the circuit. The
columns under the same letter labels are identical in entries (eg.
columns 1, 3, 5 and 7 in the table are identical). Hence if the tests
given above are applied from a test pattern generator the columns with
the same letter label may be connected to the same output of the Test
Pattern Generator (TPG). Each letter label represents a Test Input
Variable (TIV) to be generated by a TPG. The number of Distinct Test
Patterns (DTPs) in the test sequence of length 9 given above is 7
(because the 1st, 3rd and the last rows are identical to each other).
Clearly the number of DTPs is less than or equal to the length of a test
sequence. Some of the important results related to robust tests of
parity trees are concerned wtih bounds on the number of TIVs, DTPs and
test lengths. When realized by EOR gates shown in Figure 22 a parity
tree of depth d requires a minimum of d+1 TIVs, d+2 DTPs and 2d+2
tests. These bounds are achievable [15]. Similar bounds for other EOR
gate realizations have also been established [15]. The tests proposed
are given in Figure 23.

5.3 Undetectability of Stuck-open Faults in CMOS Latches and Flip-flops

Consider the cross-coupled NOR gate latch shown in Figure 24. To
detect nFET in gate G2 connected to line 2 stuck-open one must initialize
line 5 to 1 by applying input R = $\bar{Q}$ = 0 and follow with test input R = 0
and $\bar{Q}$ = 1. However, it is not possible to make $\bar{Q}$ = 1 unless Q becomes
zero, which would destroy the initialized value of line 5. Hence, stuck-
open fault of nFET connected to line 2 (as well as the nFET connected to
line 3) is not detected. In the presence of these undetectable faults,
the latch, however, becomes a dynamic latch instead of a static latch.
This and several other problems related to the detection of stuck-open
faults in CMOS latches and flip-flops are given [25]. Several designs
for single stuck-open fault detectable CMOS latch and flip-flop designs
are also given in [25].

6. TESTABLE DESIGNS

In the previous sections we have indicated several difficulties in
testing for faults in CMOS logic circuits, specially static CMOS
circuits. In this section several methods to design circuits that are
"testable" are given.

6.1 Testable Single Complex Gate Realizations

It was shown earlier that it is possible to realize any logic function using a single CMOS gate. If one desires to design single gate realizations such that all single line stuck-at and FET stuck-open and stuck-on faults are detectable by the process of applying tests and observing logic values at the output, then a modified complex gate shown in Figure 25 can be used. Note that two additional FETs and two control inputs are used. It is assumed that the nFET and pFET networks of Figure 25 are derived from logic expressions corresponding to line stuck-at fault detectable primitive gate level circuits. It can be readily shown that in this realization all single line stuck-at faults, FET stuck-open faults and FET stuck-on faults are detectable by robust tests.

If only robust testability of stuck-open faults and testability of line stuck-at faults are desired then several testable designs based on s of p and p of s logic expressions are given in [14,23]. One of the main results in [14,23] is that the FET stuck-open faults in ps-ps complex gate realization of logic functions are all robustly testable.

6.2 Robustly Testable Primitive Gate Circuits

One of the disadvantages of the testable realizations proposed in [23] is that if primitive gate circuits are used then it may be necessary to use an extra control input to realize circuits that are robustly testable for line stuck-at and FET stuck-open faults. We give below a procedure to realize primitive gate circuits that are robustly testable for these faults but do not require extra control input.

Let F be the function to be realized. We say that F is **unate** in variable x_i if a sum of products or product of sums for F exists with either only x_i or $\bar{x}_i$ present in it. If F is not unate in x_i then we say that it is **binate** in x_i. We also call x_i a unate or non-unate variable of F.

Earlier it was shown that if F is unate in all its variables then a robust testable primitive gate realization for F can be realized without the use of additional control inputs. If F is not unate in a variable x_i then decompose F as $F = x_i F_{i1} + \bar{x}_i F_{i0}$, where $F_{i1}(F_{i0})$ is obtained from F by setting $x_i = 1$ ($x_i=0$). The realization of F as shown in Figure 26 is robust testable if robust testable realizations are used for F_{i1} and F_{i0}. This suggests that by iteratively decomposing (if necessary) until implied subfunctions are unate one can realize a robust testable design for F. A somewhat different expansion of F with respect to a binate variable is also possible. Details are given in [15].

REFERENCES

1. D.K. Bhavsar, "A New Economical Implementation for Scannable Flip-flops in MOS", IEEE Design and Test, pp. 52-56, June 1986.
2. R. Chandramouli, "On testing stuck-open faults", 13th International Symposium on Fault Tolerant Computing, Milan, Italy, pp. 258-265, June 28-30, 1983.
3. K.W. Chiang and Z.G. Vranesic, "Test generation for MOS complex gate networks", Proceedings of the 12th International Symposium on Fault-Tolerant Computing, June 1982, pp. 149-157.

4. K.W. Chiang and Z.G. Vranesic, "On fault detection in CMOS logic networks", 20th Design Automation Conference, pp. 50-56, June 1983.

5. G.L. Craig and C.R. Kime, "Psuedo-Exhaustive Adjacency Testing: A BIST Approach for Stuck-open Failures", Proc. Int. Test Conference, pp. 126-137, November 1985.

6. D. Baschiera and B. Courtois, "Advances in Fault Modeling and Test Pattern Generation for CMOS", Proc. ICCD-86, pp. 82-85, October 1986.

7. S. Dasgupta and E.B. Eighelberger, "An enhancement to LSSD and some application of LSSD in reliability, availability and serviceability, Proc. of 11th Int. Symp. on Fault-Tolerant Computing, June 81, pp. 32-34.

8. R.D. Davies, "The case for CMOS", IEEE Spectrum, pp. 26-32, October 1983.

9. E.B. Eichelberger and T.W. Williams, "A logic design structure for LSI testing", Proceedings of the 14th Design Automation Conference, pp. 462-468, June 1977.

10. Y.M. El-Ziq, "Automatic test generation for stuck-open faults in CMOS VLSI", 19th Design Automation Conference, Nashville, pp. 347-354, June 1981.

11. Y.M. El-Ziq and R.J. Cloutier, "Functional-level test generation for stuck-open faults in CMOS VLSI", Int. Test Conference, Philadelphia, pp. 536-546, October 1981.

12. J. Galiay, Y. Crouzet and M. Vergnault, "Physical versus logical fault models for MOS LSI circuits: Impact on their testability", IEEE Trans. on Comp., vol. C-29, pp. 524-531, June 1980.

13. S.K. Jain and V.D. Agrawal, "Test generation for MOS circuits using D-algorithm", 20th Design Automation Conference, pp. 64-70, June 1983.

14. N.K. Jha and J.A. Abraham, "Testable CMOS logic circuits under dynamic behavior", Int. Conf. on Computer Aided Design, Santa Clara, Nov. 12-15, 1984.

15. S. Kundu, Ph.D. Thesis, Department of Electrical and Computer Engineering, University of Iowa, Iowa City, 1988.

16. S. Kundu and S.M. Reddy, "On the design of TSC CMOS combinational logic circuits", Proc. of the Int. Conf. on Comp. Design, pp. 496-499, October 1986.

17. S.R. Manthani and S.M. Reddy, "On CMOS totally self-checking circuits", Proc. of Int. Test Conf., Philadelphia, pp. 866-877, Oct. 1984.

18. V.G. Oklobdzija and P.G. Kovijanic, "On Testability of CMOS-Domino Logic", Proc. Int. Symp. on Fault-Tolerant Computing, pp. 50-55, June 1984.

19. S.M. Reddy, "Easily testable realizations for logic functions", IEEE Transactions in Computers, vol. C-21, pp. 1183-1189, Nov. 1972.

20. S.M. Reddy, M.K. Reddy and J.G. Kuhl, "On testable design for CMOS logic circuits", Proc. of the Int. Test Conf., Philadelphia, pp. 435-445, October 1983.

21. S.M. Reddy, V.D. Agrawal and S.K. Jain, "A gate level model for CMOS combinational circuits with application to fault detection", Proc. of Annual Design Automation Conf., pp. 504-509, June 1984.

22. S.M. Reddy, K.K. Saluja and M. Karpovsky, "A data compression technique for built-in self-test", Proc. of the Int. Symp. on Fault-Tolerant Computing, June 19-21, 1985.

23. S.M. Reddy and M.K. Reddy, "Testable realization for FET stuck-open faults in CMOS combinational logic circuits", IEEE Trans. on Comp., pp. 742-754, August 1986.

24. M.K. Reddy and S.M. Reddy, "Robust Tests for Stuck-open Faults in CMOS Combinational Logic Circuits," Proc. 14th Int. Symp. Fault-tolerant Comp., pp. 44-49, June 1984.

25. M.K. Reddy and S.M. Reddy, "Detecting FET Stuck-open Faults in CMOS Latches and Flip-flops," IEEE Design and Test, pp. 17-26, October 1986.

26. D.S. Ha and S.M. Reddy, "On the Design of Testable Domino PLAs," Proc. Int. Test Conf., pp. 567-573, November 1985.

27. R.L. Wadsack, "Fault modeling and logic simulation of CMOS and MOS integrated circuits", Bell System Technological Journal, vol. 57, pp. 1449-1474, May-June 1978.

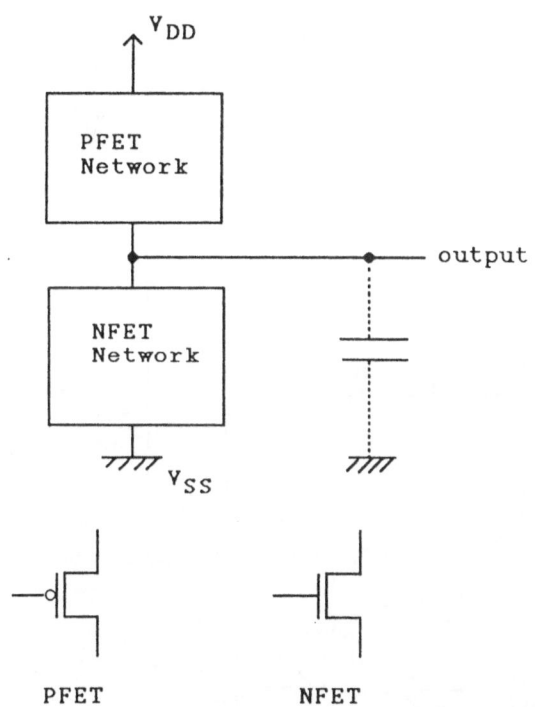

Figure 1 : Block Diagram of a CMOS Gate.

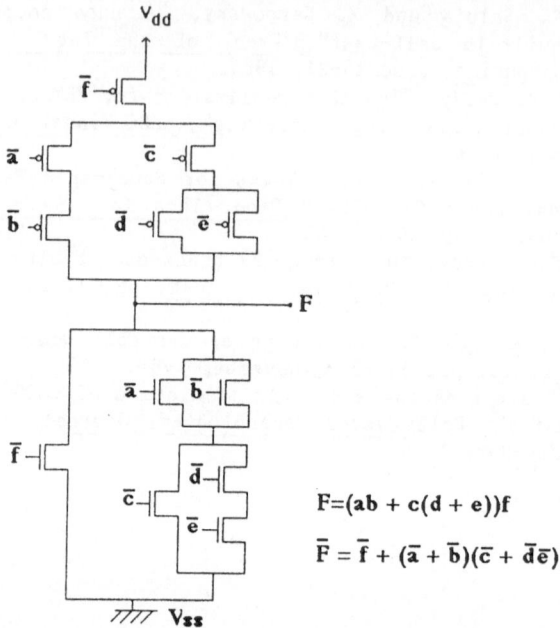

$$F = (ab + c(d + e))f$$

$$\overline{F} = \overline{f} + (\overline{a} + \overline{b})(\overline{c} + \overline{d}\,\overline{e})$$

Figure 2 : A Static CMOS Gate realizing F = (ab + c(d+e))f

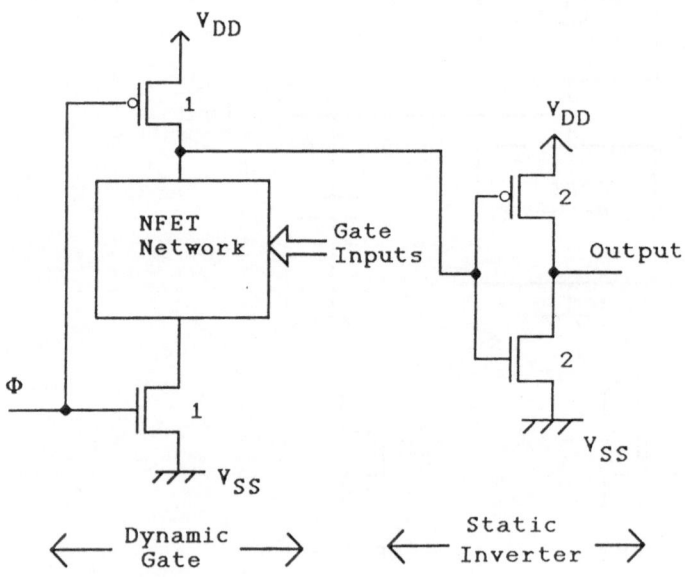

Figure 3 : Domino CMOS Gate.

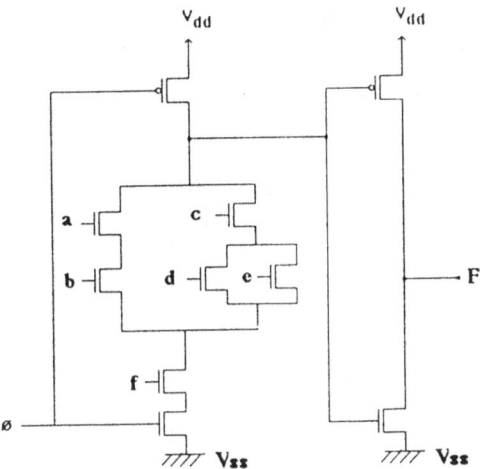

Figure 4 : Domino Gate realizing F = (ab + c(d+e))f.

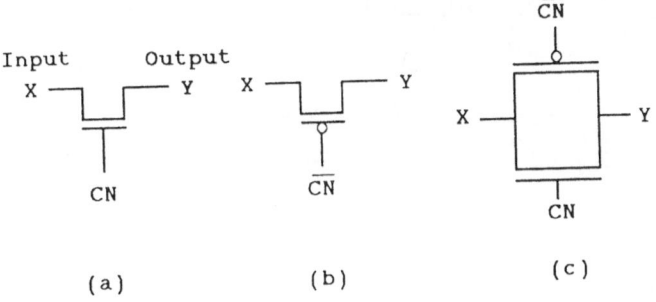

(a)   (b)   (c)

Figure 5 : MOS Transmission Gates.

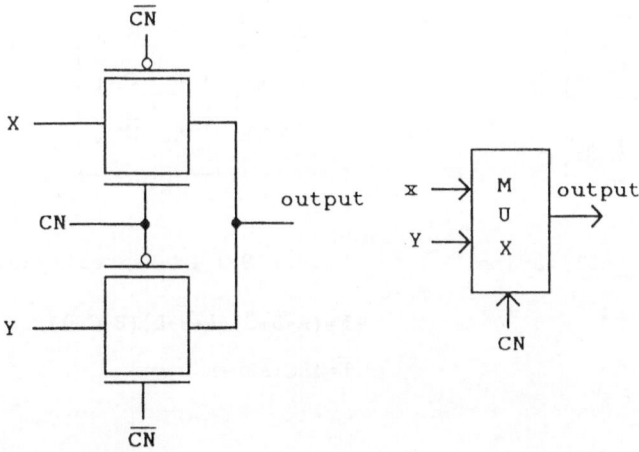

Figure 6 : A multiplexer realized by Transmission Gates.

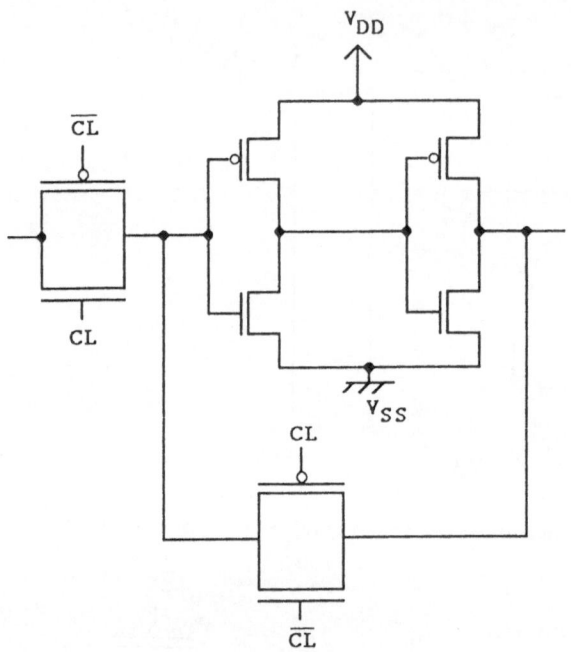

Figure 7 : A D-latch.

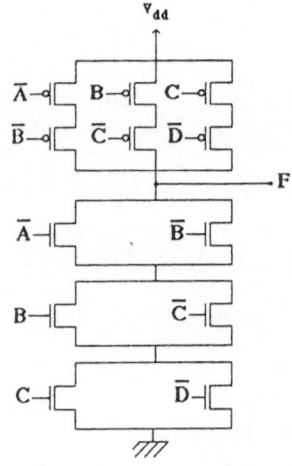

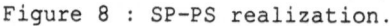

Figure 8 : SP-PS realization.

$F=AB+\overline{B}C+\overline{C}D$

$\overline{F}=(\overline{A}+\overline{B})(B+\overline{C})(C+\overline{D})$

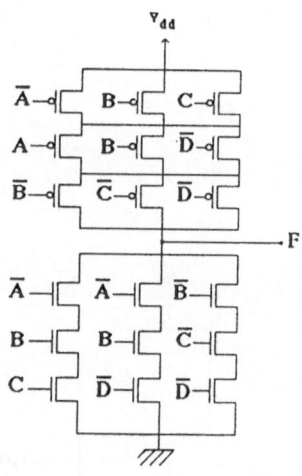

Figure 9 : PS-SP realization.

$F=(A+\overline{B}+\overline{C})(A+\overline{B}+D)(B+C+D)$

$\overline{F}=\overline{A}BC+\overline{A}B\overline{D}+\overline{B}\overline{C}\overline{D}$

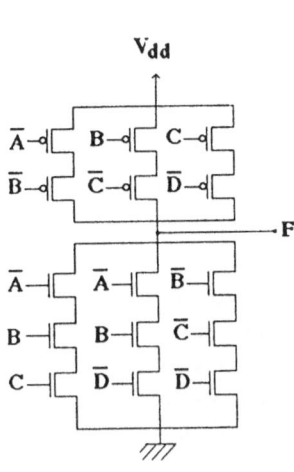

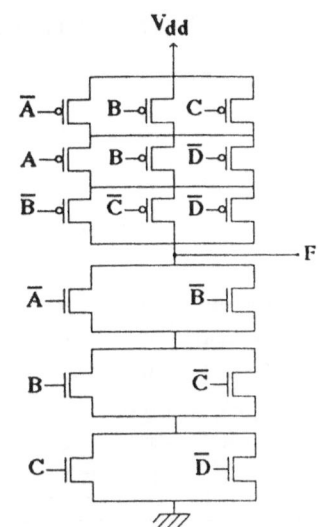

Figure 10 : SP-SP realization.

$F=AB+\bar{B}C+\bar{C}D$

$\bar{F}=\bar{A}BC+\bar{A}B\bar{D}+\bar{B}\bar{C}\bar{D}$

Figure 11 : PS-PS realization.

$F=(A+\bar{B}+\bar{C})(A+\bar{B}+D)(B+C+D)$

$\bar{F}=(\bar{A}+\bar{B})(B+\bar{C})(C+\bar{D})$

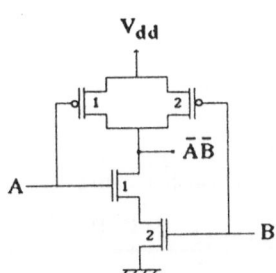

Figure 12 : A two-input NAND Gate.

Fault	Test	Fault-free output	Output with fault
A s-a-1	01	1	0
A s-a-0	11	0	1
B s-a-1	10	1	0
B s-a-0	11	0	1
pFET1 stuck-open	01	1	0 (with output initialized to 0)
pFET1 stuck-open	10	1	0 (with output initialized to 0)
nFET1 stuck-open	11	0	1 (with output initialized to 1)
nFET1 stuck-open	11	0	1 (with output initialized to 1)
pFET1 stuck-on	11	0	unknown (monitor supply current)
pFET1 stuck-on	11	0	unknown (monitor supply current)
nFET1 stuck-on	01	1	unknown (monitor supply current)
nFET1 stuck-on	10	1	unknown (monitor supply current)

Figure 13 : Tests for two-input NAND Gate.

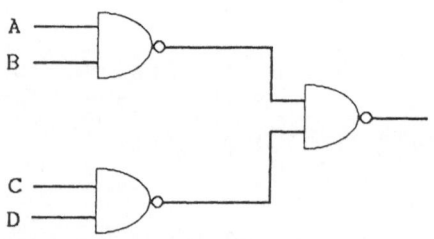

Tests for line stuck-at faults only

A B C D

1 1 0 1
0 1 1 1
1 0 1 0
0 1 0 1

tests for extended faults

A B C D

1 1 0 1
0 1 1 1
1 1 0 1
1 0 1 0
0 1 1 1
1 0 1 0
1 1 0 1

Figure 14 : Tests for line stuck-at and extended fault model.

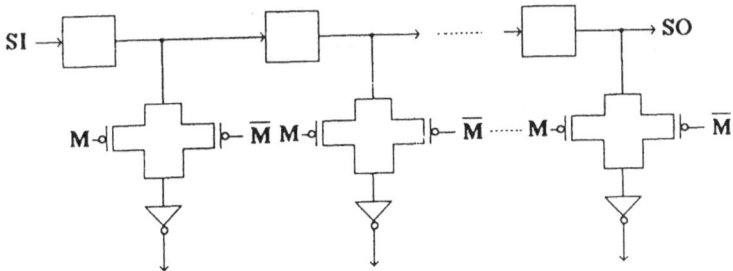

Figure 15 : A modified scan register for two-pattern
testing.

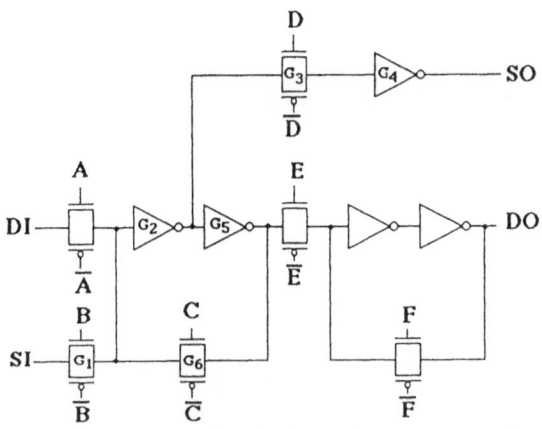

Figure 16 : A scan flip-flop for two pattern test
application in two latch sequential circuit
designs.

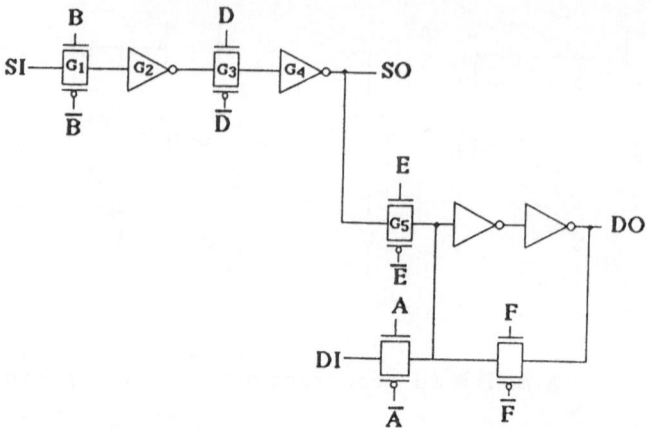

Figure 17 : A scan flip-flop for two pattern test
application in single latch sequential circuit
designs.

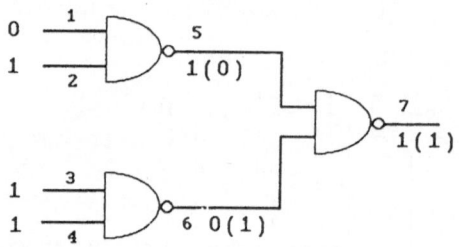

$L_5 = \{1/1, \ n1/on, \ p1/off, \ 5/0\}$

$L_6 = \{3/0, \ 4/0, \ p3/0n, \ p4/on,$
$\qquad n3/off, \ n4/off, \ 6/1 \ \}$

$L_7 = \{3/0, \ 4/0, \ p3/0n, \ p4/on,$
$\qquad n3/off, \ n4/off, \ 6/1 \ , \ n6/on, \ 7/0\}$

Figure 18 : Faults detected by test input 0111 given the
initial state of the circuit nodes.

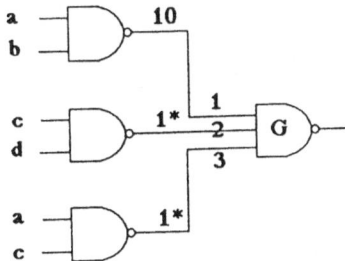

Figure 19 : Stuck-open fault test invalidation.

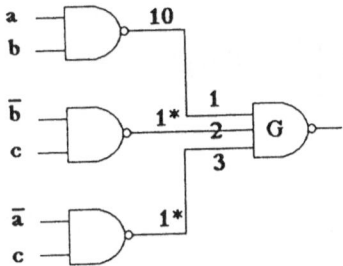

Figure 20 : Another example of test invalidation.

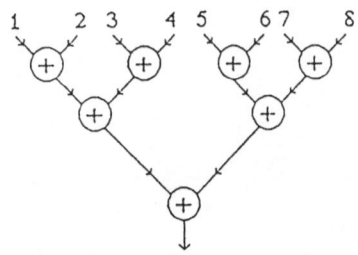

Figure 21 : A parity tree of depth 3.

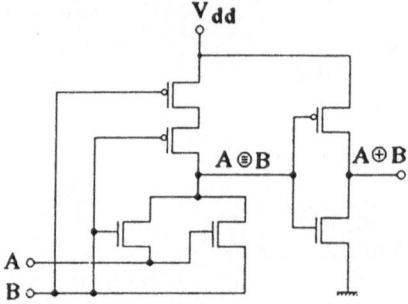

Figure 22 : CMOS EOR Gate.

1	2
1	1
1	0
0	0
0	1

1	2	3	4
1	1	1	1
1	0	1	1
0	0	1	0
0	1	0	0
0	0	0	0
0	0	0	1

1	2	3	4	5	6	7	8
1	1	1	1	1	1	1	1
1	0	1	1	1	1	1	1
0	0	1	0	1	0	1	1
0	1	0	0	0	0	1	0
0	0	0	0	0	1	0	0
0	0	0	1	0	0	0	0
0	0	0	0	0	0	0	0
0	0	0	0	0	0	0	1

$d = 1$ $d = 2$ $d = 3$

Each EOR

gate needs : < 00 or 11, 01 >,

< 00 or 11, 10 >,

< 01 or 10, 00 > .

Figure 23 : Tests for parity trees.

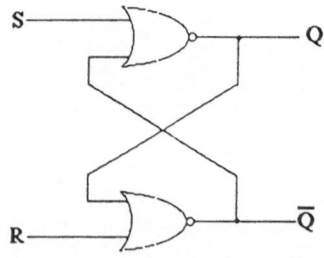

Figure 24 : An SR Latch.

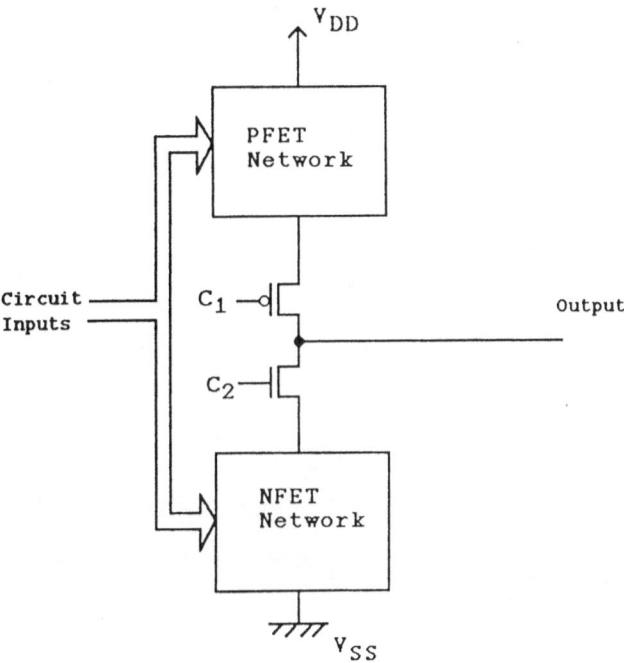

Figure 25 : A Testable Complex Gate.

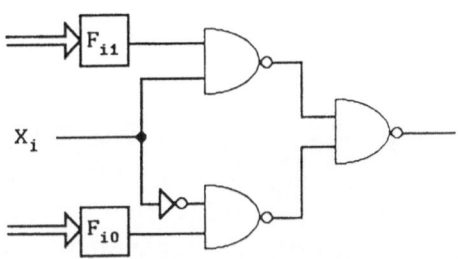

Figure 26 : Testable Primitive Gate realization.

PARALLEL COMPUTER SYSTEMS TESTING AND INTEGRATION

MIROSLAW MALEK

DEPARTMENT OF ELECTRICAL AND COMPUTER ENGINEERING
UNIVERSITY OF TEXAS AT AUSTIN
AUSTIN, TEXAS 78712

1. INTRODUCTION

The ever-growing complexity of computer systems testing is becoming more important than ever. With today's technology we are capable of building the systems that may contain in a cage or a frame several millions of circuits which, even with high level integration, may have to be packaged in several hundred thousand chips. The quest for testing in multiprocessors is a very challenging one and it is our intent to shed some light on this important, but frequently underestimated, aspect of implementing large parallel computers.

Testing is an integral part of any design process and the system life cycle. It is also necessary at every level of system packaging and integration and it needs to be complete, comprehensive, and efficient. Let us look first at the stages of the system life cycle. We know that any new system is built because the old system is obsolete or there are new needs, and/or capabilities or there may be new technology that makes building or design of a better system feasible. The system life cycle is described below:

    concept formulation
    system specification
    design
    prototype
    production
    installation
    operational life
    modification and retirement

Notice that after every step of the system life cycle the system should be tested with respect to the validity of the concept, correctness of specification, design, etc. Unfortunately, it is very difficult at the initial stage of the project to test for completeness of specification because the user very frequently is not sure exactly what he needs. There may be some constraints in the systems that the initial specification will not be able to consider, or some constraints may prove to be invalid and therefore the system with the higher performance or better design for testability will be feasible. Whatever the case may be, the bottom line is that testing should be an integral part of every stage in the design process.

Let us take another view of the system, and determine the packaging levels of system integration. We may observe that the way the system is packaged has direct impact on testing; in fact in the entire process, we can distinguish the following layers of system integration:

    integrated circuits (chips)
    printed circuit boards/cards, wafers, TCMs (thermal-conductor modules)
    boxes/cages
    cabinets/frames

F. Lombardi and M. Sami (eds.), Testing and Diagnosis of VLSI and ULSI, 93–116.
© *1988 by Kluwer Academic Publishers.*

operating systems
standard languages
special purpose languages
applications modules
applications

Notice that, similarly as in the system life cycle, the testing is needed at every level of integration or packaging. We start with the exhaustive tests that should maximize fault coverage for integrated circuits, then we test the boards, then boxes, cabinets, software, and applications. Testing should be an integral part of the system development and test and diagnostics should be written for each packaging level of the system.

Let us define yet another view, abstract view, of the system. The levels are as follows.

circuit level
logic level
register-transfer level
system hardware
operating system, languages
applications

Notice that if for each layer we define the set of tests that will maximize the coverage of the system, the same test set will also be useful for the system integration test provided tests for checking interfaces are added. It is becoming clear that in order to build reliable multiprocessor systems we need tests for every stage of the system life cycle, for every level of packaging, as well as for every level of the system functionality.

We call this concept dependability rings because at every level it is useful not only for testing, but also for fault location and fault recovery, as well as for fault tolerance.

In Figure 1 we show dependability rings for testing and fault tolerance and as you may notice every stage of the ring is surrounded by an additional circle called acceptance test. What we mean by this are both on-line and off-line tests that will verify and accept correctness of circuits or programs in the inner circles. In order to bring a multiprocessor system to life, our experience with two multiprocessor systems design and development indicate that the following staging of the test rings becomes effective. First we test diagnostic and maintenance processors that form a hardcore of the system under test. Then those diagnostic and maintenance processors test main processors of the multiprocessor system, then those processors test their memories in parallel, and then they access shared memories or pass messages to them through the network, and thus testing the network. This process can also be illustrated in a form of test rings as shown in Figure 2. This bootstrap technique, that expands from diagnostic and maintenance processors all the way to the network, needs to be executed in parallel in order to increase efficiency of tests. We will describe later in more detail how processors, memories, and networks should be tested.

2.   DEPENDABLE DESIGN AND TESTING METHODOLOGIES
We introduce here both dependable design and testing methodologies. As will be demonstrated, we strongly believe that both design for fault tolerance and design for testability should be integrated. The dependable design methodology is outlined here.
1.   Identify fault classes, fault latency and fault impact.
2.   Determine qualitative and quantitative specifications for dependability, and evaluate the design in a specific environment.
3.   Identify "weak spots" and assess potential damage.
4.   Decompose (partition) the system.
5.   Develop fault and error detection techniques and algorithms.

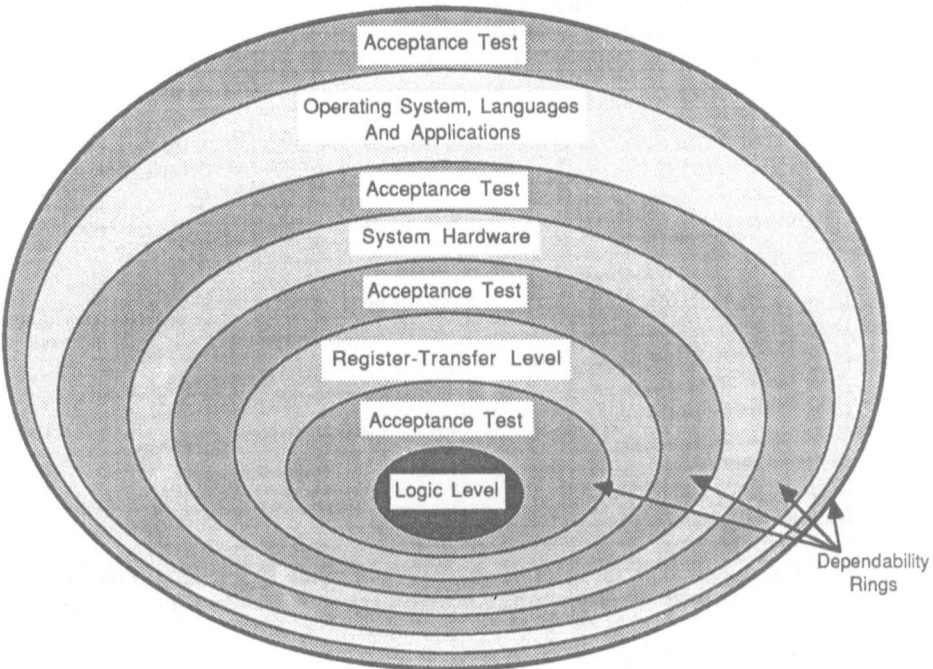

FIGURE 1. Dependability (reliability) rings for testing and fault tolerance
each dependability ring provides measures and mechanisms for testing and fault tolerance
(detection, location, testability and recovery).

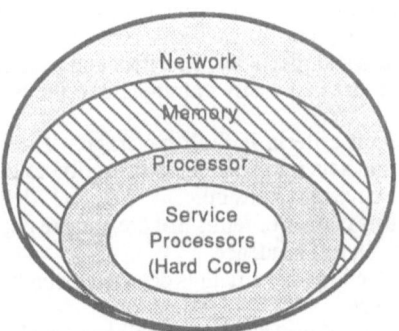

FIGURE 2. A bootstrap - test rings in a multicomputer system.

6. Develop fault isolation techniques and algorithms.
7. Develop recovery/reintegration/restart techniques.
8. Evaluate degree of fault tolerance.
9. Refine, iterate by improvement. Try to eliminate "weak spots" and minimize potential damage.

The more detailed description of every step of this design methodology for parallel systems can be found in [1].

The system testing methodology and design for testability could be summarized as follows.

1. Identify fault classes, fault latency, and fault impact.
2. Determine qualitative and quantitative specifications for testability and coverage.
3. Decompose (partition) the system.
4. Analyze the system, assess and incorporate, if necessary, design for testability during the design process at every level.
5. Generate tests for every level of system hierarchy with respect to various systems views.
6. Estimate coverage and testing time.
7. Refine and iterate for improvement.

Looking at both methodologies that we have just outlined, it is easy to notice that several steps do overlap; therefore, we strongly believe and urge you to integrate design for testability with design for dependability, performance and functionality. In fact, the integration of dependability, testing, performance, and functionality we call computer integrity. Since the topic of the paper is system testing, we will not comment further on computer integrity, but would like to point out and make the reader aware that in future multiprocessor systems, because of the built-in redundancy, trading performance for dependability will become a standard feature.

3. TESTABILITY AND DIAGNOSABILITY MEASURES

In testing at any level it is important to know and measure the objective of testing. Of course, it is clear that we would like to test for any fault in a very short time with a very small overhead in time and test equipment, but we know that in reality it is not usually possible. We have to compromise the test time or what we call fault coverage. Several measures were developed over the years, and this section will highlight a few, in our opinion, important ones and give some specifics and examples on the fault coverage at the logic level. The reader can easily extrapolate to the higher levels of previously defined hierarchies.

One of most important criteria is fault coverage. There are numerous ways of defining fault coverage and several researchers subscribe to different defintions. Here are some examples of the coverage measure.

a. Qualitative: which lists classes of faults that are and are not recoverable (testable, diagnosable) for a given test sequence.
b. Quantitative: the probability that a fault is detected given that the failure has occurred for a given test sequence.
c. Quantitative: percentage of testable/diagnosable/recoverable faults for a given test sequence.
d. Quantitative: sum of the coverages of all fault classes, weighted by the probability of occurrence of each fault class with respect to a given test sequence
$$[p_1C_1+p_2C_2+...+p_nC_n].$$

The testability measure is a combination of controllability and observability properties of the system as well as the quality of test generator. Here we give an example of logic level testability as defined in SCOAP [2]. In SCOAP, six levels of testability are distinguished.

combinatorial zero controllability (CC0)
combinatorial one controllability (CC1)
combinatorial observability (C0)
sequential zero controllability (SC0)
sequential one controllability (SC1)
sequential observability (S0)

The values of all these testabilities grow proportionally to the required testing effort, and they provide a good quantitative measure of controlling and observing logic values of internal points. So how are testability measures derived? They should be a function of circuit description and test generator description. Based on that information, after running tests, we should be able to estimate the testability results. In the fine paper by V.D. Agarwal and M.R. Mercer [3] titled "Testability Measures-What Do They Tell Us?" the authors define detection probability (fault coverage) as a function of testability measure.

Let us define $p(t)=e^{-\alpha t}$ a detection probability as a function of testability where testability ranges from 1 to ∞. α is a test generation procedure parameter that indicates how good the procedure is. In the circuit with n faults, there are n different testability measures for each point $t_1, t_2, ..., t_n$.

Having defined that, we can easily find a fault coverage after applying v test vectors:

$$f(v) = 1/n\sum_{i=1}^{n}\left\{ 1-\left[1-p(t_i)\right]^v \right\} = 1-1/n\sum_{i=1}^{n}\left[1-e^{-\alpha t_i}\right]^v$$

This type of testability measures should be applied at every level of the system design.

Frequently it is not sufficient to test the system and find that it is faulty or fault-free. It is also important to know where the fault is especially with respect to a field replaceable unit.

In a seminal paper by Preparata, Metze and Chien (PMC) [4], the self-diagnosability of systems, called t-diagnosability, was defined.

A system of n units is one step t-fault diagnosable (t-diagnosable) if all faulty units within the system can be located without replacement, provided the number of faulty units does not exceed t, where

1. $2t+1 \leq n$, and
2. At least t units test each unit.

The above represents conditions for the system to be t-diagnosable. More details on the PMC model description can be found in Section 8 of this Chapter. This work initiated extensive research in the area [5, 6, 7, 8, 9, 10].

Of course, not all systems need to be self-diagnosable. Another example of the definition of t-diagnosability can be stated as follows: A system of n units is t-fault diagnosable if under a given number of k tests up to t-faulty units, can be located with respect to given inputs and outputs (observable points).

This type of diagnosability definition is very useful for multistage interconnection networks that are frequently used in multiprocessor systems [11, 12].

Coverage, testability, and diagnosability are important test parameters, but what is also of great significance are time and cost of test development, amount of time spent for test that is equivalent of course to the cost of test, and the overhead in design for testability.

With respect to testability measures, it is up to the specifications team to decide which of them are the most important and why. How to precisely define and model a system under tests and its environment, is another difficult and important problem that has to be considered. How to partition for testability and design for testability and how to evaluate tradeoffs is highly dependent on a particular system, its environment and cost associated with design for testability.

It is clear that testing, verification, validation, and debugging have become the most consuming parts of the design process. We are familiar with the phrases like "the last 10%" in the design of the code where the last 10% (debugging) takes over 80% of the total development time or the systems that cost initially $1000 and then it costs several times more to fix or maintain them. Of course, the difficulty associated with testing is its exponential growth. So, reiterating some of the previous questions and ideas, the top-down design, partitioning, structuring, and integration are important components of good testing at the system level. The interfaces have to be clearly defined and though tests for interfaces are difficult, they are essential in proper system verification and validation.

When we begin to consider testing for a given system, several questions have to be answered. At what level exhaustive tests should be carried out; logic, functional, program, algorithmic? Our philosophy is that testing should take place at every level. Are fault models realistic? Careful definition and identification of fault classes is essential. What frequently happens is that diagnostics are highly efficient in finding solved problems, and inadequate for a majority of faults that occur in real systems. In fact, we should test for both intended and unintended functions. The two approaches are black box versus white box. The black box approach relies only on specification and the classical approach says that the testing engineer should not care what is inside, as long as inputs create anticipated outputs. The white box approach requires good knowledge of the system structure for derivation of efficient tests.

4. PROCESSOR TESTING

Several papers and books were written on digital logic testing. Techniques such as LSSD [14] and design for testability [13] and several versions of the D-algorithms are most popular among these. In a complex multiprocessor system, once the chips are tested, the functional testing seems to be the only feasible approach during system integration. One of the important contributions by Thatte and Abraham [15] was to advocate the data flow graph model for functional testing of processors. In that model nodes represent registers, or active units, while links correspond to data transfer paths. The three classes of instructions are defined: sequential control, data storage, and transfer and data munipulation. For example, a fault model for data transfer functions is such that allows any number of lines to be stuck-at-0 or 1, and bridges could be of the type "AND" or "OR". With this technique, depending on the complexity of the chip or a processor, 70-96% coverage is feasible. Functional testing is usually supplemented by additional pseudo-random tests.

Another widespread technique is a built-in test. By using coding [16, 17], LSSD [14], signature analysis [18, 19, 20] and watchdog processors [21], it is possible to develop extensive tests. There are several processors that use this technique and at several companies, scanning or LSSD are required design rules.

Let us introduce some examples that could use the functional testing approach. *Flip-flops* require that they can be set from 0 to 1, and reset from 1 to 0. We need to validate that this is possible as well as that the flip-flops can hold their state. *Counters* can be tested by checking their contents after each cycle. For a 32-bit or 64-bit counters that are common in today's implementations, this approach would take an extremely long time. Therefore, a practical approach requires that a counter is divided into a 6-bit or 8-bit sections, and that way, the testing time can be significantly decreased by counting only within subsections while maintaining high fault coverage. For example, for a 24-bit counter, an exhaustive technique would require over 16 million tests, while using the partitioning approach with 6 bits in each partition, only $4 \times 2^6 = 256$ tests are required. When we test, for example, *multiplexers*, we may observe the following: that either no source is selected, the wrong source is selected, or more than one source is selected and multiplexer output is wired-AND or wired-OR. Therefore, at least $2^n + 1$ tests are required. An additional test is required to check "enable" line. In *demultiplexer* the faults that may occur are that no destination is selected, or instead of, or in addition to, the selected correct destination, one or more other destinations are selected. To test for this kind of a fault also $2^n + 1$ tests are required. An extra test is needed to check "enable" line. In the case of *decoders* the faults that may occur are: either no output is selected or the incorrect decoding has occurred. Also, in this case to test for all these faults, $2^n + 1$ tests are required plus one test for "enable" line.

Considering a functional testing approach the following processor testing procedure may be proposed [22, 23].

1. Reset a microprocessor.
2. Divide and test the program counter PC ($2 \times 16 = 32$) by incrementing it through its states via the NOP instruction ($2 \times 2^{16} = 128K$ states of the PC; state of the PC is compared to the tester's counter).
3. Test the general purpose registers by transferring selected test patterns to each register in turn via the PC.
4. Test the stack pointer register by incrementing and decrementing it through 128K states. Access via the PC.
5. Test the accumulator by transferring selected test patterns to it via previously tested registers.
6. Test the ALU and flags by exercising all arithmetic, logical and conditional branch (flag testing) instructions.
7. Exercise all previously untested instructions and control lines.

It is clear that this functional approach to testing cuts down on the testing time, but high coverage cannot be assured. Also, some basic rules in design for testability should be followed. Here are some of them [23]: allow all memory elements to be initialized before testing begins, preferably via a single reset line; provide means for opening feedback loops during testing; allow external access to the clock circuits to permit the tester to synchronize with, or disable, the unit under test; insert multiplexers to increase the number of internal points which can be controlled or observed from the external pins; avoid logically redundant circuits if possible; use synchronous circuitry whenever possible. Recently in high performance processors built in the latest technologies, a new phenomenon has occurred: Testers are not able to run at the unit under test speeds. Comparison techniques seem to be effective in this type of environment.

With the ever-increasing system complexity, the built-in test technologies are becoming indispensable. Instead of external test, coding, self-checking circuits, and program self-testing are used more frequently than ever before.

5. MEMORY TESTING

In recent years, rapid advances in technologies allowed to develop and deliver to the markets memory chips ranging from 64K to 4 Mbit in size. This high chip density created new problems: the memories are very susceptible to α-particles and pattern sensitivity. Diversity of technologies and layouts created additional difficulty in testing memories despite their regularity. One of the biggest problems in today's memories are temporary (soft) errors. In order to have a sound operational environment, the error correction codes or, concurrent error detection with retry are becoming standard. Many of today's multiprocessor systems have random access memory with sizes in the order of gigabytes. This type of environment creates not only stressful and highly demanding testing conditions, but also a requirement for error correction becomes necessary.

Numerous failure n.ɔdes in random access memories are possible. As mentioned earlier α-particles and pattern sensitivity are predominant, but for completeness, let us list most of the failure modes that may occur in semiconductor memories [24].

1. Stuck-at-1 or stuck-at-0 hard or soft faults.
2. Open and short circuits - too much or too little metallization; also bonds not making contact with the pad (opens).
3. Input and output leakage - leakage current in excess of the specified limit.
4. Decoder malfunction - inability to address same portions of the array due to an open or defective decoder.
5. Multiple writing - data written into more than one cell when writing into one cell.
6. Pattern sensitivity - device does not perform reliably with certain test pattern(s). (e.g., 111...101...1 may be read as all 1's).
7. Refresh dysfunction - data are lost during the specified minimum refresh time.
8. Slow access time - considerable capacitive charge on the output driver circuit taking excessive time to sink current and thus making the access time slow.
9. Write recovery - access time becoming longer when a read follows immediately after a write.
10. Sense amplifier recovery - when data accessed for a number of cycles are the same and then suddenly changed. The sense amplifier tends to stay in the same state as before.
11. Sleeping sickness (memory loses information in less than the stated hold time).

In such environment the RAM test requirements are as follows.
1. Every cell of memory must be capable of storing 0 and 1.
2. The addressing circuits or decoders must correctly address every cell.
3. The sense amplifiers must operate correctly.
4. There must be no interaction between cells.
5. The cells must be capable of storing data for a specified time without being refreshed (dynamic RAMs only).

The three categories of tests can be distinguished:
1. Functional (1's and 0's).
2. DC parametric (signal timing, fall and rise).
3. AC parametric (access times, setup and hold times and cycle times).

Several RAM testing techniques have been proposed. Here is the list of some of them followed by description of each procedure with its anticipated coverage and complexity [24, 25].

TEST	COMPLEXITY
Column Bars	$4N$
Checkerboard	$4N$
Marching 1's/0's	$12N$
Shifted Diagnoal	$4N^{3/2}$
Ping Pong	N^2
Walking 1's/0's	$2N^2 + 6N$
Galloping 1's & 0's	$2N^2 + 8N$
Algorithmic Test Sequence	$8N$
Eulerian Test	$256N$

MEMORY TESTING TECHNIQUES

MEMORY SCAN [MSCAN] PROCEDURE

FOR	$i = 0, 1, 2, ..., N-1$ DO:
WRITE	$c_i \leftarrow 0$
READ	$c_i = (0)$
WRITE	$c_i \leftarrow 1$
READ	$c_i = (1)$

The MSCAN can detect any stuck-at fault in the memory array, in the memory data register, or in the read/write logic.

The MSCAN will not detect any stuck-at fault in the memory address register or in the decoder.

Complexity: 4N

TEST: COLUMN BARS (also checker board test is similar)

Test Procedure:

Step 1: Write 1's in all even columns and 0's in all odd columns.
Step 2: Read each column and row.
Step 3: Repeat steps 1 and 2 for complementary patterns (i.e., interchange 0's and 1's).

Complexity: 4N

TEST: VOLATILITY TEST PATTERN

Purpose: Checks for hold time in dynamic memories.

Test Procedure:

Step 1: Load memory with a test pattern (various different simple patterns are typically used).
Step 2: Pause T units of time (inhibit all clocks).
Step 3: Read entire memory.
Step 4: Repeat for complementary patterns.

Complexity: 4N

TEST: MARCHING 1's AND 0's

Purpose: Minimal functional testing; detection of many decoder errors; minimal check on cell interactions.

Test Procedure:

Step 1: WRITE: $c_i \leftarrow 0$ for i=0, 1,...,N-1

Step 2: FOR i=0, 1, ..., N-1 DO
    READ: c_i (=0)

    WRITE: $c_i \leftarrow 1$
    READ: c_i (=1)

Step 3: FOR i=N-1, N-2, ... 0 DO
    READ: c_i (=1)

    WRITE: $c_i \leftarrow 0$
    READ: c_i (=0)

Step 4: Repeat steps 1-3 interchanging 0's and 1's. That is repeat for the complementary pattern.

Complexity: 12N

TEST: PING-PONG (often used as a subtest procedure in other tests)

Purpose: Pattern sensitivity and the interaction between pairs of cells.

Test Procedure:

One cell c_i is designated as the test bit. Now for all $j \neq i$ some patterns of read and write operations are done on cells c_i and c_j.

Complexity: N^2

Comments: Row/column ping-pong (complexity $N^{3/2}$ to cut down the large amount of tester time).

TEST: WALKING 1's AND 0's

Purpose:
Verifies that each cell can be set to both 0 and 1; that any cell can be set to either state without causing any other cell to change its state; and that decoder addressing is correct. Slow amplifier recovery is also detected by this test.

Test Procedure:

Step 1: WRITE: $c_i \leftarrow 0$ FOR i=0, 1, ..., N-1

Step 2: FOR i = 0, 1, ..., N-1 carry out the following full ping-pong procedure on test bit c_i.

      2A: WRITE: $c_i \leftarrow 1$
      2B: READ: c_j (=0) for all $j \neq i$ (tests that no cell is disturbed).
      2C: READ: c_i (=1) (tests that test bit is still correct).

      2D: WRITE: $c_i \leftarrow 0$ (restores test bit to original value).

Complexity: $2N^2 + 6N$

TEST: GALLOPING 1's AND 0's

Purpose:
To test all possible address transitions with all possible transitions when reading.

Test Procedure:

(Galpat 1): Same as for "walking 0's and 1's" except that in step 2B each READ: $c_j(=0)$ operation is followed by a READ: c_i (=1) operation. A modified version of this test procedure, known as Galpat II, also exists.

Complexity: $2N^2 + 8N$

TEST: ALGORITHMIC TEST SEQUENCE (ATS)

Purpose:
The test detects any combination of stuck-at faults.

Test Procedure:
Step 1: Divide a memory array into three regions as follows.

$$G_0 = \left\{ c_i \mid i=0 \text{ (MODULO 3)} \right\}$$

$$G_1 = \left\{ c_i \mid i=1 \text{ (MODULO 3)} \right\}$$

$$G_2 = \left\{ c_i \mid i=2 \text{ (MODULO 3)} \right\}$$

Step 2: WRITE 0's in partitions G_1 and G_2
Step 3: WRITE 1's in G_0
Step 4: READ 0's from G_2
Step 5: WRITE 1's in G_2
Step 6: READ 0's from G_3
Step 7: READ 1's from G_1 and G_2
Step 8: WRITE 0's in G_0
Step 9: READ 0's from G_0
Step 10: WRITE 1's in G_3
Step 11: READ 1's from G_3

TEST: EULERIAN-BASED TEST METHOD

Purpose:
To detect pattern sensitive faults (effect of 0/1 and 1/0 transitions is observed).

Example:
Testing three cells of all 0/1 and 1/0 pattern generation is based on using Eulerian (every arc is traversed exactly once). See Figure 3.

In general a k-bit neighborhood would require 2^k nodes and therefore 2^k transitions arcs.

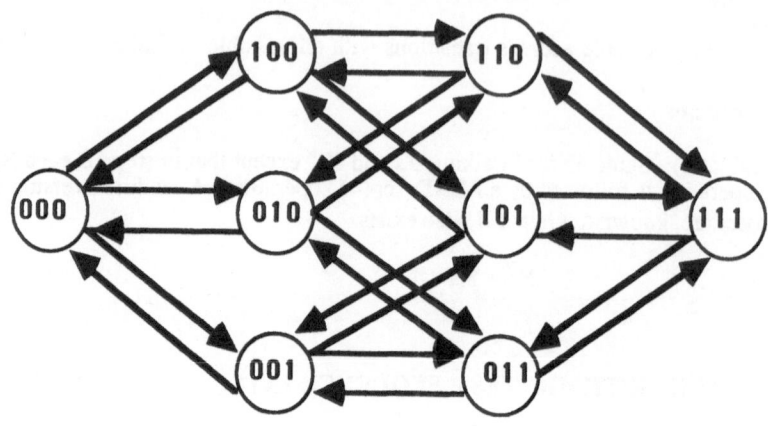

FIGURE 3. Testing three cells with Eulerian method.

The tests that we have just listed are becoming obsolete for large memories. Notice in Table 1, when we look at the time required to execute testing procedures, the marching 1's and 0's take only 1/10th of a second for 64 Kb memory provided the access time is 100 nanoseconds, while 1 Mb requires 1.6 seconds. If we use tests of the higher complexity such as $N^{1.5}$ or N^2, we may notice that the tests for even 64 Kb memory, especially with respect to GALPAT are becoming unacceptable and may take anywhere from several minutes to several hours for 1 Mb memory chips.

Test Block Size	MARCH 1/0 (14N)	GALTDIA ($6N^{1.5} + 10N$)	GALPAT ($4N^2 + 2N$)
64Kb	0.1 sec	10.1 sec	28.6 min
128Kb	0.2 sec	28.6 sec	1.9 hr
256Kb	0.4 sec	1.4 min	7.6 hr
1Mb	1.6 sec	10.8 min	122.2 hr

TABLE 1. Time required to execute memory test procedure
N = number of cells in a memory block,
Access time = 100ns.

There are a variety of approaches to cut on test cost. The ways to reduce testing are shown in Figure 4.

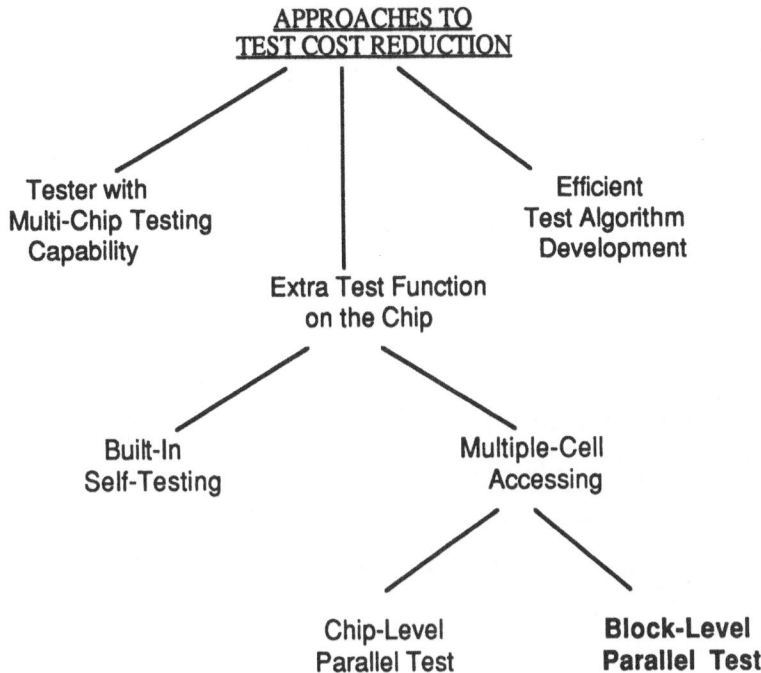

FIGURE 4. Various approaches to reduction of memory test time.

Straightforward approaches include testers with multichip testing capability, and obviously improvement in complexity of test algorithms. But, the area that is flourishing right now is the extra test functions built into the chip [26, 27, 29]. This includes built-in self-testing when test generation and evaluation function is built into the chip. Most of those techniques have used a serial test approach. The other techniques are based on multiple cell accessing and two methods can be distinguished. The first is called chip level parallel test. In this technique memory blocks are tested in parallel, and a single cell is tested in a memory cycle in each memory block. The test time, of course, is reduced. Block level parallel test is the other approach that we will describe here in more detail. When multiple cells are tested in one memory cycle in each memory block high speedup can be achieved. We propose here the method we have developed, called Multiple Access with Read Compaction (MARC) [28]. MARC is a block level parallel test method that is used in execution of a given memory test procedure to increase the test throughput per chip. For multiple cell accesses address decoder is modified such that test data can be broadcast to a set of cells and the set of cells can be read within the chip. For multiple read, a linear feedback shift register (LFSR), and then parallel signature analyzer (PSA), is built into the chip such that a set of read data can be monitored in a single memory cycle. A memory with parallel testing capability is shown in Figure 5, and a block level testing is explained in more detail in Figure 6.

The multiple selection and test data broadcast in the modified decoder allows access to k cells, and then the test data is compressed and test signature is created for evaluation. We call this scheme block level parallel test. Of course, parallel testing with MARC, may result in some masking of faults. Some of them are related to interaction faults that are

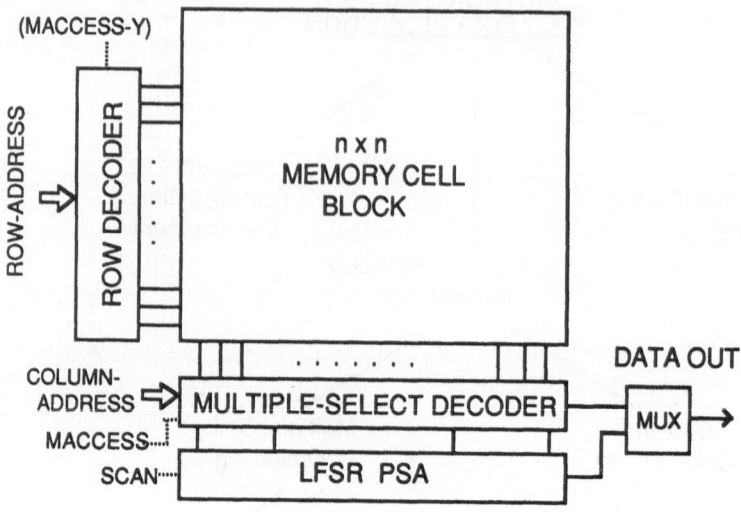

FIGURE 5. Parallel testing of RAMs.

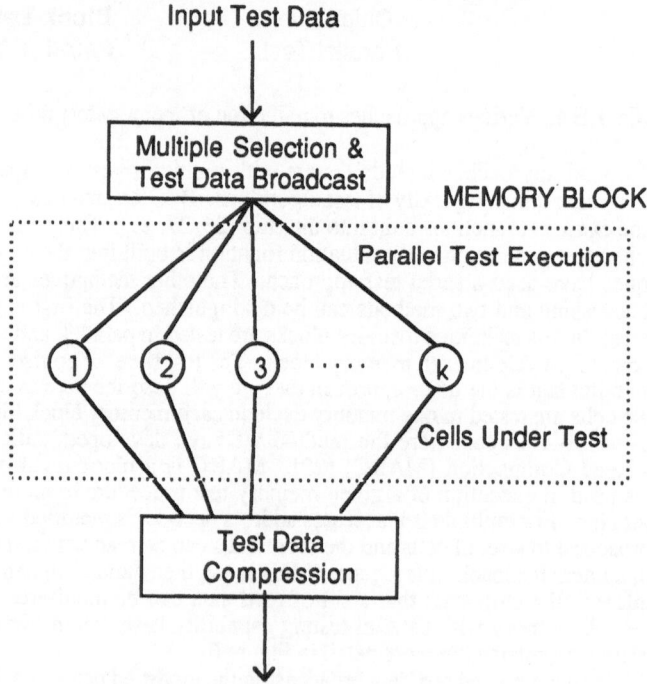

FIGURE 6. Block-level parallel test scheme.

caused by multiple accessing, the others are caused by exclusive-OR operations performed on the multiple input sequences which are applied to LFSR, PSA, or are caused by the data compression in LFSR. But on the other hand, MARC creates substantial savings with a very small additional overhead. Table 2 indicates and compares the testing time required for 16, 64, and 256 Kb chips.

N	Test Time (sec)			
	Block-based	Subblock-based	2-level Part	Normal
16Kb	3.4	0.2	0.3	53.7
64Kb	53.7	3.4	3.6	14.3 (min)
256Kb	14.3 (min)	53.7	54.7	3.8 (hour)

TABLE 2. Test time for Walking 1's & 0's test
read/write cycle=100ns.

In summary, we would like to point out that very frequently tests for memories should be customized depending on a particular memory chip because layout, technology, operating environment, size, and density have a direct impact on pattern sensitivity faults and therefore test requirements. Parallel testing will become indispensable in the future to insure complete testing of memories. As in the majority of cases, a fundamental tradeoff here is between testing time and fault coverage.

Having developed tests for processor and memory, we are ready to outline the PMS test procedure.

1. Test the microprocessor (isolate μP from data bus so the tester can supply instructions) and busses.
2. Test ROMs (RAMs are disconnected during this test; every ROM location is accessed and a fixed signature is derived by the processor.)
3. Test RAMs.
4. Test I/O using a loop-back technique. (The above tests can be implemented with only a few instructions such as load, store and NOP.)
5. Complete the processor-memory test by testing traps, interrupts and all the other special systems conditions.
6. Test the interconnection network.

The last part of a system test requires test of the interconnection network. That is what we are going to cover in the next section.

6. NETWORK TESTING

With emergence of commercial multiprocessor systems, it is becoming critically important to fully test complicated interconnection networks. The number of switches for a multiprocessor as a function of the number of processors may vary from $0(N)$ to $0(N^2)$ where N is the number of processors. The testing objective here is to verify that each switching element behaves correctly under various traffic conditions, that it properly responds to all the priority requirements and collisions, and transfers the data as required: reliably and on time. Complex switching elements, which allow a variety of broadcast modes, are very difficult to test. Imagine that you have a multistage network built of 8x8 crossbar switches. If you allow selective broadcasts, the exponential complexity of the network states makes it difficult to test in any reasonable amount of time. One can develop and build a variety of fault models. Imagine that in the 8x8 crossbar you have 64 cross points, which translates into 2^{64} of possible switch states. Of course, it is not possible to test it in any reasonable time. Therefore, the fault classes have to be limited and constrained only to particular lines and the subset of states.

Several papers cover the network testing problem [1, 11, 12, 29], but none of them adequately defines fault classes. In Table 3 we have defined the following fault classes for interconnection networks.

TABLE 3. Fault Classes.
Fault Class I - Data Link or Data Registers
* s-a-0 or s-a-1
* OR-bridge
* AND-bridge
Fault Class II - Control Lines
* Enable line(s) (data valid, request)
  (s-a-valid, s-a-broadcast
  equivalent to s-a-0 or s-a-1)
* Clear-to-send (release, acknowledge) line
  or s-a-1)

Fault Class III - Fault in Priority Logic (s-a-0 or s-a-1)
Fault Class IV - Clock and Power Lines
Fault Class V - Timing Faults (Switch Latency, Frequency)
Fault Class VI - Faults in Error Handling Logic

Notice that we distinguish between the data link faults, control lines, priority logic, clock and power lines, timing faults, and faults in error handling logic in case this type of logic is on the switching element. Also we would like to indicate that there are a variety of errors that may occur in the network from the obvious ones such as single-bit errors, burst errors, or random multiple-bit errors to more subtle that are listed in Table 4 as message errors.

TABLE 4. Message Errors

1. A message elongated at the front (head).
2. A message elongated at the end (tail).
3. Two messages merged together.
4. A false message.
5. A message truncated at the front.
6. A message truncated at the end.
7. A message partitioned into two parts.
8. An erased message.

As in processors we can derive efficient functional tests for the networks. As an example, let us take a multistage network that is used in BBNs butterfly system, Research Parallel Processor Prototype (RP3), the Texas Reconfigurable Array Computer (TRAC), and CEDAR project [1]. The banyan network is composed of small crossbars with a certain number of inputs and outputs respectively that we call fanout or spread and denote them by f. In Figure 7 an 8-processor system with 2x2 crossbars is shown. In Figure 8 a graph model of the same network is shown. Fanout and spread correspond to in-degree and out-degree of a vertex. Notice that the networks are characterized by the number of levels which, in a particular example, L=3, and fanout and spread which in Figures 7 and 8 are equal to 2. The number of processors or processor memory pairs, interconnected over the network like this is denoted by $N=f^L$.

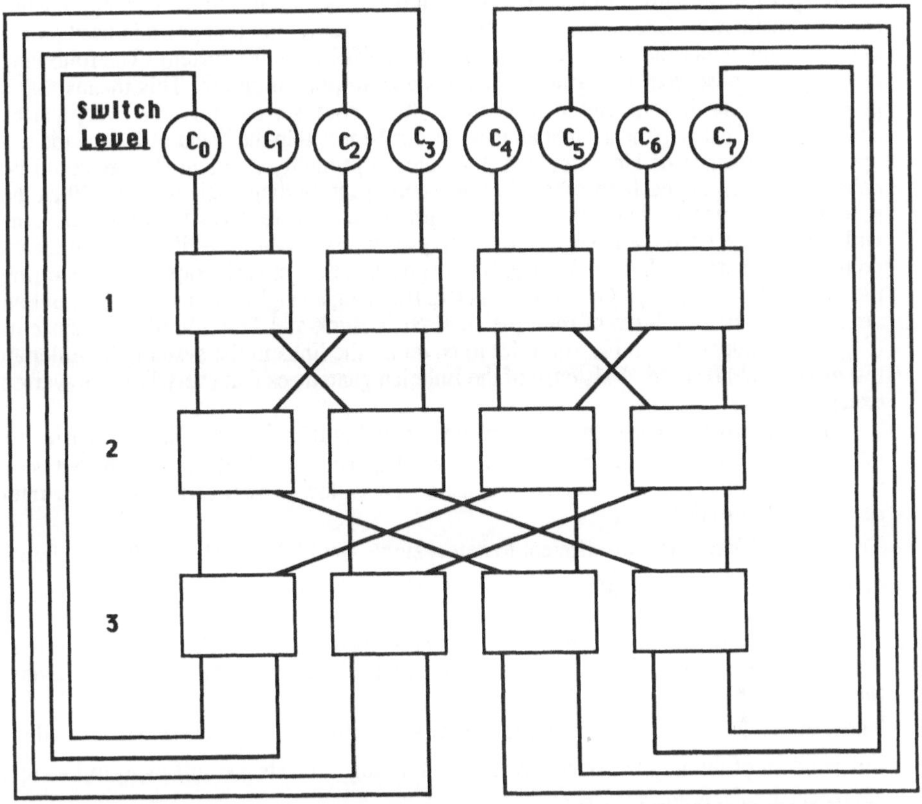

FIGURE 7. An eight processor system connected via a multistage network.

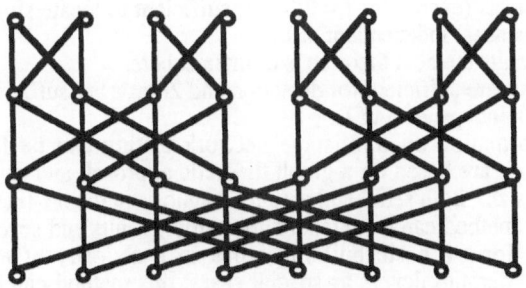

FIGURE 8. (2,3) rectangular SW-banyan.

In an actual system configuration the outputs of the network could be connected via computers to the inputs. This is an example of a unidirectional system that we denote as a system (f,L) where f corresponds to fanout and L to the number of levels. For a particular example, let us take the following fault model. Stuck-at-1 and stuck-at-0 faults on vertices and edges for the data part as well as on the vertices for the routing tag. We also allow AND/OR bridges between pairs of vertices for the data part, and we assume a single fault, though these methods could be extended for multiple faults. We also assume permanent faults. Let us analyze fault detection and location. Two approaches can be considered. One is a serial diagnosis where only one computer is used as the diagnostic computer that generates tests on-line or off-line in the system. The following property was observed: the system (f,L) has a Hamiltonian circuit. This means that a given packet that is transmitted from the testing processor can traverse every vertex exactly once. That also means that all vertices can be tested in the linear order and the test optimal. Existence of Hamiltonians guarantees optimality and actual tests could be enhanced by forcing each processor to time stamp passing diagnostic packet. When the sender sends a packet that routes via every processor, and receives the packet on time with all the time stamps and signatures of all processors, it can be verified that all the tie points on the network that are denoted by the nodes in the network graph are transferring the information properly. Only two tests 01, 10, using a single port as input and output are needed to detect single or multiple stuck-at-0, stuck-at-1 faulty vertices in systems (f,L), independent of f and L. In order to cover all the links in the system, the Eulerian test graph could be used. Existence of the Eulerian guarantees that every link is traversed exactly once.

In multiprocessor systems parallel testing is highly advised. Notice that in the system (f,L) f pairwise edge-disjoint test graphs, each with f^L disjoint paths between pairs of computers can always be constructed. This allows a very efficient test for fault detection and location.

Theorem 1: Two tests are sufficient to detect single (multiple) stuck-at-α (0 or 1) faulty vertices and four tests are sufficient to locate a single stuck-at-α faulty vertex independent of f and L.

The proof of this theorem is in [11]. This allows us to locate stuck-at faults at vertices on the data part. As far as stuck-at fault at edges on the data part, the next theorem is of use.

Theorem 2: 2f tests are sufficient to detect single (multiple) stuck-at-α faulty edges independent of the number of levels L and $2f + \lceil \log_2 L \rceil$ tests are sufficient to locate a single stuck-at-α faulty edge [11].

As far as the routing tag stuck-at-faults at vertices, the following result follows: Theorem 3: f tests are sufficient to detect a single routing tag stuck-at fault (by missing sequence) and f tests (except for f = 2,3) are sufficient to locate single such a fault (by two missing sequences) independent of L.

For bridge faults the next theorem is of interest here.
Theorem 4: f tests are sufficient for detection and 2f tests are sufficient for location of a single bridge fault independent of L.

Survey information on multistage network testing can be found in [1]. The suggested methods are based on a graph theoretic approach and can be applied to any networks of any size. Both serial and parallel diagnosis in the off-line and on-line modes is possible. The method can be extended to multiple faults and several system models. Since crossbars form a basic building block in a majority of networks including multiplexers and demultiplexers as special cases, this method can be generalized to a variety of systems including cube interconnection networks, meshes, trees and multiple busses.

7. GRAPH THEORY CONCEPTS FOR SYSTEM TESTING

Graph theory is a very powerful tool in analyzing system level diagnosis. Here are some examples that proved to be useful in actual practice of system testing and integration.

One of them is a dominating set. A dominating set in a graph G is a set of vertices that dominates every vertex v in G in the following sense: Either v is included in the dominating set or is adjacent to one or more vertices included in the dominating set. The concept of dominating set becomes essential for providing service/testing/support/I/O processors in the multiprocessor environment. Also, for recovery purposes the members of the dominating set can directly access all the processors in the system.

Other useful concepts are Hamiltonians and Eulerians. A Hamiltonian cycle is a cycle in the graph that allows traversal of every node exactly once. So if nodes have to be tested, try to build a system whose graph is a Hamiltonian, because then a test could be generated that would traverse every node exactly once.

If on the other hand, links are to be tested, try to build a system whose graph is an Eulerian because then a test could be created that would traverse every link exactly once. An Eulerian graph is very easy to build; as long as a degree of every vertex, that means number of edges incident on each vertex, is even, a graph has an Eulerian.

The next concept, test trees, search trees, decomposition trees, is also important. The test trees in adaptive testing procedures allow minimization of the test sequence. Refer to, for example, Huffman trees, where pendant nodes are weighted by probability of failure of particular component and the total optimization criteria is the sum of products of path lengths from the root to the pendant node multiplied by the weight of that node. This type of tree would allow to optimize the adaptive test sequence. Another application of trees would be to have a system partition where after every test the subsystem is divided further depending on the response of the test. Trees are also very helpful in modeling the system hierarchy. In fact, it is useful to model a multiprocessor system as a tree. In homogeneous environment, once we derive a test for a particular element of the homogeneous set, we can apply the same test to the other elements. The hierarchical approach is especially important in simulation and test development. Also, other graphs can be used to show data flow, control flow, program, or physical structure of the system. As we may see, the graphs have a very wide applicability to testing.

8. SYSTEM DIAGNOSIS

Frequently, it is not enough to test and find out whether the system is faulty or not faulty. Especially in the multiprocessor environment with several hundreds of thousands of chips, it is important to locate the fault at least within a replaceable unit, which is usually a board (card). Fault diagnosis techniques combine fault detection and fault location.

Let us outline just a few of the fault detection and location techniques. One of them are replication checks. In homogeneous environments in multiprocessor systems we can execute exactly the same tests, compare the outcome and find out which of the elements are faulty.

The next approach is by timing checks. Timing checks could be executed by incorporating timer that is reset after several tests. When the timer is not reset the alarm goes off indicating that the system did not pass the test. The timing tests are extremely important in distributed, parallel, and real-time system environment; e.g., modes of communication in the network could be tested by timing checks.

Another approach are reversal checks, which are the type of acceptance tests. The reversal checks are created by inversing or reversing a particular function. For example, if we calculate $N=2^n$ the reverse check would be $n=\log_2 N$.

There is a wide body of knowledge on coding checks. Starting with the very simple parity checks for fault detection, we may incorporate other error detecting codes all the way to error correcting codes. The popular codes for networks include Berger codes, checksumming, interlaced parity, and CRC [16].

The other type of testing is called reasonableness check. If the computer, for example, predicts the weather, if the temperature is 1000 degrees, we know that something is wrong with the machine. Structural tests verify that the structure and connectivity of the components is according to the expectations.

And the last characterized here are the algorithmic tests [30, 31, 32]. When we develop an algorithm we should have some other way of achieving the same result or execute special operation or a mathematical trick to verify that the result is correct.

Many tests should be built-in or used not only off-line, but also on-line in order to locate field replaceable units such as processor boards, memory boards, switching element boards, interface boards, I/O boards, support processor boards, and hopefully software modules. Unfortunately, in many systems, packaging determines the structure of testing. In fact packaging, testability, diagnosability, and performance instrumentation are usually afterthoughts in the design process. Since fault impact in multiprocessor environment is very severe, concurrent error detection is indispensable in multiprocessor environment. We should place, as indicated earlier, a strong emphasis on algorithmic level diagnosis by employing sophisticated acceptance tests. The complexity of state-of-the-art parallel systems is so high that concurrent error detection is becoming an absolute necessity.

With ever-growing system complexity and development of parallel computers, the pioneering work by Preparata, Metze and Chien in 1967 is becoming increasingly important [4]. The Preparata-Metze-Chien (PMC) model inspired many papers in the area of system self-diagnosis. The model is very simple; it takes into account a pair of processors where one is testing another one. If a fault-free processor tests fault-free, we say that the test is reliable and denote that particular edge by 0. If a fault-free processor has a faulty processor, we also can see that the test is reliable and indicate this by 1 that the processor is faulty. The problem begins when faulty processor tests fault-free or a faulty unit. Then, the outcome of the test is not determined. It can be 0 or 1. In the work by Preparata, Metze and Chien it was shown that in order to have a t diagnosable system, meaning that we can identify up to t units that are faulty, t must be less than half, to be precise 2t+1 should be less than or equal to n, and each unit should be tested by at least t other units. Barsi, Grandoni and Maestrini [5] developed a new model a few years later. In the new, simplified model the response in presence of a faulty unit testing a faulty unit was determined instead of being undetermined, as in the PMC model. That simplified the result and significantly increased diagnosability. Malek developed comparison models. In those models units are not testing each other; they are compared. In the first model [33] the fault-free comparator is assumed, and then, if there is discrepancy, the pair is assumed to be faulty. If there is no discrepancy, we know that both units are faulty. The conditions for locating faulty and non-faulty units were also determined in [33]. In the work by Maeng and Malek [34] three units are considered where one of them is a comparator. Then eight states can be derived if comparisons among three units are three ways. The faulty unit can be identified if only a single fault occurs within each triple. The comparison approach was implemented at AT&T [35]. It uses signature analyzers for data compression and resulting signatures are compared. The summary of models is shown in Figure 9. For large multiprocessor systems we propose partitioning [36]. For practical reasons, we propose to run diagnostics in the following manner.

For processors within a 4-processor cluster, as depicted in Figure 10, the quadruplet of processors is connected in such a fashion that it allows testing by each processor all the remaining three processors, and itself. If processors agree that a particular processor is faulty, they may automatically disconnect the processor from operation by the AND gate agreement signal as shown in the figure. In the example depicted in Figure 10, the

Testing Models

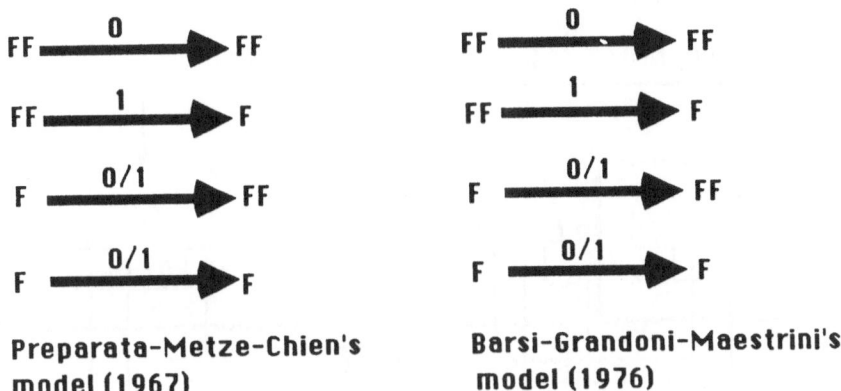

Preparata-Metze-Chien's
model (1967)

Barsi-Grandoni-Maestrini's
model (1976)

Comparison models

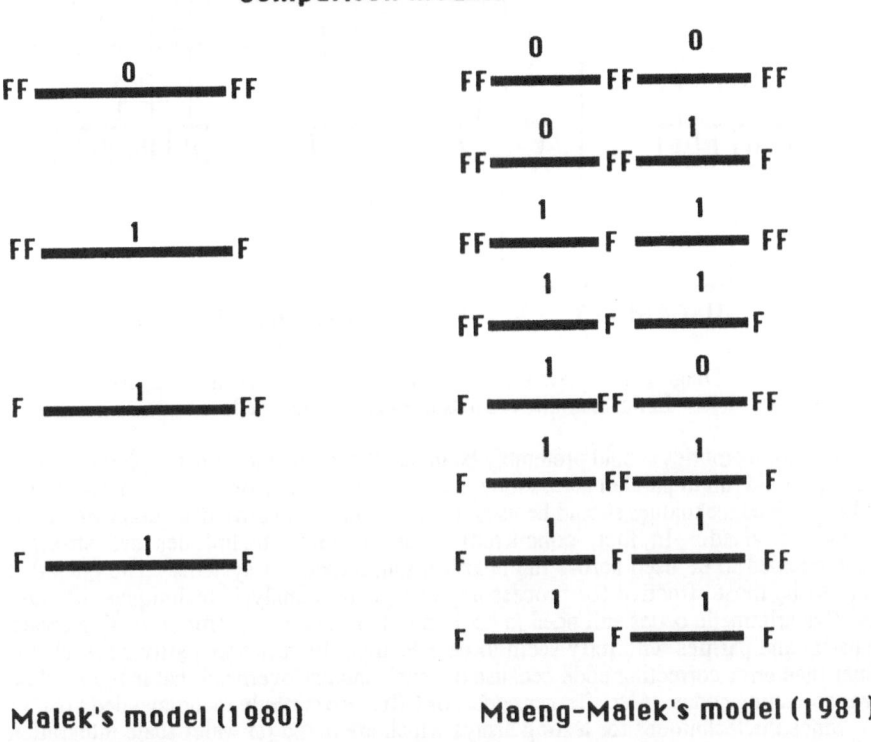

Malek's model (1980)

Maeng-Malek's model (1981)

Fig 9. System diagnosis models (FF – fault free unit; F – faulty
unit).

voting of processor P4 is irrelevant. It is going to be disconnected even if it indicates that the other processors are faulty, not it itself. In large multiprocessors all tests should be run in parallel on quadruplets.

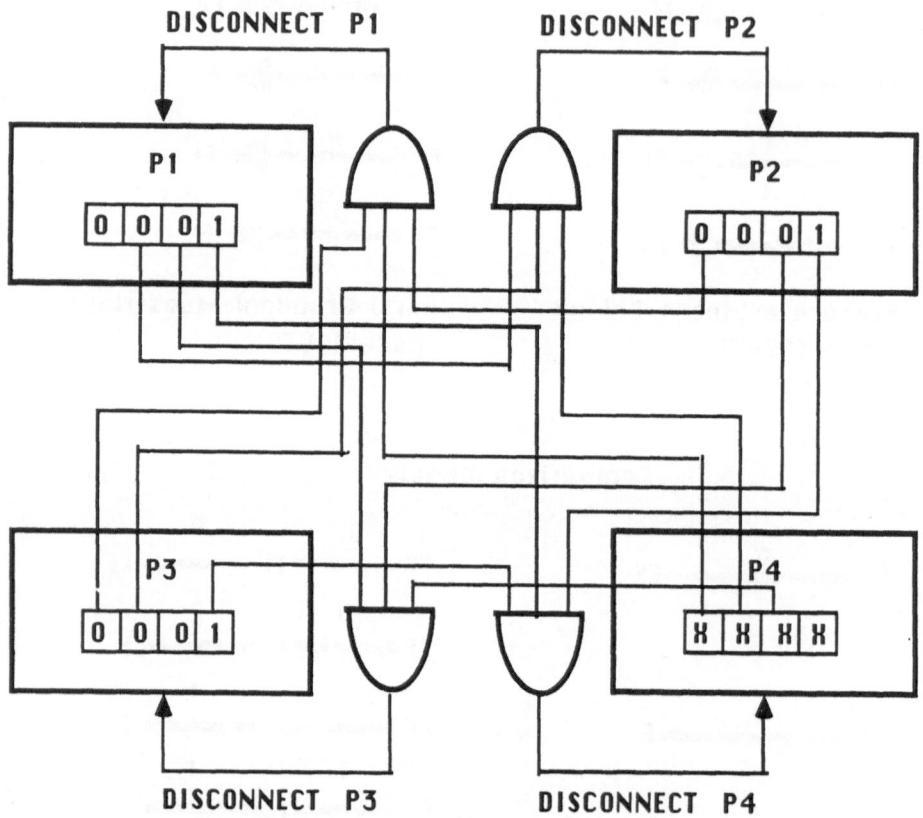

FIGURE 10. Diagnostic registers and decision logic in a quadruplet of processors. In the example processor P4 is disconnected.

Test for memories should preferably be in parallel within the chip and on top of it, of course, should run in parallel across the system, locally, or in pairs. For the network, the earlier outlined techniques should be used, but also concurrent error detection by codes is strongly advised. In fact, concurrent error detection techniques are strongly recommended to be used across the board in multiprocessor systems. The ones that seem to be most effective for processors are signature analysis techniques, because effective arithmetic codes still need to be found. For memories, error correcting codes, or parity and parities with retry seem to be effective. In networks parity with retry is better than error correcting code because of much smaller overhead that is required on error correcting codes. Also, Berger codes, or CRCs are strongly recommended [16].

Since the techniques for testing arrays which are based on wafer scale integration (WSI) are discussed by M. Sami in this book we refer a reader to that chapter.

9. SUMMARY AND TRENDS

The emergence of several commercial multiprocessors and distributed systems will demand ever more sophisticated testing and fault tolerance techniques. Testing and built-in tests at every level of abstraction and packaging will become a must for systems with even relaxed reliability requirements. It is obvious by now, that complete tests will not be feasible and living with faults will have to be an acceptable mode of operation. Therefore, fault tolerance will have to be incorporated in every multiprocessor to make it operational and trading performance for reliability will become a standard feature in the future systems. The emphasis should be put on testing and careful integration at every layer of computer systems, but algorithm-based acceptance testing, especially for special purpose architectures, will become indispensable. Parallel testing will evolve into standard and testing itself will be a major bottleneck in commercialization process of parallel systems.

REFERENCES
1. G.J. Lipovski and M. Malek, "Parallel Computing: Theory and Comparisons," John Wiley-Interscience, New York, 1987.
2. L.H. Goldstein, "Controllability/Observability Analysis of Digital Circuits," IEEE Trans. of Circuits and Systems, Vol. CAS-26, Sept. 1979, pp. 685-693.
3. V.D. Agrawal and M.R. Mercer, "Testability Measures-What Do They Tell Us?" 1982 Int. Test Conf. Proc., Nov. 1982, pp. 391-396.
4. F.P. Preparata G. Metze, and R.T. Chien, "On the Connection Assignment Problem of Diagnosable Systems," IEEE Trans. on Electronic Computers, EC16, 1967, pp. 848-854.
5. F. Barsi, F. Grandoni and P. Maestrini, "A Theory of Diagnosability of Digital Systems," IEEE Trans. on Com., C-25, 1976, pp. 885-893.
6. S.L. Hakimi and A.T. Amin, "Characterization of the Connection Assignment of Diagnosable Systems," IEEE Trans. on Comp., C-23, 1974, pp. 86-88.
7. S. Mallela and G.M. Masson, "Diagnosable Systems for Intermittent Faults," IEEE Trans. on Comp., C-27, 1978, pp. 560-566.
8. A.D. Friedman and L. Simoncini, "System-Level Fault Diagnosis," Computer, C-13, 1980, pp. 47-53.
9. M. Malek and K.Y. Liu, "Graph Theory Models in Fault Diagnosis and Fault Tolerance," Design Automation and Fault-Tolerant Computing, Vol. III, Issue 3/4, 1980, pp. 155-169.
10. S.E. Kreutzer and S.L. Hakimi, "System Level Diagnosis," Microprocessing and Microprogramming, The Euromicro Journal, Vol. 20, No. 4 and 5, May 1987, pp. 323-330.
11. M. Malek and E. Opper, "Multiple Fault Diagnosis of SW-banyan Networks," Digest of Papers of 13th Fault-Tolerant Computing Symposium, June 1983, pp. 446-449.
12. E. Opper and M. Malek, "Real-time Diagnosis of Banyan Networks," Proc. of the 1982 Real-Time Systems Symp., Dec. 1982, pp. 27-36.
13. T.W. Williams and K.P. Parker, "Design for Testability-A Survey," IEEE Trans. on Comp., Vol. C-31, Jan. 1982, pp. 2-15.
14. E.B. Eichelberger and T.W. Williams, "A Logic Design Structure for LSI Testability," Proc. of 14th Design Automation Conf., June 1977.
15. S.M. Thatte and J.A. Abraham, "Test Generation for Microprocessors," IEEE Trans. on Computers, Vol. C-29, No. 6, June 1980, pp. 429-441.
16. M. Malek and K.H. Yau, "Cost-effective Error Detection Codes in Multicomputer Networks," Microprocessing and Microprogramming, The Euromicro Journal, Vol. 20, No. 4 and 5, May 1987, pp. 331-343.

17. J.F. Wakerly, "Error Detecting Codes, Self-Checking Circuits and Applications," Elsevier North Holland, 1978.
18. R.A. Frohwerk, "Signature Analysis: A New Digital Field Service Method," Hewlett-Packard Journal, May 1977.
19. S.Z. Hassan, D.J. Lu, E.J. McCluskey, "Parallel Signature Analyzers - Detection Capability and Extensions," COMCON 83, February 1983, pp. 440-445.
20. J. Duran and T. Mangir, "Application of Signature Analysis to the Concurrent Test of Microprogrammed Control Units," Microprocessing and Microprogramming, Vol. 20, No. 4 and 5, May 1987, pp. 309-322.
21. M. Namjoo and E.J. McCluskey, "Watchdog Processors and Capability Checking," 12th Fault Tolerant Comp. Symp., June 1982, pp. 245-248.
22. A.C.L. Chiang and R. McCaskill, "Two New Approaches to Simplify Testing of Microprocessors," Electronics, Vol. 49, No. 2, Jan. 22, 1976, pp. 100-105.
23. J.P. Hayes and E.J. McCluskey, "Testability Considerations in Microprocessor-Based Design," Computer, March 1980, pp. 17-26.
24. M.A. Breuer and A.D. Friedman, "Diagnosis and Reliable Design of Digital Systems," Computer Science Press, 1976.
25. R. Nair, S.M. Thatte and J.A. Abraham, "Efficient Algorithms for Testing Semiconductor Random-Access Memories," IEEE Trans. on Comp. Vol. C-27, No. 6, June 1978, pp. 572-576.
26. T. Sridhar, "A New Parallel Test Approach for Large Memories," Proc. 1985 Int. Test Conf., Nov. 1985, pp. 462-470.
27. K.T. Le and K.K. Saluja, "A Novel Approach for Testing Memories Using a Built-in Self-Testing Technique," Proc. of 1986 Int. Test Conf., Sept. 1986, pp. 830-839.
28. S.H. Han and M. Malek, "Two-dimensional Multiple-access Testing Technique for Random-access Memories," Proc. of Int. Conf. on Computer Design, Oct. 1986, pp. 248-251.
29. T.Y. Feng and C.-L. Wu, "Fault Diagnosis for a Class of Multistage Interconnection Networks," IEEE Trans. on Comp., Oct. 1981, pp. 743-758.
30. K.H. Huang and J.A. Abraham, "Algorithm-based Fault Tolerance for Matrix Operations," IEEE Trans. on Comp., Vol. C-33, June 1984, pp. 518-528.
31. Y.-H. Choi and M. Malek, "A Fault-Tolerant FFT Processor," Digest of Papers of 15th Int. Symp. on Fault-Tolerant Computing, June 1985, pp. 266-271.
32. Y.-H. Choi and M. Malek, "A Fault-Tolerant VLSI Sorter," Proc. of Int. Conf. on Computer Design, Oct. 1985, pp. 510-513.
33. M. Malek, "A Comparison Connection Assignment for Diagnosis of Multiprocessor Systems," Proc. 7th Ann. Symp. on Computer Architecture, May 1980, pp. 31-36.
34. J. Maeng and M. Malek, "A Comparison Assignment for Self-Diagnosis of Multiprocessor Systems," IEEE Symposium on Fault-Tolerant Computing, 1981, pp. 173-175.
35. P. Agrawal, "RAFT: A Recursive Algorithm for Fault Tolerance," Proc. Int. Conf. on Parallel Computing, 1985, pp., 814-821.
36. M. Malek and J. Maeng, "Partitioning of Large Multicomputer Systems for Efficient Fault Diagnosis," Digest of Papers of 12th Int. Symp. on Fault-Tolerant Computing, June 1982, pp. 341-348.

ANALOG FAULT DIAGNOSIS

Chin-Long Wey[*], Chwan-Chia Wu[&], and Richard Saeks[#]

* Department of Electrical Engineering, Michigan State University,
East Lansing, MI 48824, U.S.A. & Department of Electrical Engineering,
National Taiwan Institute of Technology, Taipei, Taiwan, Republic of China.
Department of Electrical and Computer Engineering, Arizona State
University, Tempe, AZ 85287, U.S.A.

ABSTRACT: A simulation-after-test algorithm for the analog fault diagnosis
problem is presented, in which a bound on the maximum number of
simultaneous faults is used to minimize the number of test points required.
Based on this self-testing algorithm, an analog automatic test program
generation (AATPG) for both linear and nonlinear circuits and systems is
being developed. The AATPG code is subdivided into off-line and on-line
processes. The actual test can be run in either a fully automatic mode or
interactively. In addition, the issues of parallel processing and
testability for analog circuits and systems are also addressed.

1. INTRODUCTION

Over the past decade, the circuits and systems community has undertaken
a considerable research effort directed at the development of a viable
analog fault diagnosis algorithm [1-21]. This effort has borne fruit with
the proposal of several algorithms that alleviate the usual test
point/computer time limitations by imposing a bound on the number of
simultaneous system failures. (A maximum of two or three simultaneous
failures is typically sufficient for most applications.)

Given our experience with the digital test problem and the analog
computer-aided design (CAD) problem, one might initially assume that the
analog test problem could be resolved simply by integrating the tools and
techniques of these two well-established fields. In fact, however, the
analog problem is characterized by tolerance, modeling, and simulation
problems which have no counterpart in the digital problem, while many of
the concepts derived from the analog design problem are incompatible with
the economics of the test environment. Indeed, the major contribution of
the Analog Automatic Test Program Generation (AATPG) research effort in its
early years may very well have been the characterization of the unique
problems encountered in analog system testing. These problems are
summarized in reference [7] and [8]. These papers took orthogonal
approaches to the problem with the former emphasizing the properties of
specific implementations and the latter dealing with the generic properties
of various algorithms. The two papers, taken together, however, represent
an essentially complete summary of the problems which must be resolved in
the development of a successful AATPG.

The various proposed algorithms for AATPG may naturally be divided into
three categories characterized by the simulation occurrence, the test
inputs allowed, and the fault hypotheses employed. These include: [6,9]

    (1) Pre-Test Simulation Techniques,
    (2) Post-Test Simulation with Multiple Test Vectors, and
    (3) Post-Test Simulation with Failure Bounds.

The first is commonly employed in digital testing and is characterized by
minimal on-line computational requirements. However, the high cost of

F. Lombardi and M. Sami (eds.), Testing and Diagnosis of VLSI and ULSI, 117–150.
© 1988 by Kluwer Academic Publishers.

analog circuit simulation coupled with the large number of potential fault modes limits the applicability of this algorithm. Typically, the Post-Test simulation technique attempts to model the analog fault diagnosis problem as a nonlinear equation in which the internal variables or component parameters are computed in terms of the test data. In the case where sufficiently many test points are available, only a single test is required and will reduce the fault diagnosis problem to the solution of a linear equation [10,15]. Therefore, the on-line computational requirements are moderate. However, the test points requirement grow with circuit complexity [9]. To reduce the test point requirements one may consider using multiple test vectors to increase the number of equations obtained from a given set of test points. However, the on-line computation required to solve the resultant sets of complex nonlinear equations (even for linear systems) is extremely expensive. Alternatively one can use an upper bound on the number of simultaneous failures to reduce the test point requirements associated with the post-test simulation algorithm.

To evaluate analog diagnosis algorithms, a set of criteria has been proposed [9], in which the following criteria are considered: computational requirements, number of test points and test vector employed, robustness to tolerance effects, availability of models, and the degree to which the algorithm is amenable to parallel processing. Based on the above criteria, the above three simulation techniques are compared and the Post-Test simulation with failure bounds seems to be the closest to the "ideal" algorithm [9].

Motivated by the above considerations the authors have developed a self-test theory for analog circuit fault isolation [8,19-21] to reduce the complexity of the nonlinear fault diagnosis equations while still retaining computational simplicity. The salient features of the proposed self-test algorithm, a post-test simulation with failure bounds, are that the algorithm

(1) works with both linear and nonlinear systems modelled in either the time or frequency domain;

(2) can be used to locate multiple hard or soft faults; and

(3) is designed to locate failures in "replaceable modules" such as an IC chip, PC board, or subsystem rather than in discrete components.

Moreover, this is achieved at an acceptable computational cost and with minimal test point requirements.

Based on the proposed self-testing algorithm, an AATPG for both linear [19,20] and nonlinear [21] circuits has been implemented. The AATPG code is subdivided into off-line and on-line processes. The former is used by the test system designer to input nominal system specifications and to generate a database used by the on-line process. To implement the actual test, the field engineer invokes the on-line process inputing data describing the unit under test (UUT) and the source of the test data. The actual test can then be run in a fully automatic and/or an interactive mode.

Since the Computer-Aided Testing (CAT) problem is inherently a large scale system problem, it is essential to exploit whatever computational power is available to reduce both on-line and off-line computational requirements. In particular, digital CAT algorithms often use some degree of parallel processing in their implementation. Therefore, the degree to which an analog CAT algorithm can be implemented in parallel processing becomes a significant factor in determining its viability. Furthermore, since the on-line computation is directly related to the per unit cost, it is important to keep the post-fault computation time short and simple. Therefore, a parallel version of our self-test algorithm has also been developed.

With the rapidly increasing complexity of circuits and systems, the ability to adequately design a diagnosable circuit or system is a prime requisite for rapid fault location. In order to determine the diagnosability of a designed circuit, testability conditions are presented. The unique feature of these testability conditions is that they depend only on the topological structure of a given circuit, not on the component values. Therefore, these conditions can be implemented in both linear or nonlinear circuits or systems. With appropriate computer aids, test points may be generated automatically by the proposed algorithm and the testability of a circuit can be established.

In the following section, the simulation model known as the Component Connection model and the detailed implementation of the self-testing algorithm are discussed. Although it is reasonable to assume that at most two or three components have failed simultaneously in a given circuit containing several hundred components, we do not know which two or three. Therefore, some type of search is still required. Three decision algorithms are presented to search for the faulty components in Section III. The parallel processing algorithm and the associated parallel test system architecture are demonstrated in Section IV. Design of testability is discussed in Section V and, finally, the work is summarized in Section VI.

2. THE SIMULATION MODEL AND THE SELF-TESTING ALGORITHM

For the purpose of doing fault diagnosis, we work with an interconnection system model known as the Component Connection Model (CCM) [22]. Assume the analog circuit or system under test consists of n components, k external test inputs, and m external test points. In the linear case, the unit under test (UUT) characterizes its components and/or subsystems together with an algebraic connection equation as follows:

$$b = Z \, a \qquad (1)$$

and

$$a = L_{11} \, b + L_{12} \, u \qquad (2)$$

$$y = L_{21} \, b + L_{22} \, u \qquad (3)$$

where, $a = col(a_i)$ and $b = col(b_i)$, $i=1,2,...,n$, are the column vectors composed of the component input and output variables, respectively; while u is the vector of external test inputs applied to the system and y is the vector of system responses measured at the various test points. The L_{ij}'s are constant, generally sparse matrices, known as the connection matrices, whose dimensions conform to the given vector quantities.

For the linear case, the component equation (1) is modeled in the frequency domain, where $Z=diag(Z_i)$, $i=1,2,...,n$, is a frequency domain composite component transfer matrix. Each of such Z_i, $(=Z_i(s,r))$, describes the i-th component of the circuit or system, here $r=col(r_i)$ is the column vector of unknown component parameters and s is the complex frequency variable. Typically, the unknown component parameters take the form of resistances, capacitances, inductances, amplifier gains, etc. In particular, it is assumed that enough parameters are employed to completely characterize the performance of the device.

For the nonlinear case, a similar model is used with the component characteristics represented by a set of decoupled state models,

$$\dot{x}_i = f_i (x_i, a_i)$$
$$b_i = g_i (x_i, a_i) \qquad ; \quad x_i(0) = 0 , \quad i=1,2,...,n.$$

where x_i is the **state vector** for component i. For notational purpose we stack the individual component equations together to form the composite component equations:

$$\dot{x} = f (x, a)$$
$$; \quad x(0)=0. \qquad (4)$$
$$b = g (x, a)$$

Similar to the linear case, we append the connection equations (2) and (3) to these component equations to form the CCM for the nonlinear case.

2.1. The Self-testing Algorithm

Conceptually, at each step of the test algorithm the components (individual chips, discrete components or subsystems) are subdivided into two groups: the "Tester Group" (or, group "1") and the "Testee Group" (or, group "2"). At each step one assumes that the Tester Group is composed of good components and one uses the known characteristics of those components together with the test data (test inputs and and measured system responses at the external accessible test points), to determine if the remaining components in the Testee Group are good. In fact, the first group of components is testing the second, hence the "self-testing" algorithm. Of course, if all components in the Tester Group are actually good, then the resultant test outcome for each component in the Testee Group will be reliable. On the other hand, if any one of the testers is faulty the test outcomes on the testee will be unreliable. Consequently, one repeats the process at the next step of the test algorithm with a different subdivision of components. Finally, after a number of such repetitions the test outcomes obtained at the various steps are analyzed to determine the faulty component.

Of course, the number of components which may be tested at any step is dependent on the number of test points available; while the number of steps required is determined by the number of testee components and the bound on the maximum number of simultaneous failures. Therefore, this procedure yields a natural set of tradeoffs between the number of test points, simultaneous failures, and steps required by the algorithm.

2.2. Linear Case

In order to represent the subdivision step, a superscript is assigned to denote the Group designation, i.e. superscript 1 for Group 1 components, etc. This induces a partitioning of the equations (1) through (3), as

Component equations:
$$b^1 = z^1_2 a^1 \qquad (5)$$
$$b^2 = z^2 a^2 \qquad (6)$$

Connection equations:
$$a^1 = L^{11}_{11} b^1 + L^{12}_{11} b^2 + L^1_{12} u \qquad (7)$$
$$a^2 = L^{21}_{11} b^1 + L^{22}_{11} b^2 + L^2_{12} u \qquad (8)$$
$$y = L^1_{21} b^1 + L^2_{21} b^2 + L_{22} u \qquad (9)$$

Since, in our application, the test responses, y, and the test inputs, u, are known, and the Group 1 components are assumed to be good, the remaining objective is to   the Group 2 component input and output waveforms, a^2 and b^2. To this end we assume that L^2_{21} admits a left inverse. Under this assumption, equation (9) becomes

$$b^2 = [\ L_{21}^2\ ]^{-L}\ [\ -\ L_{21}^1\ b^1\ -\ L_{22}\ u + y\ ] \tag{10}$$

by making substitutions into equations (7) and (8). The "pseudo circuit" connection equations are obtained as,

$$a^1 = K_{11}\ b^1 + K_{12}\ u^P \tag{11}$$

$$y^P = K_{21}\ b^1 + K_{22}\ u^P \tag{12}$$

where $u^P = col(u,y)$ and $y^P = col(a^2, b^2)$ are the column vectors of the external inputs and outputs of the pseudo circuit; while the connection matrices K_{ij}'s are defined as,

$$K_{11} = [\ L_{11}^{11} - L_{11}^{12}\ [\ L_{21}^2\ ]^{-L}\ L_{21}^1\ ] \tag{13}$$

$$K_{12} = [\ L_{12}^1 - L_{11}^{12}\ [\ L_{21}^2\ ]^{-L}\ L_{22}\ |\ L_{11}^{12}\ [\ L_{21}^2\ ]^{-L}\ ] \tag{14}$$

$$K_{21} = \begin{bmatrix} L_{11}^{21} - L_{11}^{22}\ [\ L_{21}^2\ ]^{-L}\ L_{21}^1 \\ \hline -\ [\ L_{21}^2\ ]^{-L}\ L_{21}^1 \end{bmatrix} \tag{15}$$

$$K_{21} = \begin{bmatrix} L_{12}^2 - L_{11}^{22}\ [\ L_{21}^2\ ]^{-L}\ L_{22} & | & L_{11}^{22}\ [\ L_{21}^2\ ]^{-L} \\ \hline -\ [\ L_{21}^2\ ]^{-L}\ L_{22} & | & [\ L_{21}^2\ ]^{-L} \end{bmatrix} \tag{16}$$

For each pseudo circuit, substituing (5) into (11) and (12) the input-output relation

$$y^P = M\ u^P \tag{17}$$

is obtained, where

$$M = K_{21}\ Z^1\ [I - K_{11}\ Z^1]^{-1}\ K_{12} + K_{22} \tag{18}$$

upon the partitioning M to conform with the partitions of y^P and u^P this then becomes

$$a^2 = M_{11}\ u + M_{12}y \tag{19}$$

$$b^2 = M_{21}\ u + M_{22}y \tag{20}$$

Note, since equation (18) is dependent only on the nominal values of the Tester Group components, the M_{ij}'s may be computed off-line and stored in a data base to be retrieved at the time when a test is conducted. Furthermore, since only a single test vector is required, single-frequency testing can be employed. In this case, the M_{ij}'s need only be computed at a single frequency. The only on-line computation required for the fault diagnosis of a linear system is thus the matrix/vector multiplication indicated in (19) and (20) together with the computation of

$$\hat{b}^2 = Z^2\ a^2 \tag{21}$$

to determine which, if any, of the Testee Group components are faulty.

The test outcome (either "good", or "bad", for each Testee Group component) is obtained by comparing the b^2 and $\hat{b}^2$. If their i-th elements are the same, i.e., $b_i^2 = \hat{b}_i^2$, then we say that the test outcome for the i-th component in the Testee Group is good; otherwise, the component is "bad". In a more realistic environment, instead of requiring that two values be equal, one may say that a component is "good" if these two values are sufficiently close to each other in some sense. In this way, one may compensate for numerical errors and tolerances. Moreover, b_i^1 and b_i^2 are not necessarily scalars, they may be vectors, depending upon the component type with which one deals. For instance, a two-port component may require a two tuple vector to represent its input/output characteristics.

Example 1. [20]

Consider a linear circuit, a BJT small signal amplifier circuit with beta-independent bias, as shown in Figure 1.

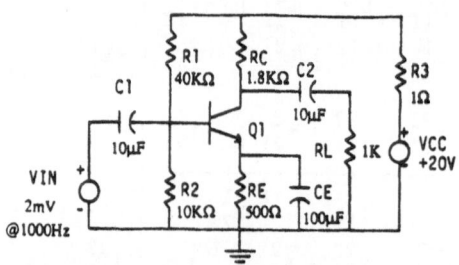

Figure 1. Linear Circuit

Connection equations:

$$
\begin{bmatrix}
IC1 \\ IR1 \\ VR2 \\ IRC \\ IBQ1 \\ VCEQ1 \\ VRE \\ VCE \\ VC2 \\ IRL \\ VR3 \\ \hline IC1 \\ IR1 \\ V45 \\ V13
\end{bmatrix}
=
\begin{bmatrix}
0 & 0 & 1 & 0 & 0 & 0 & 1 & 1 & 1 & 0 & 1 \\
0 & 0 & 0 & 0 & 0 & -1 & 0 & 0 & -1 & 0 & -1 \\
-1 & 0 & 0 & 0 & 0 & 0 & 0 & 0 & 0 & 0 & 0 \\
0 & 0 & 0 & 0 & 0 & 1 & 0 & 0 & 1 & 0 & 0 \\
0 & 0 & 0 & 0 & 0 & -1 & 1 & 1 & 0 & 0 & 0 \\
0 & 1 & 0 & -1 & 1 & 0 & 0 & 0 & 0 & 0 & 0 \\
-1 & 0 & 0 & 0 & -1 & 0 & 0 & 0 & 0 & 0 & 0 \\
-1 & 0 & 0 & 0 & -1 & 0 & 0 & 0 & 0 & 0 & 0 \\
-1 & 1 & 0 & -1 & 0 & 0 & 0 & 0 & -1 & 0 & 0 \\
0 & 0 & 0 & 0 & 0 & 0 & 0 & 1 & 0 & 0 & 0 \\
-1 & 1 & 0 & 0 & 0 & 0 & 0 & 0 & 0 & 0 & 0 \\
\hline
0 & 0 & 1 & 0 & 0 & 0 & 1 & 1 & 1 & 0 & 1 \\
0 & 0 & 0 & 0 & 0 & -1 & 0 & 0 & -1 & 0 & -1 \\
-1 & 0 & 0 & 0 & -1 & 0 & 0 & 0 & 0 & -1 & 0 \\
0 & 1 & 0 & -1 & 0 & 0 & 0 & 0 & 0 & 0 & 0
\end{bmatrix}
\begin{bmatrix}
VC1 \\ VR1 \\ IR2 \\ VRC \\ VBEQ1 \\ ICQ1 \\ IRE \\ ICE \\ IC2 \\ VRL \\ IR3
\end{bmatrix}
+
\begin{bmatrix}
0 & 0 \\ 0 & 0 \\ 1 & 0 \\ 0 & 0 \\ 0 & 0 \\ 0 & 0 \\ 1 & 0 \\ 1 & 0 \\ 1 & 0 \\ 0 & 0 \\ 1 & -1 \\ \hline 0 & 0 \\ 0 & 0 \\ 0 & 0 \\ 0 & 0
\end{bmatrix}
\begin{bmatrix}
VIN \\ VCC
\end{bmatrix}
$$

Component equations:

$$
Z =
\begin{bmatrix}
Z_{C1} & 0 & 0 & 0 & 0 & 0 & 0 & 0 & 0 & 0 & 0 \\
0 & Z_{R1} & 0 & 0 & 0 & 0 & 0 & 0 & 0 & 0 & 0 \\
0 & 0 & Y_{R2} & 0 & 0 & 0 & 0 & 0 & 0 & 0 & 0 \\
0 & 0 & 0 & Z_{RC} & 0 & 0 & 0 & 0 & 0 & 0 & 0 \\
0 & 0 & 0 & 0 & Z_{h1} & Z_{h2} & 0 & 0 & 0 & 0 & 0 \\
0 & 0 & 0 & 0 & Z_{h3} & Z_{h4} & 0 & 0 & 0 & 0 & 0 \\
0 & 0 & 0 & 0 & 0 & 0 & Y_{RE} & 0 & 0 & 0 & 0 \\
0 & 0 & 0 & 0 & 0 & 0 & 0 & Y_{CE} & 0 & 0 & 0 \\
0 & 0 & 0 & 0 & 0 & 0 & 0 & 0 & Y_{C2} & 0 & 0 \\
0 & 0 & 0 & 0 & 0 & 0 & 0 & 0 & 0 & Z_{RL} & 0 \\
0 & 0 & 0 & 0 & 0 & 0 & 0 & 0 & 0 & 0 & Y_{R3}
\end{bmatrix}
$$

The component subdivision is generated by the assumption that $[L_{21}^2]^{-L}$ exists. Therefore, 34 possible subdivisions are generated in Table I.

Table I. Subdivisions

Subdivision Number	Testee Group Components	Subdivision Number	Testee Group Components
1	1 2 3 9	18	2 8 10 11
2	1 2 3 11	19	3 4 9 10
3	1 2 7 9	20	3 4 10 11
4	1 2 7 11	21	4 7 9 10
5	1 2 8 9	22	4 7 10 11
6	1 2 8 11	23	4 8 9 10
7	1 3 4 9	24	4 8 10 11
8	1 3 4 11	25	2 3 5 6
9	1 4 7 9	26	2 5 6 7
10	1 4 7 11	27	2 5 6 8
11	1 4 8 9	28	2 5 6 9
12	1 4 8 11	29	2 5 6 11
13	2 3 9 10	30	3 4 5 6
14	2 3 10 11	31	4 5 6 7
15	2 7 9 10	32	4 5 6 8
16	2 7 10 11	33	4 5 6 9
17	2 8 9 10	34	4 5 6 11

To generate the data base for matrices K_{ij}'s, we assume the first component subdivision is chosen, or the Tester Group consists of components #1, #4, (#5,#6), #7, #8, #10, and #11, and the Testee Group contains the remaining components. The K-matrix is generated as follows,

$$
\begin{bmatrix} IRC \\ IBQ1 \\ VCEQ1 \\ VRE \\ VCE \\ IRL \\ VR3 \\ \hline IC1 \\ IR1 \\ VR2 \\ VC2 \\ \hline VC1 \\ VR1 \\ IR2 \\ IC2 \end{bmatrix} =
\begin{bmatrix}
0 & 0 & 0 & 0 & 0 & 0 & -1 & | & 0 & 0 & 0 & -1 & 0 & 0 \\
0 & 0 & -1 & 1 & 1 & 0 & 0 & | & 0 & 0 & 0 & 0 & 0 & 0 \\
0 & 1 & 0 & 0 & 0 & 0 & 0 & | & 0 & 0 & 0 & 0 & 0 & 1 \\
0 & 0 & 0 & 0 & 0 & 1 & 0 & | & 1 & 0 & 0 & 0 & 1 & 0 \\
0 & 0 & 0 & 0 & 0 & 1 & 0 & | & 1 & 0 & 0 & 0 & 1 & 0 \\
0 & 0 & -1 & 0 & 0 & 0 & -1 & | & 0 & 0 & 0 & -1 & 0 & 0 \\
1 & 1 & 0 & 0 & 0 & 1 & 0 & | & 1 & -1 & 0 & 0 & 1 & 1 \\
\hline
0 & 0 & 0 & 0 & 0 & 0 & 0 & | & 0 & 0 & 1 & 0 & 0 & 0 \\
0 & 0 & 0 & 0 & 0 & 0 & 0 & | & 0 & 0 & 0 & 1 & 0 & 0 \\
0 & 1 & 0 & 0 & 0 & 1 & 0 & | & 1 & 0 & 0 & 0 & 1 & 0 \\
0 & 1 & 0 & 0 & 0 & 0 & 0 & | & 1 & 0 & 0 & 0 & 1 & 1 \\
\hline
0 & -1 & 0 & 0 & 0 & -1 & 0 & | & 0 & 0 & 0 & 0 & -1 & 0 \\
1 & 0 & 0 & 0 & 0 & 0 & 0 & | & 0 & 0 & 0 & 0 & 0 & 1 \\
0 & 0 & 1 & -1 & -1 & 0 & 0 & | & 0 & 0 & 1 & 1 & 0 & 0 \\
0 & 0 & -1 & 0 & 0 & 0 & -1 & | & 0 & 0 & 0 & -1 & 0 & 0
\end{bmatrix}
\begin{bmatrix} VRC \\ VBEQ1 \\ ICQ1 \\ IRE \\ ICE \\ VRL \\ IR3 \\ \hline u1 \\ u2 \\ y1 \\ y2 \\ y3 \\ y4 \end{bmatrix}
$$

2.3. Nonlinear Case

Similar to the linear case, for each component partition, a pseudo circuit is generated in the form,

$$\dot{x}^1 = f^1 (x^1, a^1)$$
$$b^1 = g^1 (x^1, a^1) \quad ; \quad x^1(0) = 0 \qquad (22)$$

and

$$a^1 = K_{11} b^1 + K_{12} u^p \qquad (23)$$

$$y^p = K_{21} b^1 + K_{22} u^p \qquad (24)$$

where the notations are defined in the linear case.

Although these equations have the same form as a set of standard circuit equations, they do not correspond to a physical circuit and hence the term "pseudo-circuit" equations. Since both parameters u and y are known in our test algorithm, these pseudo-circuit equations can be solved via any standard circuit analysis code, such as SPICE [23], or NAP2 [24]. Once the component variables in Testee Group, a^2 and b^2, are computed, we calculate

$$\dot{x}^2 = f^2 (x^2, a^2)$$
$$\hat{b}^2 = g^2 (x^2, a^2) \quad ; \quad x^2(0) = 0 \qquad (25)$$

and compare b^2 and $\hat{b}^2$ to determine which, if any, of the group "2" components are faulty.

Given any component subdivision, a pseudo circuit with connection matrix K is created by evaluating equations (13)-(16). The database takes the form of SPICE code for equations (22)-(25). Since we work with an algebraic representation of the system connectivity structure rather than a graph or a schematic diagram to facilitate the use of SPICE, we first develop a "controlled source equivalent circuit" for the connection matrix from which the required SPICE code may be generated. Indeed, this greatly simplifies the database generation process for our "pseudo circuits" which have no "obvious" physical realization.

The SPICE code is generated as follows. Consider equations (22)-(25). Our goal is to compute the value of b^2 and $\hat{b}^2$ from these equations. We first solve the equations (22) and (23) for b^1 and substitute it into equation (24) to evaluate y^p, or a^2 and b^2, then substitute a^2 into (25) to compute b^2.

Mathematically, in equation (23), the i-th element of the vector a^1, a_i, is the sum of the following two products: the i-th row of K_{11} and b^1, and the i-th row of K_{12} and u^p. When the test is conducted, the internal input vector u and the external output vector y are known. Thus, these unknown values can be modelled by independent sources. Suppose that the vectors b^1 and u^p in the pseudo circuit are

$$b^1 = col[V_{b1},I_{b2},V_{b3},I_{b4},V_{b5}] \quad \text{and} \quad u^p = col[V_{u1},V_{u2},I_{u3},V_{u4}]. \quad (26)$$

Suppose also the i-th rows of K_{11} and K_{12} are respectively given by

$$P_i = [1 \ 0 \ -1 \ 0 \ 0] \quad \text{and} \quad Q_i = [1 \ 0 \ 0 \ -1],$$

if the voltage measurement at component a_1 is considered, then

$$V_{a1} = P_i b^1 + Q_i u^p = V_{b1} - V_{b3} + V_{u1} - V_{u4} . \qquad (27)$$

Here V_{b1} may be implemented as a dependent voltage source controlled by the voltage measured across component b1 and similarly for V_{b3}, V_{u1}, and V_{u4}. Physically, from equation (27), V_{a1} is the voltage measured across the serial connection of the above four dependent sources. Thus, equation (27) may be modelled as shown in Figure 2.

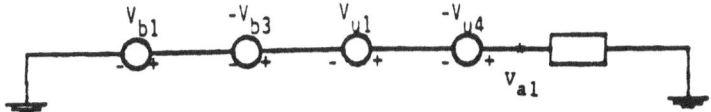

Figure 2. Circuit Modelling of Equation (27).

Therefore, b^1 may be computed from the characteristics of the components, equation (22), and the known vector value a^1, by simulating the circuit shown in Figure 2. To implement this simulation, the box of Figure 2 is replaced by the component employed. More specifically, if the component is a resistor, the resistor will be connected in series with those four dependent voltage sources, as illustrated in Figure 3. If V_R is known, by Ohm's law, the current I_R can be calculated. In SPICE code, the current flow through a component is modelled by connecting a zero-valued voltage source in series with the component [23]. The current flow through the voltage source is then the current to be measured.

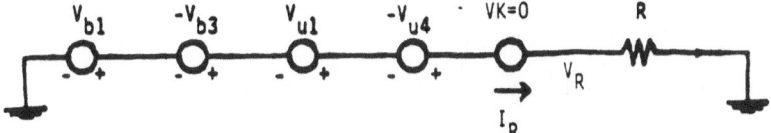

Figure 3. Circuit Modelling for Equation (27) with a Resistor.

Similarly, Consider the current measurement of an element of a^1. Suppose the i-th rows of K_{11} and K_{12} are [0 1 0 -1 0] and [0 -1 1 0], respectively, then

$$I_{a1} = I_{b2} - I_{b4} - I_{u2} + I_{u3} \qquad (28)$$

Here I_{b2} may be implemented as a dependent current source controlled by the current measured at component b2 and similarly for I_{b4}, I_{u2}, and I_{u3}. The circuit for equation (28) is modelled as shown in Figure 4.

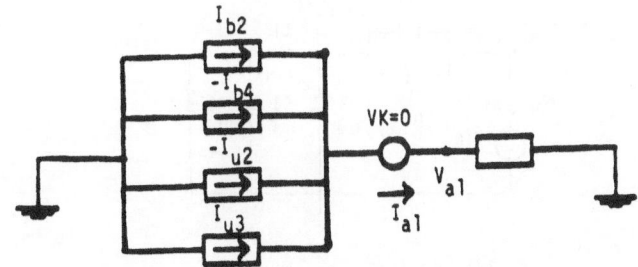

Figure 4. Circuit Modelling for equation (28).

126

The above four dependent current sources are connected in parallel to produce the sum of the currents. A zero-valued voltage source is serially connected to represent this sum, or the current I_{a1}. As in the previous discussion, a component box is connected in series so that the voltage across the component, or V_{a1}, can be measured. Again, if the component is a resistor, then the component box is replaced by a resistor as shown in Figure 5.

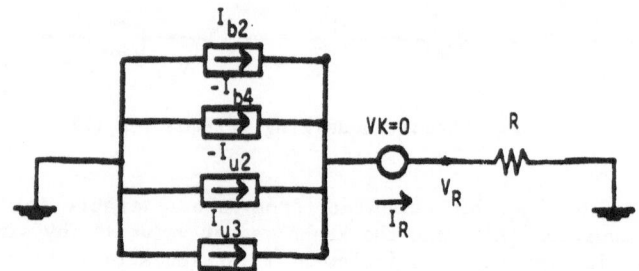

Figure 5. Circuit Modelling for Equation (28) with a Resistor.

For multi-port components, the circuit is modelled in the similar way using multiple connection and component equations.

Given that the SPICE code for equation (23) has been generated, consider equation (24) with the partitioned matrices,

$$a^2 - K_{21}^1 b^1 + K_{22}^1 u^P \qquad (29)$$

$$b^2 - K_{21}^2 b^1 + K_{22}^2 u^P \qquad (30)$$

SPICE code for equation (30) is generated in a similar manner except that, the box in Figure 2 is replaced by a zero-valued voltage source if the element of b^2 is a current measurement, or a resistor with resistance one Ohm for a voltage measurement. The SPICE code for equation (29) can be generated in the same way, however, in this case we are interested in is the vector value b^2. Therefore, the box in Figure 2 is determined by equation (25).

Once the values b^2 and $\hat{b}^2$ are computed, the test outcome can be determined for the current step of the algorithm.

Example 2. [21]

Consider the nonlinear power supply circuit shown in Figure 6.

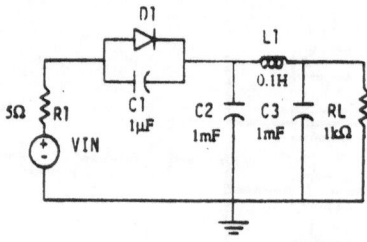

Figure 6. Power Supply Circuit

The connection matrix L and vectors a, b, u, and y are automatically generated in the form.

$$
a = \begin{bmatrix} IR1 \\ VC1 \\ ID1 \\ VC2 \\ IL1 \\ VC3 \\ VRL \end{bmatrix} \quad b = \begin{bmatrix} VR1 \\ IC1 \\ VD1 \\ IC2 \\ IL1 \\ IC3 \\ IRL \end{bmatrix} \quad u = [\ u1\ ] = [\ VIN\ ]
$$

$$
y = \begin{bmatrix} y1 \\ y2 \\ y3 \end{bmatrix} = \begin{bmatrix} ID1 \\ IL1 \\ VRL \end{bmatrix}
$$

$$
L = \begin{bmatrix} L_{11} & | & L_{12} \\ ---- & + & ---- \\ L_{21} & | & L_{22} \end{bmatrix} = \left[\begin{array}{rrrrrrr|r}
0 & 0 & 0 & 1 & 0 & 1 & 1 & 0 \\
0 & 0 & 1 & 0 & 0 & 0 & 0 & 0 \\
0 & -1 & 0 & 1 & 0 & 1 & 1 & 0 \\
-1 & 0 & -1 & 0 & 0 & 0 & 0 & 1 \\
0 & 0 & 0 & 0 & 0 & 1 & 1 & 0 \\
-1 & 0 & -1 & 0 & -1 & 0 & 0 & 1 \\
-1 & 0 & -1 & 0 & -1 & 0 & 0 & 1 \\
\hline
0 & -1 & 0 & 1 & 0 & 1 & 1 & 0 \\
0 & 0 & 0 & 0 & 0 & 1 & 1 & 0 \\
-1 & 0 & -1 & 0 & -1 & 0 & 0 & 1
\end{array} \right]
$$

12 possible component subdivisions are generated as follows:

Subdivision number	Component number			Subdivision number	Component number		
(1)	1	2	6	(7)	2	5	6
(2)	1	2	7	(8)	2	5	7
(3)	1	4	6	(9)	3	4	6
(4)	1	4	7	(10)	3	4	7
(5)	2	3	6	(11)	4	5	6
(6)	2	3	7	(12)	4	5	7

The pseudo circuits and the database are generated as follows: Suppose the first subdivision is chosen, i.e. components 3, 4, 5, and 7 test components 1, 2, and 6. The K-matrix is computed as follows,

$$
\begin{bmatrix} ID1 \\ VC2 \\ IL1 \\ VRL \\ --- \\ IR1 \\ VC1 \\ VC3 \\ -- \\ VR1 \\ IC1 \\ IC3 \end{bmatrix} = \left[\begin{array}{rrrr|rrrr}
0 & 0 & 0 & 0 & 0 & 1 & 0 & 0 \\
0 & 0 & 1 & 0 & 0 & 1 & 0 & 0 \\
0 & 0 & 0 & 0 & 0 & 1 & 0 & 0 \\
0 & 0 & 0 & 0 & 0 & 1 & 0 & 0 \\
\hline
0 & 1 & 0 & 0 & 0 & 1 & 0 & 0 \\
1 & 0 & 0 & 0 & 0 & 1 & 0 & 0 \\
0 & 0 & 0 & 0 & 0 & 1 & 0 & 0 \\
\hline
-1 & 0 & -1 & 0 & 0 & 1 & 0 & 0 \\
0 & 0 & 0 & 0 & 0 & 1 & 0 & 0 \\
0 & 0 & 0 & 0 & 0 & 1 & 0 & 0
\end{array} \right] \begin{bmatrix} VD1 \\ IC2 \\ VL1 \\ IRL \\ --- \\ u1 \\ y1 \\ y2 \\ y3 \end{bmatrix} \quad (31)
$$

where u1=VIN, y1=ID1, y2=IL1, and y3=VRL.

Using the modified circuit shown in Figure 7, the data base for the SPICE code is generated as follows,

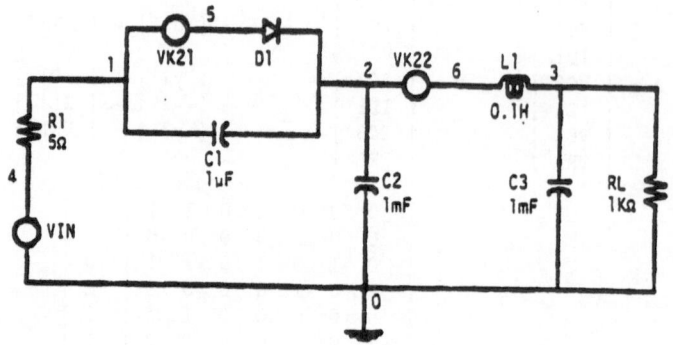

Figure 7. Power Supply Circuit with the Additional Sources.

SPICE code for the circuit in Figure 7:

```
POWER SUPPLY CIRCUIT
VIN     4     0     SIN(0 10 60)
.MODEL DMOD1 D IS=1.0E-06 N=0.97
VK21    1     5     0
VK22    2     6     0
R1      4     1     5
C1      1     2     1U
D1      5     2     DMOD1
C2      2     0     1M
L1      6     3     0.1
C3      3     0     1M
RL      3     0     1K
.TRAN  10M 200M
```

SPICE code for source description:

```
EE701   701    0      4     0    1
RR701   701    0      1
FF702     0  702   VK21     1
VK702   702    0      0
FF703     0  703   VK22     1
VK703   703    0      0
EE704   704    0      3     0    1
RR704   704    0      1.
```

SPICE code for the K-matrice shown in equation (31):

```
****   1
FF101    0  101  VK702   1
VK101  101  102     0
DD101  102    0  DMOD1
****   2
EE101  103    0  POLY(2)  106 0 704 0 0 1 1
VK102  103  104     0
CC101  104    0    1M
****   3
FF102    0  105  VK703   1
VK103  105  106     0
LL101  106    0    0.1
****   4
EE102  107    0   704   0    1
VK104  107  108     0
RR101  108    0    1K
****   1
FF103    0  109  POLY(2)  VK102 VK703 0 1 1
VK105  109  110     0
RR102  110    0     5
****   2
EE103  111    0   102   0    1
VK106  111  112     0
CC102  112    0    1U
****   3
EE104  113    0   704   0    1
VK107  113  114     0
CC103  114    0    1M
****   1
EE105  115    0  POLY(4)  102 0 106 0 701 0 704 0 0 -1 -1 1 1
RR108  115    0     1
****   2
FF104    0  116  POLY(3)  VK102 VK702 VK703 0 1 -1 1
VK109  116    0     0
****   3
FF105    0  117  POLY(2)  VK104 VK703 0 -1 1
VK110  117    0     0
.PRINT TRAN    V(115),V(110)
.PRINT TRAN    I(VK109),I(VK106)
.PRINT TRAN    I(VK110),I(VK107)
.END
```

2.4. Software Implementation

The AATPG code for both the linear and nonlinear circuits is subdivided into off-line and on-line processes. The former, corresponding to the test system design stage, is used by the test system designer to input nominal system specifications to generate a database which is used during the on-line process. To implement the actual test on a UUT (Unit Under Test), the field engineer invokes the on-line process by inputing data describing the UUT; the assumed maximum number of simultaneous failures; the type of decision algorithm to be employed; and the resources of the test data. The actual test can then be run in a fully automatic mode, or interactively.

A circuit description and test objectives are initially given in the off-line process to generate the test program. Necessary changes may be implemented if the resultant test does not satisfy all requirements. On the other hand, if the design is satisfactory, the off-line process will then generate the test program and its associated test data for the use of the on-line process. In the test package, the greatest part of the required computation is carried out by the off-line process which computes the "pseudo-internal test data". This "test data" is computed from the test measurements via a simple on-line matrix/vector multiplication (of (19) and (20)) for the linear case. To the contrary, in the nonlinear case, a circuit simulator, SPICE, is used to compute the "pseudo-internal test data" via an on-line simulation of an appropriate pseudo circuit.

In our implementation, the AATPG code resides in the host computer. When the on-line process is conducted, the ATE (Automatic Test Equipment) receives the commands from the host computer instructing the ATE to perform the measurements. After the ATE completes it transmits the test data to the host to identify the faulty component(s).

Both the off-line and the on-line processes have user-oriented interfaces6Htosimplify the process of generating a new test program. The AATPG has been implemented on a VMS operating system VAX 11/780 in FORTRAN 77 and DCL (DEC Command Language). The input syntax is a free-format style, in other words, it does not require that data be entered in fixed column locations.

3. DECISION ALGORITHMS

Since the results of the tests described in the preceding section are dependent on our assumption that the Group "1" components are not faulty, they are not immediately applicable. A decision algorithm is required to cope with this ambiguity so that the actual fault(s) can be precisely identified. Following the philosophy initiated by Preparata, Metze, and Chien, in their study of self-testing computer network [25-27], if one assumes a bound on the maximum number of simultaneous failures, it is possible to determine the actual fault(s) from an analysis of the test results obtained at the various steps of algorithm. To this end we have formulated three decision algorithms [20], namely, an Exact algorithm, a Heuristic algorithm, and a Boolean algorithm. The Exact algorithm is employed to locate a single fault, while the Heuristic algorithm is used to identify the multiple faults. The Boolean algorithm is a formal decision algorithm that can be implemented for both single and multiple faults.

3.1. The Exact Algorithm

Consider an analog system which is known to contain at most one faulty component. Assume that m test points are available, in other words, the Testee group, G_2, consists of m components. Suppose that the test outcomes obtained from a given step of the algorithm indicate that all Testee Group components are good as indicated in the following table, where 0 (1) indicates the component is good (faulty) according to the results of that test step.

```
                      "1"
        "2"  | a b c . . . k
        ---+---+-----------------
        0 | x |
        0 | y |
        . | . |
        . | . |
        0 | z |
```

In this case, we claim that the group "2" components are, in fact, good. Indeed, if a good component were actual faulty then our test outcomes are incorrect, which could only happen if one of the group "1" components was faulty. Therefore, the system would have two faulty components contradicting our assumption to the effect that a most one component is faulty.

Now consider the case where the test outcomes from a given step of the test algorithm indicates that exactly one group "2" components, say x, is faulty.

```
                          "1"
         "2"  | a b c . . . k
       ---+---+------------------
         1 | x |
         0 | y |
         . | . |
         . | . |
         0 | z |
```

In this case, the same argument used above will guarantee that the components which test good: say, y through z; are, in fact, good. On the other hand, we have no information about x. It may be faulty or, alternatively, the test outcome may be due to a faulty group "1" component.

Finally, consider the case where two or more group "2" components test bad in a given step indicated in the following table:

```
                          "1"
         "2"  | a b c . . . k
       ---+---+------------------
         1 | x |
         1 | y |
         . | . |
         . | . |
         0 | z |
```

Since, under our assumption of a single fault, it is impossible for two or more group "2" components to be faulty, this test result implies that at least one of the group "1" components is bad. Since we have assumed that there is at most one faulty component, this implies that all of the group "2" components are, in fact, good.

Table II summarizes all possible test outcomes obtained from one step of the test algorithm and the conclusions.

Table II.   Test Results for single fault.

Test Results (1 2 3 4 . . m)	Conclusions
0 0 0 0 . . 0	all components in group "2" are good
1 0 0 0 . . 0	all components in group "2" except (possibly) component #1 are good
1 1 0 0 . . 0	all components in group "2" are good
1 1 1 0 . . 0	: :
: : : . . . :	: :
1 1 1 1 . . 1	all components in group "2" are good

Consistent with the above arguments, at each step of the test algorithm, either all, or all but one, of the components in group "2" are found to be good. Therefore, if one repeats the process several times eventually arriving at a component subdivision in which only known good components are included in the Tester Group, then the test outcome obtained at that step will be reliable. Thus, an accurate determination of the faulty component(s) in group "2" at that step will be obtained and the process will terminate.

Algorithm I describes that the software implementation of the Exact algorithm for the single failure analysis.

Algorithm I. [20] (Exact Algorithm for a single failure)
Step 0. In the pre-process, a component subdivision table must be generated, and a weight, the number of components in group "2", is assigned to each subdivision.
Step 1. Retrieve the component subdivision table and the weight for each subdivision.
Step 2. Choose a subdivision with the maximum weight, and delete it from table.
Step 3. If the maximum weight is less than or equal to 1, then go to step 6. Otherwise, derive the test outcome and determine the good components.
Step 4. Reduce the weight of each subdivision by the good components it currently contains.
Step 5. Record the undetermined component number, then go to step 2 to repeat the process.
Step 6. (Case of the maximum weight equal to one)
Choose the first subdivision that contains an undetermined component and check if the component is good. Repeat this step until all undetermined components have been chosen.
Step 7. The actual faulty component is then determined.

3.2. The Heuristic Algorithm

The problem of multiple faults location can be greately simplified if one adopts an "analog heuristic" to the effect that two independent analog failures will never cancel [28]. Needless to say, this is an inherently analog heuristic since two binary failures have a fifty-fifty chance of cancelling one another. In the analog case, however, two independent failures are highly unlikely to cancel each other (as long as one works with reasonably small tolerances).

Recall from our discussion of the single fault case that whenever a test result indicates that a component is good, then it is, in fact, good. Although this is not rigorously true in the multiple fault case, it is true under the assumption of our heuristic. For instance, consider the test outcomes indicated in the following table in which x is found to be good.

```
                  "1"
       "2"  | a b c . . . k
     ---+---+-----------------
       0 | x |
       1 | y |
       . | . |
       . | . |
       0 | z |
```

Now, if x is actually faulty there must be a faulty group "1" component whose effect is to cancel the error in x as observed during this step of the test algorithm. This is, however, forbidden by our heuristic and, thus, we conclude that x is actually good.

Interestingly, our heuristic can be carried a step further than that indicated above since, under our heuristic, a bad group "1" component would always yield erroneous test results. An exception would, however, occur if some of the group "1" components are totally decoupled from some of the group "2" components. Therefore, if prior to our test we generate a coupling table (by simulation, or a sensitivity analysis) which indicates whether or not a faulty group "1" component will affect the test results on a group "2" component, our heuristic may be used to verify that certain group "1" components are good whenever a good group "2" component is located. Consider the following table,

```
                        "1"
        "2"  | a b c . . . k
        ---+---+------------------
        0 | x | 1 0 1 . . . 1
        1 | y | 1 1 0 . . . 0
        . | . | . . . . . . .
        . | . | . . . . . . .
        0 | z | 0 1 1 . . . 0
```

in which a "1" in the i-j position indicates that the test result for component i is affected by component j, while a "0" indicates that the component j does not affect the test result for component i. Now, since component x has been found to be good in this test, our heuristic implies that those group "1" components which are coupled to x in this test are also good. Similarly, since z is good the heuristic implies that b and c are also good. Thus, with a single test step, we have verified that x, z, a, b, c, and k are all good.

Since in any practical circuit the coupling table is composed mostly of 1's, it has been our experience that relatively few steps of the algorithm will yield a complete diagnosis. To implement the heuristic, however, one must assume that the maximum number of faulty components is strictly less than the number of group "2" components. If not, the test results at each step may show that all group "2" components are faulty, in which case no reliable test information is obtained. Moreover, the degree to which the number of group "2" components exceeds the maximum number of faulty components determines the number of algorithm steps which will be required to fully diagnose a circuit.

Algorithm II. [20]
Step 0. Retrieve the component subdivision table and input t (t is the number of simultaneous failures.)
Step 1. Choose a subdivision and call subroutine to derive the test results. (indicating a 0 for a good component and a 1 for a bad componen.)
Step 2. Retrieve the coupling table corresponding to this subdivision.
Step 3. If the ith test result is 0, then all components with 1 in the i-th row of the coupling table are good.
Step 4. Repeat the Step 3 until all 0 test results have been processed.
Step 5. Repeat Steps 1-4 until the actual faulty components are determined.

3.3. The Boolean Algorithm

A Boolean expression is derived from each step of the test algorithm which includes all possible fault patterns associated with the test data. The actual fault(s) can then be located by "multiplying" the Boolean expressions associated with several steps of the algorithm or equivalently comparing the fault patterns obtained from each test step and excluding the impossible fault patterns.

Consider the case where Group "1" contains five components; namely, a b, c, d, and e; and Group "2" contains three components; x, y, and z. Suppose that the test results is as follows:

```
                    "1"
      "2"  | a b c d e
    ---+---+-----------
     0 | x |
     1 | y |
     0 | z |
```

All possible faulty patterns for this test result can be expressed by a Boolean form as

$$T1 = a + b + c + d + e + \bar{a}\,\bar{b}\,\bar{c}\,\bar{d}\,\bar{e}\,x\,y\,\bar{z}$$

(Here, the letter a indicates the component is bad, and its complement $\bar{a}$ means the component a is good.) The first five terms of T1 represent the case where one of the Group "1" components is faulty and, thus, the remaining components are unknown, or "don't care"; while the last term of T1 indicates that all Group "1" components are good and the test results for the Group "2" components are as shown.

To implement this symbolic Boolean expression, a simple "array-like" data structure (a tabulated expression) is presented [20] as follows, (where $\emptyset$ denotes the "don't care" term):

a	b	c	d	e	x	y	z
1	$\emptyset$	$\emptyset$	$\emptyset$	$\emptyset$	$\emptyset$	$\emptyset$	$\emptyset$
$\emptyset$	1	$\emptyset$	$\emptyset$	$\emptyset$	$\emptyset$	$\emptyset$	$\emptyset$
$\emptyset$	$\emptyset$	1	$\emptyset$	$\emptyset$	$\emptyset$	$\emptyset$	$\emptyset$
$\emptyset$	$\emptyset$	$\emptyset$	1	$\emptyset$	$\emptyset$	$\emptyset$	$\emptyset$
$\emptyset$	$\emptyset$	$\emptyset$	$\emptyset$	1	$\emptyset$	$\emptyset$	$\emptyset$
0	0	0	0	0	0	1	0

The following rules are used to compute the production of any two Boolean expressions:

Rule 1: Let the Boolean set B=\{0,1,$\emptyset$\}, then
$x * x = x$; $x * \emptyset = \emptyset * x = x$, where x= 0, 1, or $\emptyset$, and
$0 * 1 = 1 * 0 =$ null (impossible pattern).

Rule 2: $(A + B + C) * (A + B + D) = A + B + CD$.

This effectively excludes all possible fault patterns which are not consistent with the test data. The above algorithm is termed a Regular Boolean algorithm.

Unfortunately, the following problem may be encountered: "How fast are the impossible fault patterns excluded so that the actual faulty component(s) can be determined?" To accelerate the speed of convergence one may specify the maximum number of simultaneous failures. Once this number has been specified the number of impossible patterns will be reduced. For example, if a single failure is assumed, recall from the first term of T1, a, that the component a is bad and the remaining components are "don't care". If, at most, one faulty component is allowed, however, the remaining components would not be bad. Therefore, component a is the only bad component in this pattern, and all "don't care" terms are replaced by 0.

Algorithm III. [20] (Regular Boolean Algorithm)
Step 1. Retrieve the component subdivisions table and input t (the maximum number of simultaneous failures)
Step 2. Choose a subdivision and call the routines to derive the test results and the associated tabulated expression.
Step 3. Search for patterns which contain more than t 1's, and delete the impossible patterns.
Step 4. If this subdivision is the first one, then GOTO step 2, otherwise,
Step 5. Call the routine to compute the product of the associated tabulated expression and the previous one.
Step 6. Search for the impossible patterns and delete them.
Step 7. Repeat the above steps until the actual faulty components are determined.

Since the Regular Boolean algorithm often requires a great number of steps to accelerate the speed of convergence, two additional algorithm are presented [20]: namely, Boolean Exact algorithm and Boolean Heuristic algorithm. The former is used for the single fault case, while the latter is employed for the multiple faults.

4. PARALLEL PROCESSING FOR ANALOG FAULT DIAGNOSIS
The Computer-Aided Testing (CAT) problem is inherently a large scale systems problem, it is essential to exploit whatever computational power is available to reduce the computational requirements for both on-line and off-line test processes. In particular, digital CAT algorithms often use some degree of parallel processing in their implementation. Therefore, the degree to which an analog CAT algorithm can be implemented in parallel becomes a significant factor in determining its viability.

Due to the iterative nature of the self-testing algorithm, it requires many simulation steps to validate the final decision in order to locate the faults. This deficiency can, however, be alleviated by implementing parallel processing in the self-testing algorithm. In fact, it is possible to simulate several pseudo-circuits simultaneously and to implement the test algorithm in parallel. The objective of this section is to present a parallel-type decision algorithm.

Consider the circuit of Figure 1 and its component subdivision table shown in Table I. The table consists of all allowable subdivisions that are generated according to the self-test theory.

The implementation of Algorithm I is described as follows. After the subdivision table is generated, a weight, the number of components in group "2" having unknown status, is initially assigned to each subdivision. In our implementation, a weight of the number of components in group "2" is initially assigned to each subdivision simply because all components have unknown status at the very beginning. As shown in Column 6 of Table III, a weight of four is assigned to subdivisions #1 through #24, but a weight of three, however, is assigned to subdivisions #25 through #34 because that components 5 and 6 represent the two ports of a single component, transistor Q1.

Table III.   Allowable Subdivisions with the Assigned Weight

Subdivision Number	Testee Group Components	Weights (Algorithm I)									Weights (Algorithm IV)		
		Step 0	1	2	3	4	5	6	||	0	1	2
1	1 2 3 9	4	0	0	0	0	0	0	||	4	0	0
2	1 2 3 11	4	2	1	1	0	0	0	||	4	1	0
3	1 2 7 9	4	2	1	1	1	1	1	||	4	1	0
4	1 2 7 11	4	3	1	1	1	1	1	||	4	2	0
5	1 2 8 9	4	2	2	0	0	0	0	||	4	2	1
6	1 2 8 11	4	3	2	1	1	1	1	||	4	2	1
7	1 3 4 9	4	2	1	1	1	1	1	||	4	1	0
8	1 3 4 11	4	3	1	1	1	1	1	||	4	2	0
9	1 4 7 9	4	3	1	1	1	1	1	||	4	2	0
10	1 4 7 11	4	4	0	0	0	0	0	||	4	3	0
11	1 4 8 9	4	3	2	1	1	1	1	||	4	2	1
12	1 4 8 11	4	4	2	1	1	1	1	||	4	2	1
13	2 3 9 10	4	1	1	1	1	1	0	||	4	1	0
14	2 3 10 11	4	2	1	1	1	1	1	||	4	2	0
15	2 7 9 10	4	2	1	1	1	1	1	||	4	2	0
16	2 7 10 11	4	3	1	1	1	1	1	||	4	3	0
17	2 8 9 10	4	2	2	1	1	1	1	||	4	2	1
18	2 8 10 11	4	3	2	1	1	1	1	||	4	3	1
19	3 4 9 10	4	2	1	1	1	1	1	||	4	2	0
20	3 4 10 11	4	3	1	1	1	1	1	||	4	3	0
21	4 7 9 10	4	3	1	1	1	1	1	||	4	3	0
22	4 7 10 11	4	4	1	1	1	1	1	||	4	4	0
23	4 8 9 10	4	3	2	1	1	1	1	||	4	3	1
24	4 8 10 11	4	4	2	1	1	1	1	||	4	4	1
25	2 3 5 6	3	1	1	1	1	0	0	||	4	2	2
26	2 5 6 7	3	2	1	1	1	1	1	||	4	3	2
27	2 5 6 8	3	2	2	1	1	1	1	||	4	3	3
28	2 5 6 9	3	1	1	1	1	1	1	||	4	2	2
29	2 5 6 11	3	2	1	1	1	1	1	||	4	3	2
30	3 4 5 6	3	2	1	1	1	1	1	||	4	3	2
31	4 5 6 7	3	3	1	1	1	1	1	||	4	4	2
32	4 5 6 8	3	3	2	1	1	1	1	||	4	4	3
33	4 5 6 9	3	2	1	1	1	1	1	||	4	3	2
34	4 5 6 11	3	3	1	1	1	1	1	||	4	4	2

Suppose that component #1 (C1) is faulty, and the single fault case is considered.   Among the possible subdivisions, the subdivision #1 with a maximum weight of 4 is first selected and simulated.

Subdivision number	Testee Group Component Numbers				Test Results			
1	1	2	3	9	1	0	0	0

According to the discussion in Table II, the simulation shows that the components 2, 3, and 9 are good, but the status of component 1 is unknown. other words, after completing this simulation, the status of components 1, 4, 5(6), 7, 8, 10, and 11 are still unknown. In the subsequent steps, the weight of each subdivision is reassigned according to the number of components having unknown status, and a maximum weight subdivision is selected and simulated in the next step. The detailed selection process is listed in Table III, and the following simulation steps are executed sequentially.

Subdivision number	Testee Group Component Numbers				Test Results			
1	1	2	3	9	1	0	0	0
10	1	4	7	11	1	0	0	0
5	1	2	8	9	1	0	0	0
2	1	2	3	11	1	0	0	0
25	2	3	5	6	0	1	1	1
13	2	3	9	10	0	1	1	1

The faulty component is #1 (C1).

From the simulation results of the first four steps, we know that the components 2, 3, 4, 7, 8, 9, and 11 are definitely good, and component 1 is possibly faulty. The following two steps show that components 5(6) and 10 are also good. In other words, the simulation results illustrate that all components except #1 are good. Consequently, the component 1 is found to be faulty. Otherwise, it contradicts to the test results shown in the first four steps.

Using Algorithm I for every possible single fault, the simulation results are summarized in Table IV.

Table IV. Simulation Results

Simulated Faulty Component		Required Simulation Steps	Detected Faulty Component(s)
1	(C1)	6	C1
2	(R1)	5	R1
3	(R2)	6	R2
4	(RC)	4	RC
5,6	(Q1)	4	Q1
7	(RE)	5	RE
8	(CE)	5	RE,CE
9	(C2)	6	C2
10	(RL)	5	RL
11	(R3)	5	R3

In fact, the on-line computational time and cost of a test algorithm are proportional to the number of simulation steps required to locate the faulty component. It is unrealistical, however, for a test algorithm to require an average of 5.1 simulation steps to locate a single faulty component in an 11-components circuit.

Before describing the modification of this sequential-type decision algorithm, the following definitions [29] are first considered.

A subdivision space is defined as a collection of all possible component subdivisions generated from a circuit or system based on the self-test algorithm. A subdivision basis is a minimum subset of the subdivision space, that contains all components. In other words, every component must appear at least once in group "2" of a subdivision in the subdivision basis. Moreover, in a subdivision basis, we are guaranteed that there exists at least one subdivision which contains all good components in the Tester Group of a subdivision for single fault case [29]. As a result, this subdivision will provide a reliable test for locating the faulty component. The generation of a subdivision basis is illustrated in

Algorithm IV.

Algorithm IV. [29] (Generation of Subdivision Basis)
*   Let S_i, i=1,2,..,s, be all component subdivisions,
       BASE be the number of subdivisions in the subdivision basis, and
       C be the set of all components, i.e., $C=\{C_1,C_2,..,C_n\}$.
1. DO i=1 TO s
  BEGIN
     Subdivision #i is selected;
     $C=C-\{C_i$ in subdivision #i\};
     COUNT=1;
2.   WEIGHT=number of components in both the set C and a subdivision;
     Select a subdivision that has the maximum weight, say #k;
     $C=C-\{C_k$ in subdivision #k\};
     COUNT=COUNT+1;
     IF COUNT > BASE, THEN
     BEGIN
       IF C=∅ THEN
       BEGIN
          BASE=COUNT;
          Record the selected subdivisions;
       END
       ELSE
          GO TO Step 2.
     END
  END

According to the stepwise illustration of the weight assignment for the circuit described in Table III, a subdivision basis consisting of three seed subdivisions, is generated as follows,

Subdivision number	Testee Group Component Numbers			
1	1	2	3	9
22	4	7	10	11
27	2	5	6	8

In practice, the number of seed subdivisions is approximately $\lceil n/m \rceil$, where n and m are the numbers of the circuit components and the selected test points, respectively. In fact, the ratio n/m represents a natural measure of the possible tradeoffs between test points and computational requirements. The on-line test process time may be decreased if the number of test points is increased. However, since the number of possible test points is limited due to geometrical effects [9] and fault diagnosability [12], our emphasis is then turned to how to reduce the on-line computational requirements for a given set of test points.

Suppose component #1 (C1) is faulty, the simulation results associated with these seed subdivisions are

Subdivision number	Testee Group Component Numbers				Test Results			
1	1	2	3	9	1	0	0	0
22	4	7	10	11	0	1	0	1
27	2	5	6	8	0	1	1	1

According to the discussion in Table II, we can easily locate the faulty component #1 (C1) by using these three simulation steps.

In the single fault case, if the simulation results can identify that that there is only one component having unknown status, the component is then identified as faulty. However, due to the nature of component coupling, a good component in the Testee Group may be misjudged as faulty due to a faulty component contained in the Tester Group. (This component is referred to as the **equivalent faulty component**.) The equivalent faulty components are, indeed, often encountered in our implementation. Therefore, it is necessary to apply more simulation steps to distinguish the equivalent faulty components from the actual one.

Suppose component #4 (RC) is faulty, three seed subdivisions are simulated as follows,

Subdivision number	Testee Group Component Numbers				Test Results			
1	1	2	3	9	0	0	1	0
22	4	7	10	11	1	0	0	0
27	2	5	6	8	0	1	1	1

The simulation results identify the components 1, 2, 5(6), 7, 8, 9, 10, and 11 as good, but the components 3 and 4 are unknown. If subdivision #7 is further selected and simulated,

Subdivision number	Testee Group Component Numbers				Test Results			
7	1	3	4	9	0	0	1	0

the equivalent fault component #3 is then distinguished from the actual faulty component #4.

On the other hand, however, due to the topological structure of a given circuit or system, the subdivision space may not always contain all possible combinations of components. In many cases, it may not have a subdivisions that can distinguish the equivalent faulty component(s) from the actual one. Consequently, an **ambiguity set** that consists of both equivalent and actual faulty components, results.

Suppose component #8 (CE) is faulty and the simulation results associated with the seed subdivisions are

Subdivision number	Testee Group Component Numbers				Test Results			
1	1	2	3	9	1	0	1	0
22	4	7	10	11	0	1	0	0
27	2	5	6	8	0	0	0	1

We conclude that the components 1, 2, 3, 4, 5(6), 9, 10, and 11 are good, but components 7 and 8 are unknown. Unfortunately, there exists no subdivision that can distinguishes between then. Thus, an ambiguity set consisting of the components 7 and 8 results.

Based on the above discussion, a modification of Algorithm I is given in in Algorithm V and the number of steps required for each possible single fault is listed in Table V. The number of simulation steps required for the same circuit has been reduced from an average of 5.1 steps to 3.4 steps. However, the on-line computation required for 3.4 simulation steps are still high for this 11-components circuit. A further improvement on both on-line test process time and cost is still needed.

Algorithm V. [29] (Modified Exact Algorithm)
*   Let the subdivision basis $B=\{B_1, B_2, .., B_r\}$, $Z=\emptyset$.
  1. DO i-1 TO r (r: the number of subdivisions in the basis)
     BEGIN
          Simulate the test results with each B_i.
          IF Z_i is the only element with "1" in B_i, THEN $Z - Z \cup \{Z_i\}$.
     END
  2. IF the number of components in the set Z is 1, THEN STOP
     ELSE (Eliminate the equivalent faulty components)
          BEGIN
              DO i-1 TO s (s: the size of the subdivision space)
              BEGIN
                  WEIGHT:-number of components in both Z and S_i.
              END
              COUNT-0.
              REPEAT
                  Select and simulate the maximum WEIGHTed subdivision.
                  COUNT-COUNT + 1
                  Eliminate the equivalent faulty components from the set Z
                  Reduce the WEIGHT of each subdivision.
              UNTIL the maximum WEIGHT is 1.
          END
  3. IF $|Z| \neq 1$, THEN Z is defined as the ambiguity set.

Table V.   Simulation Results

Simulated Faulty Component	Required Simulation Steps	Detected Faulty Component(s)
1 (C1)	3	C1
2 (R1)	3	R1
3 (R2)	5	R2
4 (RC)	4	RC
5,6 (Q1)	4	Q1
7 (RE)	3	RE
8 (CE)	3	RE,CE
9 (C2)	3	C2
10 (RL)	3	RL
11 (R3)	3	R3

4.1. Parallel Processing

Unlike the original Exact Algorithm that determines the next selected subdivision in terms of the previous simulation results, in the Modified Exact Algorithm the determination of the next subdivision is based on the simulation of the seed subdivisions which can be simulated independently. In other words, the seed subdivisions can be simulated by using different processors. Therefore, a parallel implementation is possible.

Let S be the subdivision space and B be the subdivision basis, the numbers s and r are respectively denoted as the cardinalities of S and B. Suppose that r processors are available, namely, P_1, P_2, .., P_r. A parallel algorithm, Algorithm VI, is presented.

Algorithm VI: [29] (Parallel algorithm)
* Load the necessary data to each processor.
1. Central processor sends the test data to each processor.
2. Perform the simulation at each processor and then transmit the test
   outcomes back to the central processor.
3. (Central processor makes decision for next iteration)
   IF the number of components with unknown status is 1, THEN STOP.
4. IF no subdivision contains two or more components with unknown
   status, THEN STOP. (These components form the ambiguity set) ELSE
5. i=0
   REPEAT
       i=i+1
       select the maximum weighted subdivision, and send the index of this
           subdivision to the i-th processor.
   UNTIL (i=r OR no subdivision with two or more of unknown components)
6. GOTO step 2.

Suppose component #1 is faulty, and three processors are used to
simulate the test results. The simulation results associated with the seed
subdivisions in the first step of the process are,

Proceesor	Subdivision number	Testee Group Component Numbers				Test Results			
P_1	1	1	2	3	9	1	0	0	0
P_2	22	4	7	10	11	0	1	0	1
P_3	27	2	5	6	8	0	1	1	1

Similar to Tables IV and V, the performance of the parallel algorithm for
each possible single component failure is shown in Table VI. The average
number of simulation steps for the same circuit has been reduced to 1.3.

Table VI.   Simulation Results

Simulated Faulty Component		Required Simulation Steps	Detected Faulty Component(s)
1	(C1)	1	C1
2	(R1)	1	R1
3	(R2)	2	R2
4	(RC)	2	RC
5,6	(Q1)	2	Q1
7	(RE)	1	RE
8	(CE)	1	RE,CE
9	(C2)	1	C2
10	(RL)	1	RL
11	(R3)	1	R3

Although the proposed parallel test system has not yet been physically
implemented on a parallel processor, a simulation program has been
developed to demonstrate it via a multiple task facilities of the VMS
operating system on a VAX 11/780.
    To obtain an experimental estimates of the performance of the proposed
parallel test algorithm for the single fault case, various linear and
nonlinear circuits have been simulated.

142

Example 3.

Consider the cascade filter with 2, 4, 10, 20, or 40 stages [19], shown in Figure 8. Suppose that four satellite processors are used, the simulation results are listed in Table VII. Table VII shows that the test process is required only one simulation step in each case.

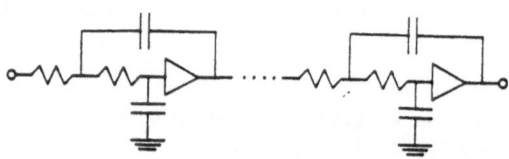

Figure 8. Linear Circuit.

Table VII. Simulation Results

# of stages		2	4	4	4	10	10	20	20	40	40
# of components		10	20	20	20	50	50	100	100	200	200
# of test points		4	8	6	5	20	15	40	30	80	60
# of inaccessive nodes		0	0	4	6	0	10	0	20	0	40
Fault coverage (%)		100	100	80	70	100	80	100	80	100	80
Size of the basis		4	4	4	4	4	4	4	4	4	4
Average of steps		1	1	1	1	1	1	1	1	1	1
Size of the ambiguity set	1	8	18	14	12	48	38	98	78	198	158
	2	2	2	2	2	2	2	2	2	2	2
	4			4							
	6				6						
	10						10				
	20								20		
	40										40

Example 4.
Consider the two nonlinear circuits shown in Figure 9 [10,21]: a power supply and an astable multivibrator. The test results are shown in Table VIII.

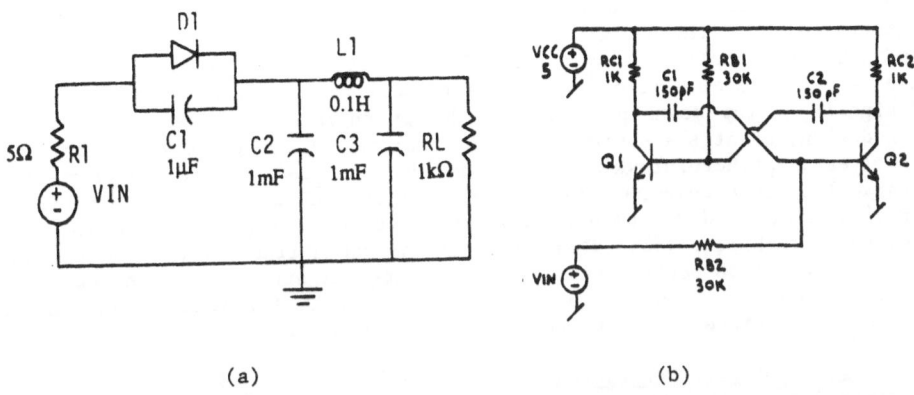

(a)   (b)

Figure 9.   Nonlinear Circuits: (a) Power Supply, (b) Astable Multivibrator.

Table VIII.   Simulation Results.

# of components		7	11
# of test points		3	4
# of inaccessive nodes		0	0
Fault coverage (%)		100	100
Size of the basis		3	4
Average of steps		1*	1**
Ambiguity set	1	4	10
	2	3	1

*: three satellite processors are used.
**: four satellite processors are used.

5. DESIGN FOR TESTABILITY

With the rapidly increasing complexity of circuits and systems, the ability to adequately design a diagnosable circuit or system is a prime requisite for rapid fault location.

Given the fact that future electronic systems will rely heavily on CAD tools to reduce design costs, increase design accuracy and reduce development times, it is clearly important that testability factors be integrated into these CAD tools. Testability as it now stands in the industry is a "bottom up" process [30]. Virtually all the known techniques need a detailed cirucit design before testability can be measured. However, design is a "top down" process. The design engineer is given requirements for design, creates a design approach, analyzes it for performance, finally involves a structural design that can be analyzed for testability. For testability to become an effective part of the CAD process the testability process will need to become more of a top-down approach instead of a bottom-up technique [31]. Therefore, if one can define a condition for the testability which depends only on the topological structure of the designed circuit, not on the component values, then design for testability can be established before analyzing the circuit performance.

5.1. Pseudo-Circuit Generation

In a linear circuit, at each step of the test algorithm, the required pseudo circuit is formed via equations (5), (11) and (12). The crucial assumption of this pseudo circuit approach is that $[L_{21}^2]^L$ exists. This implies that the Group 1/Group 2 partition must be chosen so that $[L_{21}^2]^L$ exists.

Consider the pseudo nominal tableau equations. In the linear case, they are simply a stacking of equations (5),(7-9):

$$
\begin{bmatrix} -I & 0 & z^1 & 0 \\ L_{11}^{21} & -I & 0 & L_{11}^{22} \\ L_{11}^{11} & 0 & -I & L_{11}^{12} \\ L_{21}^1 & 0 & 0 & L_{21}^2 \end{bmatrix} \begin{bmatrix} b^1 \\ a^2 \\ a^1 \\ b^2 \end{bmatrix} = \begin{bmatrix} 0 \\ -L_{12}^2 u \\ -L_{12}^1 u \\ y-L_{22} u \end{bmatrix}.
\tag{32}
$$

where

$$
T = \begin{bmatrix} -I & 0 & z^1 & 0 \\ L_{11}^{21} & -I & 0 & L_{11}^{22} \\ L_{11}^{11} & 0 & -I & L_{11}^{12} \\ L_{21}^1 & 0 & 0 & L_{21}^2 \end{bmatrix}
$$

is known as the tableau matrix [22].

Theorem 1: [32,33]
   The tableau matrix T is left invertible if and only if the matrix

$$Q = \begin{bmatrix} -I + L_{11}^{11} z^1 & L_{11}^{12} \\ L_{21}^1 z^1 & L_{21}^2 \end{bmatrix}$$

is left invertible.

Note that if L_{21}^2 is left invertible, consider the matrix

$$Q^* = (-I + L_{11}^{11} z^1) - L_{11}^{12} (L_{21}^2)^{-L} L_{21}^1 z^1$$

$$= -I + [L_{11}^{11} - L_{11}^{12} (L_{21}^2)^{-L} L_{21}^1] z^1$$

$$= -I + K_{11} z^1 \qquad (\text{ by equation (9)})$$

Since the matrix K_{11} is a connection structure, the matrix $(I-K_{11}z^1)$ is invertible [22], and Q^* is left invertible. Therefore, the left invertibility of L_{21}^2 is a sufficient but not necessary condition [32,33].

Remark 1: [32,33]
   The tableau matrix T is generically left invertible, if and only if the matrix Q' is left invertible, where

$$Q' = \begin{bmatrix} -I + L_{11}^{11} & L_{11}^{12} \\ L_{21}^1 & L_{21}^2 \end{bmatrix} \qquad (33)$$

Remark 2: [32]
   In the linear case, the pseudo circuit exists with respect to a subdivision if and only if Q' is left invertible.

   The salient feature of this condition is that Q' is a constant matrix and depends only upon the connection matrices, L_{11}^{11}, L_{11}^{12}, L_{21}^1, and L_{21}^2, but not upon the component equations. Obviously, this condition depends only upon the connection structure of a given circuit. Moreover, for this reason the development for linear circuits can be immediately extended to nonlinear circuits.
   Similarly, the tableau matrix T' for nonlinear circuits is given as follows [33],

$$T' = \begin{bmatrix} -I & 0 & [\frac{\partial g^1}{\partial x^1}]P + \frac{\partial g^1}{\partial a^1} & 0 \\ L_{11}^{21} & -I & 0 & L_{11}^{22} \\ L_{11}^{11} & 0 & -I & L_{11}^{12} \\ L_{21}^1 & 0 & 0 & L_{21}^2 \end{bmatrix}$$

where

$$P = \{d_0 I - \frac{\partial f^1}{\partial x^1}\}^{-1} [\frac{\partial f^1}{\partial a^1}].$$

Remark 3:
   The tableau matrix T' is generically left invertible, if and only if the matrix Q' is left invertible.

Remark 4:
   In the nonlinear case, the pseudo circuit exists with respect to a subdivision if and only if Q' is left invertible.

   The above development is summarized in the following theorem,

Theorem 2: (The necessary and sufficient condition)
   The pseudo circuit exists with respect to a particaulr allowable subdivision if and only if the matrix T or T' is left invertible. The matrix T or T' is generalically left invertible if and only if the matrix Q' is left invertible.

5.2. Design of a Testable Circuit
   We start with a definition of diagnosability.
Definition 1: [25]
   A system is t-diagnosable if, given the results of all allowable tests, one can uniquely identify all faulty units provided that the number of faulty units does not exceed t.

Lemma 2:
   A system is t-diagnosable under the Exact algorithm if and only if every 2t-tuple of components appears in at least one Group 2.

Definition 2: [13]
   Let B be an m by n matrix, $n \geq m$. The global column-rank of B is said to be t if every combination of t columns of B is linearly independent, and some combination of (t+1) columns of B is linearly dependent.

Definition 3:
   Given a set of m test points and the connection matrix
$$L = \begin{bmatrix} L_{11} \\ L_{21} \end{bmatrix}, \qquad (34)$$
a corresponding matrix of a given combination of t components is defined as a matrix Q', (33), whose Group "2" contains these t components.

Definition 4:
   The Q'-rank of L is said to be t, $t \leq m$, if for every combination of t components, there exists at least one corresponding matrix Q' with full column rank, and for some combination of (t+1) components, there is no corresponding matrix with full column rank.

Theorem 3:
   A system is t-diagnosable under the Exact algorithm if and only if the Q'-rank of L is at least 2t.

   Similarly, if the Heuristic algorithm is employed, the following results are summarized [32,33].

Lemma 3:
   A system is t-diagnosable under the Heuristic algorithm if and only if every (t+1)-tuple of components appears in at least one group "2".

Theroem 4:
A system is t-diagnosable under the Heuristic algorithm if and only if the Q'-rank of L is at least t+1.

In fact, Theorems 3 and 4 can be combined as the Theorem 5,

Theorem 5:
A designed circuit is t-diagnosable if and only if the Q'-rank of the connection matrix L is at least t+1.

Based on above development, an algorihm to check the t-diagnosability of a circuit is given,

Algorithm VII: [32] (Diagnosability)
```
*
*       N              : number of components.
*       M              : number of test points.
*       COMB2T(i,j): all combinations of 2t elements out of N.
*                      i-1,2,..,2t.  j-1,2,..,r.  r-C(N,2t)
*       FLAG(i)      : .true. if the COMB2T(i,*) has been selected.
*                      .false. otherwise.
*       COMBM(i,j) : all combinations of M elements out of N.
*                      i-1,2,..,m.   j-1,2,..,s.  s-C(N,M)
*       WEIGHT(i)    : number of COMB2T(k,*) whose FLAG(k) is .false.,
*                      contained in COMBM(i,*).
Step 1. FLAG(k) := FALSE,  for all k-1,2,..,r.
     2. Calculate WEIGHT(k), k-1,2,..,s, and select the maximum weight,
        say, WEIGHT(i).
     3. If WEIGHT(i)-0, Then "the system is not t-diagnosable"
        Else   (Compute the Q'-rank)
     4.    Take COMBM(i,j),j-1,2,..,M, as Group 2, and derive the matrix Q'.
     4.1  If coulmn-rank of Q' < n, Then GOTO step 5.
          Else  (reset the flag of the selected COMB2T(*,*))
     4.2    FLAG(k) := .true.  for all k, if COMB2T(k,*) is contained
     5. Do k-1 To r
           If (FLAG(k)-.false.) Then GOTO step 2.
           End of loop k
     6. "The system is t-diagnosable".
```

Theorem 5 provides a criterion to determine the dignosability of a circuit under design. However, as far as design for testability is concerned, one may be interested in not only the diagnosability of the designed circuit, but also the problems of (1) how many test points are required, and (2) how to select the test points.

To determine the number of test points required, the following aspects may be considered: First, a geometric "rule of thumb" was proposed [9,18], in which the circuit complexity is proportional to the area of a printed circuit board (if not a power theoreof), while the number of accessible test points is proportional to the edge length of the board, therefore, the number of test points should grow with the square root of the circuit complexity (or less).

Secondly, as discussed in the previous section, the number of components which may be tested at any step depends on the number of test points available, while the number of steps required (approximately n/m) is determined by the number of testee components. Since the on-line computational requirement is proportional to the number of simulation steps, more test points may reduce the number of steps and thus decrease the computational requirements.

Finally, for a t-diagnosable circuit design, by Theorem 5 and Definition 4, the number of test points, m, must be greater than, or equal to, t+1, which would be the minimum number of test points required. This conclusion may be equivalent to the (k+1) test points needed for a k-node fault [12].

From the above considerations, the first and third define upper and lower bounds on the number of test points required, while the second provides the tradeoffs between the number of test points and the on-line computational requirements.

In order to precisely identify faulty components, the remaining task is to select the test points for a diagnosable circuit design. In our application, two steps are employed to automatically generate the test points: Selection and Compaction [32,33].

<u>Rule 1</u>: If components are connected in parallel and they are all in either tree edges or co-tree edges, then the test points at some component outputs are selected such that the corresponding columns of L (of equation (34)) are not identical.

<u>Rule 2</u>: Test points are selected so that the number of nonzero entries of each column of L is not less than t+1 for a t-diagnosable circuit design.

If the test points are selected by following the Selection Rules 1 and 2, then they are called the <u>dominate test points</u>.

<u>Rule 3</u>: If the number of distinct dominate test points is less than m, then the remaining test points can be selected arbitrarily.

Finally, the Compaction step is considered after the Selction step has been performed. Of course, if the number of distinct dominate test points is not greater than m, the desired number of test points, then it is not necessary to perform the Compaction step, otherwise, the test points are compacted such that the number of test points is reduced to m. To compact the selected test points, one may select an appropriate test point whose coefficients cover that of the dominate test points and hopefully cover as many of the test points selected by Selection Rule 2 as possible. Of course, during the Compaction procedure, the Selection Rules 1 and 2 are checked recursively.

Evidently, the previous development for single-port components can be extended to the case of multiple-port components. Let

$$L = \begin{bmatrix} L_{11} \\ L_{21} \end{bmatrix} = \begin{bmatrix} M_{11} & M_{12} & \cdots & M_{1r} \\ M_{21} & M_{22} & \cdots & M_{2r} \\ \cdot & \cdot & \cdots & \cdot \\ M_{s1} & M_{s2} & \cdots & M_{sr} \end{bmatrix}$$

where the block matrices M_{ij}'s contain the columns and rows of the corresponding components. Let w_i be the number of columns in the submtrix M_{1i}, $i=1,2,..,r$, then

$$\sum_{i=1}^{r} w_i = n,$$

where $w_i=r$ for the r-port components.

The definition of Q'-rank may be generalized to the block matrix M_{ij} for the multiple-port case. Under the definition of the generalized Q'-rank, it is easy to verify that Theorem 5 is still valid for the multiple-port components. For a t-diagnosable circuit design, the number of test points required is defined as the summation of w_i, i=1,2,..,t+1, which are the (t+1)'s highest number of the component ports in the circuit. Obviously, in the case of single-port components (w_i=1), the summation is equal to t+1.

6. SUMMARY

A simulation-after-test algorithm for the analog fault diagnosis problem is presented, in which a bound on the maximum number of simultaneous faults is used to minimize the number of test points required. Based on this self-testing algorithm, an analog automatic test program generator (AATPG) for both linear and nonlinear circuits and systems is being developed. The AATPG code is subdivided into off-line and on-line processes. The actual test can be run in either a fully automatic mode or interactively.

In order to further reduce the on-line compuational requirements, the adoption of parallel processing techniques to the proposed algorithm is addressed. In addition, the important issue of design of testability for analog circuits is discussed. Since the proposed testability condition is dependent only on the topological structure of a given circuit under test, not on the component values, hence, the proposed algortihm can be applied to design a diagnosable analog circuit.

REFERENCES

1. Saeks R., and S.R. Liberty, Rational Fault Analysis. Marcel Dekker, 1977.
2. Duhamal, P. and Rault, J.C., "Automatic Test Generation Techniques for Analog Circuits and Systems: A Review", IEEE Trans. on Circuits and Systems, Vol. CAS-26, pp. 411-440, 1979.
3. Plice, W.A., "Automatic Generation of Fault Isolation Test for Analog Circuit Boards: A Survey", presented at ATE EAST '78, Boston, Sept, 1978, pp.26-28.
4. Bandler, J.W., and A.E. Salama, "Fault Diagnosis of Analog Circuits," IEEE Proceedings, pp.1279-1325, August, 1985.
5. Liu, R.-W., Analog Fault Diagnosis, IEEE PRESS, 1987.
6. Liu, R.-W., and R. Saeks, (Editors), Report on the ONR Workshop on Analog Fault Diagnosis, Univ. of Notre Dame, Notre Dame, IN., May 1981.
7. Plice, W.A., "A Survey of Analog Fault Diagnosis", presented at the workshop on Analog Fault Diagnosis, Univ. of Notre Dame, Notre Dame, IN., May 1981.
8. Saeks, R., "A self-testing algorithm for analog fault diagnosis", presented at the workshop on Analog Fault Diagnosis, Univ. of Notre Dame, Notre Dame, IN. May 1981.
9. Saeks, R., "Criteria for Analog Fault Diagnosis", in Nonlinear Fault Analysis, Technical report. Texas Tech U., Lubbock, Texas, pp.19-28.
10. Saeks, R. and Singh, S.P. and R.-W. Liu, "Fault isolation via component simulation", IEEE Trans. on Circuit Theory, Vol.CT-19, pp.634-640, Nov. 1972.
11. Sen, N. and R. Saeks, "A measure of testability and its application to test point selection - theorey", Proc. 20th Midwest Symp. on Circuits and Systems, Lubbock, TX. August 1977.
12. Huang, Z.F., Lin, C.-S., and R.-W. Liu, "Node-fault Diagnosis and a Design of Testability", IEEE Trans. on Circuits and Systems, Vol. CAS-30, pp.257-265, May 1983.

150

13. Lin, C.S. Huang, Z.F., and R.-W. Liu, "Fault diagnosis of linear analog networks: a theorey and its application", Proc. IEEE Int'l Symp. on Circuits and Systems, pp.1090-1093, May 1983.

14. Biernacki, R.M., and J.W. Bandler, "Multiple-fault Location of Analog Circuits", IEEE Trans. on Circuits and Systems, Vol.CAS-28, pp.361-366, May 1981.

15. Trick, T.N., Mayeda, W., and A.A. Sakla, "Calculation of parameter value for fault detection in analog circuits", 1980 IEEE Int. Symp. on Circuits and Systems, pp. 1057-1077.

16. Rapisarda L. and R. DeCarlo, "Analog Multifrequency Fault Diagnosis", IEEE Trans. on Circuits and Systems, Vol. CAS-30, No. 4, pp.223-234, April, 1983.

17. Flecha, E. and R. DeCarlo, "The nonlinear analog fault diagnosis scheme of Wu, Nakajima, Wey, and Saeks in the tableau context", IEEE Trans. on Circuits and Systems, Vol. CAS-31, pp.828-830., September 1984.

18. Saeks, R. and R.-W. Liu, "Fault diagnosis in electronic circuits", in Nonlinear Fault Analysis, Texas Tech Univ., Lubbock, TX., pp.3-7.

19. Wu, C.-c., Nakajima, K., Wey, C.-L., and R. Saeks, "Analog Fault Analysis with Failure Bounds", IEEE Trans. on Circuits and Systems, Vol.CAS-29, No.5, pp.277-284, May 1982.

20. Wey, C.-L., and R. Saeks, "On the Implementation of an Analog ATPG: The Linear Case", IEEE Trans. on Intrumentation and Measurement, Vol. IM-34, pp.277-284, 1985.

21. Wey, C.-L., and R. Saeks, "On the Implementation of an Analog ATPG: The Nonlinear Case", Proc. IEEE International Symp. on Circuits and Systems, Montreal, Canada, pp.213-216, May 1984.

22. DeCarlo, R.A. and R. Saeks, Interconnection Dynamical Systems, Marcel Dekker, New York, 1981.

23. Nagel, L.W., SPICE2: A Computer Program to Simulate Semiconductor Circuit, Univ. of California, Berkeley, 1976.

24. NAP2: a Nonlinear Analysis Program for Electric Circuits. Version 2, Technical University of Denmark, Lyngby, Denmark, Dec. 1976.

25. Preparata, F.P., Metze, G., and R.T. Chien, "On the connection assignment problem of diagnosable systems", IEEE Trans. on Electronic Computers, Vol.EC-16, pp.448-454, 1967.

26. Amin, T., unpublished notes, Bell Laboratories, 1980.

27. Hakimi, L.S., "Fault analysis in digital systems - a graph theoretic approach", in Rational Fault Analysis (ed. R. Saeks and S.R. Liberty), Marcel Dekker, New York, 1977, pp.1-12.

28. Liu, R.-W., unpublished notes, Univ. of Notre Dame, Notre Dame, IN, 1980.

29. Wey, C.L., "Parallel processing for analog fault diagnosis", Proc. 27th Midwest Symp. on Circuits and Systems, Morgantown, WV. pp.435-438, June 1984.

30. Breurer, M.A. (editor) Digital System Design Automation: Languages, Simulation & Database. Computer Sciences Press, Inc. Potomac, MD. 1975.

31. Fennell, T.L. and T.A. Nicolino, "Computer Aided Testability", Proc. IEEE Annual Reliability and Maintainability Symp., pp.6-10.

32. Wey, C.L., "Design of Testability for Analog Fault Diagnosis", Proc. of IEEE 1985 Internatioanl Symp. on Circuits and Systems, Kyoto, Japan, June 1985. pp.515-518.

33. Wey, C.L., "Design of Testability for Analog Fault Diagnosis", International Journal of Circuit Theory and Application. Vol.15, No.2, pp.123-142, April 1987.

SPECTRAL TECHNIQUES FOR DIGITAL TESTING

JON C. MUZIO

VLSI DESIGN & TEST GROUP, DEPARTMENT OF COMPUTER SCIENCE
UNIVERSITY OF VICTORIA, VICTORIA, B.C., CANADA

1. INTRODUCTION

The normal approach to testing a combinational network is to identify a test set which can ensure the correct behaviour of the network with respect to some specific class of faults. The most frequent fault class is restricted to single lines in the network being stuck at 0 or 1. The complexity of finding a test set increases rapidly with the size of the network. Further, both the test set and the correct response vector must be stored at test time. This means that test sets are unlikely to be of use for built-in self-test.

Data compaction methods have been proposed to reduce the response vector to a reasonable size. For example all of (28) (80) and (21) present approaches of this type. However, these methods still use a test set and have the problem of storage of this set.

An alternative to the use of test sets is exhaustive testing where all input assignments are systematically applied to the network. In this case the input generator is straightforward, avoiding the storage requirement of a test set. In this paper, we consider the use of Rademacher- Walsh spectral coefficients in the testing of combinational networks. The requirement is for exhaustive testing, with the responses being collected by a simple accumulator in a compacted form.

The use of such coefficients was first suggested by Bennetts and Hurst (7) and has since been explored by a number of authors, including Susskind (86) and Hsaio and Seth (32). In the following sections, we give an overview of the thrust of recent results in this area, although all the proofs and detailed derivations of the results have been omitted. The reader is referred to the relevant literature for more explanation of each of the topics.

Savir (71) terms a fault in a combinational network syndrome testable if it causes the faulty function to contain a different number of ones in its truth vector from that of the fault-free function. The syndrome is just one particular coefficient from the Rademacher-Walsh spectrum which can be verified to ensure the correct behaviour of a circuit.

Our initial consideration concerns single stuck-at faults and methods for guaranteeing 100% fault coverage. The results are deterministic and indicate clearly whether a circuit is syndrome testable. If it is not, a number of possibilities exist. One approach involves hardware modification as proposed by Savir (71). Alternatively we can augment the syndrome with other spectral coefficients until complete coverage results.

A third approach uses constrained syndrome testing, in which some subset of the inputs is held constant while the remainder is exhaustively exercised. Savir (72) has shown that a set of constrained syndrome tests can always be found to cover any single output combinational network against single stuck faults. His approach, developed in the Boolean domain, forces untestable lines to be unate to ensure that they are testable. The spectral conditions derived below show that this requirement is unnecessarily stringent,the condition of unateness being sufficient, but not necessary, to ensure syndrome testability. The complete characterization of the necessary and sufficient conditions are much more easily derived in the spectral

F. Lombardi and M. Sami (eds.), Testing and Diagnosis of VLSI and ULSI, 151–180.

domain than in the Boolean domain and they are included below.

The question of multiple stuck faults and spectral signatures for their detection is discussed. The results presented are primarily based on the work of Lui and Muzio (45)and they show that a complete deterministic analysis of all multiple faults in a circuit of any size is computationally unreasonable. However, it is possible to define a number of straightforward signatures which giveb very good protection against almost all multiple faults.

There has been increasing interest recently in bridging faults (or short circuit faults). We give a complete spectral characterization of the requirements for coverage against such faults, taken primarily from (57) and (58).

The area where the methods show the greatest promise is for multiple output circuits and the ability that exists to take advantage of a signature that is compressed in both time and space. The generalized weighted spectral sum that is introduced is shown to be very powerful for providing good error coverage in a multiple output circuit. These methods were first reported by Serra (79). The practical application of these results to a special class of circuits, PLA's, leads to a method that guarantees complete fault coverage against all single stuck faults, all single contact faults and all single bridging faults with reasonable coverage against multiple faults. In rare cases there is the requirement for the addition of one product line and one output line to the PLA. A more detailed treatment of these methods has recently appeared in (78)

A brief final section gives a short summary of a recent probabilistic analysis of syndrome testing and a comparison between spectral methods and other data compaction schemes, taken from (60).

2. BACKGROUND

2.1 Notation

The results presented below require some theoretical background for their understanding. Let $f(X)$ be a Boolean function where $X=\{x_1,x_2,\ldots,x_n\}$ are the input variables, (x_n is the high order variable), and let Z be its column vector. The *weight*, $W(f)$, of $f(X)$ is the number of ones in the corresponding Z. The *syndrome*, $S(f)$, of $f(X)$ is the normalized weight, $S(f) = \dfrac{W(f)}{2^n}$. The *spectrum* of $f(X)$ is given by $R = T^n \cdot Z$ where T^n is the nth order transform of size $2^n \times 2^n$. The simplest form of t is given by the recursive definition based on the Hadamard transform as

$$T^0 = [1], \qquad T^n = \begin{bmatrix} T^{n-1} & T^{n-1} \\ T^{n-1} & -T^{n-1} \end{bmatrix}.$$

The Hadamard transform is one example of transforms based on the discrete Rademacher-Walsh functions; more detail about different forms of these spectral transforms can be found in (34, 35). The spectral domain is just an alternate domain in which a function can be represented, the transformation between the two domains being carried out by the orthogonal transform of choice (see Figure 1). As an example the transform matrices for T^0, T^1 and T^2 are as follows

$$T^0 = [1], \quad T^1 = \begin{bmatrix} 1 & 1 \\ 1 & -1 \end{bmatrix} \quad \text{and} \quad T^2 = \begin{bmatrix} 1 & 1 & 1 & 1 \\ 1 & -1 & 1 & -1 \\ 1 & 1 & -1 & -1 \\ 1 & -1 & -1 & 1 \end{bmatrix}.$$

The spectrum of a Boolean function is unique. The column vector R consists of the

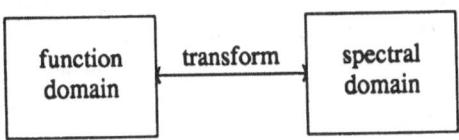

FIGURE 1: The Spectral Domain

spectral coefficients. We denote by $\mathbf{R}_f$ the spectrum of f(X). More details about spectra and their evaluation can be found in (35). The truth vector $\mathbf{Z}$ for the Boolean function is assumed to be ordered such that x_n is the high-order variable in the table defining f(X). We illustrate here an example for n=3.

Consider the function $f(x_1,x_2,x_3) = x_3x_2 + \bar{x}_3x_1$, whose truth vector is represented by $[0\,1\,0\,1\,0\,0\,1\,1]^t$, where x_3 is the high-order input. The spectrum of f is obtained by premultiplying the truth vector by the appropriate size transform $\mathbf{T}^3$. The computation and its result are shown in Figure 2, with the proper labeling of the coefficients.

As shown in Figure 2, the spectral coefficients are distinguished using subscripts identifying the variables directly correlated to them. Since the first row of $\mathbf{T}^n$ is all ones, $r_0=W(f)$. The coefficients $r_1,r_2,...,r_n$ are called first order coefficients and they measure the correlation of the function f(X) with the variables $x_1,x_2,...,x_n$ respectively. All the other coefficients measure the correlation with particular Exclusive-OR functions. For example, r_{12} compares

$$
\begin{bmatrix}
1 & 1 & 1 & 1 & 1 & 1 & 1 & 1 \\
1 & -1 & 1 & -1 & 1 & -1 & 1 & -1 \\
1 & 1 & -1 & -1 & 1 & 1 & -1 & -1 \\
1 & -1 & -1 & 1 & 1 & -1 & -1 & 1 \\
1 & 1 & 1 & 1 & -1 & -1 & -1 & -1 \\
1 & -1 & 1 & -1 & -1 & 1 & -1 & 1 \\
1 & 1 & -1 & -1 & -1 & -1 & 1 & 1 \\
1 & -1 & -1 & 1 & -1 & 1 & 1 & -1
\end{bmatrix}
\cdot
\begin{bmatrix}
0 \\ 1 \\ 0 \\ 1 \\ 0 \\ 0 \\ 1 \\ 1
\end{bmatrix}
=
\begin{bmatrix}
4 & r_0 \\
-2 & r_1 \\
-2 & r_2 \\
0 & r_{12} \\
0 & r_3 \\
-2 & r_{13} \\
2 & r_{23} \\
0 & r_{123}
\end{bmatrix}
$$

FIGURE 2: Example of Hadamard spectrum

$f(X)$ with $x_1 \oplus x_2$. A general coefficient is denoted by r_α, $\alpha \subseteq \{1,...,n\}$. Note that when the subset $\alpha = \phi$, the corresponding spectral coefficient is labeled r_0.

The transformation between the Boolean domain and the spectral domain does not change the information content; indeed, one can move freely between the two domains using the appropriate transform or its inverse. In our case the inverse transform is identical to the original except for the presence of a scaling factor. However, some properties which are very difficult to derive in the Boolean domain are straightforward in the spectral domain. The results themselves may then be moved back to the Boolean domain. Further, as we shall see below, the spectral coefficients themselves are very powerful for detecting faults in a network. The complexity of calculating a fast transform in Hadamard order is $n \cdot 2^n$, (35), by using the fast butterfly algorithm.

In order to explore the application of spectral testing some results about decompositions of spectra are necessary (see (35) for more details). The subfunctions of $f(x_1,...,x_n)$ are defined by $f_u(x_1,...,x_m) = f(x_1,...,x_m,u_1,...,u_{n-m})$ where $(u_1,u_2,...,u_{n-m})$ is the binary expansion of u, with u_1 the least significant bit. For example, for $f(x_1,x_2)$ where n=2, and for m=1, $f_0(x_1) = f(x_1,0)$ and $f_1(x_1) = f(x_1,1)$. The spectra of the subfunctions are defined analogously as $R_u = T^m \cdot Z_u$, where Z_u is the column vector for f_u. It is important to know the relationship between the spectrum of the function and the spectra of the subfunctions in the cases when m=n-1, and when m=n-2, that is for the decomposition for the nth input variable, or for the two high order variables x_n and x_{n-1}. They correspond to the Shannon decomposition in the Boolean domain.

Consider halving the vector Z of $f(X)$ as $Z = \begin{bmatrix} Z^0 \\ Z^1 \end{bmatrix}$ where Z^0 and Z^1 are the vectors for $f_0(x_1,...,x_{n-1}) = f(x_1,...,x_{n-1},0)$ and $f_1(x_1,...,x_{n-1}) = f(x_1,...,x_{n-1},1)$ respectively, so that

$$R_0 = T^{n-1} \cdot Z^0 \quad \text{and} \quad R_1 = T^{n-1} \cdot Z^1. \tag{1}$$

The spectrum R_f can be partitioned as in

$$R_f = \begin{bmatrix} R^0 \\ R^1 \end{bmatrix} = \begin{bmatrix} T^{n-1} & T^{n-1} \\ T^{n-1} & -T^{n-1} \end{bmatrix} \begin{bmatrix} Z^0 \\ Z^1 \end{bmatrix}$$

that is,

$$R^0 = T^{n-1}(Z^0 + Z^1) = R_0 + R_1 \quad R^1 = T^{n-1}(Z^0 - Z^1) = R_0 - R_1 \tag{2a}$$

which are equivalent to

$$R_0 = \frac{1}{2}(R^0 + R^1) \quad R_1 = \frac{1}{2}(R^0 - R^1). \tag{2b}$$

In the decomposition about two input variables, which is useful for the later discussion on bridging faults, the function $f(x_1,...,x_n)$ can be decomposed about inputs x_n and x_{n-1} as

$$f = \bar{x}_n \bar{x}_{n-1} f_0 + \bar{x}_n x_{n-1} f_1 + x_n \bar{x}_{n-1} f_2 + x_n x_{n-1} f_3$$

where $f_h = f(x_1, ..., x_{n-2})$, $0 \le h \le 3$, are functions of n-2 variables. This divides the truth vector into four parts, namely Z_0, Z_1, Z_2, Z_3, and the respective spectra are R_0, R_1, R_2, R_3, obtained by $R_k = T^{n-2} Z_k$, $0 \le k \le 3$. If the spectrum R of f is partitioned as

$$R = T^n \cdot Z = \begin{bmatrix} R^0 \\ R^1 \\ R^2 \\ R^3 \end{bmatrix}$$

it is straightforward to derive the following relationships; see (35, p.119) for details (the derivation uses the coding of {-1,+1} to represent {0,1} respectively).

$$R^0 = R_0 + R_1 + R_2 + R_3, \quad R^2 = R_0 + R_1 - R_2 - R_3, \tag{3a}$$

$$R^1 = R_0 - R_1 + R_2 - R_3, \quad R^3 = R_0 - R_1 - R_2 + R_3, \tag{3b}$$

and conversely

$$R_0 = \frac{1}{4}(R^0 + R^1 + R^2 + R^3), \quad R_2 = \frac{1}{4}(R^0 + R^1 - R^2 - R^3), \tag{3c}$$

$$R_1 = \frac{1}{4}(R^0 - R^1 + R^2 - R^3), \quad R_3 = \frac{1}{4}(R^0 - R^1 - R^2 + R^3). \tag{3d}$$

This partitioning process can be carried out for any value of $m < n-1$, where R and Z are partitioned into 2^{n-m} subvectors, as in

$$R = \begin{bmatrix} R^0 \\ R^1 \\ . \\ . \\ R^\beta \end{bmatrix}, \qquad Z = \begin{bmatrix} Z^0 \\ Z^1 \\ . \\ . \\ Z^\beta \end{bmatrix} \quad \text{where } \beta = 2^{n-m}-1.$$

The general result, first derived by Tokmen, (87), is

$$[R^0 R^1 \cdots R^\beta] = [R_0 R_1 \cdots R_\beta] \, T^{n-m} \quad \text{and} \quad [R_0 R_1 \cdots R_\beta] = \frac{1}{2^{n-m}} \, [R^0 R^1 \cdots R^\beta] \, T^{n-m}. \tag{4}$$

A small example may be useful.

Let $f(x_1,x_2,x_3) = x_1 x_2 x_3 + \bar{x}_3 x_2$, where x_3 is the high order variable and $Z = [0\,0\,1\,1\,0\,0\,0\,1]^t$. Its spectrum is calculated as $R = T^3 \cdot Z = [3\,{-1}\,{-3}\,1\,1\,1\,{-1}\,{-1}]^t$, (written here transposed), where $R^0 = [3\,{-1}\,{-3}\,1]^t$ and $R^1 = [1\,1\,{-1}\,{-1}]^t$. Using the Shannon decomposition about x_3 we have $f(X) = \bar{x}_3 f_0(x_1,x_2) + x_3 f_1(x_1,x_2) = \bar{x}_3(x_2) + x_3(x_1 x_2)$.

The corresponding spectra for the subfunctions are $R_0 = T^2 \cdot Z_0 = [2\,0\,{-2}\,0]^t$ and $R_1 = [1\,{-1}\,{-1}\,1]^t$, which could easily have been deduced directly from equations (2b).

2.2 Data compaction and spectral testing

It is important to be clear as to the aims of the testing process. Given a circuit with n inputs there is one correct function of n variables that is realized by the network. On the other hand, there are $2^n - 1$ other possible functions of n variables which might be realized by a faulty function. The purpose of any test is to distinguish between these two classes. Ideally, we need some mapping function that will map the single correct circuit to one value and each of the multitude of incorrect networks to a different value. All testing methods are an attempt to find a suitable mapping function. In practice, it is some approximation to a suitable function which is actually used, because otherwise the complexity of finding a suitable function and the complexity of using it for a test both become unreasonable.

The situation is further complicated in that it is obviously not possible for all $2^n - 1$ incorrect functions to actually be realized in the presence of some fault, but only a very small subset. It is this small subset that is actually of interest to us. Unfortunately, no-one has been able to provide a reasonable characterization of this subset of circuits which can be attained in the presence of a fault.

There are three different aspects of testing that we have to consider, and the relevance of the spectral approach to each one. First, there is the derivation of a test. This is a process that is carried out once, and is quite separate from the actual device to be tested. Secondly, there is the application of the test; this is the process which is carried out many times. We are prepared to put a large effort into the first operation if it might simplify the second; that is, it is reasonable to adopt a complex method to derive the test if, by so doing, we can simplify the actual test to be applied.

The third area is the investigation and analysis of the test to determine its efficacy and coverage. The spectral domain appears to be particularly useful for the first and third of these areas. In the second case, it is likely that the most useful application is in the restricted sense of syndrome testing and constrained syndrome testing, rather than the full generality of spectral testing.

The spectral coefficients were originally used in areas such as function classification and circuit design, (35), which applications are not pursued here. The first systematic discussion of their capabilities in fault detection is shown in (52) although the capabilities of two of the coefficients had been explored in (86). The coding induced by the transform gives a particular meaning to the values of the spectral coefficients. As noted above, r_0 corresponds to the weight of the function; r_1 adds the number of 1's in the part of the truth table when $x_1=0$ and subtracts the number of 1's in the part of the truth table when $x_1=1$. Similar comments apply to the other first-order coefficients, $r_2, \ldots, r_n$. By exploiting this coding, conditions have been proven by Miller and Muzio (52) for the syndrome testability, i.e. the r_0 testability of input lines or internal lines in a circuit, and for the general testability by a signature comprised of spectral coefficients. In the following sections they are presented separately with examples.

3. SPECTRAL TESTING OF STUCK-AT FAULTS

The results presented in this section come primarily from (52), and can also be found in (35). Susskind (86) first discussed the use of the last of the spectral coefficients for fault detection purposes and Carter (11) includes some results based on the parity of r_0.

We abbreviate a line b stuck-at 0 by b/0 and similarly for b/1. The results are stated here without proof.

THEOREM 1. $x_i/0$ and $x_i/1$ are testable by r_α, $\alpha \subseteq \{1,2,\ldots,i-1,i+1,\ldots,n\}$, and testable by $r_{\alpha i}$ if and only if $r_{\alpha i} \neq 0$.

The interesting special case follows when r_α is r_0, giving the result for syndrome testability.

COROLLARY 2. The input stuck-at faults $x_i/0$ and $x_i/1$ are syndrome testable (r_0 testable) if and only if $r_i \neq 0$.

As an example, consider the circuit in Figure 3, realizing $f = x_1 x_2 x_3 + x_4(\bar{x}_2 + \bar{x}_3)$. The spectrum is as follows:

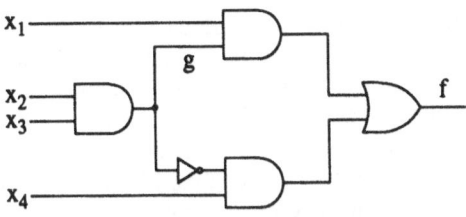

FIGURE 3: Example Circuit

r_0 r_1 r_2 r_3 r_4 r_{12} r_{13} r_{14} r_{23} r_{24} r_{34} r_{123} r_{124} r_{134} r_{234} r_{1234}

8; –2 0 0 –6; 2 2 0 0 –2 –2; –2 0 0 2; 0

For this circuit the analysis shows, without fault simulation, that stuck-at faults on inputs x_1 and x_4 are syndrome testable, while such faults on x_2 and x_3 are not. The complete results for general r_α coefficients are shown in Table 1.

This result gives a very straightforward condition to decide when an input stuck-at fault is

TABLE 1: Testability of input stuck-at faults

Faulty input	Testing Coefficients	Criterion
x_1	r_0, r_1	$r_1 \neq 0$
	r_2, r_{12}	$r_{12} \neq 0$
	r_3, r_{13}	$r_{13} \neq 0$
	r_{23}, r_{123}	$r_{123} \neq 0$
x_2	r_1, r_{12}	$r_{12} \neq 0$
	r_4, r_{24}	$r_{24} \neq 0$
	r_{13}, r_{123}	$r_{123} \neq 0$
	r_{34}, r_{234}	$r_{234} \neq 0$
x_3	r_1, r_{13}	$r_{31} \neq 0$
	r_4, r_{34}	$r_{34} \neq 0$
	r_{12}, r_{123}	$r_{123} \neq 0$
	r_{24}, r_{234}	$r_{234} \neq 0$
x_4	r_0, r_4	$r_4 \neq 0$
	r_2, r_{24}	$r_{24} \neq 0$
	r_3, r_{34}	$r_{34} \neq 0$
	r_{23}, r_{234}	$r_{234} \neq 0$

158

syndrome testable, without using the complex Boolean manipulations presented in (71). The results are general for r_α testability, so that a general spectral signature can be derived, without full fault simulation.

For internal lines, the function $f(X)$ must be rewritten to make the Boolean dependence on an internal line g explicit. Thus $f(X) = h(X, g(X))$, where $g(X)$ is the function realized by the line g, and h is a function of n+1 inputs. This is reflected in the model shown in Figure 4.

For the example of Figure 3 and the line labeled g, we have $g(X) = x_2 x_3$ and $h(X, g) = x_1 g + x_4 (\overline{x}_2 + \overline{x}_3)$. Note that, as this example shows, $g(X)$ and $h(X, g)$ may have common inputs and their realization may share gates. If the inputs to g and to the rest of the circuit are disjoint then simplified results may be used. Note that $g(X)$ and $h(X, g)$ are defined by the network in question and not by $f(X)$ alone. Either $g(X)$ or $h(X, g)$ may be independent of one or more x_i, $1 \le i \le n$. To prove that a line g is testable, by either the syndrome (r_0) or by some other spectral signature, some coefficients for the spectrum of function h are required, and the proofs rely on some properties of the decomposition of spectra as outlined above. Here only the statements of the conditions are given (see (52) for the proofs).

Let $\hat{R}$ be the spectrum of $h(X, g)$. Let $\hat{R}_0$ and $\hat{R}_1$ be the spectra of $h_0(X)$ and $h_1(X)$ respectively. The relations on decomposition of spectra given earlier apply between $\hat{R}_0$, $\hat{R}_1$, $\hat{R}^0$ and $\hat{R}^1$.

THEOREM 3. The fault g/0 is testable by r_α if and only if $r_\alpha \ne \dfrac{1}{2} (\hat{r}_\alpha + \hat{r}_{\alpha n+1})$.

The fault g/1 is testable by r_α if and only if $r_\alpha \ne \dfrac{1}{2} (\hat{r}_\alpha - \hat{r}_{\alpha n+1})$.

The statement of the Theorem depends on both R, the spectrum of the function realized by the network, and $\hat{R}$, the spectrum of the function realized by the residual network found by treating g as primary input. The factor of 1/2 arises because $h(X, g)$ is a function of n+1 variables, whereas $h(X, 0)$ and $h(X, 1)$ are functions of only n. Full details in (52).

As for input faults, it is useful to state explicitly the results for syndrome (r_0) testability.

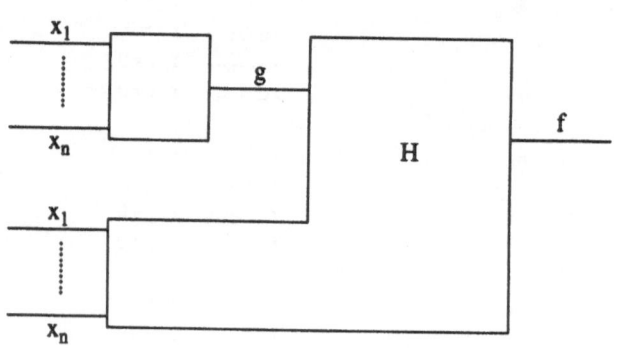

FIGURE 4: Internal Line Model

We follow common usage in using the term syndrome testable although the actual verification by a tester is of the weight (r_0), rather than the syndrome. Clearly syndrome testability and r_0 testability are isomorphic.

COROLLARY 4. The fault g/0 is syndrome testable if and only if $r_0 \neq \frac{1}{2}(\hat{r}_0 + \hat{r}_{n+1})$.

The fault g/1 is syndrome testable if and only if $r_0 \neq \frac{1}{2}(\hat{r}_0 - \hat{r}_{n+1})$.

For the same example of Figure 3, $r_0=8$, while $\hat{r}_0=17$ in the function h(X,g), and $\hat{r}_{n+1}=-5$ corresponding to the input g. Thus g/0 and g/1 are syndrome testable. Note that the detailed analysis of the spectral coefficients is needed only at design time when the testability of a line is evaluated. When the test is applied, only a counter and a comparator are necessary, as only the weight of the function is needed. This aspect is discussed in more detail below.

Two faults are functionally equivalent if they cause the network to realize the same faulty function. Only one fault from any class of functionally equivalent faults need be explicitly considered. When possible, it is advantageous to choose an input fault, single or multiple, in preference to an internal stuck-at fault since the testability of an input fault depends only on R. For an equivalence class containing no input fault, it is best to consider the fault closest to the network output since this will, in general, involve a simpler $\hat{R}$.

Of course, the evaluation of the relevant g and h functions for a large number of lines in a network would be very time consuming. Luckily, the evaluation is only required in a comparatively small number cases for most networks. This area is discussed after we have given a brief discussion of the coverage of multiple faults.

4. SPECTRAL TESTING OF MULTIPLE STUCK-AT FAULTS

A multiple-input stuck-at fault is a fault where one or more input lines are stuck. Single input stuck-at faults are a special case. In a network with p lines, there are 3^p-1 possible multiple faults. This large number makes any exhaustive analysis of all faults impractical. The fault is termed *unidirectional* if all stuck lines take the same value; otherwise it is termed mixed. The research on spectral signatures for multiple stuck-at faults is based primarily on (45), and a summary is given here without proofs.

Consider a k-multiple-input fault in a network realizing f(X) with spectrum **R**. Suppose $x_\lambda = \{x_{\lambda_1} \cdots x_{\lambda_k}\}$ are the stuck-at inputs with x_{λ_i} stuck at u_i, $u \leq i \leq k$, $u_i \in \{0,1\}$. Denote this fault by "x_λ s-a-u", where $u = \sum_{i=1}^{k} u_i 2^{i-1}$. This class of multiple-input stuck-at faults is sufficient, since all others are equivalent up to permutation of variables. Let α be a possibly empty string of variable labels from $\{1...n\}$ such that α and $\lambda = \{\lambda_1 \cdots \lambda_k\}$ are disjoint ($r_\alpha = r_0$ when α is empty). Let λ' be a nonempty subset of λ.

THEOREM 5. x_λ s-a-u is

a) $r_{\alpha\lambda'}$ testable if and only if $r_{\alpha\lambda'} \neq 0$.

b) r_α testable if and only if $r_\alpha \neq [r_\alpha \ r_{\alpha\lambda_1} \ r_{\alpha\lambda_2} \ r_{\alpha\lambda_1\lambda_2} \cdots r_{\alpha\lambda_1\cdots\lambda_k}] \ T^k_u$ where T^k_u is the (u+1)th column of T^k.

This result is not easy to apply in general and the more useful practical results come from the Corollaries. Unfortunately, these only apply in a restricted set of cases.

COROLLARY 6.

a) If $r_\alpha \neq 0$, r_α covers all multiple faults involving any x_i if $i \in \alpha$.

b) If $r_{1...n} \neq 0$, it is a signature for all 3^n-1 multiple-input faults.

c)   A sufficient condition for r_α to cover a k multiple-input fault is that the fault-free value of r_α is not divisible by 2^k.

d)   If r_0 is odd, any coefficient covers all multiple-input faults.

Note that although the condition in (c) is sufficient, it is not necessary. It can be shown by induction that a fan-out free network, without Exclusive-OR gates, with two or more inputs, must realize a function with an odd number of minterms. Since r_0 is odd for such a function, all spectral coefficients are odd. It is a property of the spectrum that all coefficients have the same parity. Since all stuck-at faults, single or multiple, in a fan-out free network, are functionally equivalent to single or multiple stuck-at input faults, it follows that they are all testable by r_α. It also known (11) that all input faults are testable by r_0 if r_0 is odd. The results above generalize this result for any r_α and for cases when r_0 is even.

The analysis of r_α testability for multiple-input faults outside the conditions of Corollary 6 can be tedious. However simplification is possible in some cases leading to more efficient results.

COROLLARY 7.

a) x_i s-a-0 and x_i s-a-1 are either both r_α testable or r_α untestable.

b)   The 2^n n-multiple faults involving all of $x_1 \cdots x_n$ are either all r_α testable or all r_α untestable.

For the same example from above, in Figure 3, each of $\{r_1, r_4, r_{12}, r_{13}, r_{24}, r_{34}, r_{123}, r_{234}\}$ test all multiple faults involving two or more inputs, while r_0 tests those faults involving all four inputs. While it is possible to derive some elegant results for particular networks, such as the following for fanout free tree circuits, the treatment of internally fanout free networks and general irredundant circuits becomes increasingly complex. The interest in the result below is its inclusion of XOR/XNOR gates in the structure and we note that the fault(s) need not be functionally equivalent to input stuck-at faults. The reader is referred to (45) for more details.

THEOREM 8.   For a tree network composed of NAND, NOR, AND, OR, NOT and k two-input XOR/XNOR gates $(k \geq 0)$, we have

a) $r_{1 \cdots n} \neq 0$

b) $r_{1 \cdots n}$ is a multiple of 2^k but not of 2^{k+1}

c) $r_{1 \cdots n}$ covers all multiple faults in the network.

5. ALGORITHMS AND IMPLEMENTATION

We consider the testability of complete networks and the alternatives for dealing with untestable faults. Unate lines play an important role. Any line in a network can be expressed in the form $f(X) = A(X)g(X) + B(X)\bar{g}(X) + C(X)$, where $A(X)$, $B(X)$ and $C(X)$ are independent of $g(X)$, the function realized by the network line labeled g. In this context, g may be either an input or an internal line. If g is a positive line then $f(X) = A(X) g(X) + C(X)$. g/0 yields $f^*(X) = C(X)$, while g/1 yields $f^*(X) = A(X) + C(X)$. Both these faulty functions have different weights from $f(X)$, unless g is redundant, and thus g/0 and g/1 are syndrome (r_0) testable. A single stuck-at fault on a unate line is always syndrome testable.

Consequently, it is useful to identify, a priori, the unate lines. Complete results are in (53). Here we present the algorithm. The following procedure identifies the unate lines in an irredundant single-output combinational network. The algorithm labels each line as P, positive, N, negative, or C, a candidate for r_0 untestability. If there is redundancy, the procedure may label certain unate lines as being candidates for syndrome untestability. These lines

161

should correctly be identified as syndrome testable.
   Procedure:

(1) Label the network output P.

(2) For each gate whose output line is labeled but whose input lines are not, do one of the
     following:
     a) AND, OR gates: label all input lines identical to output line;
     b) NOT, NAND, NOR gates: label the input lines P if the gate's output line is labeled
        N, N if the gate's output line is labeled P, and C if the gate's output line is labeled C;
     c) EXOR, EXNOR gates: label each input C.

(3) For each fanout point whose branches are labeled but whose stem is not, construct the
     stem label from the branch labels using the following commutative and associative
     operator:

	P	N	C
P	P	C	C
N	C	N	C
C	C	C	C

(4) Repeat steps 2 and 3 until all network lines are labeled.

Consider the network of Figure 5, first presented in (68), here labeled according to the marking algorithm. This circuit realizes a function with an odd number of ones, so all input
stuck-at faults, both single and multiple, are r_0 testable. There are only 6 internal lines
which are candidates; in the figure they are labeled as c1 ... c6. Of these, c1/1, c3/1 and c5/1
are r_0 testable since they are functionally equivalent to input stuck-at faults. So are c1/0, c3/0
and c5/0. Faults on c4 and c6 are syndrome testable since they result in a faulty function
independent of some x_i. c/0 is syndrome testable since it is functionally equivalent to c3/0.
c/1 is the only fault which must be explicitly considered using Corollary 4. It is found to be

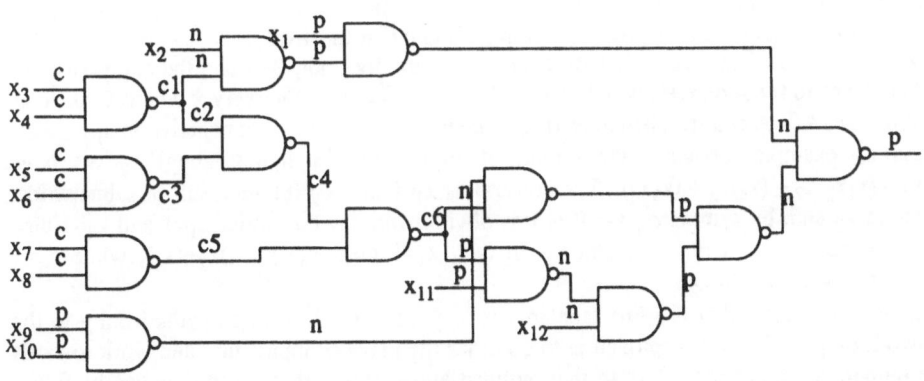

FIGURE 5: Marking Algorithm

syndrome testable. Hence the network of Figure 5 is syndrome testable for all single stuck-at faults.

The discussion so far implies that $h(X,g)$ must be explicitly determined before $\hat{R}$ is found. This is not the case. $\hat{R}$ can be determined directly from the network. In addition, while the various $\hat{R}$ required are distinct, properly ordering their computation allows the sharing of partial results (see (35, 55)).

Obviously, the preference is for a network that is syndrome testable. When a network fails to be syndrome testable, there are three possible courses of action. First, is the construction of a spectral signature by augmenting the syndrome with other coefficients, using the results of Theorems 1 and 3. Of course, each such additional coefficient increases the testing time required, unless some parallel arrangement is designed for accumulating the coefficients. Otherwise, each new coefficient added to the signature requires a complete pass with all inputs exhaustively exercised. A second possibility is the hardware modification of the circuit to embed it in a larger network which is syndrome testable. This requires the addition of a number of extra control inputs and some gating to the network. The third alternative is probably the most promising and is explored in more detail in the next section. It is based on constraining a subset of the input variables to fixed values while exhaustively exercising the remainder.

6. CONSTRAINED TESTING OF STUCK-AT FAULTS

Constrained syndrome testing takes the opposite approach to hardware modifications for testability. A network containing syndrome untestable faults is partitioned and subsequently tested as a number of, not necessarily disjoint, subnetworks. The network need not be modified to make it testable. In particular, the number of network inputs does not increase and, consequently neither does the length of time taken to exhaustively exercise all of them. Embedding the circuit in a larger network by the addition of extra control inputs doubles the test time for each such input added. A mechanism for controlling the constraints is, of course, required.

Consider a network realizing $f(X)$ with a syndrome untestable input x_i. Expanding about x_i yields $f(X) = Ax_i + B\bar{x}_i + C$, where A, B, and C are Boolean expressions independent of x_i. Since faults on x_i are syndrome untestable, $f(X)$ is nonunate in x_i, so $A \neq 0$ and $B \neq 0$. Let $\hat{x}$ denote a constraint on x found by setting certain x_i inputs to constant values. Let $\hat{A}$, $\hat{B}$, and $\hat{C}$ be the expressions which $\hat{x}$ produces from A, B and C respectively.

There must exist an $\hat{x}$ such that $\hat{B}=0$, since, if there does not, $B=1$ and the expression for $f(X)$ reduces to $f(X)=A+\bar{x}_i+C$ which is negative in x_i. Suppose for every $\hat{x}$ where $\hat{B}=0$, $\hat{A}=0$. In this case, $AB=A$ and $f(X)$ becomes $f(X)=AB+B\bar{x}_i+C$ which is again negative in x_i.

As an example, consider the circuit of Figure 6. Expanding about x_2 we have $f(X) = (x_1x_3)x_2 + (x_4)\bar{x}_2 + (\bar{x}_3x_4)$. The constraints $x_1=0$ and $x_4=0$ both result in subnetworks unate in x_2 and, by symmetry, x_3. It is less obvious that the constraint $x_1=1$ and $x_4=1$ also yield subnetworks unate in x_2 and x_3. In fact, $x_1=1$ gives $f(\hat{X})=(x_3)x_2+(x_4)$, while $x_4=1$ makes $f(\hat{X})$ negative in x_2.

It should be clear that if a line labeled g is unate, so are all lines on paths from g to the network output. The best approach is to consider the network inputs first and work towards the output. A similar procedure to that outlined above shows that constraints can be found when one consider a syndrome untestable internal line (53). It is important to know how to select an input constraint. In practice it appears to be rather straightforward by inspection, but difficult to formalize into an algorithmic process without resorting to exhaustive search. This quickly becomes infeasible as the complexity of the network increases.

One can develop a less restrictive and more flexible spectral requirement. A constrained

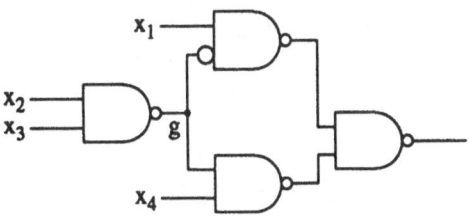

FIGURE 6 : Example for Constrained Testing

syndrome test is a syndrome test of a subfunction of f(X). The derivation and proofs of the spectral characterization for constrained testing are based on the decomposition of spectra induced by the testing subfunctions, leading to the following.

THEOREM 9. $x_t/0$ and $x_t/1$ are r_0 (syndrome) testable with respect to the input constraint $x_{i^s}=u_s$, $1 \le s \le n-k$, with no $i^s=t$, if and only if

$$[r_t \, r_{\gamma^1,t} \, r_{\gamma^2,t} \cdots r_{\gamma^\beta,t}] \, T^{(n-k)}{}_u \ne 0$$

where γ^p, $1 \le p \le \beta$, are the nonempty subsets of $\{i^1,i^2,...,i^{n-k}\}$ ordered so that $\gamma^1=\{i^1\}$, $\gamma^2=\{i^2\}$, $\gamma^3=\{i^1,i^2\},..., \gamma^\beta=\{i^1,i^2,...,i^{n-k}\}$.

Consider the example in Figure 6, with $f(X)=x_1x_2x_3+x_4(\bar{x}_2+\bar{x}_3)$. We found earlier that single stuck-at faults on x_2 and x_3 are not r_0 testable. Consider the constraint $x_1=0$ with respect to faults on x_2. In this case, the left hand side of Theorem 9 becomes $[r_2 r_{12}] \, T^{(1)}{}_0$ which from the spectrum listed earlier yields $[0 \ 2] \, [1 \ 1]^t=2$. Hence by Theorem 9, $x_2/0$ and $x_2/1$ are syndrome testable with respect to the input constraint $x_1=0$. The constraint $x_1=1$ gives $[r_2 r_{12}] \, T^{(1)}{}_1$ yielding $[0 \ 2] \, [1 \ 1]^t=-2$. Hence faults on x_2 are syndrome testable for the constraint $x_1=1$.

The corresponding theorem for internal lines is of a similar style with both results being rather difficult to use extensively. However, following from the general result there are direct results that show showing the connection between constrained syndrome testing and r_α testing. These are of much more use in deriving useful constraints for testing circuits.

Let $q_\alpha = \hat{r}_\alpha + \hat{r}_{\alpha(n+1)} - 2r_\alpha$.

THEOREM 10. $g/0$ is syndrome testable with respect to the contraint $x_{i^s} = u_s$, $1 \le s \le n-k$, if and only if $[q_0 \, q_{\gamma^1} \, q_{\gamma^2} \cdots q_{\gamma^\beta}] \, T^{(n-k)}{}_u \ne 0$, where the γ^i are defined as in Theorem 9.

Having established a spectral characterization of constrained syndrome testability, we now consider the connection between constrained testing and spectral coefficient testability. Suppose x_i is not a syndrome testable input. Let r_α be a nonzero spectral coefficient which does cover x_i such that $|\alpha|$, the cardinality of α, is minimal. Such a coefficient must exist since if it does not, faults on x_i are not testable and x_i is redundant. Using the conditions of Theorem 9, we can state the following.

THEOREM 11. If $x_i/0$ and $x_i/1$ are testable by r_α with $|\alpha|$ minimal, $i \notin \alpha$, they are also syndrome testable for every constraint on all the variables included in α.

A similar condition can also be stated for an internal line g.

The theorem provides a simple method to identify which inputs to constrain to make a particular fault syndrome testable. On an individual basis the choice of constraint is arbitrary. In combination the constraint must be selected so that all single stuck-at faults are tested. As an example consider the circuit in Figure 7, which realizes $f(X) = x_1\bar{x}_3\bar{x}_4 + x_2x_3\bar{x}_4 + \bar{x}_1\bar{x}_3x_4 + \bar{x}_2x_3x_4$, with the spectrum:

$$r_0 \ r_1 \ r_2 \ r_3 \ r_4 \ r_{12} \ r_{13} \ r_{14} \ r_{23} \ r_{24} \ r_{34} \ r_{123} \ r_{124} \ r_{134} \ r_{234} \ r_{1234}$$

$$8; \ 0 \ 0 \ 0 \ 0; \ 0 \ 0 \ -4 \ 0 \ -4 \ 0; \ 0 \ 0 \ -4 \ 4; \ 0$$

Since $r_i = 0$, $1 \le i \le 4$, all single input stuck-at faults are syndrome untestable. Given that $r_{24} \ne 0$ and $r_{14} \ne 0$, we can use a constraint on x_4 to test faults on x_1 and x_2. The situation can be summarized as shown below.

constrain	to be tested	condition
x_4	x_1, x_2	$r_{24} \ne 0, r_{14} \ne 0$
x_1 and x_4	x_3	$r_{134} \ne 0$
x_1 or x_2	x_4	$r_{14} \ne 0, r_{24} \ne 0$

Each application of a constraint makes a line unate and thus testable. We choose actual assignments of values for the constraints to ensure that all internal lines are also covered by this set if possible. In this case a reasonable set of constraints to apply (which test for all stuck-at faults in the circuit) are $x_4=0$, $x_1=0$; $x_4=0$, $x_1=1$; $x_4=1$, $x_1=1$. Larger examples and proofs can be found in (53).

7. SPECTRAL TESTING OF BRIDGING FAULTS

The fault model adopted in practice is usually restricted to stuck-at faults, and mostly to single ones. The introduction of new technologies has presented test engineers with different problems, not covered by the stuck-at fault model. In the class of permanent faults, bridging faults or, more restrictively, stuck-at-neighbour faults, are acquiring new importance. Here we give consideration to such faults and motivation for their study; we derive

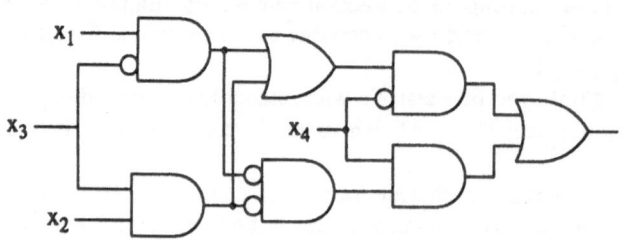

FIGURE 7 : Example 2 for Constrained Testing

some theoretical results for their testability by compaction, specifically by syndrome or spectral testing.

A single *Bridging Fault* (BF) is defined as the fault occurring when two normally uncon-nected lines x, y, in a circuit are shorted together. The function Z(x,y) is performed at the connection by the BF, and Z assumes either the function AND(x,y) or OR(x,y) depending on the technology. This is an assumption based on most technologies (61), and it appears to exclude CMOS, where the effect of a short can produce an indeterminate effect. There are two main types of bridging faults:

(1) Non-Feedback Bridging Faults (NFBF): there exists no path between the lines x and y and thus no feedback loop is created. This is the type of bridging fault on which we focus in this research.

(2) Feedback Bridging Faults (FBF): if there exists at least one path between the lines x and y, then the BF(x,y) creates a feedback loop and changes the combinational circuit into a sequential one.

It is important to determine the size of the testing problem itself. Given a circuit with k lines, there are 2k possibilities for single stuck-at faults, and testing simulators must expli-citly cover this number, or the subset obtained after fault collapsing. The number k is a count of all line segments, including separate segments for stems and branches in a fan-out situation. A bridging fault does not need this separation since the signal performed by the wired logic on any part of the fan-out is carried both forward and backward to stems and branches. Thus the number q of segments to be considered is less than k; q = k in a fan-out free network. The number of single bridging faults is $\binom{q}{2}$. This number is usually bigger than the number 2k of single stuck-at faults. If the number of fan-outs is minimized in a cir-cuit, q becomes much closer to k, and the number of single bridging faults becomes much bigger than the number of single stuck-at faults. The possible number of multiple BF's (i.e. involving more than two lines) is proven to grow faster than $q^{q/2}$ and a more accurate analysis is given in (81). Based on these numbers, it is infeasible to do an explicit simulation to cover all BF's and generate appropriate test sets.

It is useful to define *stuck-at-neighbour faults* as a subset of single bridging faults: these are single bridging faults between two adjacent line segments. The definition assumes that the actual layout of the circuit is known. In a stuck-at-neighbour fault two neighbouring lines x,y are seen to evaluate to the same result induced by the Z(x,y). If the set of single bridging faults to be considered is restricted to stuck-at-neighbour faults, the number is much smaller, depending entirely on the layout. Although it remains large, it is computa-tionally feasible. In this paper only stuck-at-neighbour faults are explicitly considered; the name "bridging fault", or BF, is used to mean a single bridging fault between adjacent lines.

In the current testing environment, the coverage of bridging faults is often left to the optimistic belief that the effect of a BF is usually covered by some vector in the stuck-at fault test set. The only BF's explicitly examined may be the few between adjacent input pins on an IC. While this attitude is defensible considering the difficulty of the problem, it is only reasonable if the coverage does carry over. Unfortunately, this is often not the case. A further problem arises as it can be shown (42) that undetectable BF's can invalidate a test set for stuck-at faults. A bridging fault is defined to be *undetectable* when the faulty circuit pro-duces the same output vector as the fault-free circuit. However, it is possible for an undetectable bridging fault to change the coverage of the test vectors for stuck-at faults.

We introduce some new work for the compaction testing of BF's with spectral testing, and in particular syndrome testing. The criteria which are developed below provide deterministic conditions for the testability of NFBF's. As in the case of stuck-at faults, the conditions are to be evaluated at design time only and their overhead in computation is contained at that stage. At test time only a counter with the signature of choice is used. These conditions can be applied in parallel with the ones for stuck-at faults. Thus a signature for stuck-at faults can be augmented if necessary, or it can be verified whether an existing one provides adequate coverage for bridging faults as well. The simpler case of input lines is considered first, followed by a number of criteria for internal lines. The results for internal lines are computationally expensive if a large number of adjacent lines have to be checked. Hence the second part of the discussion focuses on reducing the number of lines to be examined and how to apply such results to guidelines for design for testability. It is shown that the marking algorithm presented in (35) can be extended to predetermine which pairs of lines are susceptible to testing problems.

Testability criteria for bridging faults between two primary input lines, x_i and x_j, are proved in (35), and Theorem 12 is taken from that source.

THEOREM 12. For input lines x_i, x_j, the AND–BF(x_i,x_j) is:

(a) testable by r_α and testable by $r_{ij\alpha}$ if and only if $r_{i\alpha} + r_{j\alpha} + 2r_{ij\alpha} \neq 0$

(b) testable by $r_{i\alpha}$ if and only if $r_{i\alpha} - r_{j\alpha} \neq 0$,
where i,j $\notin \alpha$.

COROLLARY 13. For input lines x_i, x_j, the AND–BF(x_i,x_j) is syndrome testable (r_0 testable) if and only if $r_i + r_j + 2r_{ij} \neq 0$.

THEOREM 14. For input lines x_i, x_j, the OR–BF(x_i,x_j) is:

(a) testable by r_α and testable by $r_{ij\alpha}$ if and only if $r_{i\alpha} + r_{j\alpha} - 2r_{ij\alpha} \neq 0$

(b) testable by $r_{i\alpha}$ if and only if $r_{i\alpha} + r_{j\alpha} \neq 0$,
where i,j $\notin \alpha$.

COROLLARY 15. For input lines x_i, x_j, the OR–BF(x_i,x_j) is syndrome testable (r_0 testable) if and only if $r_i + r_j - 2r_{ij} \neq 0$.

The importance of these results is in their simplicity and direct similarity to the comparable results for stuck-at faults (35). They follow the philosophy mentioned earlier; at design time we want to minimize the amount of extra work to verify the testability of bridging faults by a particular coefficient. After having found a signature for stuck-at faults, composed of one or more spectral coefficients, it is straightforward to evaluate its coverage for input bridging faults. In particular no simulation is required and the results are completely deterministic.

The above results are, of course, only concerned with the testability of the bridging faults. As shown earlier, some BF's are undetectable, that is the presence of the fault has no effect on the output vector produced by the circuit. If a BF is undetectable, then the conditions above lead to an untestability result. However, it is not known whether the fault is untestable by the specific coefficient considered or if it is functionally undetectable. Since the undetectability of a BF has been shown to pose problems for test sets, we deduce here straightforward results to detect this situation.

THEOREM 16.

(a) The AND–BF(x_i,x_j) between input lines x_i, x_j is undetectable if and only if $R^1 = R^2 = -R^3$.

(b) The OR–BF(x_i,x_j) between input lines x_i, x_j is undetectable if and only if $R^1 = R^2 = R^3$.

As expected, the testability of internal lines for bridging faults is much more complex than that for input lines. We limit our discussion to NFBF, that is adjacent lines with no paths between them. A functional model is required in order to express the output of the circuit explicitly in terms of the lines under evaluation.

Consider two internal lines labeled g_1 and g_2. The function $f(x_1,...,x_n)$ may be written as $f(X) = h(X,g_1(X),g_2(X))$, where $X=(x_1,...,x_n)$ and $h(X,g_1,g_2)$ is formally considered as a function of n+2 inputs. We can partition h into four subfunctions:

$$h_0 = h(x_1,...,x_n,0,0), \quad h_1 = h(x_1, \ldots , x_n,1,0),$$

$$h_2 = h(x_1,...,x_n,0,1), \quad h_3 = h(x_1, \ldots , x_n,1,1),$$

where $h = h(x_1, \ldots , x_{n+2})$, and $x_{n+1}=g_1(X)$, $x_{n+2}=g_2(X)$ are the high order inputs. The truth vectors for h_0, h_1, h_2, h_3 are denoted by H_0, H_1, H_2, H_3, and for g_1, g_2 by G_1, G_2 respectively. Note that each of these vectors contains exactly 2^n entries.

For the derivation of the testability conditions for internal lines a number of spectral results are required. The interested reader is referred to (58, 59, 79), where all the Lemmata and proofs are given in full detail. We use '$*$' to denote the convolution sum, defined below, and J the 2^nx1 column vector such that $J_0 = 2^n$ and $J_i = 0$ ($1 \le i \le 2^n-1$). For convenience the abbreviations $N=2^n$, $N'=\dfrac{1}{2^n}$, $N''=(N')^2$ are used. Lastly, '$\circ$' is the *Hadamard product* of two matrices i.e. the componentwise product of the entries so that if $A_i = R_iS_i$ for all i, then we write $A = R \circ S$. The *convolution sum*, or '$*$' product (44), is defined for two N-place vectors R and S by

$$P = R * S, \quad \text{where} \quad P_i = \sum_{j=0}^{N-1} R_{i \oplus j} S_j, \quad i=0,...,N-1.$$

The aim of this section is to obtain some form of result for the testability of a bridging fault of the type

$$\hat{f}_\alpha \ne l\,(r_\alpha,r_\beta,...)$$

that is, a spectral coefficient calculated for the circuit under fault, $\hat{f}_\alpha$, being different from some linear combination l of spectral coefficients for the fault-free circuit. If l is not too complex, the resulting testability criteria would be computationally easier than simulation or Boolean testing. We derive a number of different forms of the results.

THEOREM 17. $R_f - R_{\hat{f}} = T^nW$ where

$$W = [G_1 \circ G_2 \circ (2H_0 - H_1 - H_2) + G_1 \circ (H_1 - H_0) + G_2 \circ (H_2 - H_0)].$$

This is the first form of a testability condition of the type $R_X \ne 0$, where X is not a Boolean function. Note that when the left hand side of Theorem 17 is 0, then $H_0 = H_1 = H_2$ and the bridging fault is undetectable as expected. For syndrome testability all that is required is for the first row of the result to be nonzero. For this the transform need not be evaluated; the relevant value is just the sum of the entries of W. We state here explicitly the result for syndrome testability.

COROLLARY 18. The AND–BF(g_1,g_2) is syndrome testable if and only if $\sum_{i=1}^{2^n} W_i \ne 0$.

It is useful at this point to look at an example where Corollary 18 and the input testability criteria are applied. Consider the circuit of Figure 8, implementing

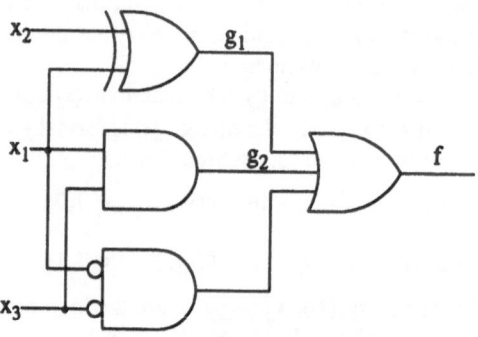

FIGURE 8: Example Circuit

$F(x_1,x_2,x_3)=x_1\bar{x}_2+\bar{x}_1x_2+x_1x_3+\bar{x}_1\bar{x}_3$, whose spectrum is

$$\begin{bmatrix} r_0 & r_1 & r_2 & r_{12} & r_3 & r_{13} & r_{23} & r_{123} \\ 6 & 0 & 0 & -2 & 0 & 2 & 2 & 0 \end{bmatrix}$$

For the input bridging fault between x_1 and x_2,

$r_1+r_2+2r_{12} = -4 \neq 0$, and so it is syndrome testable (Corollary 13).

$r_{13}-r_{23} = 0$, and it is not r_{13} testable (Theorem 12).

For a bridging fault between internal lines labeled g_1 and g_2, we have

$f(x_1,x_2,x_3) = h(x_1,x_2,x_3,x_4,x_5)$ where

$h(x_1,x_2,x_3,x_4,x_5) = x_4+x_5+\bar{x}_1\bar{x}_3$

$x_4 = g_1(x_1,x_2,x_3) = x_1\bar{x}_2+\bar{x}_1x_2$

$x_5 = g_2(x_1,x_2,x_3) = x_1x_3$.

Note that while the proof of the results is carried out in the spectral domain, the application is entirely in the integral domain. For syndrome testability we need the sum of the entries in the computed vector W to be nonzero. To simplify the computation, let $A = G_1 \circ G_2 \circ (2H_0 - H_1 - H_2)$, $B = G_1 \circ (H_1 - H_0)$, $C = G_2 \circ (H_2 - H_0)$, and $W = A + B + C$. In the following table the truth vectors for x_1, x_2, x_3, are shown, together with the computed vector for $H_0, \ldots, H_3$, and for G_1, G_2, A, B, C.

x_1	x_2	x_3	H_0	H_1	H_2	H_3	H_1-H_0	H_2-H_0	G_1	G_2	A	B	C	W
0	0	0	1	1	1	1	0	0	0	0	0	0	0	0
0	0	1	0	1	1	1	1	1	0	0	0	0	0	0
0	1	0	1	1	1	1	0	0	1	0	0	1	0	1
0	1	1	0	1	1	1	1	1	1	0	0	1	0	1
1	0	0	0	1	1	1	1	1	1	1	-2	1	1	0
1	0	1	0	1	1	1	1	1	0	0	0	0	0	0
1	1	0	0	1	1	1	1	1	0	0	0	0	0	0
1	1	1	0	1	1	1	1	1	0	1	0	0	1	1

For syndrome testability, Corollary 18 gives $\sum_{i=1}^{8} W_i = 3 \neq 0$, and the AND–BF$(g_1,g_2)$ is syndrome testable. The full transform can also be calculated as $R_f - R_{\hat{f}} = T^n \cdot W = [\ 3\ -1\ -1\ 3\ -1\ -1\ -1\ -1\ ]^t$; in fact the AND–BF$(g_1,g_2)$ is also testable by any other r_α coefficient.

There are a number of alternative forms for giving essentially the same result, of which the form below is interesting in that it illustrates a result which is in the Boolean domain although it was derived in the spectral domain. It is interesting to compare this result with that derived by Savir (71) in the Boolean domain for single stuck-at faults.

COROLLARY 19. An AND bridging fault between g_1 and g_2 is syndrome testable if and only if $S(h_1 g_1 \bar{g}_2) + S(h_2 \bar{g}_1 g_2) \neq S[h_0(g_1 \oplus g_2)]$.

In this criterion, S(f) is the normalized syndrome; in practice only the unnormalized weights need to be computed by the testing hardware, with the proviso that all subfunctions are formally considered to have n inputs. Note that Corollary 19 could have been derived directly from the Shannon decomposition by working in the Boolean domain only. As a last check, we note that if lines g_1, g_2 are replaced by primary input lines, Corollary 19 reduces to Corollary 18 as expected.

A very different form of the result is the one below, which is entirely in the spectral domain.

COROLLARY 20. $R_f - R_{\hat{f}} = R_{g_1 g_2 + g_2 h_2 + g_1 h_1} - R_{g_1 g_2 + g_1 h_0 + g_2 h_0}.$

The derivation of these conditions, their direct applicability, their computational complexity are discussed in (58, 59).

8. MULTIPLE-OUTPUT SPACE COMPACTION

The focus in this section is on testing multiple-output circuits in parallel, that is with a single signature computed at the same time for all outputs, as opposed to having separate signatures to be compared for each of them. We call this *space compaction* to distinguish from *time compaction* which is the actual compaction to produce a signature of length k bits from a stream of up to 2^n bits. The only existing results within this framework are based on multiple-input feedback shift registers (MISR), (26, 27). The research presented here can be found with all the relative proofs and derivations in (77, 79). We also show that the space compacted counting signature is excellent for certain types of circuits, specifically PLA's and ROM's. For them, full testability of all single faults is guaranteed.

Space compaction implies using a compaction method on the output vectors both in time and in space. In a circuit with n inputs and m outputs, there are m output vectors each of length 2^n to be tested. The *time compaction* applied by an LFSR or by a syndrome counter reduces them to m vectors of length r each, where $r \ll 2^n$, after 2^n time cycles. For an LFSR, r is usually 16 bits, for a syndrome counter r=n+1 bits. The *space compaction* testing proposed here reduces the m vectors of length r to one vector of length k, where the upper bound on k depends on the chosen method. Typically we would expect k to be of the order of m+n. Figure 9 gives a schematic view of the processes which are applied in parallel in order to conserve both time and space for the tester.

The space compacted signature proposed here tries to achieve full testability with the simplest coding possible. Specifically the aim is to use as a tester a weighted sum of function weights. This can be considered as the most general case, with other counting schemes, like syndrome testing, are a subset of it.being special cases. As we are considering networks of many outputs, it is necessary to identify the vectors and spectra for the individual outputs. Square brackets $[Z]_j$ and $[R]_j$ indicate respectively the truth vector and the spectrum for the jth output in the network. We define the *weighted spectral sum (WSPS)* for a network of m outputs as

$$K = \sum_{j=1}^{m} w_j\, T^n\, [Z]_j = \sum_{j=1}^{m} w_j\, [R]_j$$

with individual entries in K denoted by k_α. The weights w_j are positive nonzero integers. Note that the first coefficient of the weighted spectrum, k_0, is the weighted sum of the first coefficients of the individual spectra for each function $[f]_j$, since the first row of T^n is all 1's. Thus k_0 is the weighted sum of the weights of all functions in the network. Formally the *weighted syndrome sum* (WSS) is the normalized value of this coefficient,

$$WSS = \frac{1}{2^n}k_0 = \frac{1}{2^n}\sum_{j=1}^{m} w_j[W(F)]_j.$$

The WSPS as defined earlier is the unnormalized weighted sum of the spectra, and thus its first coefficient, k_0, is the unnormalized weighted sum of the function weights. Often however the two terms "WSS" and "syndrome sum" are also informally used to denote k_0. In practice, of course, it is this latter value that is computed by the tester. The testability properties of the WSS and k_0 are isomorphic, and the conditions hold for both. Formally, for all derivations, we must always consider all functions to depend on n inputs, and all $[Z]_1, \ldots, [Z]_m$ to be vectors of 2^n entries. In fact not all corresponding functions may functionally depend on all input variables. In this setting, $2^n \cdot WSS = k_0$, so that the WSS is a

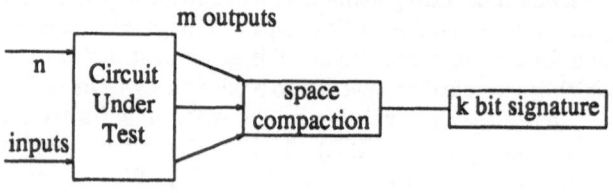

FIGURE 9: Space Compaction

real number while k_0 remains a larger integer.

The testability analysis for the space compaction of the WSPS is different from the single output case as masking is more complex to detect. When a fault occurs on a line in the circuit, we denote by $[f]_j^*$ the function produced by the faulty circuit at output j. The syndromes or some spectral coefficients, at the outputs which depend on the faulty line, change if the fault is testable at those outputs. When a set of coefficients from the WSPS is used as a signature for a circuit, there may be faults which are considered untestable for two reasons: (a) they do not change the signature at any of the outputs, (b) there is cancellation induced by the weighted summation process inherent in the WSPS definition.

In the former case the problem is the same as that for separately testing all the single outputs. If the signature aliases at all outputs for a particular fault, then the design must be changed to achieve testability, or the signature augmented. A signature from a space-compacted WSPS is very useful in this case; it is shown in section 9 that one straightforward solution to the problem is tp introduce an extra output in the network which acts as a test point for untestable lines. Since the testing is done in parallel for all outputs, the extra output does not increase the test time and only increases the storage of the space-compacted signature by a small amount.

For case (b), the aliasing is a new problem introduced by the space compaction itself. The weighted summation may cause the cancellation of two faulty signatures which do not alias on their own. However, different weights can be selected to avoid such a cancellation; the design for testability changes are straightforward and can be deterministically checked. In the implementation, this is achieved by changing the layout with no extra hardware being required.

We consider initially the case for a single stuck-at fault on an input line for a network N with n inputs and m outputs.

THEOREM 20. Let $k_{\alpha i}$ be a coefficient from a WSPS with some set of weights, where $k_{\alpha i} = \sum_{j=1}^{m} w_j [r_{\alpha i}]_j$. The input stuck-at faults $x_i/0$ and $x_i/1$ are k_α and $k_{\alpha i}$ testable iff $k_{\alpha i} \neq 0$.

It is useful to state explicitly the condition for k_0 (WSS) testability.

COROLLARY 21. Input stuck-at faults $x_i/0$ and $x_i/1$ are k_0 testable (WSS testable) iff $k_i \neq 0$.

As an example, consider the circuit of Figure 10, with 4 inputs and 2 outputs. This circuit is the same example used in (5); there exhaustive simulation had to be performed in order to show it to be WSS untestable for their chosen weights. We show below a set of weights to make it testable. The functions are $[f]_1 = (x_1 x_4 + x_3) x_2$ and $[f]_2 = \bar{x}_1 x_2 + x_1 \bar{x}_2 \bar{x}_3$. The weights chosen in (5) are $w_1 = 2$ and $w_2 = 3$ giving $K = T^4 (2[Z]_1 + 3[Z]_2)$. giving the WSPS

k_0	k_1	k_2	k_{12}	k_3	k_{13}	k_{23}	k_{123}	k_4	k_{14}	k_{24}	k_{124}	k_{34}	k_{134}	k_{234}	k_{1234}
28	4	−16	−16	0	−8	12	−4	−2	2	2	−2	−2	−2	2	−2

Consider the fault $x_3/0$. Since $k_3 = 0$, then $x_3/0$ is k_0 untestable (WSS untestable). However, $k_{13} \neq 0$ so $x_3/0$ is k_1 testable. However, if the weights for the WSPS are changed to $w_1 = 1$ and $w_2 = 2$, (as it is also done in (5)), we have the WSPS

k_0	k_1	k_2	k_{12}	k_3	k_{13}	k_{23}	k_{123}	k_4	k_{14}	k_{24}	k_{124}	k_{34}	k_{134}	k_{234}	k_{1234}
17	3	−9	−11	1	−5	1	−3	−1	1	1	−1	−1	1	1	−1

Note that $k_3 = 1$ and $x_3/0$ is k_0 (WSS) testable; moreover since all $k_i \neq 0$ all input stuck-at faults are k_0 (WSS) testable. Further since all coefficients are nonzero, all input stuck-at faults are

172

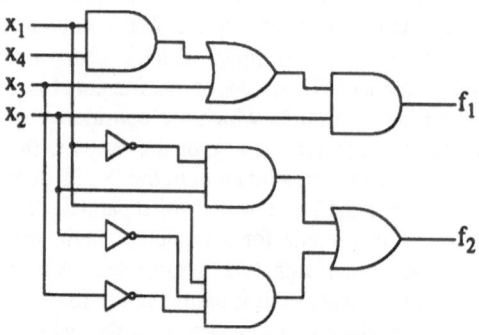

FIGURE 10: Example Circuit

k_α testable for any α. Only one computation is necessary to find the fault-free vector to which the transform is applied. All input faults are checked for testability directly from the WSPS coefficients.

Consider an internal line g; the appropriate functional model is given by $[f(X)]_j = h_j(X,x_{n+1})$ where $x_{n+1}=g(X)$, for each j, $1 \le j \le m$. Each h_j is formally considered as a function of n+1 inputs and measures the Boolean dependence of an output $[f]_j$ on the line g. Let $[\hat{R}]_j$ be the spectrum of h_j, $[V]_j$ the vector representation of h_j, and let $\hat{K} = \sum_{j=1}^{m} w_j[\hat{R}]_j$. As before, we do not need to calculate the $[\hat{R}]_j$ separately; we construct the superfunction $H = w_1[V]_1 + \cdots + w_m[V]_m$ where "+" is over the integer field, and only apply the transform once to obtain $\hat{K} = T^{n+1}H$. The check for the WSPS testability of an internal line requires the construction of the Boolean h_j functions from the circuit description. In some cases only a subset of the m outputs may depend on a line g and only that subset of corresponding h_j functions need be explicitly evaluated. The remaining h_j functions are equal to the corresponding $[f]_j$, although they must formally be considered as functions of n+1 variables.

THEOREM 22 The fault g/0 is k_α testable iff $k_\alpha \ne \frac{1}{2}(\hat{k}_\alpha + \hat{k}_{\alpha n+1})$.

The fault g/1 is k_α testable iff $k_\alpha \ne \frac{1}{2}(\hat{k}_\alpha - \hat{k}_{\alpha n+1})$.

Again it is useful to state explicitly the condition for k_0 (WSS) testability.

COROLLARY 23. The fault g/0 is k_0 testable (WSS testable) iff $k_0 \ne \frac{1}{2}(\hat{k}_0 + \hat{k}_{n+1})$.

The fault g/1 is k_0 testable (WSS testable) iff $k_0 \ne \frac{1}{2}(\hat{k}_0 - \hat{k}_{n+1})$.

In general, if all the weights chosen for the WSPS are positive, then all internally unate lines are automatically k_0 (WSS) testable. This is particularly relevant for all two-level circuits and for special circuit such as PLA's as discussed below. Larger examples and a more thorough discussion on the methods and proofs involved in space compaction can be found in (77, 79).

9. SPECIAL CIRCUITS

The first application of the general theory presented above is for Programmable Logic Arrays (PLA's). We summarize the primary results from (78). For testing PLA's various new approaches are discussed in the literature. The best surveys can be found in (3, 84, 88). Most such methods are still based on classical testing using test sets. There the thrust of improvements lies in minimizing the work for test set generation and changing the PLA functions in order to incorporate parity check lines. Moreover for built-in test, the area over-head must be reduced for storing the test generation hardware as in (9, 22, 88).

Other schemes are very efficient but not suitable for built-in self-test (BIST) (17, 41, 64). Probabilistic testing has also been proposed as a solution (26, 82) which results in less than full fault coverage. However the newer approaches of random pattern testing have proven to be not very effective (91). Notwithstanding the simple regular structure of a PLA, its testing is a difficult process for which efficient automated test pattern generators cannot guarantee full coverage (9). The common ground has been to increase the testability of the PLA by introducing extra lines, in order to achieve greater controllability of the two arrays.

The option of syndrome testing was explored by Yamada in (96). It is perhaps the closest approach to that presented here. In Yamada's method the outputs are tested separately and a very complex algorithm is required to determine the extra hardware necessary to test input faults. By introducing space compaction testing, using k_0 from the WSPS, both problems are overcome.

In (78) it is shown that the coefficient k_0 is a complete test for all internal single faults (stuck-at, cross-point and bridging). The testability conditions for primary input stuck-at faults are covered by Corollary 21 proved above. In the case of input bridging faults, any special cases can be detected using Corollary 13 or 15. This means that a single coefficient, k_0 which corresponds to a weighted syndrome sum, is sufficient as a complete signature for all single faults in a PLA. Only one signature is required for all outputs, and it is computed in parallel at the same time. There are two great advantages to this signature scheme: there exists a deterministic algorithm to check its testability at the design stage, without fault simulation; and, when changes for testability are required, they are easily achived in a PLA as described below. Any verification of testability is performed directly at the design stage, without fault simulation.

Moreover, the conditions for testability of input faults lead to a direct and simple hardware modification in the few cases where there is a problem. This would apply only in the cases when a fault on a primary input lead proves to be untestable after the direct verification of the corresponding k_i coefficient. There are two cases: one is the selection of scalar multipliers in the WSPS such that the resulting k_i are nonzero. The second possibility arises when, for all choices of scalar weights, the WSPS contains first order coefficients equal to zero.

One of the properties of the Hadamard spectrum of a particular function is that the coefficients are either all odd or all even. This, of course, carries through to the WSPS. Hence, if at least one function in the network contains an odd number of 1's in the output truth vector (and thus its weight is odd), we can assign arbitrarily a scalar multiplier of 1 to that output in the WSPS computation, while all other outputs receive scalar multipliers which are even numbers. In this way we ensure that the weighted sum of all the weights is odd, that is k_0 is odd. Consequently all coefficients in K become odd and are thus nonzero.

This ensures that all stuck-at faults on input lines x_i are k_0 (WSS) testable (by Corollary 21).

A problem occurs when a stuck-at fault on x_i is not testable at any output, that is, it aliases for each signature computed separately on each output. In the case of PLA's, the solution is still very simple. Assume that, after the checking, the k_i coefficient is equal to 0, as it is the weighted sum of $[r_i]_j$, $1 \le j \le m$, each of which is equal to 0. The required hardware change is only the addition of one product line and one output line. The extra function realized in the PLA must have one product term covering only a single minterm, $y_1 y_2 \cdots y_n$ where each $y_i \in \{x_i, \bar{x}_i\}$, $1 \le i \le n$. Any minterm suffices; indeed any function of odd weight could be used.

This extra output, $[f]_{m+1}$, is assigned a scalar multiplier of $w_{m+1}=1$ in the WSPS, while $[f]_1,\ldots,[f]_m$ are assigned values of w_1 to w_m which are even. In this way the value of k_i becomes $+1$ or -1 (depending on the minterm chosen) and obviously all other k_α coefficients become odd and nonzero. This happens solely because of the introduction of the $(m+1)$th function with odd weight. In some cases it may be more appropriate to use an extra output line connected to some already existing product lines, to achieve a similar effect. Since, in practice, with the use of standard size PLA's, there are often unused lines in the unmasked layout, this proposed solution of utilizing an extra output may not add any overhead. A similar solution can be used to solve the analogous input bridging fault testability problem.

This hardware addition for PLA's appears to be the simplest proposed so far in the literature. The only real addition is for the tester which requires one more bit, since the extra function has a weight of 1, and must be incorporated in the signature.

A similar analysis can be applied to other special circuits, namely Read Only Memories (ROM's) and Decoders. For both ROM's and decoders, it is shown that all single faults, stuck-at and bridging, are covered by the k_0 coefficient. The examination of these special circuits, which are nevertheless in wide practical use, proves that the theoretical results for WSPS testability have a general applicability. Space compaction testing may not be the best solution for all circuits, but it has been shown to be very advantageous for these modules. The next step is to provide guidelines for design for testability and/or an algorithm such that the conditions proved above plus the testing specifications can be easily implemented. A more extensive discussion on design for testability using space compaction techniques can be found in (77, 79).

10. PROBABILISTIC ANALYSIS

In (60), an analysis is undertaken of the aliasing probability of a number of common data compaction techniques. This discussion is based on the performance over the set of all possible functions and it derives a general theoretical framework for the analysis leading to general results for worst case and average case aliasing for any data compaction scheme. It is useful to have a framework in which to compare all the possible data compaction schemes, even if the theoretical assumptions of the uniformity of distribution of faulty circuits are open to some question. The results appear to indicate that the performance of syndrome testing, and similar methods, varies between very good and very bad. This is very different from the comparable results for linear feedback shift registers, where the performance is independent of the function being realized. In the paper, the results are applied to linear feedback shift register testing, syndrome testing, accumulator testing and spectral coefficient testing. Following the theoretical analysis, there is a detail evaluation of the performance of a set of 41 single and multiple output circuits to see how closely they match the average behaviour predicted in theory. In this section, we give a brief sampling of some of the relevant results as applied to spectral testing and syndrome testing.

Note that the theoretical results are derived on a probabilistic basis, which require a number of assumptions to be made. The most serious of these is that of a uniform distribution of the error patterns resulting from a faulty circuit, that is, in the presence of a fault all other bit patterns are equally likely. In practice the set of possible faults in a circuit only produces a tiny fraction of all possible bit patterns. Consequently, the assumption of uniform distribution depends on these faulty patterns being randomly distributed amongst all the possible bit patterns.

Let $N = 2^n$. For syndrome counting the worst case aliasing, W, for a function is shown to be $(\pi N/2)^{-1/2}$, while the average case aliasing is $(\pi N)^{-1/2}$. If we look at the syndrome plus one other spectral coefficient, the worst case and average case aliasing become $4/(\pi N)$ and $2/(\pi N)$ respectively. These four figures are evaluated below for 3 values of n.

	syndrome		two coefficients	
	Worst	Average	Worst	Average
n = 10	2.493 -02	1.763 -02	1.243 -03	6.217 -04
n = 15	4.408 -03	3.117 -03	3.886 -05	1.943 -05
n = 20	7.792 -04	5.510 -04	1.214 -06	6.071 -07

11. CONCLUSION

This paper has given a brief overview of the relevant results in the application of the theory of Rademacher-Walsh transforms to fault detection and analysis in combinational analysis. The results that are included are relevant both for the derivation of practical signatures for the testing of circuits and for the theoretical analysis of testing methods. If we are to develop eficient methods for testing, and, in particular, for built-in testing it is critical for us to develop a good theoretical approach to analyse the behaviour of the methods and give deterministic values for the fault coverage without relying solely on simulation studies. The bibliography below includes, in addition to the references explicitly cited in the text above, a listing of the majority of the recent work in the area.

REFERENCES.

1. Abramovici, M. and Menon, P. R.: A Practical Approach to Fault Simulation and Test Generation for Bridging Faults. IEEE Trans. Comp., C-34, July 1985.

2. Agarwal, V. K.: Multiple Fault Detection in Programmable Logic Arrays. IEEE Trans. Computers, C-29, June 1980.

3. Agarwal, V. K.: Easily Testable PLA Design. in VLSI Testing, Williams, T. W. ed.: North-Holland: Elsevier Science Publishers B.V. 1986.

4. Agrawal, V. D. and Mercer, M. R.: Testability Measures - What Do They Tell Us?. Digest 1982 IEEE Test Conference, Philadelphia, November 1982.

5. Barzilai, Z., Savir, J., Markowsky, G. and Smith, M.: The Weighted Syndrome Sums Approach to VLSI Testing. IEEE Trans. Comput., C-30, December 1981.

6. Barzilai, Z., Coppersmith, D. and Rosenberg, A. L.: Exhaustive Generation of Bit Patterns with Applications to VLSI Self-Testing. IEEE Trans. Computers, C-32, 1983.

7. Bennetts, R. G. and Hurst, S. L.: Rademacher-Walsh Spectral Transforms: a New Tool for Problems in Digital-Network Fault Diagnosis?. Computers and Digital Techniques, 1, 1978.

8. Bennetts, R. G.: Design of Testable Logic Circuits, Reading, Mass.: Addison-Wesley Publishing Co., 1984.

9. Bozorgui-Nesbat, S. and McCluskey, E. J.: Lower Overhead Design for Testability of Programmable Logic Arrays. IEEE Trans. Computers, C-35, April 1986.

10. Brzozowski, J. A.: Testability of Combinational Networks of CMOS Cells. in Developments in Integrated Circuits Testing, Miller, D. M. ed.: New York and London: Academic Press 1987.

11. Carter, W. C.: The Ubiquitous Parity Bit. Proc. FTCS-12, 1982.

12. Carter, J. L.: The Theory of Signature Testing for VLSI. Proc. 14th ACM Symp. Theory Computing, 1982.

13. Carter, W. C.: Improved Parallel Signature Checkers/Analyzers. Proc. FTCS-16, July 1986.

14. Chen, T. and Breuer, M.: Automatic Design for Testability Via Testability Measures. IEEE Trans. Computer-Aided Design, CAD-4, January 1985.

15. David, R.: Signature Analysis for Multiple-Output Circuits. IEEE Trans. Computers, C-35, September 1986.

16. Eichelberger, E. B. and Williams, T. W.: A Logic Design Structure for LSI Testability. Proc. 14th Design Automation Conf., New Orleans, June 1977.

17. Eichelberger, E. B. and Lindbloom, E.: A Heuristic Test Pattern Generator for Programmable Logic Arrays. IBM Journal Research & Development, 24, January 1980.

18. Eris, E. and Miller, D. M.: Syndrome-Testable Internally-Unate Combinational Networks. Electronics Letters, 19, 1983.

19. Friedman, A. D.: Diagnosis of Short-Circuits Faults in Combinational Circuits. IEEE Trans. Computers, C-23, 1974.

20. Frohwerk, R. A.: Signature Analysis: A New Digital Field Service Method. Hewlett-Packard Journal, May 1977.

21. Fujiwara, H. and Kinoshita, K.: Testing Logic Circuits with Compressed Data. Proc. FTCS-8, 1978.

22. Fujiwara, H.: A New PLA Design for Universal Testability. IEEE Trans. Computers, C-33, August 1984.

23. Goldstein, L. H.: Controllability/Observability Analysis of Digital Circuits. IEEE Trans. Circuits and Systems, CAS-26, 1979.

24. Goldstein, L. H. and Thigpen, E. L.: SCOAP: Sandia Controllability/Observability Analysis Program. Digest 17th Design Automation Conf., Minneapolis, Minn., June 1980.

25. Golomb, S. W.: Shift Register Sequences, Holden-Day, 1967.

26. Hassan, S. Z. and McCluskey, E. J.: Testing PLA's Using Multiple Parallel Signature Analyzers. Proc. FTCS-13, June 1983.

27. Hassan, S. Z., Lu, D. J. and McCluskey, E. J.: Parallel Signature Analyzers - Detection Capability and Extensions. Digest COMPCON, San Francisco, March 1983.

28. Hayes, J. P.: Transition Count Testing of Combinational Logic Circuits. IEEE Trans. Comput., C-25, June 1976.

29. Hayes, J. P.: Check Sum Methods for Test Data Compression. J. Design Automation and Fault Tolerant Computing, 1, 1976.

30. Hayes, J. P.: Fault Modeling. IEEE Design & Test, April 1985.

31. Hlawiczka, A.: Parallel Multisignature Analysis of Faults in the Multi-Output Digital System. Proc. FTCS-16, July 1986.

32. Hsaio, T. and Seth, S.: The Use of Rademacher-Walsh Spectrum in the Design and Testing of Digital Circuits. Proc. ICCC-82, 1982.

33. Hsaio, T. and Seth, S.: An Analysis of the Use of Rademacher-Walsh Spectrum in Compact Testing. IEEE Trans. Computers, C-33, 1984.

34. Hurst, S. L.: The Logical Processing of Digital Signals, New York: Crane Russak, 1978.

35. Hurst, S. L., Miller, D. M. and Muzio, J. C.: Spectral Techniques in Digital Logic, New York and London: Academic Press, 1985.

36. Iosuposicz, A.: Optimal Detection of Bridging Faults and Stuck-at Faults in Two-Level Logic. IEEE Trans. Computers, C-27, 1978.

37. Karpovsky, M. G.: Finite Orthogonal Series in the Design of Digital Devices, New York: John Wiley, 1976.

38. Karpovsky, M. G. and Trachtenberg, E. A.: Fourier Transforms over Finite groups for Error Detection and Error Correction in Computational Channels. Information and Control, 40, 1979.

39. Karpovsky, M. G. and Su, S. Y. H.: Detection and Location of Input and Feedback Bridging Faults in Combinational Networks. IEEE Trans. Computers, C-29, 1980.

40. Karpovsky, M. G.: Detection and Location of Errors by Linear Inequality Checks. IEE Proc. Part E: Computers and Digital Techniques, 129, 1982.

41. Khakbaz, J.: A Testable PLA Design with Low Overhead and High Fault Coverage. IEEE Trans. Computers, C-33, August 1984.

42. Kodandapani, K. L. and Pradhan, D. K.: Undetectability of Bridging Faults and Validity of Stuck-at Faults Test Sets. IEEE Trans. Comp., C-29, January 1980.

43. Kohavi, I. and Kohavi, Z.: Detection of Multiple Faults in Combinational Logic Networks. IEE Trans. Computers, C-30, 1981.

44. Lechner, R. J.: Harmonic Analysis of Switching Functions. in Recent Developments in Switching Theory, Mukhopadhyay, A. ed.: New York and London: Academic Press 1971.

45. Lui, P. and Muzio, J. C.: Spectral Signature Testing of Multiple Stuck-at-Faults in Irredundant Combinational Networks. IEEE Trans. Computers, C-35, December 1986.

46. Markowsky, G.: Syndrome-Testability Can be Achieved by Circuit Modification. IEEE Trans. Computers, C-30, 1981.

47. McCluskey, E. J.: Built-In Self-Test Techniques. IEEE Design & Test, April 1985.

48. McCluskey, E. J.: Built-In Self-Test Structures. IEEE Design & Test, April 1985.

49. McCluskey, E. J.: Logic Design Principles, Englewood Cliffs, New Jersey: Prentice-Hall, 1986.

50. McCluskey, E. J.: Comparing Causes of IC Failures. in Development in Integrated Circuits Testing, Miller, D. M. ed.: New York and London: Academic Press 1987.

51. Miczo, A.: Digital Logic Testing and Simulation, New York: Harper & Row, 1986.

52. Miller, D. M. and Muzio, J. C.: Spectral Fault Signatures for Single Stuck-at-Faults in Combinational Networks. IEEE Trans. Comput., C-33, August 1984.

53. Miller, D. M. and Muzio, J. C.: Spectral Techniques for Fault Detection in Combinational Logic. in Spectral Techniques, Karpovsky, M. ed.: London: Academic Press 1985.

54. Muzio, J. C.: Evaluation of the Spectra of Sum and Product Functions. Computers and Digital Techniques, 1, August 1978.

55. Muzio, J. C.: Composite Spectra and the Analysis of Switching Circuits. IEEE Trans. Comp., C-29, August 1980.

56. Muzio, J. C. and Miller, D. M.: Spectral Fault Signatures for Internally Unate Combinational Networks. IEEE Trans. Computers, C-32, 1983.

57. Muzio, J. C. and Serra, M.: Built-in-Testing and Design for Testability of Stuck-at-Neighbour Faults. Canadian Conference on VLSI, Toronto, Ont., 1985.

58. Muzio, J. C. and Serra, M.: Spectral Criteria for the Detection of Bridging Faults. 2nd International Workshop on Spectral Techniques, Montreal, P.Q., October 1986.

59. Muzio, J. C. and Serra, M.: Data Compaction for Bridging Faults. Submitted, 1987.

60. Muzio, J., Ruskey, F., Aitken, R. and Serra, M.: Aliasing Probabilities of Data Compression Techniques. in Developments in Integrated Circuits Testing, Miller, D. M. ed.: New York and London: Academic Press 1987.

61. Pradhan, D. K., ed.: Fault-Tolerant Computing: Theory and Techniques, New Jersey: Prentice-Hall, 1986.

62. Rajski, J. and Tyszer, J.: Easily Testable PLA Design. Euromicro, 1984.

63. Ratiu, I. M., Sangiovanni-Vincentelli, A. and Pederson, D. O.: VICTOR: A Fast VLSI Testability Analysis Program. Proc. 1982 IEEE Int. Test Conference, 1982.

64. Reddy, S. M. and Ha, D. S.: A New Approach to the Design of Testable PLA's. IEEE Trans. Computers, C-36, February 1987.

65. Robinson, J. P. and Saxena, N. R.: A Unified View of Test Compression Methods. IEEE Trans. Computers, C-36, 1987.

66. Roth, C. H.: Fundamentals of Logic Design, New York: West Publishing Co., 1979.

67. Roth, J. P.: Computer Logic and Testing, Computer Science Press, 1980.

68. Roth, J. P. and Savir, J.: Testing for and Distinguishing Between Failures. Proc. FTCS-12, 1982.

69. Saluja, K. K., Kinoshita, K. and Fujiwara, H.: An Easily Testable Design of Programmable Logic Arrays for Multiple Faults. IEEE Trans. Computers, C-32, November 1983.

70. Saluja, K. and Karpovsky, M.: Testing Computer Hardware Through Data Compression in Space and Time. Proc. 1983 IEEE Int. Test Conference, 1983.

71. Savir, J.: Syndrome Testable Design of Combinational Networks. IEEE Trans. Comput., C-29, June 1980.

72. Savir, J.: Syndrome Testing of 'Syndrome Untestable' Combinational Circuits. IEEE Trans. Computers, C-30, 1981.

73. Savir, J., "Good controllability and observability do not guarantee good testability," IBM Research Report RC 9432 (#41597), June 1982.

74. Savir, J. and McAnney, W. H.: On the Masking Probability with One's Count and Transition Count. IEEE Int. Conf. on Computer-Aided Design, Santa Clara, CA, November 1985.

75. Saxena, N. and Robinson, J.: Accumulator Compression Testing. IEEE Trans. Computers, C-35, April 1986.

76. Serra, M. and Muzio, J. C.: Testing Programmable Logic Arrays by Sum of Syndromes. 8th IEEE Workshop on Design for Testability, Beaver Creek, CO, April 1985.

77. Serra, M. and Muzio, J. C.: Space Compaction for Multiple-Output Circuits. Submitted, 1987.

78. Serra, M. and Muzio, J. C.: Testing Programmable Logic Arrays by Sum of Syndromes. IEEE Trans. Comput., C-36, Sept. 1987.

79. Serra, M., "New Methods for the Compaction Testing of Multiple-Output Digital Circuits," Ph.D. Dissertation, University of Victoria, Dept. of Computer Science, 1987.

80. Seth, S. C.: Data Compression in Logic Testing an Extension of Transition Counts. J. Design Automation and Fault Tolerant Computing, 1, 1977.

81. Sinha, B. P. and Bhattacharya, B. B.: On the Numerical Complexity of Short-Circuit Faults in Logic Networks. IEEE Trans. Comp., C-34, February 1985.

82. Smith, J. E.: Detection of Faults in Programmable Logic Arrays. IEEE Trans. Comput., C-28, November 1979.

83. Smith, J. E.: Measures of Effectiveness of Fault Signature Analysis. IEEE Trans. Comput., C-29, June 1980.

84. Somenzi, F. and Gai, S.: Fault Detection in Programmable Logic Arrays. Proc. IEEE, 74, May 1986.

85. Spencer, T. H. and Savir, J.: Layout Influences Testability. IEEE Trans. Computers, C-34, March 1985.

86. Susskind, A. K.: Testing by Verifying Walsh Coefficients. IEEE Trans. Computers, C-32, 1983.

87. Tokmen, V. H.: Disjoint Decomposition of Multi-Valued Functions by Spectral Means. Proc. 10th Int. Symp. on Multiple-Valued Logic, 1980.

88. Treuer, R., Fujiwara, H. and Agarwal, V. K.: Implementing a Built-In Self-Test PLA Design. IEEE Design & Test, April 1985.

89. Tzidon, A., Berger, I. and Yoeli, M.: A Practical Approach to Fault Detection in Combinational Networks. IEEE Trans. Computers, C-27, 1978.

90. Williams, T. W. and Parker, K. P.: Design for Testability - A Survey. IEEE Trans. Computers, C-31, 1982.

91. Williams, T. W.: VLSI Testing. IEEE Computer, October 1984.

92. Williams, T. W.: Test Length in a Self-Testing Environment. IEEE Design & Test, April 1985.

93. Williams, T. W., ed.: VLSI Testing, Amsterdam: North-Holland, 1986.

94. Williams, T. W., Daehn, W., Gruetzner, M. and Starke, C. W.: Comparison of Aliasing Errors for Primitive and Non-Primitive Polynomials. Proc. 1986 IEEE Int. Test Conference, 1986.

95. Xu, S. and Su, S.: Detecting I/O and Internal Feedback Bridging Faults. IEEE Trans. Comp., C-34, June 1985.

96. Yamada, T.: Syndrome-Testable Design of Programmable Logic Arrays. Proc. 1983 IEEE Int. Test Conference, October 1983.

97. Yamada, T. and Nanya, T.: Stuck-at Faults Tests in the Presence of Undetectable Bridging Faults. IEEE Trans. Comp., C-33, August 1984.

LOGIC VERIFICATION, TESTING AND THEIR RELATIONSHIP TO LOGIC SYNTHESIS

Srinivas Devadas, Hi-Keung Tony Ma and Alberto Sangiovanni-Vincentelli

Department of Electrical Engineering
and Computer Sciences
University of California, Berkeley, CA 94720

1. INTRODUCTION

Much work has gone into automating the integrated circuit (IC) design process over the past few years (e.g. [1] [2] [3]). A variety of Computer-Aided Design (CAD) tools for the logic [4] [5] and physical design [6] of integrated circuits have been developed. It is clear that an integrated set of computer design aids coupled with an unified approach to data management is essential for VLSI design. To this end, research has focused on IC *synthesis systems* [7] i.e. systems which can automatically generate functionally correct mask-level layout of integrated circuit chips from high level, programming language-like specifications.

The goal of this paper is to review the problems of verification and testing of IC designs in the context of logic synthesis. The context of logic synthesis provides a different point of view to solving these problems. Rigorous (formal) verification techniques have become very important as synthesis systems become a reality. Testing of automatically synthesized circuits may become easier if testability considerations are embedded in the synthesis algorithms used.

In particular, verification techniques for both combinational and sequential circuits will be reviewed. The relationship between testing and verification will be underlined and the impact of logic minimization evaluated. Recently developed algorithms for sequential testing will be presented.

1.1. Need for synthesis systems

There are integrated circuit applications, such as speech synthesis, bandwidth compression and recognition, modems and digital data transmission and digital control systems, where in order to achieve a complete and efficient integration of the system functions, it is necessary to design special-purpose chips which are to perform only a single task. Unfortunately, even though this can yield enormous savings in the size of the system and its power consumption, the design cost in both time and money can often be prohibitively expensive. In addition, there are not many designers who have the expertise to design these system chips, which often require knowledge in both analog and digital circuit design as well as digital signal processing and computer architecture.

F. Lombardi and M. Sami (eds.), Testing and Diagnosis of VLSI and ULSI, 181–245.
© 1988 by Kluwer Academic Publishers.

Semi-custom design techniques such as gate-arrays and standard cells offer an environment where faster turnaround can be guaranteed by design tools which can place and route complex functions in a short time. However, the task of logic design, i.e. specifying the gates and the interconnections which implement a certain behavior of the system, may still consume a large amount of design time.

Application-Specific Integrated Circuit (ASIC) synthesis systems have been proposed as a solution to the problem of automatic integrated circuit generation from a high-level behavioral or algorithmic description of the functions of the system to be implemented (e.g. [8] [9] [10] [7]). The demand for and use of ASIC synthesis systems is increasing at a rapid rate in the IC industry today.

1.2. Problems with Designs generated automatically

Present day ASIC synthesis systems use complex optimization and mapping strategies during the synthesis process. For example, to produce automatically a design from its register-transfer level description in the Berkeley synthesis environment, programs performing logic extraction, net-list translation, state assignment and encoding, logic optimization, technology mapping, layout generation, floor-planning, placement, channel definition, global routing, detailed routing and symbolic compaction are invoked. As behavioral synthesis systems evolve, more optimization tools will be used at the higher levels of design.

It is conceivable that one or more of the synthesis tools could introduce errors into the design. These errors could occur in the gate-level description or layout-level description and make the fabricated chip functionally incorrect. To ensure functionally correct chips and error-free optimization strategies, verification steps are required at and across each stage of the synthesis process. A prototype behavioral synthesis system developed at Berkeley is shown in Figure 1.1.1.1, which incorporates a verification subsystem.

After a chip has been fabricated, regardless of how its design was produced, automatically or manually, a cost-effective way of determining whether the chip has been manufactured correctly is necessary. The fabricated chip has to be *tested* against its functional specification.

1.3. Verification

In the design of integrated circuits, at all levels of abstraction, verification tools compare the design at different levels to make sure that in the synthesis process, the designers or optimization tools have not introduced errors, particularly *logic errors*. Due to the high complexity of VLSI design and the complexity of synthesis tools, this has become increasingly important. *Logic verification* detects any discrepancy in the function implemented by the two compared logic designs.

In the synthesis system shown in Figure 1.1.1.1, the verification subsystem checks to see if the generated layout correctly implements the register-transfer level description produced by the hardware allocation phase. As behavioral

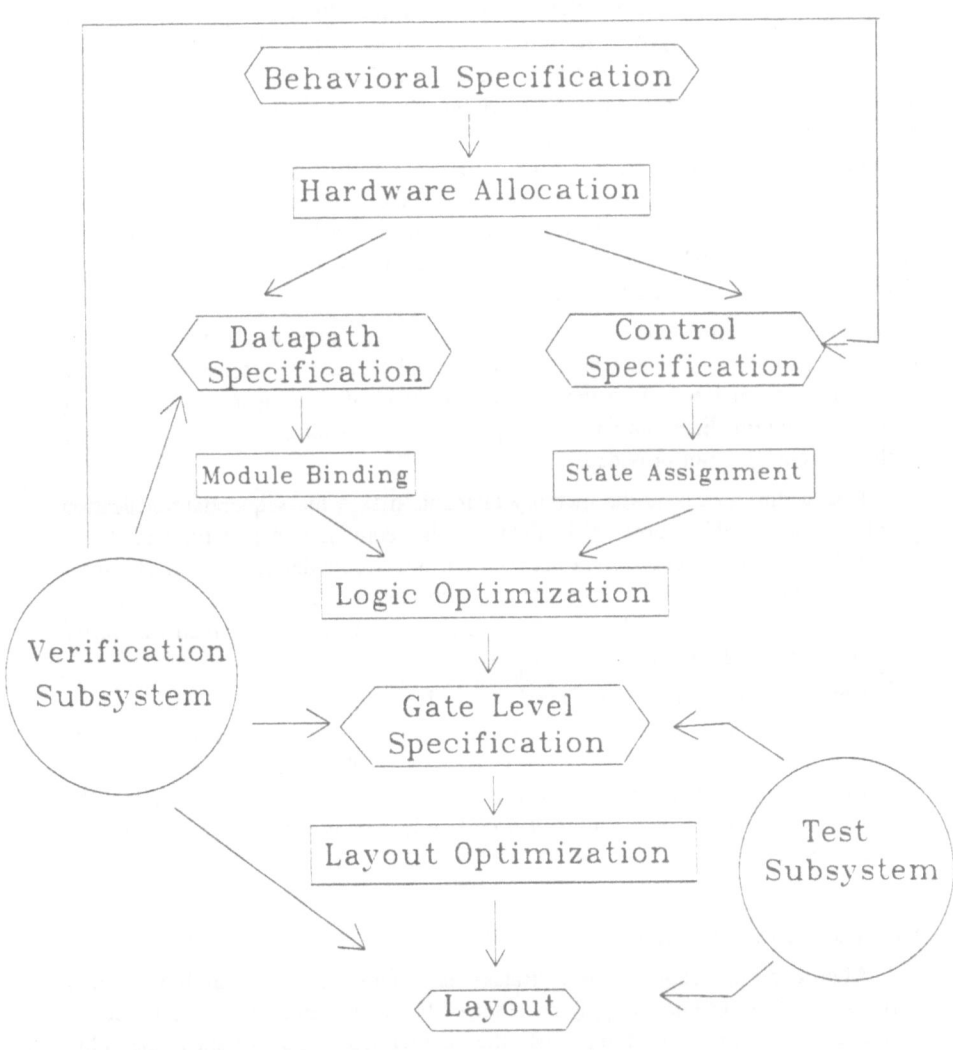

Fig. 1.1.1.1   Flowchart of A Behavioral Synthesis System

verification techniques mature, verification subsystems will be capable of verifying layout against the behavioral specification (shown by a dashed arrow in Figure 1.1.1.1).

1.3.1. Sequential and Combinational Logic Verification

More formally, logic verification refers to the Boolean equivalence check between two logic designs. The problem is known to be NP-complete [11]. Its actual complexity is reflected by the *Boolean complexity* rather than the *structural* (number of gates, wires) *complexity* of the compared designs. For example, a parity checker with a small number of gates is still very difficult to verify algorithmically. Logic verification can be carried out heuristically by non-exhaustive simulation which offers no guarantees about the functional equivalence since simulation is input pattern dependent, and a difference between the two designs may not be uncovered. *Formal verification* techniques are input-pattern independent and can guarantee functional equivalence.

ICs typically implement sequential circuits. A sequential circuit has memory elements associated with it, which are often implemented by latches or flip-flops. In general, external access to the inputs or outputs of these memory elements (which are intermediate outputs and inputs of the combinational logic circuitry within the chip) is not provided.

Not having access to the memory elements makes the sequential verification problem considerably more difficult than the combinational verification one. Equivalence has to be established between the two sequential circuits for all possible infinite binary sequences, rather than all possible binary patterns. Not surprisingly, very few approaches to sequential verification exist today, which can verify even medium-sized circuits.

Combinational logic verification techniques are relatively mature. Approaches capable of verifying large designs with thousands of gates exist today. Verifying sequential circuits can be reduced to a combinational verification problem if a one-to-one correspondence can be made between the internal states of the two circuits. Several verification systems rely on being able to make this correspondence.

1.4. Testing and Test Generation

After chip fabrication, a test strategy is required to ascertain functionality correctness. Testing entails applying a set of binary patterns to the inputs of the fabricated chip and checking that the output response satisfies the chip specification. The input binary patterns are called test sequences or tests for short. Test generation is a process of generating the input stimuli (tests) for a circuit, such that when the tests are applied to the circuit, the existing faults in the circuit are sensitized and a faulty response propagated to an observable output of the circuit.

Effective test generation requires first selecting a good descriptive model for the circuit under consideration. For digital circuits, the model is usually the gate-level model. Once the model has been selected, a fault model is then developed to define the types of faults that will be considered during test generation. Many fault models, such as the classical *stuck-at* fault model, have been introduced for digital circuits described by at the gate level. The stuck-at fault model assumes that a logic gate input or output, when faulty, is fixed to either a logic 0 or a logic 1 (is stuck at a logic 0 or a logic 1). For example, a signal line when stuck-at-1 remains 1 no matter what the correct value is.

The *single stuck-at fault* model further assumes that a faulty machine can only have a single fault. This is the prevailing model used in the industry today.

1.4.1. Combinational and Sequential Testing

Combinational logic test generation has traditionally been carried out using the D-algorithm, developed almost two decades ago. More recently, an algorithm called PODEM (Path Oriented DEcision Making) was proposed which outperforms the D-algorithm for a large class of circuits. A number of combinational test generation systems are in use in industry today.

Like in the verification case, the testing problem for sequential circuits is much harder than that for combinational circuits, and fewer test generation systems exist for sequential circuits. Extensions to the D-algorithm have been proposed for sequential test generation [12] [13]. Recently, a new algorithm for sequential testing based on the PODEM algorithm was proposed [14].

The sequential testing problem is converted into a combinational testing one using Scan Design techniques [15]. The Scan Design style, widely used in industry today, makes all the sequential elements observable and controllable from the outside and transforms the sequential testing problem into a combinational logic testing problem. Unfortunately, Scan Design has a substantial area penalty associated with it − a design constrained by Scan rules can be 20-30% larger than an unconstrained design. Test generation algorithms for sequential circuits are thus very attractive.

1.4.2. Testability and Design for Testability

A variety of different structural designs can be generated to implement a given functional specification. *Testability* refers to the ease with which a fabricated design can be tested. The testability of a circuit has a great effect on the cost of generating and applying tests to the circuit.

The synthesis process can strongly influence the testability of the final fabricated chip. Conflicting trade-offs may exist between the performance and testability of the chip during synthesis. A complex relationship exists between logic synthesis and testability (Section 1.5, 4.2).

Design for testability involves adopting testability as a design parameter. Widely used design for testability techniques are the Scan Design style described in Section 1.4.1 and built-in self-test (BIST) [16]. In BIST, logic is partitioned

186

into various modules and built-in pattern generators test each partitioned block exhaustively. BIST is used for avoiding the problem of test generation for complex combinational circuits.

1.5. Relationship between Logic Synthesis, Verification and Testing

Logic optimization is a critical and difficult task in IC synthesis. Logic minimization tools modify the Boolean structure of the circuit, in order to reduce area and increase performance.

Verification and test generation for circuits are dependent upon the Boolean structure of the logic. Since logic synthesis tools modify this Boolean structure, they strongly influence the test and verification processes.

One relationship between test generation and logic minimization deals with *redundancies* in the circuit. The absence of a test for a fault in a circuit is associated with redundancies in the circuit. Logic minimization algorithms can, in principle, remove all redundancies in a circuit, making it 100% testable. In practice, test generation algorithms like the D-algorithm are used to identify redundancies in a circuit [17]. Interestingly, the D-algorithm can also be used for logic minimization [18].

The verification problem can be formulated as a testing problem. Test generation algorithms can thus be used to verify two circuit descriptions.

The relationships between these three problems are described in greater detail in Sections 3.3 and 4.2.

1.6. Organization of this paper

This paper is organized as follows. In the next chapter, basic definitions and notations used are given. The logic verification problem is dealt with in Chapter 3. Approaches taken for combinational and sequential logic verification are reviewed. The testing problem for VLSI circuits is described in Chapter 4. The relationship between testing and logic minimization is discussed and an algorithm for sequential test generation is described. Finally, conclusions are drawn and directions for future work are presented in Chapter 5.

2. BASIC DEFINITIONS

In this chapter, a framework of concise definitions is provided for use in the following section. In the definitions, the object being defined appears in bold type.

Let $B = \{0,1\}$, $Y = \{0,1,2\}$. A **logic** (Boolean, switching) **function** ff in n input variables, $x_1, x_2, ..x_n$, and m output variables, $y_1, y_2, ..y_m$, is a function

$$ff: B^n \rightarrow Y^m$$

where $x = [x_1, ..x_n] \in B^n$ is the **input** and $y = [y_1, ..y_m] \in Y^m$ is the **output** of ff. B^n is the Boolean n-space associated with the function ff. Note that in addition to the usual values of 0 and 1, the outputs y_i may also take the **don't care value** 2 (or -). Such functions are called **incompletely specified** logic functions. A **completely specified function** f is a logic function taking values in $\{0,1\}^m$, i.e., all the values of the input map into 0 or 1 for all the components of f. For each component of an incompletely specified logic function ff, $ff_i, i = 1, ..m$, we can define: the **ON-set**, X_i^{ON} (also denoted by $FF_i^{ON}(x)$) $\subset B^n$, the set of input values x such that $ff_i(x) = 1$, the **OFF-set**, X_i^{OFF} (also denoted by $FF_i^{OFF}(x)$), the set of values such that $ff_i(x) = 0$ and the **don't care** set X_i^{DC}, the set of values such that $ff_i(x) = 2$. A logic function with $m=1$ is called a **single-output** function, while $m>1$, it is called a **multiple-output** function.

The **complement** of a completely specified logic function f, called $\bar{f}$, is another completely specified logic function, such that its components, $\bar{f}_1, ..\bar{f}_m$, have their ON-sets equal to the OFF-sets of the corresponding components of f. The **intersection** of two completely specified logic functions, $f \cap g$, is defined to be the completely specified logic function h, whose components h_i, have ON-sets equal to the intersection of the ON-sets of the corresponding components of f and g. The **difference** between two completely specified logic functions, $h = f - g$, is a completely specified logic function h given by the intersection of f with the complement of g. The ON-sets of the components of h are the elements of the ON-sets of the corresponding components of f that *are not* in the ON-set of the corresponding components of g. The **union** of two completely specified logic functions is a completely specified logic function $h = f \cup g$, such that the ON-sets of the components of h, h_i, are the union of the ON-sets of f_i and g_i.

A completely specified logic function f is a **tautology**, written $f \equiv 1$, if the OFF-sets of all its components are empty. In other words, the outputs of f are 1 for all inputs. For example, for any completely specified function f, $f \cup \bar{f}$ is a tautology.

A **cube** in a Boolean n-space associated with a logic function, f, can be specified by its vertices and by an index indicating to which components of f it belongs. An input cube c is specified by a row vector $c = [c_1, ..c_n]$ where each input variable takes on one of three values 0, 1 or 2 (or -). A 2 in the cube is a don't care input, which means that the input can take the values of either 0 or 1. For example, the cube 002 is equal to the union of the cubes 001 and 000. A cube with only 0 and 1 values of inputs is called a **minterm**. A cube, c, is defined to cover (contain) another cube, d, if each entry of c is equal to the corresponding entry of d or is equal to 2.

A **Boolean network** is a multi-level structure for representing an incompletely specified logic function. A Boolean network, η, is a pair (F, PO), where $F = \{F_j, j=1,2,..m\}$ is a set of m given representations of the ON-sets f_j of incompletely specified functions $(X_j^{ON}, X_j^{DC}, X_j^{OFF})$. With each F_j is associated a "local output" logic variable y_j in the set $IV = \{y_1,..y_m\}$. The specified primary output set is $PO \subset IV$.

The **graph** $G(V,E)$ of a Boolean network is an acyclic directed graph in which there is a vertex $v \in V$ corresponding to each function F_i and each primary input. The edges correspond to the fan-in relationships between the functions F_i: an edge $e(j,i) \in E$ exists if F_j fans into F_i. An example of a graph of a Boolean network is given in Figure 3.3.2.2.

3. THE LOGIC VERIFICATION PROBLEM

3.1. Introduction

Logic verification refers to the Boolean equivalence check of two logic designs. The designs in question may be purely combinational logic or sequential finite state machines. These designs may be described at different levels – a combinational logic design may be represented by a gate-level description or a Boolean truth table, a sequential circuit may be described by a State Transition Graph, as an interconnection of gates and flip-flops or in a register-transfer (RT) level language.

Verifying the equivalence of logic circuit descriptions at differing levels of abstraction is an important problem and has many possible applications (Section 1.3). For example, after the synthesis of a logic level finite automaton from a higher level register-transfer description, it is essential to be able to verify that the optimization tools during synthesis have not introduced any design errors in the circuit and the synthesized description and the original specification actually represent the same machine.

One approach to the general verification problem is exhaustive simulation. Unfortunately, the number of simulations required grows exponentially with the number of inputs for even a purely combinational logic circuit, and grows even faster for sequential circuits since all possible input vector *sequences* have to be simulated to prove equivalence. A different approach is to use *formal verification* techniques which are input pattern independent and can guarantee functional equivalence.

Many formal verification approaches have been taken to prove/disprove the equivalence of two combinational logic circuits, at the gate level and at differing levels [19] [20] [21] [22] [23]. Some of these approaches [20] [21] transform the verification problem into a testing problem. A package of programs called PROTEUS [24] incorporates several efficient algorithms for verifying combinational logic circuits. In particular, an algorithm called LOVER [24] in PROTEUS has successfully been used on circuits with a large number of gates.

Boolean equivalence checks can also be performed at the switch level [25] [26]. Using *symbolic simulation* and heuristically efficient Boolean function manipulation algorithms, the program MOSSYM [25] performs equivalence checks between combinational logic descriptions at the switch level against a specification of Boolean equations.

Sequential circuit verification is a considerably more difficult problem, in the general case when there is no correspondence between the latches (states) of the two circuits[1]. The approaches taken to solve the sequential verification problem include the use of temporal logic [27] and PROLOG [28]. The use of temporal logic helps for asynchronous circuits [29] but is not necessary in the synchronous

---

In the special case of a one-to-one correspondence between the two circuits, the problem reduces to a combinational circuit verification problem.

circuit case. Algorithms have been proposed for formally verifying the equivalence of two gate-level sequential circuit descriptions with differing numbers of latches using symbolic boolean manipulation [30]. However because of the intractability of the problem, most of the approaches taken so far have been restricted to small to medium sized circuits with a small amount of memory elements.

In [31] algorithms were presented for formally verifying the equivalence of two sequential machines described at the register-transfer, state table or logic level. By exploiting the don't care information available at the various levels (e.g. invalid input and output sequences) the complexity of the verification problem was reduced significantly and equivalence checks for large finite automata were shown to be feasible. For verifying equivalence between two logic level finite automata, a two-phase enumeration-simulation verification algorithm efficient both in terms of memory and CPU time usage was given [31].

This chapter is organized as follows – In Section 3.2 the formal definitions for the equivalence of two combinational logic designs are given. Combinational logic verification algorithms are reviewed in Section 3.3. Definitions of equivalence for two sequential circuits and approaches taken for the verification of sequential machines are reviewed in Section 3.4. Work that remains to be done in the logic verification area is discussed in Section 3.5.

3.2. Combinational Logic Design Equivalence

Verifying the equivalence of two combinational logic designs has been shown to be NP-complete [11]. This means that there is little hope of finding an algorithm whose running time is bounded by a polynomial in the number of inputs to the designs. Given two n-input logic designs, in the worst case, 2^n possible input combinations may have to be verified. Combinational logic designs may be represented by two-level or multi-level logic functions or by gate-level circuits.

Two completely specified (single output) logic functions, f and g, are **Boolean equivalent** if and only if

$$f = 1 \iff g = 1. \qquad (3.2.2.1)$$

Given two incompletely specified logic functions f and g, their ON and OFF-sets and the don't care set $F^{DC}(x)$ ($F^{DC}(x)$ could represent input combinations that cannot occur), f and g are Boolean equivalent under $F^{DC}(x)$ if and only if

$$(F^{ON}(x) \cap G^{OFF}(x)) \cup (F^{OFF}(x) \cap G^{ON}(x)) \subset F^{DC}(x). \qquad (3.2.2.2)$$

Given two functions F and G and a Boolean function D representing the don't care inputs (e.g. inputs which cannot occur or for which equivalence is not required) checking the condition

$$D(x) \cup (F^{ON}(x) \cap G^{ON}(x)) \cup (F^{OFF}(x) \cap G^{OFF}(x)) = 1 \qquad (3.2.2.3)$$

amounts to checking F and G for equivalence.

Given a gate-level circuit Y, every input combination evaluates the circuit to some known value, either a 0 or a 1. So, Y represents a completely specified logic function.

The Boolean equivalence of two gate-level circuits A and B implementing f, given that $F^{DC}(x) = \phi$, can be verified by checking that $A^{ON} = B^{ON}$ or $A^{OFF} = B^{OFF}$.

3.2.1. Segmentation: Single-Output Cone Extraction

A decomposition method widely adopted in performing logic verification checks for multi-output circuits is *segmentation*. Segmentation decomposes a multi-output circuit into many single output segments called *cone* circuits. Cone circuits are verified separately. In the sequel, unless otherwise specified, all circuits are assumed to be cone circuits.

3.3. Combinational Logic Verification Methods

In this section, approaches to verifying combinational logic designs are reviewed. Five different kinds of methods are described in the following subsections – verification by exhaustive simulation, verification using multi-level tautology algorithms, verification using test generation techniques, verification using symbolic simulation and verification using enumeration- simulation techniques. Comparisons between some of these methods have been drawn using the logic verification framework called PROTEUS [24]. The results obtained using PROTEUS are discussed in Section 3.3.5. Parallel implementations of logic verification methods are presented in Section 3.3.6.

3.3.1. Verification by Exhaustive Simulation

Verification using exhaustive simulation entails applying all possible input combinations to the circuits, simulating them and checking to see if the output responses are identical. This can be carried out by a hardware accelerator or a software simulator.

These simulations can be performed in parallel in both cases – using multiple processors in a hardware simulation engine or using parallel vector simulation in a software simulator. However, current capabilities of hardware or software simulators are not enough to carry out the entire simulation in parallel (at once). Hence, it makes sense to reduce time spent in simulation by exploiting the nature of the exhaustive simulation problem. A degree of freedom is the sequence in which inputs are applied to the circuit which is to be simulated.

For the sake of simplicity, assume that a single processor is being used. Simulators typically use leveling and event-driven selective trace techniques [13]. The efficiency of an event-driven simulator is in general determined by the number of events generated during simulation. The **event frequency** at any level in a circuit is defined to be the number of gates at that level that are re-evaluated during simulation. For a given circuit, the total number of events generated during simulation is the sum of all event frequencies at all levels in the circuit. This is directly influenced by the event frequency at the inputs of the circuit. The event frequency at the inputs is a function of the *distance* between two consecutive patterns being simulated (the distance between two patterns is defined to be equal to the number of bits in the two patterns that have different values).

In exhaustive simulation, input patterns can be simulated in any order, as long as all combinations are enumerated. Given a set of patterns $P_1, P_2, \cdots P_n$, the *average distance* is defined as

$$D_{av} = \frac{\sum_{i=1}^{n-1} distance(P_i, P_{i+1})}{n}$$

A good heuristic to speed-up event-driven simulation then, given a set of input patterns to simulate, would be to order the input patterns so to minimize D_{av}.

Given two logic circuits with n inputs each, assuming $C^{DC} = \phi$, all the 2^n patterns can be arranged so that the distance between any two consecutive patterns is 1 (a *Gray* code). Then D_{av} is minimum ($= 1$). The average distance in a sequential binary sequence (e.g. 000, 001, 010, ...), with n bits can be shown to be $2 - 2^{-(n-1)}$ [24].

Exhaustive simulation is impractical in most cases. Given a twenty input circuit the number of binary patterns to be simulated is over a million. Hardware accelerators can alleviate the problem to a certain extent, but more intelligent verification techniques are required for large circuits.

3.3.2. Verification using Multi-Level Tautology

Algorithms for tautology checking can be used for verification purposes. As indicated in Section 3.2.3, given two functions, F and G, to be checked for equivalence, under a don't care set D, checking that

$$D(x) \cup (F^{ON}(x) \cap G^{ON}(x)) \cup (F^{OFF}(x) \cap G^{OFF}(x))$$

is a tautology amounts to checking equivalence of F and G.

Tautology checking is one of the important operations of a two-level logic minimizer. For example, ESPRESSO [5], a two-level logic minimizer, uses tautology checking for two-level logic functions to produce an irredundant cover and to maximize the "sparsity" of the input/output plane of a PLA realization of the logic function. Even though in the worst case, this operation has exponential complexity, in practice it is quite fast if implemented as in [5].

However, verifying whether a multi-level logic function is a tautology is a much more complex task. In [23], a framework of algorithms for tautology checking for multi-level logic functions was presented. In this section, we will review the most interesting methods.

3.3.2.1. Flattening

Flattening is a method for verifying that a given multi-level logic network is a tautology by transforming the multi-level network into a two-level equivalent one and then by applying a well developed two-level tautology algorithm. [5].

Flattening involves *pushing* all intermediate nodes (nodes which receive inputs from other nodes or produce inputs for other nodes) in the Boolean network into the nodes they feed into. This process is illustrated in Figure 3.3.2.1 using a gate-level representation of a Boolean network for clarity. The original multi-level network in shown in Figure 3.3.2.1(a). The resulting two-level network after flattening is shown in Figure 3.3.2.1(b). The number of gates in the network has increased because the intermediate nodes have to be duplicated.

The order in which the intermediate nodes are pushed into the nodes they feed into has a definite effect on the overall complexity of the method. In particular, the nodes can be pushed starting from the primary inputs up until the nodes producing the primary outputs are reached (bottom-up flattening method) or from the nodes feeding the primary output down until the primary inputs are reached (top-down). Intermediate strategies are also possible, where the nodes are chosen according to some heuristics [23]. Experimental results indicate that the bottom-up method is superior.

Flattening a circuit may require exponential time and memory requirements. Certain functions like parity checkers blow up when collapsed to two-level form. However, due to the high efficiency of two-level tautology check algorithms, in practice large circuits can be verified quite quickly using the flattening method.

3.3.2.2. The Don't Care Method

In this method the network is not flattened, but the efficient two-level tautology algorithm [5] is still used.

Let $\eta = (F,PO)$ be a Boolean network on which the tautology check is to be performed. A Boolean network as defined in Chapter 2 is illustrated in Figure 3.3.2.2. A multi-level network can be seen as a structure that imposes some constraints on the inputs that feed the nodes connected to primary outputs. In fact, if these nodes are seen in isolation, their inputs are free to span the corresponding Boolean space. However, when seen in the context of a multi-level network, the input values have to satisfy the logic functions that are between the primary inputs and the nodes under consideration. We could check whether the two-level functions corresponding to these nodes are tautologies, if we had a compact way of representing the constraints that the network poses on the inputs. A possible approach is to determine the values of the variables that *will not* satisfy these constraints and declare these values as don't-cares for the tautology checker acting on the output nodes. These values can be determined by observing that if y_i is the

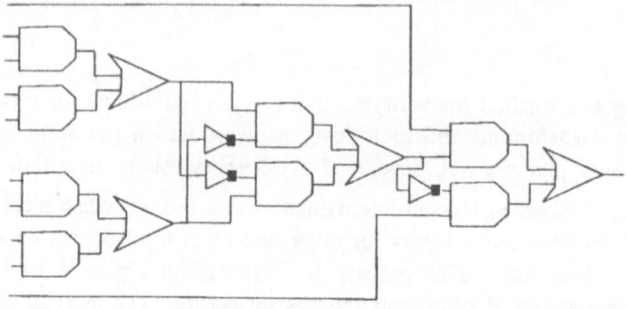

(a) Multi-level logic network

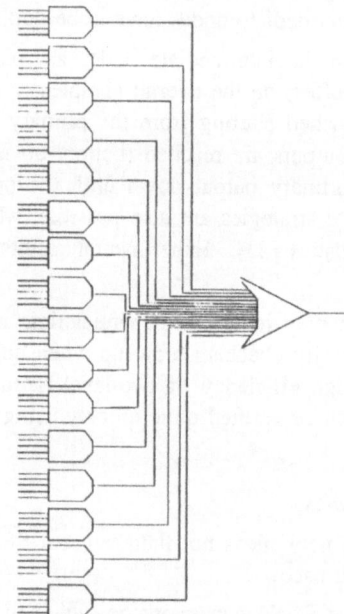

(b) After flattening to Two-level form
Fig. 3.3.2.1

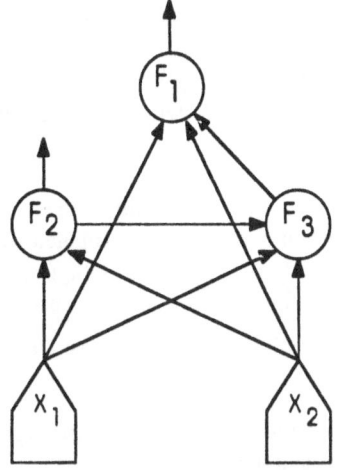

$$F_1 = \bar{x}_1\bar{x}_2 + y_3$$
$$F_2 = x_1\bar{x}_2 + \bar{x}_1x_2$$
$$F_3 = x_1x_2\bar{y}_2 + \bar{x}_1\bar{x}_2$$

$$PO = \{y_1, y_2\}$$

Fig. 3.3.2.2   A Boolean network

output of function F_i, then the network forces y_i to be identical to F_i thus creating a don't-care set of the form:

$$DI_i = \bar{y}_i F_i + y_i \bar{F}_i.$$

called the intermediate don't-care set due to F_i. Now if these don't-care set are summed over all the intermediate nodes except the ones that are providing the primary outputs, the constraints due to the structure of the Boolean network are fully taken into consideration.

This method seems quite appealing since it is very simple to build the don't-care set and the two-level tautology algorithm is already available. Unfortunately, the don't-care set DI can be huge and the tautology check has to be run over a Boolean space which includes both the n primary inputs and m intermediate variables. Experimental results confirm this fact and the don't-care set method is mainly of theoretical interest.

3.3.2.3.   The Co-factoring Method

This method is based on an idea derived from the Shannon decomposition for two-level logic. In two-level logic, the cofactor of a logic function F with respect to a variable x_i, F_{x_i} is defined to be the function obtained from F by evaluating x_i

to 1. The Shannon decomposition states that

$$F = F_{x_i} x_i + F_{\bar{x}_i} \bar{x}_i .$$

It can be easily proven that F is a tautology if and only if F_{x_i} and $F_{\bar{x}_i}$ are tautologies. The cofactor of a function F with respect a cube c F_c can also be defined. F_c is computed by computing successively the cofactors of F with respect to all the care variables of c. In particular, if J denotes the set of indices of the variables appearing uncomplemented in c and with $\bar{J}$ the set of indices of the variables appearing complemented in c, then F_c is computed by successively computing the cofactor of F with respect to $x_j, j \in J$ and with respect to $x_j, j \in \bar{J}$. The tautology theorem can be generalized to a set of cubes C such that:

$$\bigcup_{c^i \in C}^{n} c^i = 1 \text{ and } c^i \cap c^j = \phi , \ c^i, c^j, \in C$$

i.e. the cubes in C are disjoint and span the entire space. Then F is a tautology if and only if $F_c, c \in C$ are tautologies.

In the multi-level case, the definition of cofactor as well as the tautology theorem hold as above. $\eta^c = (F_c, \phi_c)$ of a Boolean network is defined to be the cofactor of $\eta = (F, \phi)$ with respect to a cube c. Each of the two-level functions in the original multi-level Boolean network are individually cofactored, one variable at a time, as in the two-level case. The variables in the cube c are then deleted from the corresponding fan in sets of the cofactored functions.

The Shannon decomposition provides a mean of answering the tautology question by answering it on a number of simpler functions.

Two key issues have to be resolved:

(1) How to compute quickly the cofactor of the Boolean network;

(2) How to select the set of cubes C so that the computation carried out to verify whether the cofactors are tautologies is simplified.

The cofactors can be computed efficiently by *simulating* the network for the appropriate values of the variables appearing in the set of cubes C. By evaluating these variables the network simplifies since some of the intermediate nodes disappear, thus yielding simpler tautology problems.

The selection of the set of cubes is not easy to perform. Several heuristics have been proposed for this task [23].

The recursion implemented by the Shannon decomposition stops at a point where either all the primary outputs of the cofactored network evaluate to one or one of them evaluates to zero. If one of the outputs evaluate to zero, the multi-level function is not a tautology and the entire process stops. If all the primary outputs of a cofactored function evaluate to one, then the recursion halts the decomposition of that cofactored function and goes on to decompose other cofactored functions that do not evaluate to one or zero. When all the cofactored functions evaluate to one, the process stops since the multi-level function has been proven to be a tautology.

The complexity of the method depends on the number of decompositions that have to be performed before an answer is obtained. Providing an effective way of pruning the decomposition tree speeds up the entire process by a large amount.

One such pruning strategy is implemented by using symbolic manipulations. Assume that the algebraic expression of all the nodes of the decomposition tree is compute and stored in a compact form such as a Polish reverse notation. The idea is to record whenever a node of the decomposition tree is known to be a tautology. The corresponding expressions can be compared with the ones that are not known to be a tautology as yet. If the comparison is successful then the corresponding nodes are not decomposed and the tree is pruned.

Note that this method can be seen as the abstraction of many methods proposed in the past among which is Bryant's method reviewed in Section 3.3.4.

3.3.2.4. A General Framework for Multi-level Tautology

A recursive procedure called ML_TAUT was presented in [23] for answering the multi-level tautology question. This recursive procedure contains all the methods, but the don't-care method, presented in the previous subsections as special cases. The methods correspond to specific choices of the subroutines. The pseudo-code for ML_TAUT is given in Figure 3.3.2.3.

There are 3 basic processes in ML_TAUT, called T_FLATTEN, FLATTEN and ML_COFACTOR. Subprocedure T_FLATTEN examines the multi-level functions, F_i, in the Boolean network and checks for trivial cases such as $F_i \equiv 1$ (function F_i has a cube of all don't cares in its cover) or $F_i \equiv \phi$ (function F_i has no cubes). Such trivial functions are flattened up into their fanouts. This may simplify the fanout functions to the point where they too become trivial. T_FLATTEN continues until no trivial functions are left in the multi-level cover. This process is in effect event-driven logic simulation, with trivial functions treated as events.

The second basic procedure in ML_TAUT is that of subprocedure FLATTEN which is driven by the edge selection subprocedure E_SELECT. E_SELECT selects a directed edge, e, from the graph $G = (V,E)$ of the Boolean network. For each edge (j,i), function F_j is flattened into its fanout F_i. E_SELECT selects edges which minimally increase the number of cubes in the fanout function F_i after flattening.

The third process in ML_TAUT is procedure ML_COFACTOR which is driven by the heuristic cube selection subprocedure CUBE_SELECT. In CUBE_SELECT a set of cubes C is generated, for which

$$\bigcup_{c^i \in C}^{n} c^i = 1 \text{ and } c^i \bigcap c^j = \phi, \ c^i, c^j, \in C$$

i.e. the cubes in C are disjoint and span the entire space of the Boolean network. The multi-level cover is a tautology only if and only if it is a tautology on each of the cubes in C. ML_TAUT is called for each of these cubes recursively. ML_COFACTOR computes the multi-level cofactor [23] $\eta^c = (F_c, \phi_c)$ of a Boolean

Procedure ML_TAUT(F, D)
Begin
($Done$, F) = T_FLATTEN(F, D)   /* *Flattens simplified functions* and */
If($Done$ = 1) **Return**(x)   /* *returns their value in vector* x */
Else
  Begin
   e = E_SELECT(F)   /* *Selects one edge flatten* */
   While($e \neq \phi$)
    Begin   /* *Flatten while appropriate* */
     FLATTEN(F, e)
     e = E_SELECT(F)
    End While
   C = CUBE_SELECT(F, D)   /* *Selects expansion cover* */
   For($c \in C$)
    Begin
    (F_c, D_c) = ML_COFACTOR(F, D, c)
    If(ML_TAUT(F_c, D_c) = 0) **Return**(0)
    End For
   Return(1)
  End Else
End ML_TAUT

Fig. 3.3.2.3 Description of ML_TAUT

network $\eta = (F, \phi)$ with respect to a cube c.

3.3.3. Verification using Testing Methods

The verification problem can be formulated as a redundancy identification problem as shown in Figure 3.3.3.1. The two circuits which are to be checked for equivalence A and B are connected by an exclusive-nor gate. Establishing the fact that F-stuck-at-1 is a redundant fault is equivalent to verifying the two designs for equivalence. If a test can be found to detect F-stuck-at-1 the test vector is an input combination which differentiates the two circuits.

Test pattern generation algorithms like the D-algorithm [32] and PODEM [33] can be used for identifying redundant faults in circuits and thus can be used to perform equivalence checks on logic circuits. VERIFY [21], a verification algorithm based on the D-algorithm finds counter-examples by performing line justification. DIVER is a verification technique built on top of VERIFY in the PROTEUS system [24] which exploits certain degrees of freedom during line justification and is thus many times faster than VERIFY. Another algorithm

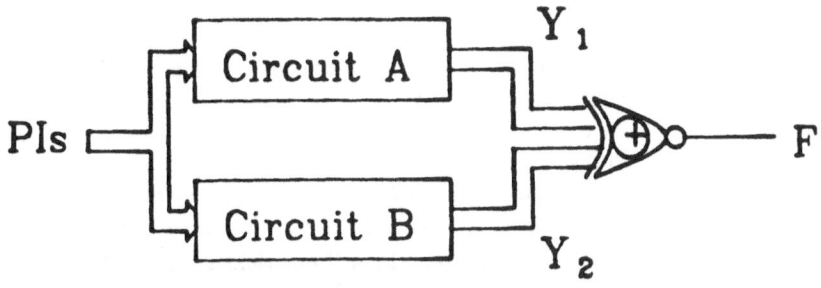

Fig. 3.3.3.1 Logic Verification via Redundancy Identification

called POVER [24] is based on the PODEM test generation algorithm. VERIFY, DIVER and POVER will be described in the rest of the section.

3.3.3.1. Justification in VERIFY

Justification is a technique which tries to find an input combination which satisfies or justifies a given output or intermediate line value in a logic circuit exploring only a small subset of the entire input space. Consider Figure 3.3.3.1. A 0 is first assigned to F and an attempt is then made to find any input combination that produces a 0 or a 1 at F. Either such an input combination is found or proof of equivalence is established. For F to be 0, we can have two possibilities of $Y1 = 0$ and $Y2 = 1$ or $Y1 = 1$ and $Y2 = 0$. For the first choice, it is necessary to justify both $Y1 = 0$ and $Y2 = 1$. The algorithm recursively justifies the values that are pending for justification using the line justification algorithm normally used in the D-algorithm. If the algorithm finds that a justification is not possible, it *backtracks* and chooses a different justification objective. For example, if the algorithm could not justify $Y1 = 0$ and $Y2 = 1$, it would choose the second choice of $Y1 = 1$ and $Y2 = 0$. The backtracking mechanism is recursive, the choices made last are changed first if justification fails.

3.3.3.2. Justification in DIVER

DIVER employs the concept of disjoint justification. Simply put, justification objectives for sets of lines are never repeated. For example, using VERIFY, three justification objectives of $c1 = 11-$, $c2 = -11$ and $c3 = 1-1$ for some three lines may be attempted during verification. When $c1$ fails to yield a solution, $c2$ is attempted as a result of backtracking. However, the objectives $c1$ and $c2$ are *not* disjoint, the min-term 111 is common to them ($c1 \cap c2 = 111$). Advance knowledge thus exists that the objective 111 will not yield a solution because it is in the coverage of $c1$ which has failed to yield a solution. So if $c2$ can be successfully justified, it must be from the coverage of $c2 - c1$, where – stands for the Boolean difference (Section 3.2.1) between $c1$ and $c2$, in this case 011. So the justification process in DIVER would attempt the choices of 11-, 011 and 101 instead of $c1$, $c2$, $c3$ in VERIFY.

Theoretical results and extensive experimental evidence presented in [24] indicate that DIVER performs significantly better than VERIFY, performing equivalence checks in much faster time. Some of these results will be presented in Section 3.3.5 in a comparison of different combinational logic verification strategies.

3.3.3.3. Justification using PODEM - POVER

Given the composite circuit which has been constructed as shown in Figure 3.3.3.1, the backtrace algorithm of PODEM [33] can be used to perform line justification. In PODEM, given an output signal and a desired value on the output ($F = 0$), a path is traced from the signal to the primary inputs (PI) to obtain a PI assignment. This PI assignment is simulated to see if the desired value of the signal has been set up. If so, the procedure terminates. If the opposite value has been set, an opposite value is assigned to the PI and this value is propagated. If the signal remains unspecified, path tracing is repeated. The above procedure continues until either a successful PI assignment has been found (a counter-example has been found and the circuits are not equivalent) or all the PI assignments have been exhausted (the circuits are equivalent).

An efficient justification algorithm tries to justify an output value by setting a minimum subset of input values. In PODEM, justification of input combinations which are cubes rather than minterms is attempted in order to cover a large portion of the input space. For example, justification could be attempted for an input combination, 1-1-, representing four minterms.

3.3.4. Symbolic Verification

Algorithms for the symbolic manipulation of Boolean functions using a graphical representation have been proposed in [26] [25]. Verification is performed by extracting the Boolean functions from switch or gate-level circuits, representing them as directed, acyclic graphs and performing Boolean operations like *and*, *or* and *complement* to check for equivalence.

The data structure used resembles the binary decision diagram proposed by Lee [34] and Akers [35]. However, further restrictions are placed on the ordering of decision variables in the vertices. These restrictions enable the development of algorithms for manipulating the representations in a more efficient manner. The representation is in terms of *reduced* graphs and is a *canonical* form, i.e. every function has a unique representation.

Several function graphs are shown in Figure 3.3.4.1. The first graph represents a single input variable x_i which can take on the values of 0 and 1. The odd parity function of n variables is denoted by a graph containing $2n+1$ vertices. A flattened two-level representation of a n-input parity function would require 2^n terms. The third example is a graph denoting the function $x_1 x_2 + x_4$ and contains five vertices. As can be seen in graphs 2 and 3 several of the subgraphs are shared by different branches. This sharing yields efficiency not only in the size of the function, but also in the performance of the manipulation algorithms.

The time complexities of the Boolean manipulation algorithms proposed in [25] are bounded by the product of the graph sizes for the functions being operated on. Complementing a function requires time proportional to the size of the function graph, while combining two functions with a binary operation (e.g. intersection, subtraction) requires at most time proportional to the product of the graph sizes. Since every function has a unique representation, checking for equivalence simply involves testing whether the two graphs match exactly. The identity check is performed by a graph isomorphism algorithm which requires time at most proportional to the sum of the two graph sizes.

The program MOSSYM [26] performs symbolic verification by extracting the Boolean equations specifying the behavior of a circuit described at the switch level. The extraction of equations is performed by *symbolic simulation*. A symbolic simulator resembles a conventional simulator except that input patterns may consist of Boolean variables in addition to the constants of 0 and 1. The simulator computes the Boolean functions describing the behavior of the circuit for the set of all possible data represented by these variables. The set of Boolean equations extracted from the switch-level circuit is represented as a directed acyclic graph (DAG) like those shown in Figure 3.3.4.1. The DAG obtained via symbolic simulation is compared against the Boolean equation specification of the circuit behavior to check for equivalence. Using symbolic simulation and graph-based Boolean manipulation algorithms, MOSSYM can verify circuits quite quickly if

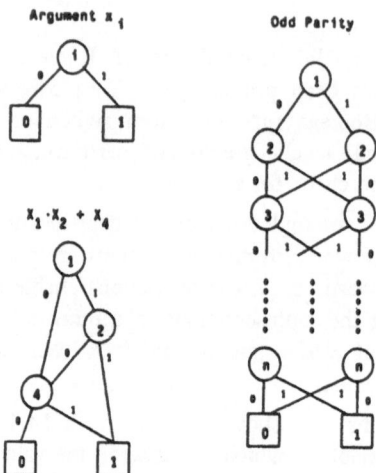

Fig. 3.3.4.1 Example Function Graphs

the size of the DAGs generated is not very large. The size of the DAG depends quite strongly upon the ordering of variables chosen and heuristics are used to find an order which minimizes the size of the graph. The program has successfully verified a 64-bit ALU against its behavioral specification. However, certain circuits like multipliers blow up when represented by DAGs [25].

The time required to validate logic designs is directly related to the size and number of the Boolean equations to be checked for equivalence. Verifying unstructured logic designs for equivalence is harder than verifying structured logic designs like ALUs. This is because the ALU can be specified by a compact set of Boolean equations (for example, the carry bit of an ALU can be compactly represented as a chain of exclusive-or's) whereas given a large unstructured control logic block, there may exist no compact Boolean representation for it. It then becomes very difficult to extract the Boolean functions the logic block implements via symbolic simulation. The logic in a multiplier is much less structured than an ALU and has no compact Boolean representation. Hence, the multiplier is much more difficult to verify using a program like MOSSYM.

3.3.5. The LOVER Approach

LOVER (LOgic VERification) is an enumeration-simulation approach first proposed in [24] and is part of the PROTEUS system.

Let the two circuits to be checked for equivalence be A and B. First, a cube c from A^{ON} is generated (*enumerated*) and then *simulated* on B to check if B produces a 1 at the output. If so, the enumeration process continues to cubes from A^{ON}. If a 0 appears the circuits are not Boolean equivalent. If an x (unknown) appears c is split (cube-split) into smaller cubes and re-simulated until a known value appears at the output of B for each of the smaller cubes. Cube-splitting and simulation are implicitly exhaustive. The process continues until all the cubes from A^{ON} have been simulated. A similar process for A^{OFF} is then performed.

This framework does not specify which enumeration and simulation algorithm to use. The next section illustrates how the justification algorithms reviewed in Section 3.3.3 can be extended to become enumeration algorithms.

3.3.5.1. Enumeration as an Extension of Justification

In general, to perform equivalence checks, all values of input combinations that produce a 1 (0) at the output of a circuit A have to be found. That is, A^{ON} (A^{OFF}) has to be found. Of course, it is of interest to find the most compact representation of A^{ON} (A^{OFF}). If setting a *subset* of input values generates a 1 (0) at the output, it can be inferred that the input combinations produced by setting the remaining unset inputs in all possible ways will produce a 1 (0) at the output. Setting a small subset of input values to justify a 1 (0) to the output results in determining a large subspace of A^{ON} (A^{OFF}).

The difference between enumeration and justification is that in enumeration one is **not** satisfied when **a single** assignment to primary inputs which creates a 1 (0) at the output has been found. **All** possible input combinations have to be found. To make sure that the entire space is examined, backtracking at the decision points is performed recursively, beginning from the deepest decision point, and an alternative assignment is tried. The process continues until all decision alternatives are examined.

Thus, most justification algorithms, especially those presented in Section 3.3.3 become enumeration algorithms after proper modification of the termination criterion. Consider a decision node in the decision tree shown in Figure 3.3.5.1(a) that a justification algorithm follows. Three choices are available at node A: c_i, i = 1, 2, 3. c_1 is chosen and eventually leads to a successful justification denoted by a "T" (Termination) leaf node. In a justification problem the goal is to reach a successful justification; the process backtracks and tries the next available choice only when the current choice of decision does not yield a solution. There is no need for the process to return to the current decision node when a solution can be found for the present choice. So c_2 and c_3 are not tried and the dashed lines reflect this. However, in the enumeration application, where the aim is to enumerate, implicitly and exhaustively, all the possible justifications, even when a successful justification is found, the process still needs to return and try all the remaining choices. This is illustrated by Fig. 3.3.5.1(b) where both c_2 and c_3 are followed regardless of the outcome of c_1.

LOVER incorporates enumeration algorithms based on PODEM backtrace algorithm, the VERIFY justification algorithm and the DIVER justification algorithm.

3.3.5.2. Comparisons between Verification Techniques

The table below compares results obtained using different verification algorithms in the PROTEUS verification system [24]. ONESIM is an exhaustive simulation algorithm using one-distance patterns. DIVER and VERIFY are verification algorithms based on line justification (described in Section 3.3.3). LOVER-SDIJUST and LOVER-PODEM are enumeration simulation algorithms based on the disjoint justification method in DIVER and the PODEM backtrace algorithm respectively. C432 and C880 are publicly available benchmark circuits from [36]. All CPU times are on a VAX 11/8600.

As can be seen, the enumeration-simulation methods in LOVER are significantly faster than the exhaustive simulation and testing methods. The enumeration methods succeed in generating a reasonably compact cover representation (ON and OFF-sets) of the logic function and are therefore quite efficient. In fact for the largest example, C880 with 60 inputs, 1.1×10^7 cubes were generated for both the ON and OFF-sets as opposed to 1.15×10^{18} ($= 2^{60}$) minterms. One major advantage the LOVER algorithms have over the flattening and symbolic verification algorithms is related to memory usage. In flattening/symbolic verification the size of the cover/DAG can become unmanageable for large circuits. However, in LOVER, only a single cube needs to be stored at any stage while verifying any circuit.

3.3.6. Parallel Implementations of Logic Verification Algorithms

Though combinational logic verification algorithms are well developed, current state-of-the-art programs sometimes take inordinate amounts of time to verify large circuits. Implementations of verification algorithms on powerful multi-processors can make equivalence checks on these large circuits feasible and speed up verification for all kinds of circuits. Efficient parallel implementations are thus very attractive.

A parallel implementation of a logic verification algorithm was first presented in [37]. The algorithm is reviewed here.

In the LOVER framework as described in Section 3.3.5, the two main tasks performed in the verification process are enumeration and simulation. Cubes are

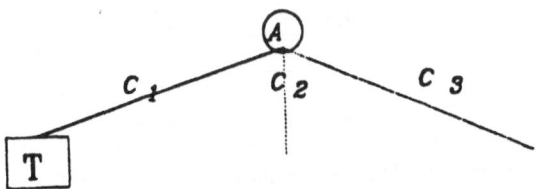

------ choices not tried in the
justification problem

(a)

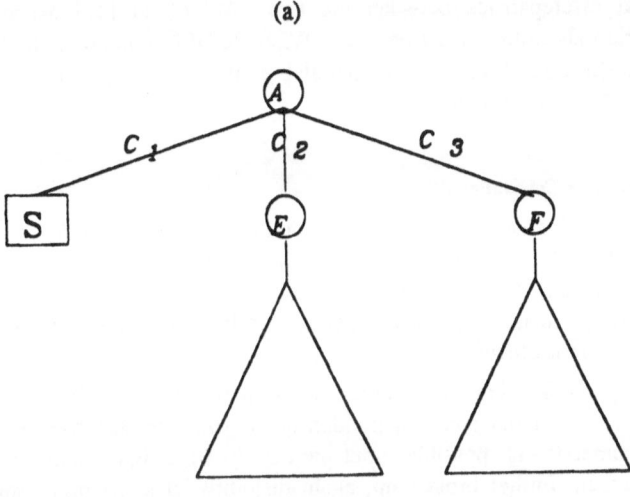

(b)
Fig. 3.3.5.1 Justification versus Enumeration

206

ckt	PLOVER	LOVER-SDIJUST	DIVER	VERIFY	ONESIM
alu4	8.3 s	4.4 s	15.1 s	131.0 s	1.3 s
c432*	27.1 s	28.9 s	182.7 s	1944.0 s	-
c432	13.67 hrs	1.56 hrs	2.2 hrs	>25 hrs	-
c880	-	4.3 hrs	12.51 hrs	-	-

c432*   : Only the first two output segments of c432
             : Job aborted due to excessive run-time

Table 3.3.5.1 Comparisons using the PROTEUS system

continuously enumerated on one circuit and simulated on the other to check any functional discrepancies between the two. Ma et. al [37] describe a parallel enumeration algorithm based on the LOVER-PODEM enumeration algorithm mentioned in Section 3.3.5.1. The parallel enumeration algorithm is based on a *dynamic scheduling* scheme.

3.3.6.1. Dynamic Scheduling

The chief goal of dynamic scheduling is to *continually distribute equal amounts of work among processors* to avoid wasteful idling and achieve high processor utilization. Good processor utilization during enumeration can be achieved by repeatedly breaking up the enumeration task(s) into smaller ones and assigning them to different processors.

The parallel enumeration algorithm is based on the LOVER-PODEM algorithm [37]. The input space is divided up into *disjoint sub-spaces* and each processor enumerates all possible input patterns in an assigned sub-space in parallel. Sub-spaces are further broken up, again disjointly, if some processors finish their assigned enumeration in the input sub-space before the others. Thus, even if the initially assigned sub-spaces are very different in enumerative complexity, processors which complete their tasks early don't remain idle but help other processors in completing their enumeration task.

Cube simulation on the cone circuit can be performed by any processor whenever the accumulated number of cubes generated by a processor is equal to the number of cubes that can be simulated in parallel by a parallel simulation algorithm. By proceeding in such fashion, an equal amount of verification work is assigned to each available processor and full utilization of processor time is achieved by continuously keeping all processors at work in parallel.

3.3.6.2. A Parallel Enumeration Algorithm

The enumeration algorithm used in LOVER-PODEM mentioned in Section 3.3.5.1 is well suited for a parallel application. Whenever a new PI assignment is made in PODEM, two disjoint input spaces are implicitly developed by the decision tree. These two input spaces correspond to the 0 and 1 values of the newly assigned input and the old values of all the previously assigned inputs. Some input values may still be unknown. Since these two input spaces are disjoint, they can be enumerated by two different processors in parallel with the guarantee that the resulting two sets of enumerated cubes will also be disjoint. Thus no redundant enumeration work is done using this technique - *each processor enumerates on a different branch of the decision tree.*

Disjoint input spaces are continually generated by all the processors doing the enumeration every time a new PI assignment is made. After a processor performs a PI assignment, it picks one of the disjoint spaces and continues enumeration on that space. As soon as a processor completes enumerating its present input space it then picks up another branch which corresponds to previously generated input spaces by other processors which have not yet been enumerated. This process continues till the entire input space has been enumerated. The selection of a new input space by a processor on the completion of its initially assigned task (this input space would have been generated by some other processor) entails an initialization overhead. It is therefore desirable to select the largest unenumerated input space available which corresponds to the space with the minimum number of assigned primary inputs.

3.3.6.3. Results

Results for five examples using dynamic scheduling are given in Table 3.3.6.1. In the table h, m and s stand for hours, minutes and seconds respectively. The first two examples are benchmark circuits from [36]. The number of outputs for the five examples are 3, 26, 2, 1 and 8 respectively. High processor utilizations were achieved [37].

3.4. Sequential Logic Verification Methods

3.4.1. Introduction

Much less work has gone into verifying sequential designs as compared to combinational designs. Sequential circuit verification is a considerably harder problem. If a one-to-one correspondence can be made between the latches/states of two sequential circuits, then the problem reduces to one of combinational logic verification, but this is not always possible.

A sequential machine can be described at register-transfer, State Transition

CKT	Number of processors							
	1		2		4		8	
	ABS. time	speed up	ABS. time	speed up	ABS. time	speed up	ABS. speed time	up
C432*	10.9h	1	5.49h	1.99	2.78h	3.92	1.38h	7.92
C880	33.9h	1	17.0h	1.99	8.54h	3.98	4.28h	7.92
ex1	69.7m	1	35.0m	1.99	17.7m	3.94	9.10m	7.68
ex2	95.4m	1	48.4m	1.99	24.2m	3.96	12.2m	7.84
alu4	104s	1	57.5s	1.81	35.3s	2.95	23.2s	4.49

C432* : only the first three outputs

Table 3.3.6.1 Results using dynamic scheduling

Graph or logic levels. Verification can be performed across the same or differing levels of abstraction. In this section, we review three different approaches to the sequential verification problem. All these methods use the State Transition Graph (STG) of the finite automata to be verified to a greater or lesser extent. Indeed, the STG represents in a compact way the structure of a finite automaton and is such a good representation to verify equivalence between two finite automata. The methods differ on how the graph is extracted from the given automaton description or how the comparison is performed.

The first approach proposed by Supowit [30] verifies two logic level sequential machines for equivalence. This approach has been used to verify finite state machines with a small number of memory elements (4-6 latches). The State Transition Graphs of the finite automata are extracted from the gate-level description using symbolic simulation techniques and compared for equivalence. This approach is restricted to Moore machines (machines whose output is purely a function of the state).

A design correctness checker for finite state machines was proposed in [38]. Given the state graph and a logic level implementation of a finite automaton, the program checks the logic level implementation for correctness using three-valued simulation. Paths in the state graph starting from the reset state are simulated on the logic description. If the reset state of the logic level machine is not known, the program finds a set of possible reset states by inspection and tries each possibility.

In [31], algorithms for verifying sequential circuits across different levels of abstraction were presented. The descriptions represent general finite automata at the differing levels – a finite automaton can be described in an ISP-like language and its equivalence to a logic level implementation can be verified using this technique. Two logic level automata or two state graphs can be similarly verified for equivalence. The efficiency of this algorithm lies in the *exploitation of don't care* information derivable from the RTL or logic level description (e.g invalid input and output sequences) during the verification process. Efficient cube enumeration procedures are used at the logic level to extract state graphs of machines from gate-level descriptions. A two-phase enumeration-simulation algorithm for verifying the equivalence of two logic level finite automata with the same or differing number of latches was also developed [31]. This algorithm is as efficient as the general approach for verifying sequential machines but is much less memory intensive. Machines with over 15 latches were verified successfully using these techniques. The equivalence problem for sequential circuits is formally defined and the three approaches are reviewed in the sequel.

3.4.2. Sequential Circuit Equivalence

A **finite automaton** (FA, finite state machine, FSM) consists of a finite set of states and a set of transitions from state to state that occur on input symbols chosen from an alphabet Σ. For each input symbol there is exactly one transition out of each state (possibly back to the state itself). One state, usually denoted q_0, is the initial state, from which the automaton starts. Some states are designated as final or accepting states.

A deterministic finite automaton (DFA) is formally denoted by a five-tuple $(Q, \Sigma, \delta, q_0, F)$, where Q is a finite set of *states* Σ is a *finite input alphabet*, δ *is the transition function* mapping from $Q \times \Sigma$ to Q, q_0 is the *initial* state and $F \subset Q$ consists of states designated as *final* or *accepting* states. That is, $\delta(q, a)$ is a state for each state q and input symbol a.

A directed graph, called a **State Transition Graph** or State Transition Diagram, is associated with an DFA as follows. The vertices of the graph correspond to the states of the DFA. If there is a transition from state q to state p on input a, then there is an arc labeled a from state q to state p in the transition diagram. The input symbol associated with this arc or edge is usually represented by a minterm (specifying a combination of primary input values). Multiple edges may exist between two states on different input minterms producing the same output. A subset (or all) of these edges may be represented more compactly by a single edge whose input symbol is a cube which is a union of the corresponding input minterms. The DFA accepts a string x if the sequence of transitions corresponding to the symbols of x leads from the start state to a final or accepting state.

The State Transition Graph of the DFA may also be equivalently represented by a State Transition Table. This table has as many rows as edges in the State Transition Graph. Each row consists of four parts – the input symbol to the corresponding edge, the state from which the edge evolves, the state which the edge enters, and the output combination asserted by the edge.

More formally, a string x is said to be **accepted** by a finite automaton $M = (Q, \Sigma, \delta, q_0, F)$ if $\delta(q_0, x) = p$ for some $p \in F$. The **language** accepted by M denoted $L(M)$, is the set $\{x \mid \delta(q_0, x) \in F\}$.

Finite automata with multiple outputs fall into two categories. The output is associated with the state in a **Moore** machine, and with the transition in a **Mealy** machine. A Moore machine is a six-tuple $(Q, \Sigma, \Delta, \delta, \lambda, q_0)$, where Q, Σ, δ and q_0 are as in the DFA. Δ is the *output alphabet* and λ *is a mapping from Q to Δ giving the output associated with each state.*

A Mealy machine is also a six-tuple $(Q, \Sigma, \Delta, \delta, \lambda, q_0)$, where all is as in the Moore machine except that λ maps $(Q \times \Sigma)$ to Δ giving the output associated with each transition.

Furthermore, a finite automaton may be described at the register-transfer (RT) or logic levels. In the former, the transitions occurring due to different input symbols (i.e. the λ and δ mappings) are specified in a high-level programming language-like description using control constructs. In the latter, the machine is described by an interconnection of logic gates and flip-flops. The logic gates implement the δ and λ mapping functions and the flip-flops (memory elements) store the state of the machine $q \in Q$. That is, the state of the machine is specified by the binary values stored in the flip-flops.

Here, we will be dealing with synchronous finite state machines. Synchronous finite state machines have the property that the inputs to the machine are sampled only at integral multiples of a duration of time called a *clock period*. The outputs of the machine also change only at integral multiples of a clock period. It is possible to define the equivalence of sequential circuits with different or same clock periods.

Each sequential circuit, C, considered here has the following properties. The circuit may be described at the register-transfer, State Transition Graph (STG) or logic level. C is assumed to be completely specified, i.e. every transition from every possible state is specified.

(1) C has exactly one output.

(2) If C is implemented by logic gates and flip-flops, it will be assumed that logic gates are delayless and that latches have unit delays (one clock cycle).

(3) All primary inputs arrive simultaneously, after $\rho(C)$ cycles, for some integer $\rho(C)$. The input lines may change values only at time $k\rho(C)$ for $k = 1, 2 \dots$ Thus the circuit is when-determinate [39].

If C is an m-input circuit, w a (possibly infinite) binary sequence, z an integer, and q_0 a state of C, then $C(w, z, q_0)$ denotes the output of C after z cycles, starting from state q_0, where the first m bits of w are input to C initially, followed by the second m bits of w, where in general bits $(r-1)m+1$ through rm of w constitute the rth set of inputs to C.

We are given sequential circuits, C_1 and C_2 along with $\rho(C_1)$ and $\rho(C_2)$ and a start state (an initial assignment to the latches for a logic level circuit, or an identified state in the given STG), q_1 and q_2 respectively.

The equivalence of C_1 and C_2 amounts to checking the condition

$$C_1(w, r\ \rho(C_1), q_1) = C_2(w, r\ \rho(C_2), q_2) \tag{3.4.2.1}$$

for all infinite binary sequences w, for all integers $r \geq 1$.

Eq. 3.4.2.1 represents a very general form of equivalence checking between sequential circuits with different input sampling rates, $\rho(C)$. In fact, one can check a combinational logic circuit for equivalence with a sequential circuit. One circuit may be a 4-bit parallel combinational adder and the other a serial single-bit sequential implementation. In this case, $\rho(C_1) = 1$ and $\rho(C_2) = 4$.

3.4.3. Difficulties in Sequential Logic Verification

Sequential logic verification is even more difficult than combinational logic verification. A trivial algorithm based on Eq. 3.4.2.1, checking all possible input sequences for equivalence, would incur huge CPU time expenditure. The State Transition Graph of a finite automaton is a more compact representation of the machine. Checking the equivalence of two machines by checking the equivalence of their State Transition Graphs is a comparatively easier task than verification by Eq. 3.4.2.1. However, given a description of a machine in the RT or logic levels, the State Transition Graph of the machine or an equivalent representation has to be extracted from the description. Extraction from logic level descriptions can be a very time and memory intensive operation.

In fact, given a logic level sequential circuit with N_S latches and N_I inputs, up to 2^{N_S} possible states can exist for the machine (some of these states may not be reachable from the given reset state of the machine, in which case they become irrelevant during verification). Each of these states has 2^{N_I} edges fanning out of it, if the edge is represented by a min-term. The number of edges in the State Transition Graph of the sequential circuit thus is $O(2^{N_I + N_S})$. The State Transition Graph of the circuit has to be extracted from the logic description to check for equivalence. Even if the cpu-time required to extract all the edges in the State Transition Graph (which is required for equivalence check) is affordable, memory requirements for storing these edges may be too expensive.

3.4.4. Supowit's approach

Supowit's approach [30] involves a formal symbolic comparison between the two given finite state machines. The algorithm assumes that the two given circuits C_1 and C_2 are Moore machines with a single output (which is also the output of a latch). A start or reset state is assumed given for each automaton. A deterministic finite automaton (DFA), given by $(Q, \Sigma, \delta, q_0, F)$, is derived from each of these circuits. Then a check is performed to see if these two DFA's accept the same language.

The DFA's, (basically State Transition Graphs of the finite automaton) are extracted from the logic description using symbolic Boolean simulation. Given an m-input circuit with start state q_0 and k latches, the extraction algorithm proceeds as shown below.

(1) The members of Q are in one-to-one correspondence with the 2^k states (i.e. assignments to the latches) of C.

(2) F is the set of states in which the output latch has the value 1.

(3) Using symbolic simulation each of the next state line Boolean expressions are extracted from the gate-level description. The next value of the i-th latch $i=1,...k$, is expressed as function $NEXT_i(x_1,x_2,...x_m,y_1,y_2,\cdots y_k)$ where $x_1,x_2,..x_m$ are the primary inputs and $y_1,y_2,..y_k$ are the current latch values. For each pair of states $v,w \in Q$, $\delta(v,w)$ is computed as follows: Let $v_1,v_2,..v_k$ $(w_1,w_2,..w_k)$ be the values of the latches corresponding to v (w). Then

$$\delta(v,w) = \prod_{i \in I^+}^{k} NEXT_i(x_1,..x_m,v_1,..v_k) \wedge \prod_{i \in I^-}^{k} \overline{NEXT_i(x_1,..x_m,v_1,..v_k)}$$

where $I^+ = \{i:w_i=1\}$ and $I^- = \{i:w_i=0\}$. If $\delta(v,w)$ is satisfiable, the edge is included in E. An expression is defined to be satisfiable if a value can be found for the input variables in the expression which asserts the value of the expression to a 1.

An example of the evaluation of $\delta(v,w)$ is illustrated in Figure 3.4.4.1.

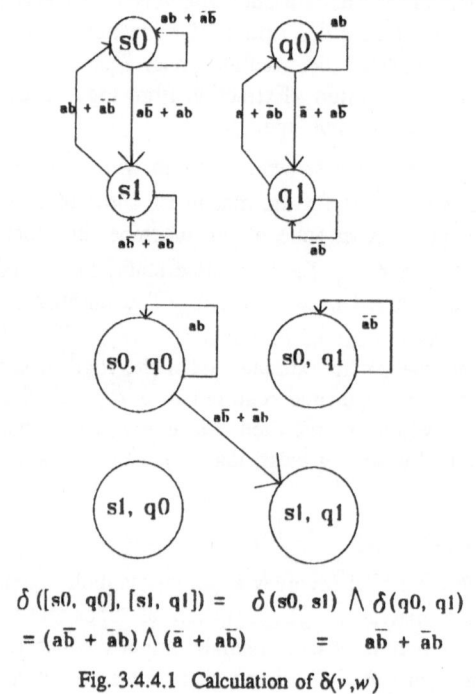

$$\delta ([s0, q0], [s1, q1]) = \delta (s0, s1) \wedge \delta(q0, q1)$$
$$= (a\overline{b} + \overline{a}b) \wedge (\overline{a} + a\overline{b}) \qquad = \qquad a\overline{b} + \overline{a}b$$

Fig. 3.4.4.1 Calculation of $\delta(v,w)$

After the extraction of the DFA's, A_1 and A_2 from C_1 and C_2, the two DFA's can be checked for equivalence using a generalization of the method used to check traditional DFA's for equivalence [40]. If $A_1 = (Q_1, \Sigma, \delta_1, s_1, F_1)$ and $A_2 = (Q_2, \Sigma, \delta_2, s_2, F_2)$, then the DFA

$$A' = (Q_1 \times Q_2, \Sigma, \delta', [s_1, s_2], F'),$$

where

(i)   for all $v_1, w_1 \in Q_1, v_2, w_2 \in Q_2$,
$$\delta'([v_1, v_2], [w_1, w_2]) = \delta_1(v_1, w_1) \wedge \delta_2(v_2, w_2),$$

(ii) E' is, the set of pairs $(v, w) \in (Q_1 \times Q_2)^2$ such that $\delta'(v, w)$ is satisfiable, and

(iii) F' is $(F_1 \times (Q_2 - F_2)) \cup ((Q_1 - F_1) \times F_2)$.

Thus, A' accepts the language
$$(L(A_1) \cap \overline{L(A_2)}) \cup (L(A_2) \cap \overline{L(A_1)}),$$

where $L(A_1)$ and $L(A_2)$ are the languages accepted by DFA's A_1 and A_2 respectively. A check is made to see if the language is empty, by determining whether there is a directed path from the start vertex of A' to one of its final vertices. If such a path exists the language accepted by A' is not empty and the machines C_1 and C_2 are not equivalent. If a path does not exist, the machines are equivalent.

This approach is efficient for machines with a reasonable number of inputs and a small number of memory elements. The State Transition Graphs of the two finite automata have to be stored as does the composite State Transition Graph. Memory requirements thus escalate rapidly with the number of latches in the two circuits. Also, the State Transition Graph extraction step entails checking the satisfiability of Boolean equations representing the edges which is an NP-complete problem. In practice, however, the satisfiability check can be performed quite quickly [30] [25]. This approach has been used to verify sequential circuits with 4-6 latches and the order of a hundred gates.

3.4.5. A Design Correctness Checker

Given a state graph of a Mealy or Moore machine and a gate-level implementation, the design correctness checker in [38] verifies that the gate-level description satisfies the state graph specification by depth-first simulation of paths from the start state in the State Transition Graph on the logic implementation. The simulation begins from a specified reset state (an assignment to latches in the logic description). In general, however, the required correspondence between the two starting states in the logic implementation and State Transition Graph is not given.

This problem is solved by finding all possible initial (reset) states for the logic description and simulating the State Transition Graph iteratively from each possible state. Three-valued (0, 1 and x) logic simulation is used. If an initial state is found for which all the paths in the State Transition Graph simulate correctly, then the logic implementation satisfies the specification for that initial state.

The number of possible initial states may explode for large circuits, but in practice this checker performs verifications quite quickly. However, it should be noted that the verification task performed by this checker is considerably easier than that of checking two logic-level finite automata for equivalence – the program assumes that a compact State Transition Graph of the machine is already given.

3.4.6. Verifying across Differing Levels

In [31], a framework for logic verification across optimization tools in a finite state machine synthesis system was described. The algorithms described can verify the equivalence of register-transfer (RT), State Transition Graph and logic level finite automata.

The State Transition Graph of a finite automaton can be extracted from the RT-level description or the logic level description. A check between descriptions at the RT and logic levels is performed by extracting the State Transition Graphs of the finite automaton from both these levels and checking the two State Transition Graphs for equivalence using the DFA equivalence algorithm described in Section 3.4.4.

Extraction from the logic level description uses cube-enumeration algorithms as opposed to the symbolic simulation techniques used by Supowit (Section 3.4.4). During the extraction the don't care information available at the RT-level is used to detect *invalid* states and edges in the logic level description's State Transition Graph and its size is reduced.

Two logic level descriptions are checked for equivalence using an enumeration-simulation approach. For large finite state machines, storing the State Transition Graphs of the machines is not viable. A technique is described where paths in the State Transition Graph of one logic-level machine are dynamically cube-enumerated and simulated on the second machine to check for equivalence.

Extraction from RT-level descriptions as performed in [31] is described in the next subsection. The enumeration-simulation algorithm is described in Section 3.4.6.2-5.

3.4.6.1. Extraction from RT-level Descriptions

The input description is at the register-transfer level, and has the following main constructs.

(1)   Procedures and functions

(2)   If and Select for control/branching

(3)   Loops - While and For.

The description is ISP-like [41] except that clock boundaries are explicitly delineated using a *wait* statement on the rising/falling edge of the clock (or clock phase) ϕ_1. A sample input description is shown in Figure 3.4.6.1.

The extraction process is only concerned with the control flow in the RTL description – a DFA controller is to be generated for the input specification. The

DFA will have the control variables as the inputs (e.g. instruction bits, ALU status bits) and assert outputs (e.g. register load, ALU add) depending on the present state.

The following steps are carried out during the DFA extraction:

```
MAIN()
BEGIN
  run = 1;
  WHILE run DO
    BEGIN
      fetch_instruction();
      effective_address();
      execute();
      IF interrupt.enable EQL 1 THEN
        IF interrupt.request EQL 1 THEN
        BEGIN
          MBR = PC ;
          MP[0] = MBR ;
          PC = 1;
          wait(φ₁);
        END
    END
END


! subroutine for effective address calculations

ROUTINE effective_address()
BEGIN

SELECT pb FROM
  [0]: BEGIN MA = 0 @ pa; END
  [1]: BEGIN MA = last.pc<0:4> @ pa; END
ENDSELECT ;

wait(φ₁);

IF ib EQL 1 THEN
BEGIN
 MA = MP[MA] ;
 wait(φ₁);
END

END  ! end of routine effective_address()
```

Fig. 3.4.6.1 Sample Input RTL Description

216

(1)　In the first pass, a one-to-one correspondence between the controlling input variables and output signals of the RTL description and the logic level description is made. For example, in the description shown in Figure 3.4.6.1, the variables *run* and *pb* are two inputs. The output signals associated with each micro-instruction are specified along with the RTL description, e.g. the micro-instruction $MA = 0 @ pa$ (where @ denotes concatenation of bit vectors) may require (a) the load signal of the MA register be high and (b) the load signal of the ALU be high with the ALU operation code 111.

(2)　Given the inputs and outputs, the description is parsed starting from the routine *MAIN* and entering and exiting all procedures in the order they are called in. If a micro-instruction is encountered then the corresponding outputs of the micro-instruction are asserted in the output of the present state. If a *wait* (ϕ_1) statement is encountered a new state is generated.

(3)　If a branch statement i.e. *IF/SELECT* is encountered, two or more states are generated depending on the number of branching conditions, a transition edge between the previous state and each possible present state is created with the corresponding input pattern. The extraction process continues with each possible present state recursively enumerating all the possible combinations. The recursion may terminate at the end of the *MAIN* routine or terminate if any input condition is violated.

The State Transition Graph for the routine *effective_address()* is shown in

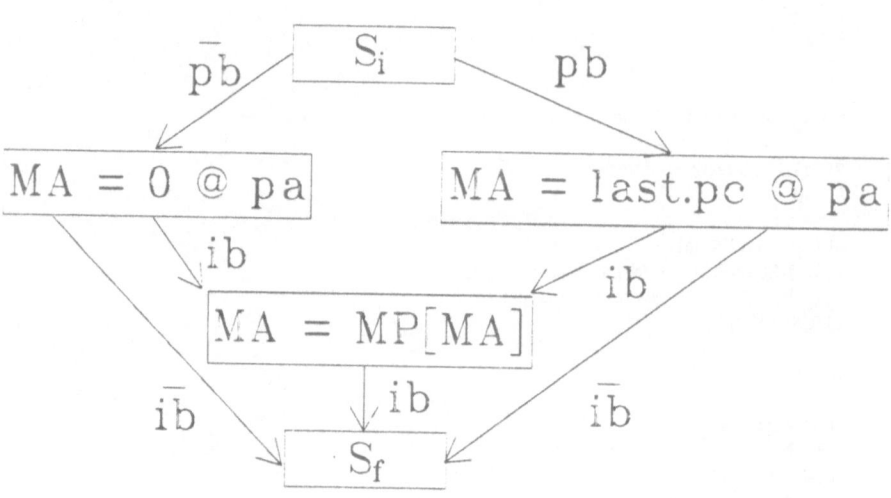

Fig. 3.4.6.2 State Transition Graph for effective_address()

Figure 3.4.6.2. Only the local inputs and outputs are shown.

3.4.6.2. The Enumeration-Simulation Approach

Two combinational logic circuits can be verified using an enumeration-simulation approach – The ON and OFF-sets of the outputs of the first circuit can be enumerated (as described in Section 3.3.5) and simulated on the other circuit to check if all input combinations produce the same values. The efficiency of this approach lies in the fact that cubes are enumerated and not min-terms, i.e. implicit but exhaustive enumeration on the input space is performed.

The same approach can be generalized to sequential circuits where no correspondence exists between memory elements. Paths starting from the reset state of the first finite automaton (usually the one with the fewer number of latches) are enumerated. Each path consists of a sequence of inputs and asserted output values representing a sequence of edges in the STG of the automaton. (The input patterns are generally cubes and not min-terms). These paths are acyclic, i.e. no state appears more than once in the edges in the path. The sequence of input cube patterns in the path is simulated on the second automaton to see if the same output values are asserted. If so, the path enumeration continues. If not, the machines are declared non-equivalent. The attractive feature of this approach is that the entire STG of the finite automaton is not stored, at any given point of time, merely a single path is stored. However, all possible paths have to be implicitly but exhaustively enumerated.

3.4.6.3. Enumerating paths in the STG

The inputs to the path enumeration program is the combinational logic block of the finite state machine and information about latch inputs and outputs, i.e. present and next state lines. This approach deals uniformly with Moore or Mealy machines.

The STG enumeration proceeds in a depth-first fashion beginning with sequential cube-enumeration of all fanout edges from the given reset state, q_0. Whenever a new edge is found, it is added to the current path if the next state it fans into does not exist in the STG. Each next state is then picked as a new starting state and the procedure repeated until no more distinct valid states can be found. All the edges in the complete STG will be implicitly and collectively enumerated in the paths. There is a hard limit, LIMIT, on the number of states in each path to restrict memory requirements. Because of this limit un-enumerated states have to be "remembered" and placed on a stack for later enumeration. At any stage all the valid states in the STG and a single path is stored. The pseudo-code shown below illustrates the global strategy used. EnumerateDfs() is initially called with the reset state, q_0, of the machine and with an empty valid state set, VAL_STATES.

This algorithm does not have to deal with the invalid states in the STG of the logic-level finite automaton. Only the states reachable from the reset state of the machine are visited during the enumeration.

```
EnumerateDfs(State, VAL_STATES)
{

UN_ENUM_STATES = φ;
PathEnumerate(State, φ, UN_ENUM_STATES, VAL_STATES);
foreach( state q ε UN_ENUM_STATES ) {
  EnumerateDfs(q, VAL_STATES) ;
  }
}
```

```
PathEnumerate(State, Path, UnEnumStates, ValStates )
{

  ValStates = ValStates ∪ State ;
  while (State.Enumerated is FALSE) {

    /* using PODEM, enumerate the next Edge fanning out from State */
    Edge = EnumerateStateFanout(State) ;

    /* check if the fanin state of the edge exists in the STG already */
    if (Edge.FaninState ε ValStates) {
      Path = Path + Edge ;
      OutputPath(Path);
    }
    else {
      ValStates = ValStates ∪ Edge.FaninState ;
      Path = Path + Edge ;
      if (Path.Length ≥ LIMIT ) {

        /* limit exceeded, push fanin state on un-enumerated stack */
        UnEnumStates = UnEnumStates ∪ Edge.FaninState ;

      }
      else {

        /* depth first enumeration continues */
        PathEnumerate(Edge.FaninState, Path, UnEnumStates, ValStates);

      }
    }
  }
}
```

The algorithm used to enumerate the fanout edges from a state, EnumerateStateFanout(), is an extension to the implicit enumeration algorithm of PODEM [33]. Initially, values of all primary inputs and next states of the logic level finite state machine are set to unknown. The logic level circuit is simulated with the present state lines fixed at their specified values. An unknown next state line is then picked and a path is backtraced from it to an unknown primary input with the objective to set the value of the chosen next state line to a known one. A 1 or 0 is assigned to that primary input. The circuit is then simulated again. The setting of primary inputs and simulation of the circuit is continued until all next state lines are set to known values – a fanout edge is enumerated. Whenever an edge is found, we backtrack to where a primary input is first set to a known value and assign it an opposite value. We then repeat the simulation and primary input setting. When no more backtracking can be done, all the edges from a state are implicitly, but exhaustively enumerated.

3.4.6.4. Simulating paths

Every time a path is produced by the routine OutputPath(), the path is simulated on the second finite automaton. Since the input patterns in the path are in general cubes and not min-terms, *cube-splitting* [24] may be necessary on the input lines to produce known output and next state values. In practice, since parallel vector simulation is used, 16 or 32 paths may be stored and simulated simultaneously on the second automaton.

The paths generated by depth-first search may be long (limited by LIMIT) and may differ only in the last few edges. So edges in the STG may be simulated a number of times if each path is separately simulated. In order to eliminate redundant work, a large set of paths (100-200) are generated and processed to identify the "tree structure" associated with them. The top level branches of the tree are simulated once and the states in the second automaton corresponding to the nodes in the tree (states of the first automaton) are "remembered" so as to simulate each branch in the tree only once.

3.4.6.5. Results using Enumeration-Simulation

Results for verifying logic level automata with known reset states for equivalence using the enumeration-simulation approach described in the previous section are given here from [31]. Table 3.4.6.1 gives both the statistics and the cpu times required for six examples which have been obtained from various industrial and university sources. The examples sbc.1 and sbc.2 are single cone output circuits from a 28 latch FSM in the Snooping Bus Controller of the Berkeley SPUR chip set.

Verification was performed between implementations of the same circuit with different encodings, implying no correspondence between the latches of the two circuits. The largest example, sbc.2, has 17 latches and 2764 valid states (total number of states is 131072) and verification was possible in 5.36 hours on a VAX 11/8650. Over 2.6 million edges were enumerated. However, these edges did not have to be stored, since the paths were dynamically generated and simulated on the second finite automaton. Most existing approaches to sequential verification are unable to deal with such large finite automata. Memory usage was restricted to less than 1M for all these examples. Cube enumeration drastically reduced the number of edges generated – min-term enumeration on each state would have

EXAMPLE	#inp	#out	#gates	#latcheS	#valid states	#edges in STG	CPU time
cse	7	7	192	4	16	167	1.2s
sand	11	9	555	6	32	237	4.9s
planet	7	6	606	6	48	182	4.2s
scf	27	56	959	8	115	393	15.6s
sbc.1	31	1	465	13	2040	4846383	8.94h
sbc.2	27	1	492	17	2764	2662236	5.36h

s denotes CPU-seconds and h denotes CPU-hours on VAX 11/8650

Table 3.4.6.1 Verification using Enumeration and Simulation

resulted in over 3.7×10^{11} (= 2764 * 2^{27}) edges.

3.5. Future Work in Logic Verification

As synthesis systems evolve, more and more complex optimization tools are used in the synthesis pipeline. The importance of verification algorithms which can ensure that these optimization tools do not introduce any design errors increases.

Presently, combinational logic verification algorithms can perform equivalence checks for circuits with thousands of gates. For instance, successful equivalence checks of gate or switch-level implementations of 64-bit ALU's against their specifications are not uncommon. However, verifying even a small circuit may result in a combinatorial explosion due to the Boolean structure of the circuit and the inherent intractability of the verification problem. For example, the verification of two gate-level exclusive-or functions using the verification by justification algorithms results in exponential run-times. Algorithms tailored for exclusive-or functions perform badly for others. Verification of larger and larger circuits is required in a synthesis environment. The search for more efficient algorithms continues.

Sequential logic verification algorithms are less developed. A few approaches have been taken to solve the problem (e.g. [30] [38] [31]. Much more work is required. Verifying sequential machines across differing levels is an important problem. Presently, algorithms exist for verifying register-transfer-level descriptions against logic implementations. However, as high-level synthesis tools (e.g. datapath synthesis tools) develop, *behavioral verification* techniques will be required. In behavioral verification, the functional specification of the design in the high-level programming language is verified against the structural specification which has been synthesized using hardware resource allocation and control synthesis steps. Verification of software programs has been attempted using theorem-proving techniques [42]. Hardware verification at the behavioral level is wide open.

Parallel computers can alleviate these problems to a certain extent. A parallel combinational logic verification algorithm was first presented in [37]. Others have been proposed [43]. Parallel sequential logic verification algorithms are very attractive. The enumeration-simulation approach to sequential verification described in Section 3.4.6.2 is eminently suitable for a parallel implementation. The efforts in parallel logic verification have concentrated on implementations on shared memory computers with a small number of powerful processors (e.g. the 12 processor Sequent Balance 8000). Massively parallel computers (e.g. the Connection Machine) could increase the capabilities of verification algorithms a great deal.

4. THE TESTING PROBLEM

4.1. Introduction

Testing of VLSI circuits is a process to ensure that a chip satisfies its functional specification. For testing a circuit, binary patterns, called test patterns or tests, are applied to the inputs of the circuit and the response of the circuit is compared with the expected one. Application of all possible input patterns, for combinational circuits, will guarantee that the chips passing the test are functionally correct. However, this exhaustive method becomes infeasible, in terms of testing time, when the number of inputs is large. In practice, a set of test patterns that are aimed to detect a high percentage of modelled faults is used. The most widely used fault model has been the *stuck-type* model [13]. Physical failures are assumed to correspond to a line in the gate-level description of the circuit stuck at a 0 or 1 value and an assumption is made that only one fault can occur at a time. It has been empirically shown that a high percentage of the chips passing the set of test patterns for stuck-type faults are correct working chips.

Test generation for combinational circuits has traditionally been considered to be a search problem [32] [33]. A test pattern for a fault is generated by searching through the input space to find an input pattern that excites the fault and propagates its effect to one of the primary outputs. The cost of test generation can be very high and it has been proved that the problem of test generation is NP-complete [11]. It is especially expensive to generate tests for circuits that contain a large number of redundant faults. Redundant faults are faults for which no test can be found after searching, implicitly or explicitly, across the entire input space. The cost for trying to generate tests for redundant faults can be more than 90% of the total test generation time. Redundant faults are due to redundancies in a circuit. Redundancies may be introduced intentionally for reliability, performance or other reasons; but often they are due to unoptimized designs. It can therefore be conjectured that designs which have gone through logic minimization, which in general aims to reduce the overall size of a circuit by removing logic redundancies, are more testable and easier to generate tests for.

The relationship between logic minimization and test generation [44] is strong. Test generation algorithms can be, in general, modified for logic minimization to identify redundancies in a circuit. However, removing all the redundancies of a circuit does not always produce an optimal solution. For example, a 2-level network without any redundancies may have an equivalent multilevel representation that is smaller in some sense. Logic minimization can, in turn, guarantee that the optimized network is 100% testable. A multi-level logic minimization algorithm using implicit don't cares [44] is described in Section 4.2. This algorithm not only produces networks that are 100% testable but also generates all the test vectors as a byproduct of the minimization. Logic optimization and test generation problems can thus be solved simultaneously.

Generating tests for sequential circuits is considerably harder than for combinational circuits. Even if the combinational part of a sequential circuit is made fully testable via logic minimization, it may still be impossible to obtain a high fault coverage for the sequential circuit. Some of the inputs and outputs of the combinational part are outputs and inputs respectively of the memory elements, i.e. flip-flops. Test patterns generated considering only the combinational part cannot be readily applied and fault effects cannot be observed directly at the inputs of these memory elements. A great amount of literature on test generation for sequential circuits has been published. A detailed review of all of them would run into pages of descriptions. In the sequel, only a few representation techniques will be outlined. In Section 4.3, the problem of generating tests for sequential circuits is discussed and a new sequential testing technique based on the concept of state space enumeration is presented.

4.2. Relationship between testing and logic minimization

4.2.1. Primality and Irredundancy

Given a Boolean network, η, a cube c of the 2-level representation of F_j is **prime** if no literal of c can be removed without causing the resulting network η' to be **not** equivalent to η. Similarly, a cube c of F_j is **irredundant** if c cannot be removed from F_j without causing the resulting network η' to be **not** equivalent to η. A Boolean network η is said to be **prime if** all the cubes in each of the representations F_j of η are prime, and **irredundant if** all of these cubes are irredundant.

These two concepts are assoicated with local minima of a cost function which is nondecreasing in the total number of cubes and literals required to represent the incompletely specified logic functions, realized by the given Boolean network.

An **internal** stuck fault is a fault in which v_k (or $\bar{v}_k$) of cube c of representation F_j or a node n_j of a boolean network is stuck at either its existing value v_k (or $\bar{v}_k$) or its opposite value $\bar{v}_k$ (or v_k). If each 2-level function F_j is physically implemented with an And-Or complex gate, each internal fault would correspond to an input stuck-at fault in the gate-level representation of the Boolean network.

Theorem 4.2.1.1. A Boolean Network is prime and irredundant if and only if it is 100% testable for internal stuck faults.

This theorem relates directly the testability of the Boolean network with concepts of logic minimization. The proof of the theorem for a Boolean network can be found in [44]. Similarly for a gate-level implementation of a Boolean network, the corresponding concept of **primality** and **irredundancy** can be defined for primitive gates AND, OR, NAND, NOR, NOT. An input stuck-at fault of a primitive gate is equivalent to an internal stuck fault of a node of a boolean network corresponding to that primitive gate.

A primitive gate is **prime** if none of its inputs can be removed without causing the resulting circuit to be functionally different. A gate is **irredundant** if its removal causes the resulting circuit to be functionally different. A gate-level circuit is said to be **prime** if all the gates are prime, and **irredundant** if all the gates are irredundant.

Corollary 4.2.1.1. A gate-level circuit is prime and irredundant if and only if it is 100% testable for all single stuck-at faults.

4.2.2. Mutlilevel logic minimization using implicit don't cares

Each node in the Boolean network, when considered in isolation, has its own inputs and output and is, in general, easily testable; but, embedded in a circuit, it may be very difficult to test. This is because, situated within a circuit, some node-input patterns can never occur and, only under certain node-input patterns the output of the node can be observed at or contributes to some primary outputs of the circuit. This can be related to the controllabilities and observability of the inputs and output respectively of the node. The set of node-input patterns that can never occur is called *fan-in don't care set* of the node. The set of node-input patterns, at which the node output does not contribute to any of the primary outputs of the circuit, is called *fan-out don't care set* of the node. The union of the two sets is simply called the *don't care set* D_j of the node.

If a fault in the node has a test set, when the node is considered in isolation, which is a subset of the don't care set D_j of the node, then it is a redundant fault. This is because neither the required test pattern can be generated nor the effect of the fault, even if excited, can be propagated to one of the primary outputs. This indicates that the don't care set D_j of a node not only plays a prominent role in logic minimization, but also is of paramount importance in making the whole circuit 100% testable. It can be proved that a circuit is prime and irredundant if the following conditions are satisfied for every node.

1) if $c \in F_j$, then $c \subseteq (F_j - \{c\}) \cup D_j$
2) if $v_l(\overline{v_l})$ appears in c, then $c' \subseteq F_j \cup D_j$, where c' is the cube obtained by deleting literal $v_l(\overline{v_l})$ from cube c.

Condition 1 guarantees that there is no redundant cubes in any of the representation F_j, and Condition 2 guarantees that each cube is prime.

The multilevel logic minimization algorithm in [44] with an optimization procedure aimed at satisfying conditions 1 and 2 and using the don't care sets produces reasonable minimization results. The procedure computes for each node n_j its associated don't care set D_j and minimizes each F_j in sequence with respect to its associated D_j by calling an efficient 2-level minimizer (the ESPRESSO program [5]). The minimization of F_js is iterated until, on one complete pass through it, no F_j changes from what it had been on the previous pass. The test vectors for the circuit also fall out as a byproduct during the minimization of F_js. Another concept, called R-Minimality, that further reduces the prime and irredundant Boolean network to a smaller one is incorporated in the procedure. Briefly stated, R-Minimality means that no one of the individual two-level functions in Boolean network can be re-expressed in terms one or more of the others to map the given prime and irredundant Boolean network into another one with less logic cost.

An example is illustrated in Figure 4.2.1.1 which shows 3 equivalent Boolean networks. The network of Figure 4.2.1.1(a) has 3 nodes, and is neither prime nor irredundant, and has non-testable stuck-at-1 faults at node x_1, and x_2, and a nontestable stuck-at-0 fault at node y_2. The equivalent Boolean network of Figure 4.2.1.1(b) is prime, irredundant and 100% testable, yet requires 9 literals and 5 product terms, and is not R-Minimal. The equivalent network of Figure 4.2.1.1(c) is similarly prime, irredundant, and testable but is R-Minimal, and requires only 2 nodes, 5 literals and 3 product terms.

Generation of all the complete don't care sets D_js can be very expensive. In practice, only subsets are used as approximations which can reduce the testability of the circuit and result in non-optimal logic implementation. The results of running the multilevel logic minimizer ESPRESSO_ML are given in Table 4.2.1.1. In the Table, "initial literals" refers to the number of literals in the original multilevel network and "literal saved" refers to the number of literals saved after logic minimization. Runtimes in the table are in CPU seconds on a Pyramid 90X, which is about twice as fast as a VAX 11/780. The CPU time requirements may become exorbitant for large circuits.

4.3. Sequential Test Generation

4.3.1. Introduction

Test generation for sequential circuits has long been recognized as a difficult task [45] [46] [13] [47]. Unstructured random sequential designs are very difficult to test. One common approach to improve the testability of a sequential circuit is to add test points to the circuitry so that tests can be applied more readily and fault effects can be observed better. But this method is not systematic and relies on designer ingenuity.

A popular approach to solving the problem of test generation for sequential circuits is to make all the memory elements controllable and observable, i.e. Complete Scan Design [15] [48], which was reviewed in Section 1.4.1. Scan design approaches have been successfully used to reduce the complexity of the problem of test generation for sequential circuits by transforming it into a combinational one which is considerably less difficult. The design rules of Scan Design also constrain the sequential circuits to synchronous ones so that the normal operation of the sequential circuit is free of critical races. However, there are situations where the cost in terms of area and performance of complete scan design is unaffordable. In addition, even though the general sequential testing problem is very difficult, there may be cases where test generation can be effective. Simply making all the memory elements scannable in a sequential circuit without first investigating how difficult is the problem of generating tests for it could unduly incur unnecessary area cost.

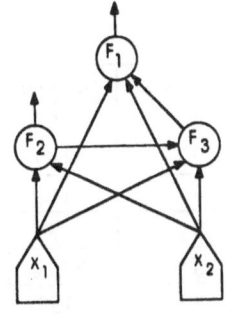

$F_1 = \bar{z}_1\bar{z}_2 + y_3$
$F_2 = z_1\bar{z}_2 + \bar{z}_1z_2$
$F_3 = z_1z_2\bar{y}_2 + \bar{z}_1\bar{z}_2$

$PO = \{y_1, y_2\}$
Number of literals = 12.
Number of cubes = 6.
Number of functions = 3.

Non-testable input stuck-at faults:
    y_2 SA0 in F_3
    z_1 SA1 in F_3
    z_2 SA1 in F_3.

Circuit is not prime, irredundant, or R-minimal.

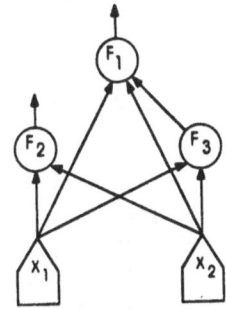

$F_1' = \bar{z}_1\bar{z}_2 + y_3 = F_1$
$F_2' = z_1\bar{z}_2 + \bar{z}_1z_2 = F_2$
$F_3' = z_1z_2$

$PO = \{y_1', y_2'\}$
Number of literals = 9.
Number of cubes = 5.
Number of functions = 3.

Non-testable input stuck-at faults: None.

Circuit is prime and irredundant.
However, it is not R-minimal.

(a) Prior to Multi-level minimization (b) Prime Irredundant equivalent

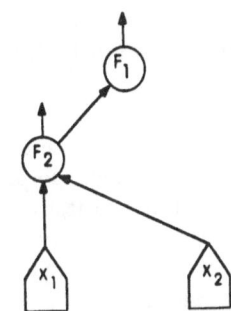

$F_1'' = \bar{y}_2$
$F_2'' = z_1\bar{z}_2 + \bar{z}_1z_2 = F_2$
$F_3'' = \emptyset$(deleted)

$PO = \{y_1'', y_2''\}$
Number of literals = 5.
Number of cubes = 3.
Number of functions = 2.

Non-testable input stuck-at faults: None.

Circuit is prime and irredundant and R-minimal.

(c) Prime Irredundant R-Minimal equivalent
Fig. 4.2.1.1

name	initial literals	literals saved	runtime
f3	73	2	57
f4	75	2	130
f5	75	5	118
insex	79	5	57
plac	191	28	3733
8fun	83	3	152
exam2	73	3	73
adder	48	4	87
dec2	149	4	7852
plab	119	20	1850
z4	58	20	77

Table 4.2.1.1 ESPRESSO_ML Multilevel Minimization Results

The difficulty in generating a test usually lies with: 1) setting the states of the memory elements into a certain combination so that the fault under test is excited; 2) propagating the fault effect to the primary outputs. An input sequence is usually required in both cases (if such a sequence exists). In general, the longer the length of the shortest input sequence needed to perform steps 1) and 2), the more difficult it is to find an input sequence to test the circuit. Both approaches mentioned above attempt to shorten the length of the input sequence. In the scan design approach, the length of the input sequence is reduced to one when all memory elements are made scannable.

Several approaches [49] [50] [12]][51] [52] [53] have been taken in the past to solve the problem of test generation for sequential circuits. They are either extensions to the classical D-Algorithm or based on random techniques [50] [52]. When the number of states of the circuit is large and the tests demand long input sequences, they can be quite ineffective for test generation. This is because no *a priori* knowledge of the length of the test sequence is available. In the extended D-Algorithm methods, a large amount of effort may be wasted in trying to find short sequence tests for faults that require long ones. Random testing techniques are based on continuous simulations and grading of test vectors according to simulation results. They can be very time consuming for difficult faults that have only a few long test sequences.

In this section, a few representative sequential test generation approaches will be reviewed and a recently proposed approach which alleviates problems mentioned above is described [14].

4.3.2. The Extended D-algorithm for Synchronous Circuits

A Synchronous sequential circuit S can be modeled as an iterative array C^p, shown in Figure 4.3.2.1, by cutting the feedback loops at the clocked flip-flops. The combinational circuits C_i, where $i = i,...p$, are all identical to the combinational portion C of the original sequential circuit and all flip-flops are modeled as combinational elements, referred to as pseudo flip-flops. The iterative array is logically equivalent to the sequential circuit - the temporal response of the sequential circuit is mapped into a spatial response of the iterative array. If an input sequence $x(1),x(2),...x(p)$ is applied to S in initial state $y(I)$ generating output sequence $z(1),z(2)...z(p)$, then the iterative array will generate the output z_i from cell C_i in response to the input x_i to cell C_i and $z_i = z(i)$ if $x_i = x(i)$ and $y_I = y(I)$. A single fault in S corresponds to the multiple fault f^p consisting of the same fault in every cell C_i of C^p. The test generation procedure for a self-initializing test sequence based on the iterative array model is as follows:

[1)] Determine the maximum number of time frames, p, allowed for test generation.

[2)] Choose an initial value for p and construct the iterative array model with y_I of unknown value. (If a reset state is given, y_I will be assigned the value of the reset state.)

[3)] Choose the time frame q from which the D-drive must be organized. Apply the D-algorithm to find a test for the multiple fault f^p so that a D or $\overline{D}$ appears at one of the outputs $z_1, z_2, ... z_p$. If a test is found, exit; otherwise, continue.

[4)] If possible, increment p by 1 and return to step 3); otherwise, exit with no test.

Since the length of test sequence cannot be determined *a priori*, a large amount of effort may be wasted in trying to generating tests with inappropriate choice of p.

4.3.3. Weighted random test-pattern generator

In a random test-pattern generator, sequence of random patterns are applied to the cirucuit. In general, all primary inputs (PI's) of the circuit have the same weights, i.e. each PI is exercised approximately the same number of times averaging over a long period of time. However, not all the PI's have the same functional importance, some being more instrumental in exciting faults and propagating their effects. In the weighted random test-pattern generator [50], different weights are assigned to the PI's in proportion to their relative importance, i.e. some PI's are exercised more often than others. Single input change between two consecutive patterns is assumed.

One way to determine the weight assigned to each PI is to measure the amount of gate switching activity produced inside the circuit as the result of exercising that PI. A set of random patterns is simulated on the circuit. The number of gates changing for the first time from a logic 1 to 0, and vice versa, due to the switching of any of the PI's, is counted. The switching activity count is then accumulated over the complete set of random patterns. By comparing the

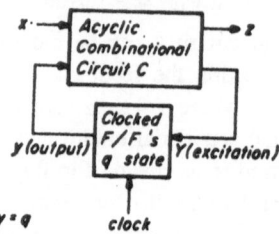

(a) Synchronous Sequential circuit

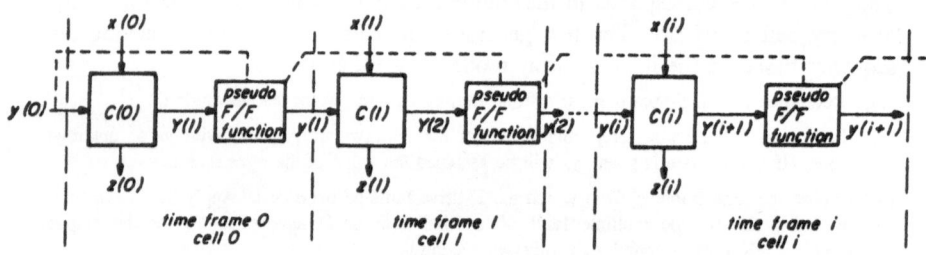

(b) Equivalent Combinational Iterative Array
Fig. 4.3.2.1

activity created by all PI's, different weights can be determined for all PI's. However, this method suffers from the fact that the importance of the order of patterns applied to detect a fault is ignored. Furthermore, test sequences consisting of more than one change between consecutive patterns cannot be generated.

A dynamic adaptive technique has been used to partially alleviate the problem of ignoring the order of test patterns mentioned above in weighted random test-pattern generation. This technique introduces the rates of changes of activity into the function of determining the weight for each PI. The switching activity count for each PI is accumulated over a subset of the random patterns simulated instead of the complete set. This is done by disregarding activity counts for the first few random patterns. The point at which the recording of the activity counts is started is determined by some heuristics. Results show that this technique achieves a significant improvement in fault coverage over the static weighted random test-pattern generators. A reduction technique is used to reduce the total number of random patterns generated as random approaches usually create a large number of test patterns. Random pattern techniques offer no guarantees of test coverage/redundancy identification unlike deterministic test pattern generators.

4.3.4. A new approach to sequential test generation

This approach to test pattern generation for sequential finite state machines represents a significant departure from previous methods and is largely based on the concept of *state space enumeration*. The problem of generating tests for faults that require a lengthy input sequence is handled efficiently by the intelligent use of information contained in a partial State Transition Graph (STG) and the integration of a few new algorithms based on the concept of state space enumeration.

The sequential circuit under test is assumed synchronous and free of races under simple design rules. A reset state for the synchronous sequential machine is assumed to be given and memory elements such as D flip-flops are identified and represented as logical primitives to facilitate loop cutting in transforming the synchronous sequential circuit into an iterative array. First, a part of the State Transition Graph (STG) of the finite state machine is extracted using purely structural information, i.e. the gate-level description of a sequential circuit. The construction of the partial STG is based on an efficient state-enumeration algorithm that finds paths from the reset state to different valid states (states reachable from the reset state) in the STG. For circuits with relatively few states, a partial STG including all valid states is built. For circuits with a large number of states, only a subset of valid states is included in the partial STG. The partial STG is then used in conjunction with efficient enumeration-based fault excitation-and-propagation and state justification algorithms for generating tests for line stuck-at faults. Using this technique, tests have been successfully generated tests for finite state machines with a large number of states and close to maximum possible fault coverages have been obtained using reasonable amounts of CPU time.

The test generation process is outined in Section 4.3.4.1 Extraction of the fully or partially connected state transition graph from the logic level finite state machine is described in Section 4.3.4.2 The enumeration-based fault excitation-and-propagation and state justification algorithms are described in Section 4.3.4.3 and 4.3.4.4 respectively. Section 4.3.4.5 describes the detection of a special class of redundant faults. Results for a number of finite state machines are presented in Section 4.3.4.6.

4.3.4.1. The Test Generation Process

Assuming the complete State Transition Graph (STG) of a sequential circuit is available, test generation for a fault under test can be done by first finding an input sequence $T1$ and an initial state $S0$ that excite and propagate the effect of the fault to the primary outputs within 4^n time frames, where n is the number of latches in the sequential circuit. A reset state for the machine is assumed to be given from which all test sequences begin. Then every path from the reset state to any state $S1$ that covers $S0$, a potential setup sequence, in the complete STG is fault simulated. If a path $T0$ (setup sequence) to a state $S1$ that covers $S0$ can be found under fault conditions, a test sequence $T2$ is generated by concatenating the path $T0$ with $T1$. Even though a setup sequence $T0$ may not be found, the fault may still be detected by one of the potential setup sequences through fault simulation. If this is the case, that particular potential setup sequence itself can serve as a test sequence $T2$. If no test sequence can be found, a new input sequence $T1$ and a new initial state $S0$ which is disjoint from all previously generated ones is searched and the procedure is repeated.

The algorithm is complete, i.e. if a fault is testable, a test will be found given sufficient time. The main drawbacks of this method are: (1) the memory storage for the complete STG may be unreasonably large and the generation of the complete STG may demand astronomical CPU time; (2) fault simulation of all potential setup sequences is extremely time consuming. A remedy to (1) is to generate the potential setup sequences on-the-fly using a backward justification algorithm that searches for paths from the reset state to the $S0$'s under fault-free conditions. No information of the STG is required/used.

A test generation algorithm following the ideas presented above is as follows.

Algorithm Structure 1

(1) Find an (new) input sequence $T1$ and an (new) initial state $S0$ that will excite and propagate the effect of the fault under test to the primary outputs within 4^n time frames using the state-enumeration-based test generation algorithm (described in Section 4.3.3.3). If no solution exists, exit without a test.

(2) Find a (new) path $T0$ (potential setup sequence) from the reset state to the initial state $S0$ using a backward justification algorithm. If no solution exists, go to (1).

(3) Fault simulate the potential setup sequence $T0$. If it detects the fault, generate the test sequence $T2$ from $T0$ and go to (5). Else if it is a valid setup sequence, go to (4). Else if $T0$ neither detects the fault nor is a setup sequence go to (2).

(4) Concatenate the input sequence $T0$ that represents the path from the reset state to the initial state $S0$ with $T1$ to form $T2$ which is the test sequence for the fault under test.

(5) Exit with a test sequence.

Even though this algorithm is potentially effective, backward justification in general is difficult when the setup sequence is long. In addition, some states may need to be justified more than once. Therefore, an important enhancement is to generate *a partial STG containing as many valid states* (and paths from the reset states to them) as possible provided that the partial STG extraction process (through forward enumeration as described in Section 4.3.4.3) is carried out efficiently. Note that the partial STG may contain all the valid states in the complete STG but contains much fewer edges. States and edges may be *added* to the partial STG via backward justification during test generation.

The second drawback mentioned above, i.e. that fault simulation of all potential setup sequences is very time consuming, does not actually pose a problem. Observations in experiments conducted in [14] indicate strongly that if $T0$ is an invalid setup sequence, it is very likely to be a test sequence. Therefore, there is rarely the need for fault simulation of more than one potential setup sequence for a fault.

Finally, an efficient test generation algorithm combining the advantages of forward enumeration and backward justification by using the partial STG is as follows.

Algorithm Structure 2

(1)   Find an (new) input sequence $T1$ and an (new) initial state $S0$ that will excite and propagate the effect of the fault under test to the primary outputs within a prescribed number of time frames using the state-enumeration-based test generation algorithm (described in Section 4.3.4.3). If no solution exists, exit without a test.

(2)   Search for a path (potential setup sequence) $T0$ from the reset state to $S0$ in the partial STG. If it is found, go to (5).

(3)   If the partial STG includes all valid states, go to (1).

(4)   Find a path $T0$ from the reset state to the initial state $S0$ using the state justification algorithm (described in Section 4.3.4.4). If no solution exists, go to (1).

(5)   Fault simulate the potential setup sequence $T0$. If it detects the fault, generate the test sequence $T2$ from $T0$ and go to (7). Else if it is a valid setup sequence, continue. Else go to (1).

(6)   Concatenate the input sequence $T0$ that represents the path from the reset state to the initial state $S0$ with $T1$ to form $T2$ which is the test sequence for the fault under test.

(7)   Exit with a test sequence.

The initial state $S0$ can be a cube containing don't care bits or a min-term with every state bit specified. In the case of a cube, a path from the reset state to a minterm covered by $S0$ can serve the purpose of a setup sequence.

4.3.4.2. State Transition Graph Extraction

The inputs to the logic level extraction program is the combinational logic block CLB of the finite state machine and information about latch inputs and outputs, i.e. present and next state lines illustrated in Figure 4.3.2.1(a). The output is a partial State Transition Graph (STG) of the finite state machine. A node in the STG represents a distinct state and an edge between two nodes represents an input combination (cube) that drives the finite state machine from one specific state to another.

The STG extraction first sequentially cube-enumerates all fanout edges from the given reset state. Whenever a new edge is found, it is added to the current STG if the next state it fans into does not exist in the STG. Each next state is then picked as a new starting state. The procedure is repeated until no more distinct valid states can be found. All the edges in the complete STG will be implicitly, but exhaustively enumerated. The partial STG constructed is a tree, i.e. there is only a single path from the reset state to any other state. This is to restrict the storage space for the partial STG so that synchronous sequential machines with very large number of states can be handled.

The algorithm used to enumerate the fanout edges from a state is an extension to the implicit enumeration algorithm of PODEM [33]. Initially, values of all primary inputs and next states of the logic level finite state machine are set to unknown. The logic level circuit is simulated with the present state lines fixed at their specified values. An unknown next state line is then picked and a path is backtraced from it to an unknown primary input with the objective to set the value of the chosen next state line to a known one. A 1 or 0 is assigned to that primary input. The circuit is then simulated again. The setting of primary inputs and simulation of the circuit is continued until all next state lines are set to known values - a fanout edge is enumerated. Whenever an edge is found, the algorithm backtracks to where a primary input is first set to a known value and assigns it an opposite value. Then the simulation and primary input setting is repeated. When no more backtracking can be done, all the edges from a state are implicitly, but exhaustively enumerated.

The extraction process can proceed in either a depth-first or a breadth-first fashion. In the breadth-first fashion, the path from the reset state to any state in the partial STG is the shortest one. The test sequences generated are shorter but the total number of test sequences is greater than using a depth-first algorithm. There are hard limits, $L1$ and $L2$, for the total number of states to be included in the final STG and the number of states at each level from the given initial state. $L1$ is used to restrict the memory usage and $L2$ restricts the maximum length of the test sequence. The pseudo-code below illustrates the partial STG extraction process in a depth-first fashion. Extract() is initially called with the reset state of the sequential circuit and the level equal to 0.

```
Extract(State, level)
{
  PresentState = State;
  PrimaryInput = unknown;
  simulate the circuit;

  while (not all edges have been enumerated) {

    if ((TotalNumStates ≥ L1) ||
      (NumStates[level] ≥ L2)) break;

    if (not all NextState lines are set) {
      find_new_pi_assignment();
      simulate with current set of pi assignments;
    }
    else {
      if (NextState is not in the partial STG) {
        add NextState to partial STG;
        TotalNumStates = TotalNumStates + 1;
        NumStates[level] = NumStates + 1;

        Extract(NextState, level + 1);
      }
      else {
        backtrack to the last set primary input
        and assign an alternative value to it;

        simulate with current set of pi assignments;

      }
    }
  }
}
```

An alternative to the backtracing/backtracking approach to STG enumeration described above is forward simulation on the input space given a starting present state. The forward simulation process begins with all the input lines set to unknown values. Inputs are set randomly to 0 or 1 in a pre-specified order till all the next state lines are all set to known values. Backtracking on primary input values is done after setting all next state lines. However, this approach is less efficient than the approach described earlier because a primary input value may be unnecessarily set in order to set the next state lines. This can lead to a great amount of redundant simulations. On the contrary, in the backtracing/bracktracking approach, the backtracing process makes sure that the next primary input to be set and the simulation following the value-setting always contribute to the setting of the next state lines.

4.3.4.3. The Fault Excitation-and-Propagation Algorithm

The Fault Excitation-and-Propagation algorithm (FEP) is based on the decision tree concept of the test pattern generation algorithm PODEM. FEP uses the conventional iterative array model for generating an input sequence $T1$ and an initial state $S0$ to excite and propagate the effect of the fault under test to the primary outputs within a prescribed number of time frames. The iterative array is considered wholly as a combinational circuit with primary inputs of different time frames time-indexed and the present state lines of the first time frame treated as pseudo inputs. The initial state $S0$ is specified by the pseudo inputs values. FEP first tries to propagate the fault effect to the primary outputs of the first time frame. If it fails, it will use the primary outputs of the second time frame for fault propagation and so on until the prescribed number of time frames is reached.

FEP uses *two decision trees*, one for the primary inputs of different time frames and the other for the initial state $S0$, as opposed to only one in PODEM. The two decision trees are built in a similar way through the backtracing and backtracking processes as used in PODEM. The present state lines of the first time frame are treated similarly as the primary inputs during the fault excitation-and-propagation process. Values of the present state lines and primary inputs of different time frames are continuously set one at a time through the backtracing process and the iterative array is simulated whenever a primary input or a pseudo iput is set to a known value. The value-setting-and-simulation process continues until the effect of the fault under test is excited and propagated to the primary outputs of at least one of the time frames or when the backtracking limit is reached. Backtracking takes place whenever it has established that under the current set of primary input and pseudo input assignments, the effect of the fault under test cannot be excited and/or observed at the primary outputs of the specified time frame with further input assignments. Backtracking during the search for $T1$ and $S0$ is done on both decision trees.

FEP employs the concept of disjoint state enumeration to make sure that all the tests it generates for a specific fault will have disjoint initial states $S0$'s; this is necessary because of the loop in the test generation process described in Section 4.3.4.1. Whenever the search for a new test is begun, the primary input decision tree ($D1$) for the previous test is scratched completely but the present state decision tree ($D2$) of the initial state $S0$ is retained. Immediately backtracking is done on $D2$. Then the value-setting-and-simulation process is carried out as described above. The reason that tests generated for a specific fault by FEP should all have disjoint $S0$'s is related to how FEP is used in the test generation process as described in Section 4.3.4.1. For a specific fault, a new test is requested only if the path from the reset state to the $S0$ in the previous test cannot be found, either in the extracted STG or through the state justification algorithm described in Section 4.3.4.4. Therefore, all tests generated for a specific fault should have disjoint $S0$'s.

A single decision tree could have been used instead of two separated ones as described above. And instead of completely resetting all primary input values to unknown, i.e. scratching the entire primary input decision tree, when a new search is started, one can simply backtrack on the single decision tree to where a pseudo input is first set to a known value and assigned it an opposite value. But due to the inherent characteristics of the enumeration approach of PODEM, it is more efficient to begin a search with as small a number of preset inputs as possible. Therefore the double decision tree method is used.

4.3.4.4. The State Justification Algorithm

Given a goal state $S0$, the state justification algorithm (SJ) attempts to find a path (setup sequence) from the reset state to it. $S0$ can be a cube containing don't care state bits or a minterm with every state bit specified. In the case of a cube, SJ needs only to find a path to any minterm state that is covered by $S0$.

First, SJ sequentially enumerates all the fanin edges to $S0$. It then checks whether any state the edges fanout from covers the reset state. If such a state exists, a path is found. Otherwise, SJ picks each fanin state as a new goal state and carries out fanin edge enumeration again. The procedure is repeated until a path is found or no path can be found. SJ actually proceeds in a depth-first fashion and there is a limit on the maximum length of the justification sequence.

The edge enumeration algorithm is an extension to the enumeration algorithm LOVER-PODEM in [37]. The difference is that here multiple line (the next state lines) values have to be justified simultaneously rather than a single output line as in LOVER-PODEM. The concept of state enumeration is also employed in SJ. There are two decision trees to be maintained as in Section 4.3.4.3, i.e. one ($D1$) for the primary inputs and the other ($D2$) for the present state lines. All the present state lines and primary inputs are set to unknown values initially. Through backtracing and backtracking processes, the primary inputs and present state lines are continuously set to some known values, 1 or 0, until all the next state lines are found to be set to their specified values through simulation. Whenever the search for a new fanin edge is begun, $D1$ is completely scratched but $D2$ is retained. Immediately backtracking is done on $D2$. Then the enumeration procedure is repeated again. All edges (with disjoint fanin states) fanning out of a state are enumerated when no more backtracking is possible. The pseudo code below illustrates the state justification algorithm proceeding in depth-first fashion. Breadth-first search is an alternative.

```
Justify_State(State)
{

    PresentState (ps) = unknown;
    PrimaryInput (pi) = unknown;
    while (not all fanin states to State are enumerated) {
      while (not all the NextState lines are justified) {

        find_new_pi/ps_assignment();
        simulate circuit with current set of pi/ps assignments;

        if (there are conflicts on NextState line values) {

          backtrack to the last set pi in D1 or ps in D2
                and assign an alternative value to it;
          simulate with current set of pi and ps assignments;
        }
      }
      if (a fanin state is found) {
        if (fanin state covers reset state) {
          a path is found;
          return;
        }
      }
      else Justify_State(fanin state);

      if (a path is not found) {

          /* scratch D1 */
        scratch all pi assignments;

        backtrack to the last set ps in D2 and
        assign an alternative value to it;
        simulate with current set of ps assignments;
      }
    }
}
```

4.3.4.5. Detection of Redundant Faults

The difficulty in test generation for sequential circuits does not just lie with finding tests for the difficult testable faults. The determination of redundant faults is equally formidable if not more difficult. Obtaining a low fault coverage does not necessary mean the test generator is inadequate if the fault coverage is close to the maximum achievable value. However, to determine whether faults, that no test has been generated for, are redundant or testable may demand astronomical CPU times. For the purpose of judging how close the fault coverage obtained by the test generator is to the maximum possible value, redundant faults based on Theorem 4.3.4.1 (given below) are found. Other undetected faults are treated as possibly testable faults. This gives a worst-case estimate of the number of redundant faults in a given circuit.

Definition 4.3.4.1. An edge in the State Transition Graph is said to be *corrupted* by a stuck-at fault if the effect of the fault can be excited and propagated to the primary outputs and/or next state lines by the input vector corresponding to the edge with the present state lines values set to the fanin state of the edge.

Theorem 4.3.4.1. In order for a stuck-at fault to be detected, the fault should at least corrupt one fanout edge from a valid state that is reachable from the reset state in the state transition graph.

Proof: In order to detect a fault we need a test sequence starting from the reset state and ending with a corrupted edge in the STG. If a fault does not corrupt any fanout edge from a valid state in the STG, no test sequence can detect the fault since no corrupted edge can be reached from the reset state.

Determining this special class of redundant faults requires the extraction of a partial STG containing all valid states reachable from the reset states. The procedure to find these redundant faults is based on the FEP algorithm described in Section 4.3.4.3. A single time frame is used and all next state lines are treated as primary outputs. All tests are generated for a potential redundant fault, with disjoint initial states. If none of the initial states exists in the partial STG, the fault under test is redundant.

4.3.4.6. Results

Results obtained for seven finite state machines are given in Table 4.3.4.1. Time profiles of test generation time for the examples are given in Table 4.3.4.2. In the tables m and s stand for minutes and seconds respectively. For each example in Table 4.3.4.1, the number of inputs (#inp), number of outputs (#out), number of gates (#gate), number of latches (#lat), number of equivalent faults (#eqv. faults), the number of test sequences (#test seq.), total number of test vectors (#vect), maximum test sequence length (max. seq. len.), fault coverage, percentage of provably redundant faults (using Theorem 4.3.4.1), total fault coverage including detected and provably redundant faults (tfc), and CPU time on a VAX

11/8800 are indicated. CPU times for extracting the partial state transition graph, test sequence generation, fault simulation, miscellaneous setup and for the entire test generation process are given in Table 4.3.4.2.

The test generation technique obtains close to the maximum possible fault coverage in all the examples. The extraction of the STG consumes a relatively small amount of CPU time with respect to the total TPG time in all cases. Fault simulation constitutes a large percentage of total TPG time in most cases except in "sse", as can be seen in Table 4.3.4.2. The fault simulator used implements the parallel-fault event-driven technique and a more sophisticated one using concurrent techniques would significantly speed up the test generation process. The reason that test generation time is the dominant constituent in the total CPU time in "sse" is because a great amount of time is consumed in trying to find tests for the large number of redundant faults.

The first five examples are finite state machines obtained from various industrial sources. The example "sbc" is the snooping bus controller [54] in the SPUR chip set. It was synthesized using the multiple level logic optimization system MIS [55], as was the largest example "stage", which is the finite state machine controller for a data encryption chip.

CKT	#inp	#out	#gate	#lat	#eqv. fault	#test seq.	#vec	max. seq. len.	fault cov. (%)	red.* fault (%)	tfc§ (%)	CPU† time
cse	7	7	192	4	680	96	472	8	99.71	0.29	100.0	53.2s
sse	7	7	130	6	486	46	284	10	84.57	15.23	99.8	69.9s
planet	7	19	606	6	2028	80	1191	26	97.39	2.56	99.95	12.6m
sand	9	6	555	6	1932	165	1077	24	94.36	5.18	99.54	22.4m
scf	27	54	959	8	3338	136	2238	21	94.37	3.86	98.23	83.0m
sbc	40	56	1011	28	3008	168	1063	24	95.68	2.66	98.34	62.1m
stage	131	64	2700	64	9155	139	425	26	93.97	6.03	100.0	125.2m

* percentage of provably redundant faults
§ total fault coverage including detected and provably redundant faults
† All times are obtained on a VAX 11/8800

Table 4.3.4.1 Results for 6 example circuits

CKT	STG Extraction	Test Generation	Fault Simulation	Miscell.	Total
cse	0.9s	8.3s	43.8s	0.2s	53.2s
sse	0.4s	52.2s	17.1s	0.2s	69.9s
planet	3.2s	1.2m	11.4m	0.7s	12.6m
sand	4.6s	10.7m	11.6m	0.6s	22.4m
scf	13.9s	11.5m	71.2m	1.2s	83.0m
sbc	12.4m	28.3m	21.4m	1.3s	62.1m

Table 4.3.4.2 Time profiles for example circuits

5. CONCLUDING REMARKS

In this paper, hardware verification and test generation algorithms were reviewed and their relationship to logic minimization discussed.

In Chapter 3, various combinational and sequential logic verification algorithms were described. Combinational logic verification techniques like exhaustive simulation, tautology checking, line justification, symbolic Boolean manipulation and enumeration-simulation were reviewed. The relationship between the test generation and logic verification problem was discussed. Parallel implementations of logic verification algorithms have been proposed. One approach was described in Section 3.3.6.

Sequential verification was also reviewed in Chapter 3. Three different approaches involving verification of sequential machines specified at the same of different levels of abstraction were described.

In Chapter 4, the testing problem for VLSI circuits was described. In particular, the relationship between the testability of the circuit and multi-level logic minimization was discussed. Logic minimization algorithms have the capability of making combinational circuits 100% testable.

Sequential testing is a very difficult problem. A algorithm recently developed for test generation of sequential circuits based on the PODEM justification algorithm was described in Chapter 4.

Verification and test generation remain "rich" areas of research. Techniques for combinational logic verification by *selective collapsing* of the circuit followed by enumeration and simulation algorithms show promise [56]. Implementations of combinational and sequential verification algorithms on massively parallel computers are very attractive. Very little work has gone into hardware verification at the behavioral level. This remains a wide open area of research. As behavioral synthesis techniques mature, the need for verification systems with the capability of verifying behavioral specifications against structural implementations will increase.

Test generation for combinational circuits is well developed, but remains a difficult problem for complex circuits. Logic minimization algorithms can alleviate this problem by making circuits easily testable. Effective sequential test generation techniques can result in higher performance in ASIC design, removing constraints posed by testability issues. The relationship between logic minimization and sequential testing should be investigated (What are the conditions on the Boolean structure of a sequential circuit for it to be 100% testable ?). Incomplete Scan Design approaches coupled with efficient test generation methods can be used to find a good compromise between performance and testability for a circuit.

REFERENCES

1. S. W. Director, A. C. Parker, D. P. Siewiorek and D. E. Thomas, A Design Methodology and Computer Aids for Digital VLSI Systems, *IEEE Transactions on Circuits and Systems CAS-28*, (July 1981), 634-645.

2. A. R. Newton, D. O. Pederson, A. Sangiovanni-Vincentelli and C. H. Sequin, Design Aids for VLSI: The Berkeley Perspective, *IEEE Transactions on Circuits and Systems CAS-28*, (July 1981), 666-680.

3. J. Allen and P. Penfield, VLSI Design Automation Activities at M.I.T., *IEEE Transactions on Circuits and Systems CAS-28*, (July 1981), 645-665.

4. R. K. Brayton and C. T. McMullen, Synthesis and Optimization of Multistage logic, *Proc. ICCD*, , October 1984.

5. R. K. Brayton, C. T. McMullen, G. D. Hachtel and A. L. Sangiovanni-Vincentelli, *Logic Minimization Algorithms for VLSI Synthesis*, Kluwer Academic Publishers, 1984.

6. J. Soukup, Circuit Layout, *Proc. of the IEEE*, , October 1981, 1281-1305.

7. A. Sangiovanni-Vincentelli, An Overview of Synthesis Systems, *Proc. of Custom Integrated Circuits Conference*, , May 1985.

8. D. Johannsen, Bristle Blocks: A silicon compiler, *Proc. of 16th Design Automation Conference*, , 1979, 310-313.

9. H. E. Shrobe, The Datapath Generator, *Proc. Conf. Adv. Res. in VLSI*, MIT, Cambridge, MASS., January 1982.

10. J. R. Southard, MACPITTS: An Approach to Silicon Compilation, *IEEE Computer*, , December 1983.

11. O. H. Ibarra and S. K. Sahni, Polynomially complete fault detection problems, *IEEE Transactions on Computers C-24*, (March 1975), 242-249.

12. R. Marlett, EBT: A Comprehensive Test Generation Technique for highly sequential circuits, *Proc. of 15th Design Automation Conference*, Las Vegas, June 1978, 332-338.

13. M. A. Breuer and A. D. Friedman, *Diagnosis and Reliable Design of Digital Systems*, Computer Science Press, 1986.

14. H. K. T. Ma, S. Devadas, A. R. Newton and A. L. Sangiovanni-Vincentelli, Test Generation for Sequential Finite State Machines, *Proc. of Int'l Conference on Computer-Aided Design (ICCAD)*, Santa Clara, November 1987.

15. E. B. Eichelberger and T. W. Williams, A logic design structure for LSI testability, *Proc. 14th Design Automation Conference*, , June 1977, 462-468.

16. B. Konemann, J. Mucha and G. Zweihoff, Built-in Self Test for complex digital integrated circuits, *IEEE journal of Solid State Circuits SC-15*, (June 1980), 315-319.

17. D. Brand, Redundancies and don't cares in logic synthesis, *IEEE Transactions on Computers C-32*, (October 1983), .

18. J. P. Roth, Minimization using the D-Algorithm, *IEEE Transactions on Computers 35*, (May 1986), .

19. W. E. Donath and H. Ofek, Automatic identification of equivalence points for Boolean Logic Verification, *IBM Technical Disclosure Bulletin 18*, (January 1976), .

20. J. P. Roth, VERIFY: an algorithm to verify a computer design, *IBM Technical Disclosure Bulletin 15*, (1973), 2646-2648.

21. J. P. Roth, Hardware Verification, *IEEE Transactions on Computers C-26*, (1977), 1292-1294.

22. G. Odawara, M. Tomita, O. Okuzawa and T. Ohta, A Logic Verifier based on Boolean Comparison, *Proc. 23rd Design Automation Conference*, , June 1986.

23. G. D. Hachtel and R. M. Jacoby, Verification Algorithms for VLSI Synthesis, *Design Systems for VLSI Circuits*, , 1986, 264-300.

24. R. S. Wei, Logic Verification and Test Generation for VLSI Circuits, *Ph.D Dissertation*, , U. C. Berkeley, September 1986.

25. R. E. Bryant, Graph-Based Algorithms for Boolean Function Manipulation, *IEEE Transactions on Computers*, , 1986.

26. R. E. Bryant, Symbolic Verification of MOS Circuits, *1985 Chapel Hill Conference on VLSI*, , December 1985.

27. M. Browne, E. Clarke, D. Dill and B. Mishra, Automatic Verification of Sequential Circuits using Temporal Logic, *Technical Report CMU*, Pittsburgh, PA-CS-85-100, 1985.

28. F. Maruyama and M. Fujita, Hardware Verification, *IEEE Computer*, , Feb. 1985.

29. D. Dill and E. Clarke, Automatic Verification of Asynchronous Circuits using Temporal Logic, *1985 Chapel Hill Conference on VLSI*, , 1985.

30. K. J. Supowit and S. J. Friedman, A New Method for verifying Sequential Circuits, *Proc. of 23rd Design Automation Conference*, , June 1986.

31. S. Devadas, H. K. T. Ma and A. R. Newton, On the Verification of Sequential Machines At Differing Levels of Abstraction, *Proc. of 24th Design Automation Conference and submitted to IEEE Transactions on CAD*, , 1987.

32. J. P. Roth, Diagnosis of Automata Failures: a calculus and a method, *IBM journal of Research and Development 10*, (July 1966), 278-291.

33. P. Goel, An Implicit Enumeration Algorithm to generate tests for combinational logic circuits, *IEEE Transactions on Computers C-30*, (March 1981), 215-222.

34. C. Y. Lee, Representation of Switching Circuits by Binary Decision Diagrams, *Bell Syst. Tech. J 38*, (July 1959), 985-999.

35. S. B. Akers, Binary Decision Diagrams, *IEEE Transactions on Computers C-27*, (June 1978), 509-516.

36. F. Brglez and H. Fujiwara, A neutral netlist of 10 combinational benchmark circuits and a target translator in Fortran, *Proc. 1985 IEEE Int. Symp. Circuits and Systems*, Kyoto, Japan, June 5-7, 1985.

37. H. K. T. Ma, S. Devadas and A. L. Sangiovanni-Vincentelli, Logic Verification Algorithms and their Parallel Implementation, *Proc. of 24th Design Automation Conference*, Miami Beach, June 1987.

38. S. Hwang and A. R. Newton, An Efficient Design Correctness Checker for Finite State Machines, *Proc. of Int'l Conference on Computer-Aided Design*, Santa Clara, November 1987.

39. J. D. Ullman, *Computational Aspects of VLSI*, Computer Science Press, Rockville, Maryland, 1984.

40. J. E. Hopcroft and J. D. Ullman, *Introduction to Automata Theory, Languages and Computation*, Addison-Wesley, Reading, Mass., 1979.

41. M. Barbacci, G. Barnes, R. Cattell and D. P. Siewiorek, *The Symbolic Manipulation of Computer Descriptions: ISPS Description Language*, Carnegie Mellon University, Research Report, 1979.

42. J. A. Robinson, A Review of Automatic Theorem Proving, *Proc. Symp. Appl. Math. Soc. 19*, (1967), .

43. G. D. Hachtel and P. H. Moceyunas, A Parallel Implementation of a Tautology Algorithm, *Proc. of Int'l Conference on Computer-Aided Design*, Santa Clara, November 1987.

44. K. A. Bartlett, R. K. Brayton, G. D. Hachtel, R. M. Jacoby, C. R. Morrison, R. L. Rudell, A. L. Sangiovanni-Vincentelli and A. R. Wang, Multi-level logic minimization using implicit don't cares, *submitted IEEE Transactions on CAD*, , August 1986.

45. F. C. Hennie, Fault detecting experiments for sequential circuits, *Proc. of 5th Annual Symp. on Switching Circuit Theory and Logical Design*, Princeton, N. J., November 1964, 95-110.

46. W. G. Bouricius and al, Algorithms for Detection of Faults in Logic Circuits, *IEEE Transactions on Computers C-20*, (November 1971), .

47. A. Miczo, The Sequential ATPG: A Theoretical Limit, *Proc. of 1983 International Test Conference*, Philadelphia, PA, October 1983, 143-147.

48. V. D. Agarwal, S. K. Jain and D. M. Singer, Automation in Design for Testability, *Proc. of Custom Integrated Circuits Conference*, Rochester, NY, May 21-23 1984.

49. M. A. Breuer, A Random and an Algorithmic technique for fault detection and Test generation for sequential circuits, *IEEE Transactions on Computers C-20*, (November 1971), 1366-1370.

50. H. D. Schnurmann, E. Lindbloom and R. G. Carpenter, The Weighted Random Test-Pattern Generator, *IEEE Transactions on Computers C-24*, (July 1975), 695-700.

51. S. Mallela and S. Wu, A Sequential Test Generation System, *Proc. of International Test Conference*, Philadelphia, PA, October 1983, 57-61.

52. S. Nitta, M. Kawamura and K. Hirabayashi, Test Generation by Activation and Defect-Drive (TEGAD), *INTEGRATION, the VLSI Journal 3 (1985)*, (1985), 2-12.

53. S. Shteingart, A. W. Nagle and J. Grason, RTG: Automatic Register Level Test Generator, *Proc. of 22nd Design Automation Conference*, Las Vegas, June 1985, 803-807.

54. M. Hill and al, Design decisions in SPUR, *IEEE Computer 19*, (November 1986), 8-22.

55. R. K. Brayton, R. Rudell, A. Sangiovanni-Vincentelli and A. Wang, MIS: A Multiple Level Logic Optimization System, *IEEE Transactions on CAD*, , November 1987.

56. A. R. Wang, Logic Verification by Selective Collapsing, *EECS 219 Final Report*, , December 1986.

PROVING
THE NEXT STAGE FROM SIMULATION

G. Musgrave
The work presented here is that of the Brunel Abstract Hardware Workshop,
in particular: M.P. Fourman, W. Palmer and R. Zimmer.

Simulation as a means of design verification is now well established in the VLSI design environment. It still has yet to be fully accepted in the PCB environment. Essentially this modelling tool is used to allow the designer to exercise his creation with a viewpoint of establishing 'that it works'. Invariably he has made many assumptions using simulation tools, and it can only be tested against these pre-conceived notions of its behaviour. Many of the problems today are that the behaviour of the good machine is reasonably well understood by the designer, but the behaviour of the same machine under slightly faulty conditions is not understood. Yet it can perform as if it is operating in a good state.

Engineering has been described as the application of scientific laws to the design and implementation of useful systems. Laws allow the engineer to proceed from a specification of requirements to an implementation satisfying those requirements. The law is described properly as a model of the system. The better our models, the better our calculus for reasoning about them, the less we need to resort to trial and error. Two good examples could be

1. The design of an electric heater, the model, the heater element as a pure resistance ohms law allows us to calculate the current which will flow for a given voltage in resistance.

2. To implement a state machine using a PLA we model the PLA as a boolean function, we use boolean algebra to arrive at a suitable function and to optimise PLA.

Most such models are so ingrained that we are liable to forget that they are only models. But it is important, particularly when we are dealing with the design of digital systems (hardware and software) and the laws they obey are somewhat more difficult and need to be put on a more formal basis if we can abstract from these models meaning for the behaviour of our designs. It could be asked why we need to bother with this when we are only just becoming familiar with logic tools in terms of simulation. This seems an unnecessary move, but if we examine the fundamentals of logic simulation then we will see that there are very distinct limits as to the processing capability we have. Currently we may have two-million transistors on a chip and simulation requires of hundreds of instructions per simple gate evaluation. If we take account of the rate of complexity increase, typically technology density doubling every year, then the computational task is very significant with the need of billions of instructions in order to do simple evaluations. This can be further multiplied by the requirements under fault conditions to handle the necessarily more complex models for the gates. Of course the technology is improving but just as the computational speeds increase so we move into the next generation of chips. A good evaluation of these numbers was given by Robert Smith [1986].

SPECIFICATION

The design engineer may well consider many aspects of system design. Three particular levels could be the electrical, physical and behavioural. If one looks at these individual aspects there are basic rules that help one understand and reason about them. But in essence these rules are only means of helping the construction to satisfy a specification. There are basically two forms of specification but in today's integrated world these are becoming much more intertwined.

Firstly let us consider the contractual specification. This is the essential central feature for the engineer. His task is to produce a design which meets a given specification, and therefore this forms part of the contract between the engineer and customer. Thus for a contract to be effective it must be possible to check objectively whether the design produced satisfies this requirement. Of course in many traditional engineering areas acceptance criteria are well established, the designer can formulate a solution, and the customer can check the calculations to make sure this design is verified. In the case of digital systems it is impossible to establish by testing that the design meets the

F. Lombardi and M. Sami (eds.), Testing and Diagnosis of VLSI and ULSI, 247–255.
© 1988 by Kluwer Academic Publishers.

specification. There are too many possible combinations of input sequences. We need to use verification as part of this technique. But in this form of verification we are already establishing that the specification provides the interface between the customer and designer and must be understood by both. In other words the customer and designer have to have a common language and not just the vocabularly they share. Hence the constructing statements about the concepts and the reasoning about them must also be understood by both parties. It is a very difficult matter.

The second form of specification is in respect of the methods of working. For example the specification of a component should allow us to use that component without worrying about, or even knowing about, its internal structure. In other words, when we are dealing with the design of large systems we have to use a hierarchical design to manage the complexity. From this perspective therefore, it is essential that the design is specified at its highest level. Because at its highest level of abstraction we have the fewest number of primitives, and therefore the least chance of being misunderstood by the designer and the customer. But in examining this automatic high-level design entry, we must provide tools for design validation at the same level. However, before we can build such tools we must formalise the abstractions the designer uses, and to be useful such abstractions must appear natural to the designer, because a tool which requires lengthy retraining will not be used.

Of course many different abstractions are employed in system design. Uncovering formalism and integrating these will be a difficult but necessary task in the development of future tools. It is essential to look at this as an experimental system, finalising the behaviour of systems formed by synchronous composition of simpler sub-systems. This abstraction is suitable for the analysis of a wide range of systems, for example controllers, communication protocols and microcode. From a point of view of synthesis, this area is relatively well understood: there are many tools for the automatic implementation of finite state machines. Our concern to provide verification for the state machines specified really meets some more abstract system requirement. This is viewed as a prototype of tools which must accompany more sophisticated automated synthesis.

Providing design verification at the post-design manufacturing stage, is not what is required. It is essential that we provide tools which the designer can use in his very earliest conceptual thoughts, which will enable him to have feed-back at that particular level, so that his design can be refined at the system level, before such synthesis tools are applied to get a given implementation. Unfortunately, to even attempt this view, one can only illustrate at a relatively simple level. To cope with the realities of today's complexity becomes a very difficult task and requires a highly automated computational system to support the necessary symbolic manipulation. To gain an appreciation of this method, two simple but different examples will be given.

SIMPLE PROBLEM 1:

The traffic light controller is a well known problem. However, the task is non-standard because we are not trying to design the traffic light controller *per se* , our task is to specify the traffic light controller. Essentially we are designing a mechanism whereby the traffic can continue to flow rapidly but collisions are avoided, rapid safety. Firstly the problem must be described formally, and there are two parts to this description. The first gives a formal model of the possible behaviour of the crossing and the cars. This model must tell us which events in the system are subject to our control. The problem is to restrict these behaviours in an orderly fashion. The second component of this description specifies what properties we require for the controlled system. The unconstrained system is described as a Petri net which is built from smaller Petri nets representing individual cars and lights. The controller is also represented as a Petri net. We have to verify the behaviour, resulting from imposing the controller on the system, against the specification.

THE TRAFFIC-LIGHT CONTROLLER.

The problem is familiar.

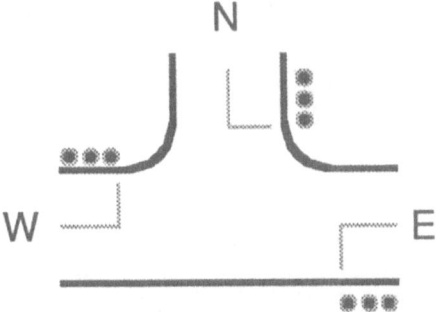

We must design a mechanism whereby traffic can continue to flow (liveness) but collisions are avoided (safety). Firstly, the problem must be described formally. There are two parts to this description. The first gives a formal model of the possible behaviours of the crossing and cars. This model must tell us which events in the system are subject to our control. The problem is to restrict these behaviours in an orderly fashion. The second part of the description specifies what properties we require of the controlled system. (In our example, safety and liveness conditions.) The unconstrained system is described as a Petri Net which is built from smaller Petri Nets.

The arrival and departure of cars at each light is described by

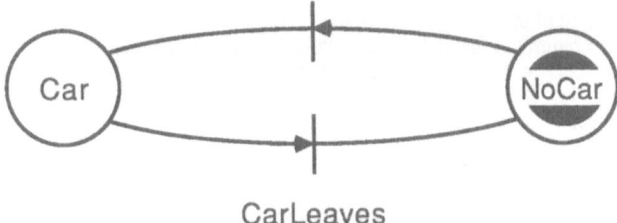

CarLeaves

The behaviour of a traffic light is given by the Petri Net

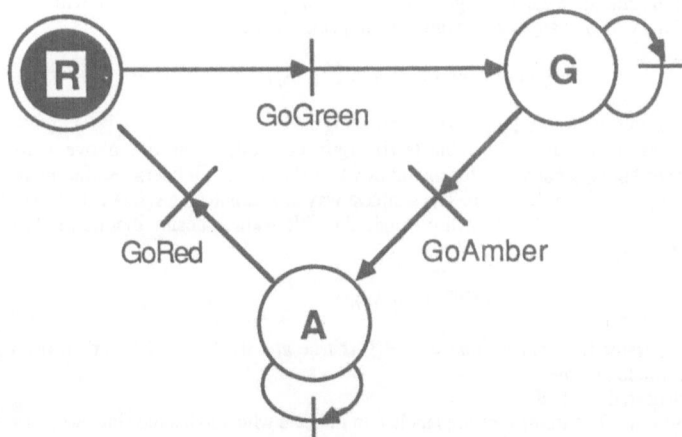

250

The traffic regulations (which we assume are obeyed) tell us how these two nets must synchronise. A car can only leave if the light is green. Otherwise, transitions can happen independently. We specify this by introducing a higher-level event, CarLegallyLeaves, which may be viewed as the synchronisation of the lower-level events CarLeaves and IsGreen. We also allow any of the lower-level events except CarLeaves to occur asynchronously. This gives the Petri Net

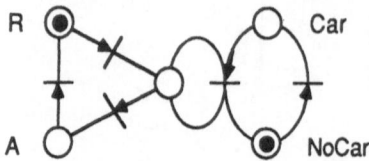

The physical system which must be controlled is modelled as three such Petri Nets, which are initially independent. Our system allows us to create parametrised instances of a given Petri Net automatically. We use the prefixes N, E, W, to distinguish the three copies. In fact, to be complete, we must also specify which parts of the system can be controlled directly - we have no direct control over the arrival and departure of cars (bringing in a policeman is not an admissible solution), we can only control the changing of the lights. Thus the designer must produce a control Petri Net and say how it synchronises with the events,
N_GoGreen E_GoGreen W_GoGreen
N_GoAmber E_GoAmber W_GoAmber
N_GoRed E_GoRed W_GoRed,
which change the various lights:.

(This is to be a very naïve controller which does not -and in this statement of the problem may not- look to see if cars are waiting. If we wished to allow this possibility we should make events SeeCar and SeeNoCar and make them visible to the controller.)

To specify the behaviour required of the final system, we use Clarke's Temporal Logic. We give an informal introduction here and return to it later. A CTL formula expresses some property of a state of a system which takes account of possible subsequent behaviour.

A first attempt at expressing part of the safety requirement might be

$$G \text{ (green\_W} \Rightarrow \text{red\_N).}$$

CTL formulae are , in general, true in some states and false in others. The "G" says "globally". G(j) is true in some state, s, iff j is true in every state reachable from s. The rest of this formula should be familiar. (Classical propositional logic is a part of CTL). We really want to say more, however. We want to ensure that if in some state the west light is green then the north light is red and it will still be red in the next state (this will give any car crossing time to get safely across).

$$G \text{ (green\_W} \Rightarrow \text{red N \& A (red\_N))}$$

"A" stands for "all" and means "for all immediately following states".
There are two other safety conditions for the traffic light controller. The one above with E replacing W and a corresponding condition for the times when N is green. Of course, so far, there is no guarantee that any light will ever be green. The simplest way of guaranteeing safety is to leave all lights always red! The liveness conditions will preclude this. We want to ensure that no light ever gets stuck on red. For example:

$$G \text{ (EF green\_W).}$$

This introduces the operator EF, "exists finally". EF(j) is true at s iff there exists a path from s which reaches a state where j is true.
The problem has now been described.
Here is a putative controller. The transitions are labelled to indicate which lights they change:

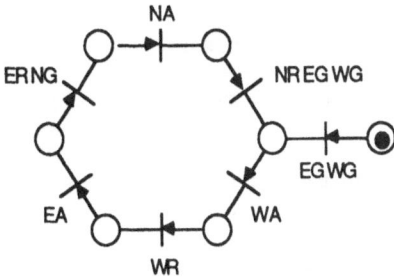

We synchronise this with the three Petri Nets representing the intersection, and verify automatically that the specification is satisfied. We can then go on to implement the controller using various standard methods. (This is intended as a pedagogic rather than a practical example. In real life, we would have a more demanding requirement and a more complex implementation.)

COMMENT:
So far, we have given an idealised picture of a part of the high-level design process of a simple controller. What is normally missing in current practice is a formal expression of the design requirement.

The value of higher-level specification may not be appreciated by all. Therefore, let us consider a lower level specification dealing with a four bit sampler. We use this example to illustrate how the design process can be driven from the specification.

The process we use is the formal procedure of goal directed proof within higher order logic both interpreted as a formalisation of the design process. Therefore, any design produced by this method is *already verified* formally. Rules that express how a complex module may be added to a design are derived by combining simple rules. In this framework *modular design is using rules as lemmas*. An extended example, illustrating these points, follows.

SIMPLE PROBLEM 2 : A FOUR-BIT SAMPLER

The four-bit sampler specification is

```
For i = 1 to 4,
   outi(t+i-1) = if sel(t+i-1) is HIGH then LOW
                                      else in(t).
```

After unfolding and a bit of boolean algebra, the specification is.
```
SPEC   : out1(t)   = in(t) * ~(sel(t))    ∧
         out2(t+1) = in(t) * ~(sel(t+1)) ∧
         out3(t+2) = in(t) * ~(sel(t+2)) ∧
         out4(t+3) = in(t) * ~(sel(t+3))
```

Clearly, this specification can be implemented modularly due to its essential symmetry. A detailed account of a design of this circuit is given below.

Step 1 : Add a delay - There are many ways of resolving with delayRule; we choose the branch that rewrites the 2nd,3rd and 4th instances of in(t)
as a(t+1). This results in
```
SPEC2          out1(t)   = in(t)  * ~(sel(t))    ∧
               out2(t+1) = a(t+1) * ~(sel(t+1)) ∧
               out3(t+2) = a(t+1) * ~(sel(t+2)) ∧
               out4(t+3) = a(t+1) * ~(sel(t+3))
```

and the derived rule

$$\frac{\text{H} \vdash \text{SPEC2}}{\text{DELAY(in,a)}, \ \text{H} \vdash \text{SPEC}}$$

Step 2 : Add an inverter - use invRule with the interpretation which rewrites the first `~sel(t)` of SPEC2 as `v1(t)`. This results in

```
SPEC3    out1(t)   =   in(t)  *    v1(t)       ∧
         out2(t+1) = a(t+1)  * ~(sel(t+1)) ∧
         out3(t+2) = a(t+1)  * ~(sel(t+2)) ∧
         out4(t+3) = a(t+1)  * ~(sel(t+3))
```

and the derived rule

$$\frac{\text{H} \vdash \text{SPEC3}}{\text{INV(sel,v1)}, \ \text{DELAY(in,a)}, \ \text{H} \vdash \text{SPEC}}$$

Step 3 : Add an and gate - use andRule with the alternative that rewrites the first boolean conjunction. The result is

```
SPEC4    out1(t)   =   out1(t)              ∧
         out2(t+1) = a(t+1)  * ~(sel(t+1)) ∧
         out3(t+2) = a(t+1)  * ~(sel(t+2)) ∧
         out4(t+3) = a(t+1)  * ~(sel(t+3))
```

and the derived rule

$$\frac{\text{H} \vdash \text{SPEC4}}{\text{AND(in,v1,out1)}, \ \text{INV(sel,v1)}, \ \text{DELAY(in,a)}, \ \text{H} \vdash \text{SPEC}}$$

Step 4 : Resolve SPEC4 with the rule:

$$\frac{\text{H} \vdash \text{P}}{\text{H} \vdash \text{x=x} \ \wedge \ \text{P}}$$

This produces

```
SPEC5    out2(t+1) = a(t+1)  * ~(sel(t+1)) ∧
         out3(t+2) = a(t+1)  * ~(sel(t+2)) ∧
         out4(t+3) = a(t+1)  * ~(sel(t+3))
```

and results in the derived rule

$$\frac{\text{H} \vdash \text{SPEC5}}{\text{AND(in,v1,out1)}, \text{INV(sel,v1)}, \ \text{DELAY(in,a)}, \ \text{H} \vdash \text{SPEC}}$$

Step 5 : Similarly, applications of steps 2-5 will reduce the specification to
```
SPEC6    out3(t+2) = b(t+2)  * ~(sel(t+2)) ∧
         out4(t+3) = b(t+2)  * ~(sel(t+3))
```

and produce the derived rule

$$H \vdash SPEC6$$

```
AND(in,v1,out1),INV(sel,v1), DELAY(in,a),
AND(a,v2,out2),INV(sel,v2), DELAY(a,b),  H ⊢ SPEC
```

Step 6: Another two applications of the sequence, 2-5, produces the specification
> TRUE

and the final rule

$$H \vdash TRUE$$

```
AND(in,v1,out1),INV(sel,v1), DELAY(in,a)
AND(a, v2,out2),INV(sel,v2), DELAY(a,b)
AND(b, v3,out3),INV(sel,v3), DELAY(b,c)
AND(c, v4,out4),INV(sel,v4), DELAY(c,d), H ⊢ SPEC .
```

H can now be instantiated to the empty list, and the premiss of the rule discharged to yield our theorem:

$$\text{implemented\_behaviour} \vdash \text{specified\_behaviour}$$

The first four non-trivial steps (steps 2 - 5) are in effect the design of a module and could be combined into resolution with one derived rule:

$$H \vdash F(z=z \ \& \ P(c(t+1)))$$

```
AND(x,v,z),INV(s,v),DELAY(x,c),  H ⊢ F(z(t)=x(t)*~s(t) ∧ P(x(t)))
```

Using this rule corresponds to adding the module,

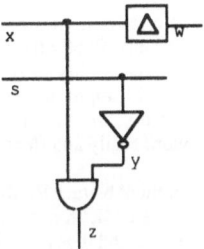

Four resolutions of the original identity rule with this rule produce the specification:

$$\text{out1}(t) \ = \text{out1}(t) \quad \wedge \ \text{out2}(t+1) \ = \ \text{out2}(t+1) \ \wedge$$
$$\text{out3}(t+2) \ = \ \text{out3}(t+2) \ \wedge \ \text{out4}(t+3) \ = \ \text{out4}(t+3)$$

which is easily proved.

254

The end result is the circuit seen on the bottom line of step 7 or in diagramatic form:

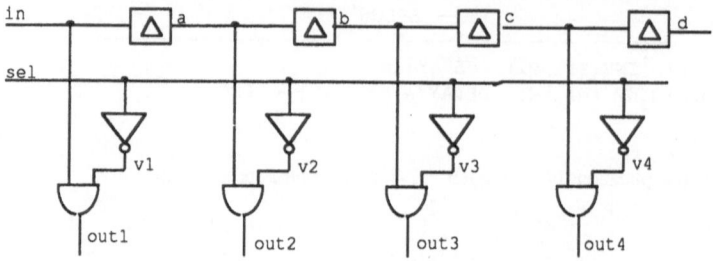

This is a somewhat redundant design : the fourth delay and three of the inverters are superfluous. It is sometimes worth keeping superfluous components in a design just for the sake of modularity. In the sampler, for example, by allowing the superfluous fourth delay, we only needed to design one module. But, we could have a postprocessing system that would recognise that only one port of that delay is connected and would query the designer. The extra three inverters can be removed by straightforward optimisations.

The real design should be

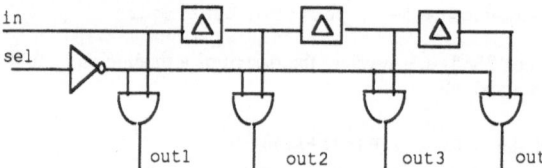

FINAL COMMENTS :

It is easy to dismiss these two examples as being trivial. They are. But they illustrate a number of attributes that are necessary for our future CAD tools. Firstly they show the formal representation at various levels of the design, thus allowing the capability of providing machine assistance for transformations, either at one level for the purpose of improving existing design, or between levels for purposes of implementing specification. It is essential that this implementation methodology should be interactive to enable the designer to work easily and flexibly with various CAD functions and extendable by the user.

In order to ensure a realistic implementation the Abstract Hardware Workshop has reviewed a number of programming languages, and has chosen ML in order to achieve the implementation. This is so that assistants can be developed at all levels, and thus make full use of AI techniques. Hence the traditional imperative languages like PASCAL have been rejected in favour of the functional programming languages which have developed from LISP, of which ML is one. The tremendous value of this lies in the fact that ML is a statically scoped programming language, which supports higher order objects, namely functions, which are first class objects, but can be passed as parameters, returned as values and embedded in the data structures. Useful examples of this language have been demonstrated in this presentation. One of its greatest benefits however, is the ability to handle modules which will allow the maintenance of large software systems by the system of hierarchy of these modules. As changes to a module will not affect the interfaces then there are no further requirements for modification or recompilation of other parts of the system.

Clearly, not all the answers are here. This is just the beginning. There are significant questions such as efficiency, but utilisation of modern machines which will be of the non Von Neumann structure, together with object orientated graphics, should enable considerable progress to be made to support the designer in his quest for design verification, which is essential if the modern systems are going to continue to have the integrity that all the users demand.

BIBLIOGRAPHY

1. Barrow H: Proving the Correctness of Digital Hardware Designs. VLSI Design, July 1984.

2. Clarke E, Emerson E & Sistla A: Automatic Verification of Finite Stase Concurrent Systems Using Temporal Logic Specifications: A Practical Approach. Carnegie-Mellon University, CMU-CS-83-152, Pittsburgh, 1983.

3. Cole N: Frontend to Simulation. Electronic Design Automation 1987.

4. Fourman M, Holte R, Palmer W, and Zimmer R: Top-Down Design as Bottom-Up Proof. Electronic Design Automation, 1987.

5. Fourman M, Palmer W, and Zimmer R: Using Higher-Order Functions to Describe Hardware. Electronic Design Automation 1987.

6. Gordon M: How to Specify and Verify Hardware Using Higher-Order Logic. Lecture Notes, University of Cambridge, 1984.

7. Hanna FK and Daeche N: Specifications and verification of digital systems using higher-order predicate logic. IEE Proceedings, Vol 133, Pt. E, No.5, September 1986.

8. Smith R: Fundamentals of Parallel Logic Simulation. Design Automation Conference, 1986, ISBN 0-8186-0702-5.

9. Wikstrom A: Functional Programming using ML, Prentice Hall 1987

There is a great deal more literature available but this tends to be in private communications/internal research reports.

PETRI NETS AND THEIR RELATION TO DESIGN VALIDATION AND TESTING

G. MUSGRAVE

BRUNEL UNIVERSITY, UK

INTRODUCTION

With complex systems of today, the need to design 'Right First Time' is an
increasingly heard slogan which has a very significant cutting edge of cost
penalties if you get it wrong. All too often, this fact is only attributed
to 'chip design', yet the cost penalties of system and module design errors
are often greater; it is just the accounting cost that is not as visible as
the cost of a new set of masks. If we have the correct environment for a
design automation system we still have to ensure that the tools and
methodology are compatible with the overall objective of the design.
Invariably the design objective is set at the behavioural level, the 'top
level' of the design; a wish for an artefact to DO a particular job-
control the pressure in a vessel say. But the design validation process is
often at a 'lower level' of the design, namely the module is tested to
perform a function. Immediately there is a dichotomy do we do a 'top-down'
or a 'bottom-up' approach to design? There are two schools of thought, or
perhaps more, because there are those who start in the middle and work in
both directions.

Considering Figure 1 then one may define the levels of design as system,
logic and physical, and when passing through these stages procedures
covering the spectrum of behavioural, functional, structural and silicon
levels are required. The important aspect of this 'staircase' is that the
design aids should allow one to descend and ascend these levels increasing
and abstracting the data respectively. The silicon compiler is not with us
such that the behavioural level can be mapped directly to the silicon, nor
is it desirable. This is because at each stage a great deal of information
is gathered and the designer makes decisions and thus modifies the
specification before proceeding. The local iterations are very important
in the learning process, they allow the parochial tools to provide the
analytical capability without dominating the design. The problem is that
many of these tools have not been designed to be part of a 'patchwork'
system and until software suppliers appreciate the wider implications of
design then it is difficult to see the specification languages coping with
the spectrum.

The question then arises as to how can we define the system in the initial
specification to ensure that it can be developed upon at subsequent stages
and retain some rigour. In other words, what are the characteristics that
are required? A wish-list may be formulated as follows:

a.   Capable of representing both hardware and software
b.   Easily assimilated by the designer, simple to modify and edit
c.   Unambiguous and concise - used by designers, implementers and
     customers

F. Lombardi and M. Sami (eds.), Testing and Diagnosis of VLSI and ULSI, 257–272.
© 1988 by Kluwer Academic Publishers.

design procedure / design stage	SPECIFICATION AND ANALYTIC TOOLS			
	behavioural level	functional	structural	silicon level
system	formal methods: directed graphs petri nets etc. — algorithmic evaluation	system design languages: PMS SARA LOGOS etc. — system simulation		
logic		hardware descriptive languages: RTL's ISP etc. logic eqns state-machines — logic simulation at function level	logic diagrams assigned machines — logic simulation at gate level: test generation	
physical			MOS transistor logic circuits etc. — circuit analysis	layout, sticks diagrams etc. — design rule checker

increasing level of abstraction

increasing degree of detail

FIGURE 1 DESIGN TOOLS FOR DIGITAL SYSTEMS

d. From description to realisation in either hardware or software-indication of which is best
e. Parallel operations in synchronous and asynchronous mode
f. Large data flows must be handled at the macro level
g. Systems evaluations performed by formal analysis
h. Hierarchical block structured.

Unfortunately, even today, after numerous promising projects and papers, this largely remains a wish list. Of course there are many techniques which map into some of these characteristics, but none of them are comprehensive. The functional description languages such as register-transfer languages (RTL) [1] and their related simulation languages such as DDL [2], HILO [3] and even the more computationally effective APL [4], have only satisfied a few of the characteristics. Basically this group lack the formal mathematical rigour to enable effective synthesis, nor can they easily relate to hardware implementation. On the other hand, the finite state machine tools do have the formalism which is necessary for an algorithmic solution. However, state tables, regular expressions [5] require infinite storage if realistic sequential circuits are to be considered. By far the most promising class of languages are the graph-theoretics. The fulcrum of this group is the Petri Net which Holt [6] and Patel [7] brought to the hardware world in the early seventies with project MAC at MIT. At the same time the LOGOS project [8] at Case Western University offered a specification technique based upon directed graphs. This certainly was easily understood by designers and offered some formalism although the concept of having a control flow graph and a data flow graph tended to present the designer with an initially partitioned system which does not lend itself to hardware/software tradeoffs. Certainly the use of directed graphs in handling large complex concurrent software systems is encouraging. However, the concept of having an interactive computing of graphs seems further away than ever when considering complexity of technology today. Nevertheless, researchers continue to work in this area and there is more mathematical rigour being given to what was traditionally the hardware description languages field. This, together with the more limited disciplined cellular structure of modern chips could mean that directed graphs could be computationally more practicable.

In this area, like many others, the need for discipline and engineering compromise appears to be the order of the day. The work of Clare [9] and its adoption by Hewlett Packard proves the point.

The over-riding requirment remains that we need to specify our systems in terms which will enable all those involved in the design process to evaluate against, and that means test criteria. Historically the development of specification linguistics has been valuable.

HISTORICAL PERSPECTIVE
There are three classes of specification languages, namely

a. Functional Description Programming Languages
b. Algorithmic Techniques
c. Graphical Theoretic Expressions.

The first group contains the simulator languages and has been widely used by designers mainly because they relate to hardware design concepts of registers etc. to which they can relate. Their main disadvantage is that they provide no effective analysis capability but merely a description. The algorithmic group attempts to redress this shortfall, and have such tools as finite state machine theory and regular expression. Very definite analytical attributes but rather impractical to use because the computational explosion of even a system with say thirty input/output variables cause enormous problems of state assignment etc. The third group has been slowly emerging and as graphics tools become more available has been adopted because it provides the synthesis route together with the conceptual visual appreciation to the designer.

An example of the first group of languages would be Schorr Register Transfer Language where some attempt has been made to give the standard R.T.L. some timing information. This is illustrated by the simple example of the Shift Instruction Figure 2.

Schorr R.T.L. Shift Instruction Description

Timing	Operation
$\|t_1 Start\|$:	$1 \rightarrow t_2$
$\|t_2\|$:	$V(K) \rightarrow T_0; 1 \rightarrow t_3$
$\|t_3 T\|$:	END
$\|t_3 T\|$:	$L_1(AC) \rightarrow X; 1 \rightarrow t_4$
$\|t_4\|$:	$X \rightarrow AC; 1 \rightarrow t_5$
$\|t_5\|$:	$K \rightarrow K-1; 1 \rightarrow t_2$

Figure 2

The next most significant development in this area was that of PMS/ISP the work of Bell G. and Newell [10]. Essentially this was a means to enable formal comparison of performance of complex machines. In practice it lead to the fundamental structure for Digital Equipment Corporation machines.

The processor Memory Switch (PMS) uses the fact that all digital systems can be described most generally as systems that at any time exist in one of a discrete set of states and that undergo discrete changes of state with time. A very abstract view which leads to the excessive state exposition for real systems. However, the tangible asset is that each state must reside in a component. Hence the PMS system consists of a number of structural primitives connected together in such a way as to form a network of components, each performing data processing operations on the information flow through the network. Information occurs in packets called i-units and is measured in bits (or other base value). The PMS offers seven basic component types, each distinguished by the kinds of operations it performs.

Memory, M A component that holds or stores information (i.e. i-units) over time.
Link, L A component that transfers information (i.e. i-units) from one

place to another in a computer system.

Control, K A component that evokes the operations of other components in
the system.

Switch, S A component that constructs a link between other components.

Transducer, T A component that changes the i-unit used to encode a given
meaning (i.e. a given referent).

Data-operation, D A component that produces i-units with new meanings.

Processor, P A component that is capable of interpreting a program in
order to execute a sequence of operations.

Components of the seven types can be connected to make stored-program
digital computers (C). For example:

a)   C : = Mp - Pc - T - X

Pc central processor, Mp primary memory, T transducer connected to the
external environment, represented by X.

Actually the classic diagram had four components, since it decomposed the
Pc into a control (K) and an arithmetic unit or data-operation (D):

$$Mp - K - T \mid Ms^1 - X \qquad\qquad or \qquad\qquad Mp - D - T \mid Ms - X$$
$$\mid$$
$$D \qquad\qquad\qquad\qquad\qquad\qquad\qquad\qquad\qquad\qquad\qquad K$$

where the solid information-carrying lines are for instructions and their
data, and the dotted lines signify control.

If we associate local control of each component with the appropriate
component, we get

Pc:=

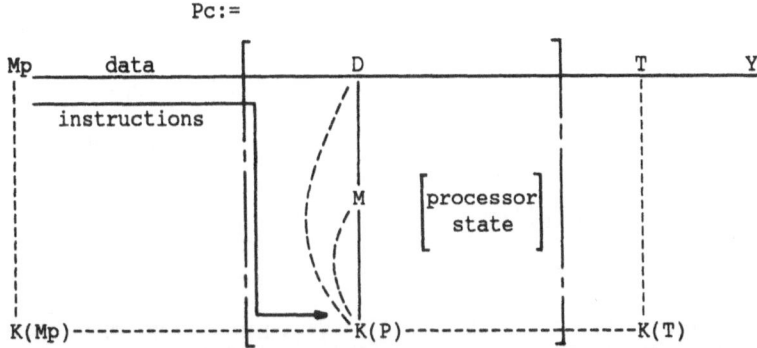

where the solid lines carry the information in which we are interested, and
the dotted lines carry information about when to evoke operations on the
respective components. The solid information carrying lines between K and
Mp are instructions. Now, suppressing the K's, then lumping the processor
state memory, the data operators, and the control of the data-operations,
and processor state memory to form a central processor, we again get

$$Mp - Pc - T - X$$

---

[1]The "|" expresses mutually exclusive alternatives. Here, a T or Ms exists
at the periphery.

By using suffixes particular properties of the components may be denoted

thus enabling a more detailed description to be achieved.
For example:

P_C (operation times : add : 2 μs; store : 2 μs; multiply : 18 μs;...)
In which the times for each operation are listed independently.
Alternatively, it is possible to give an abbreviated description by simply
stating a range of operating times, i.e.

Pc (operation time : 2~18μs).
Again, a core store may be defined as:

M (function : primary; technology : core; operation time : 1.5μs;
size : 4096 W; word : 16 bits)
but it is also possible to convey the same information to an informed
reader by the expression

M core (1.5 μs; 4kW : 16 bits).
(These figures all refer to a PDP8 machine - real history.)

Of course this is only half the picture, namely the data structured part,
and in order to complete the system model it is necessary to model the
control structure which is done with Instruction Set Processor (ISP). The
function of the ISP notation is to describe the programming level (that is
register-transfer level) of a computer in terms of memory, instruction
format, data types, data operations, etc. Thus ISP allows the
specification of an instruction set and the rules for its interpretation to
be precisely defined. In effect ISP is an ALGOL-based register-transfer
language with facilities for defining an instruction set in terms of basic
operations, registers, and data types. Concurrent operations and the
sequencing of activities can also be handled in a limited manner.

Using ISP the memory structure of a machine may be specified by declaring a
name and the number of storage bits associated with it. For example, an
accumulator register may be defined as

AC <0 : 11>
which signifies a 12-bit register, with bits 0-11 labelled from left to
right, and called AC. Similarly a core store would be specified as

M [0 : 7777_8] <0 : 11>
which is a primary memory consisting of 7777 octal words of 12 bits each.
The basic data operators include the following classes:
Concatenation
Boolean (AND/OR/EXOR/INV/EQIV)
Arithmetic (+, -, *, ÷)
Relational (=, ≠, <, ≤, >, ≥)
Transfer ←
These classes are used to specify data-processing operations between
registers. For instance, the conventional fetch operation in a digital
computer could be expressed as

Instruction ←M[PC];PC←PC+1;next
which signifies that the primary memory M is addressed with the contents of
the program counter PC and the result placed in the instruction register;
the contents of PC are then incremented by 1.

In other words the schema offers a set of tools which has proved its worth
in modelling IBM 360 versus CDC machines, but still falls short of the many
requirements set out in the initial part of the paper. In terms of
addressing the testing issues they have only been of help in the basic
portion of data flow and control flow and then the partial functional
partitioning afforded by the P.M.S. structure. The key features that are
needed are the time concurrency requirements, abstraction and analysis

criteria.

In the final group there is the potential to redress some of the
shortcomings.   The work of C.A. Petri [11] around 1965, laid the
foundations which were built on in the electrical system world by Holt [6]
and Patel   [7], while Wolfgang Reisig [12] has presented a useful
introduction. The nets have been widely used by computer scientists to
handle complex parallel data processing programs, a vital feature for
complex sequential concurrent hardware of today.   The   readers are
recommended to read some of the literature if they want the full
understanding of this diversely applied and customized theory.   Only the
essential points can be presented here.

Basic Petri Nets
There are four components of the nets:
  Places   represented by O
  Transitions   represented by | (alternative ☐)
  Arcs   represented by ——————————▶
  Markers   represented by ● (which only reside in a place)

The basic structure is:
  An input place connected by an arc to a
  transition connected via another arc to
  an output place - diagrammatically

    I/P Place   Transition   O/P Place

The rules of operation of a structure are known as the firing rules,
namely:
  i)   A transition is enabled if all of its input places hold a token
  ii) Any enabled transition may be fired if the output places have no
       tokens
  iii) A transition is fired by transferring tokens from input places to
       output places.
There is a set of mathematics which can relate these firings of the net
which may have the following properties:
The net is:-
  SAFE-   when each place is restricted to containing only one token
                 at the same time
  CONFLICT FREE- a conflict arises when two transitions share at least
                 one input place (indeterminant)
  LIVE-   the net is live when it is possible to fire any transition
                 of the net by some firing sequence, irrespective of the
                 marking that has been reached.

A typical net is shown in Figure 3. It is actually a model of a D type
flip-flop and most engineers would rightly say how complex for a simple
circuit. The key is that this is a generic model which has inherent
timing, and illustrates the behaviour. To actually work through the whole
firing sequence is possible and can be compared to the full set of state
tables, however, the net can be represented in matrix format which is more
easily computed.

264

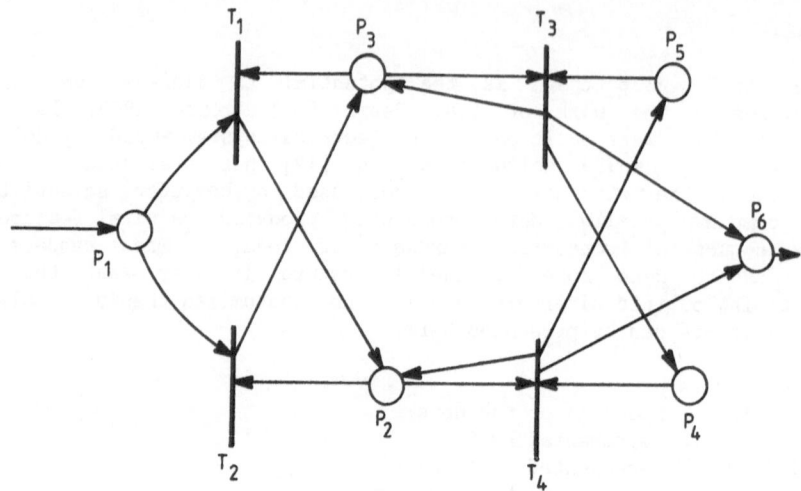

Figure 3
PEtri-Net TEst Generation ON Systems (PENTEGONS)

In order to cope with the complexity the simplicity of the straightforward net has had to give way to a series of net definitions which allow multi-level device modelling. The device modelling procedure is based upon hierarchical structure where each cell is cocooned with its own interconnection cell. These structures may be nested/recursive to any level.

An automatic Cubical Petri Net (CPN) modelling procedure, illustrated in a flow chart Figure 3, has been established, based on model partitioning that allows the modelling of more complex circuits. The following is a classification of cells:
i)   Input and Output Outer Cells (IOC and OOC)
ii)   Input and Output Memory Cells (IMC and OMC)
iii) Clock Cell (CC)
iv)   Feedback Cell (FC)
v)   Reflective Cell (RC)
vi)   Output Sequential Cell (OSC).

All these cells have their own internal activity which is dictated by their Cell Tokens (CT). In order to reflect this local activity to the higher level the cells are linked by places to provide paths for the Active Token (AT) movements. An example is illustrated by the 2-input NAND gate model given in Figure 5.

This model appears more complex than is necessary for functional modelling but it has all the necessary features to be a universal model for both design verification and test pattern generation.

The basic definition of the standard Petri Net (PN) is the starting points of the Cubical Petri Net. Essentially transitions, places, arcs and tokens are the attributes of the net, with the appropriate marking vectors enabling the process of transition firing to be mathematically computed. The CPN form a sub-class of the PN developed specifically to satisfy the needs of modelling and testing of digital circuits. It has been necessary to increase the simple single type of place to nine different types,

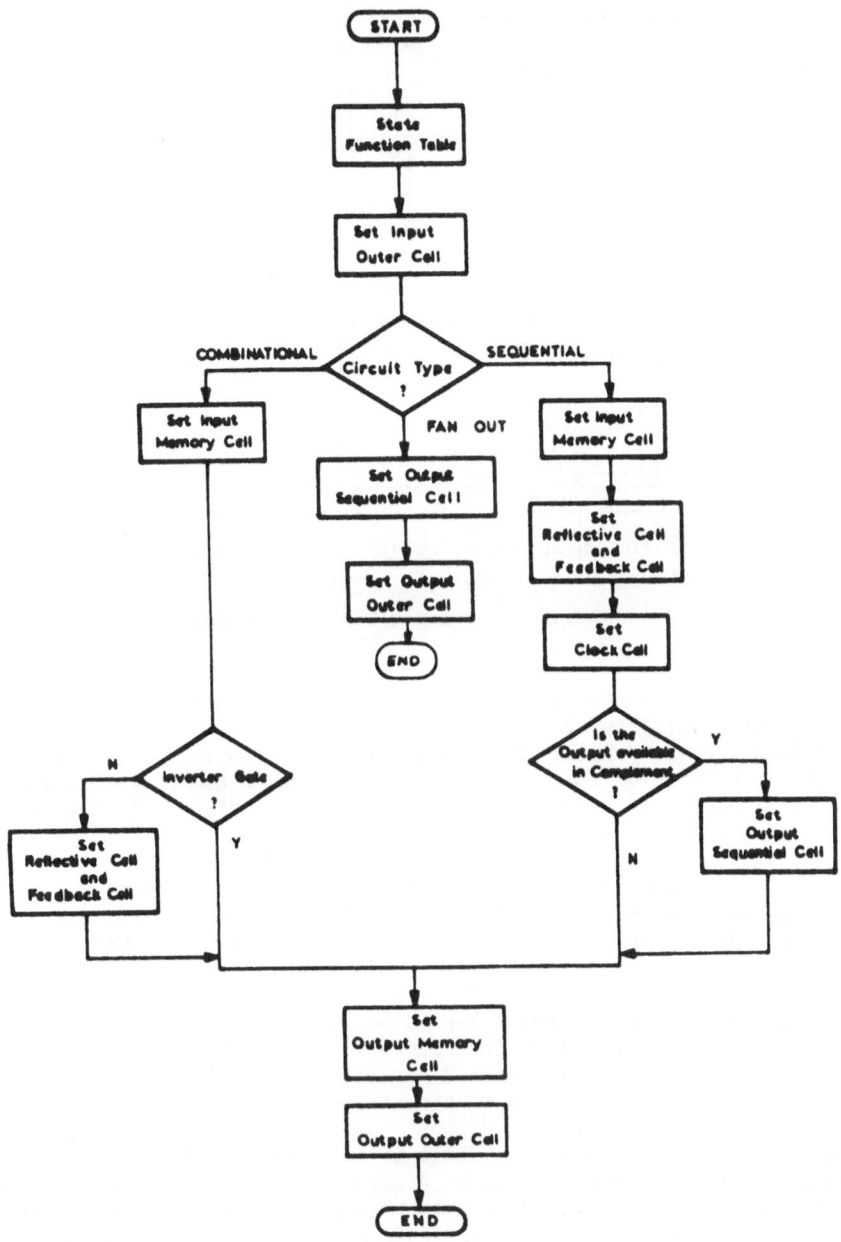

CPN Devices Modelling Procedure

Figure 4

266

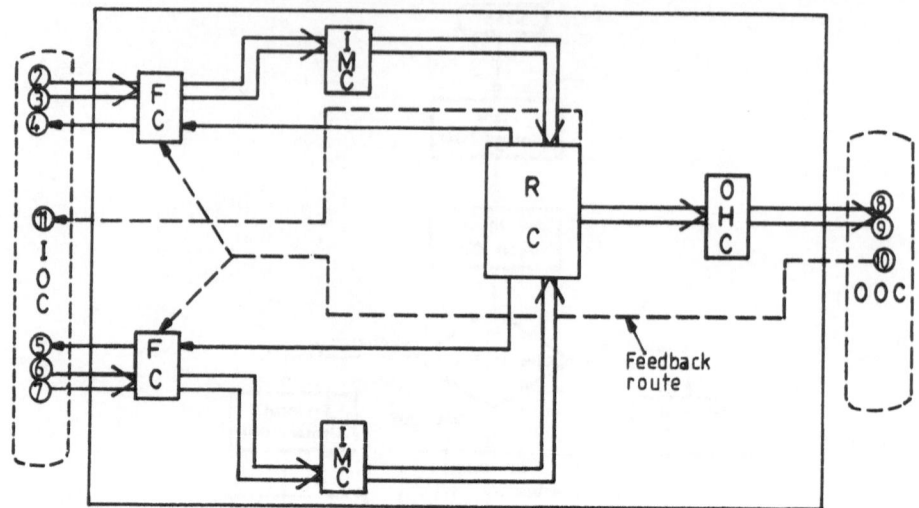

FM: Feedback Cell IMC: Input Memory Cell RS: Reflective Cell
OMC: Output Memory Cell IOC: Input Outer Cell OCC: Output Outer Cell

Figure 5 2-Input Nand Cell Model

together with two different types of transition and two types of token:
Types of Transition; Primary Transitions (PT), and Normal Transitions (NT).
Types of Place; Primary Place (PP), Outer Transit Places (OTP), Rejective
Place (RP), Reflective Cell Places (RCP), Coupled Memory Places (CMP),
Feedback Cell Places (FCP), Output Sequential Cell Places (OSCP), Inner
Transit Places (ITP), and Pseudo Place (PsP).

There was the further need to handle the multi-components in a
computationally efficient mathematical way which resulted in two matrices,
the Device Descriptive Matrix and Circuit Descriptive Matrix (see
references [13/14] for more details).

Hence any circuit to be considered has to be transformed to a CPN model, a
procedure which is indicated in Figure 6, and if this is applied to a
typical sequential circuit of Figure 7 it will result in a CPN model of
Figure 8. The main aspects to be considered are:

1. When a fan out exists: this is solved by creating a fictitious
    block, called the Fan Out Block (FOB) which can have any size, for
    example 2-outputs, or 3-outputs, ---etc. The OSC (introduced
    earlier) in fact exactly represents a 2-output FOB. Higher order
    FOBs can be designed in a similar manner.
2. The bus case: this is overcome by first spotting the type of bus
    used. A wired-AND type is replaced by an AND gate, and similarly a
    wired-OR type is replaced by an OR gate. A tri-state type, however,
    is treated differently by the use of a special CPN block called 'Tri
    State Block', which simply satisfies the Boolean function of the bus.
    These blocks actually encompass the same gates that are attached to
    the bus and therefore act favourably in reducing the total number of
    the PN elements in the PN model.
3. Each logic wire in the Circuit Under Test CUT is replaced by 3 PN

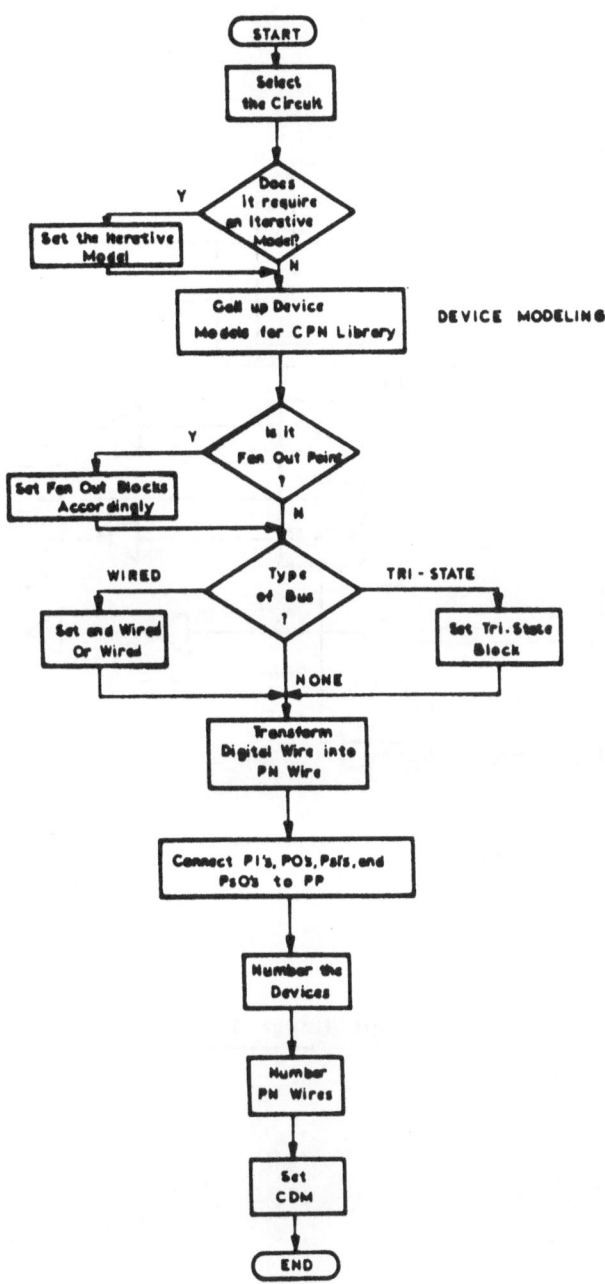

Procedure of Transformation of a Circuit from Logical to
CPN Domain

Figure 6

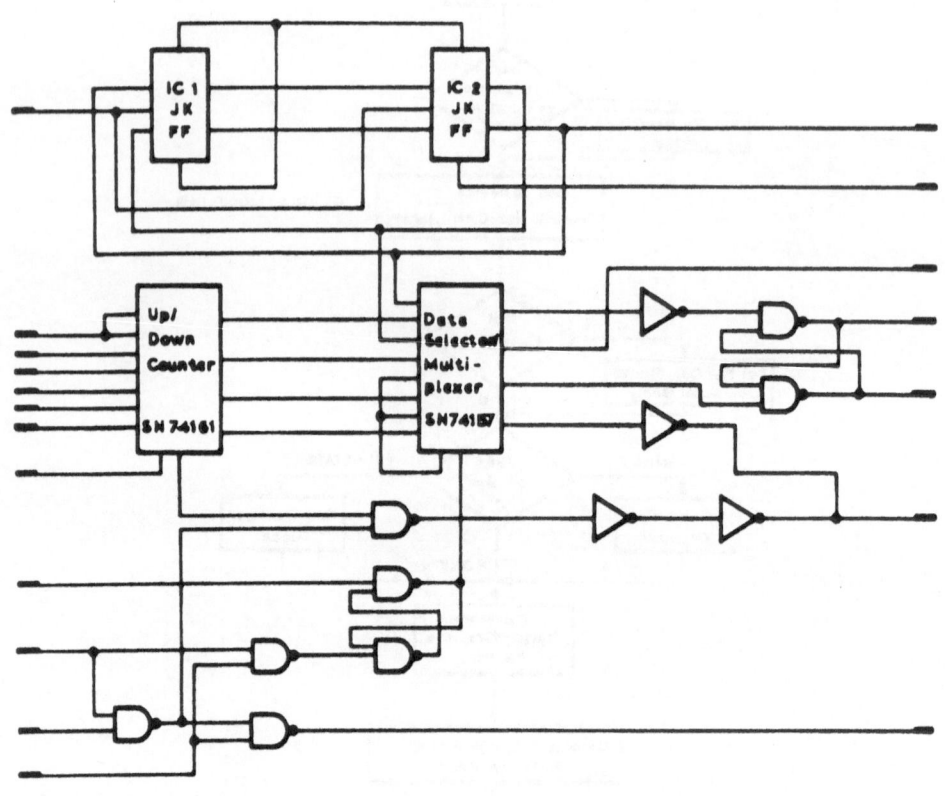

Circuit Under Test

Figure 7

269

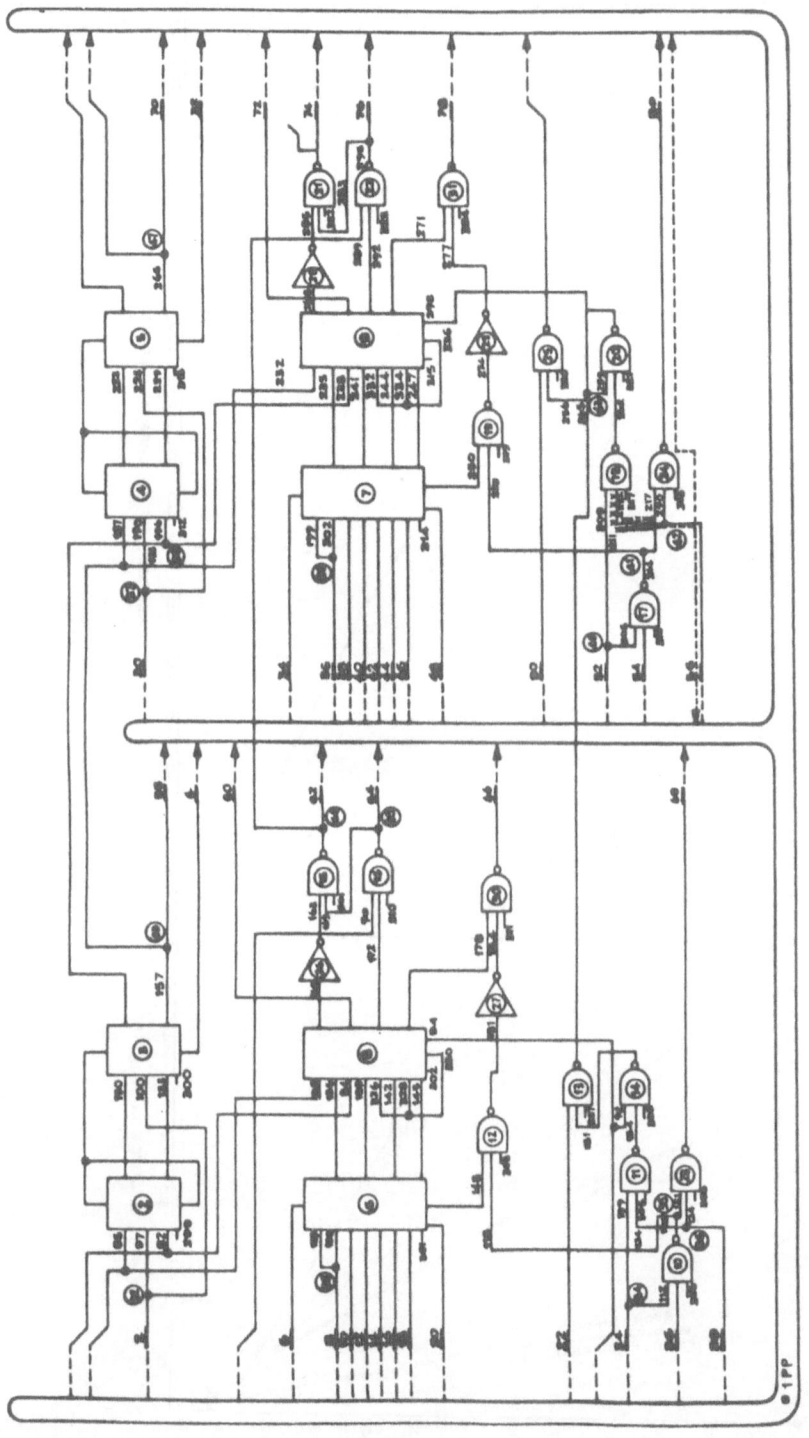

Complete Circuit CPN Model

Figure 8

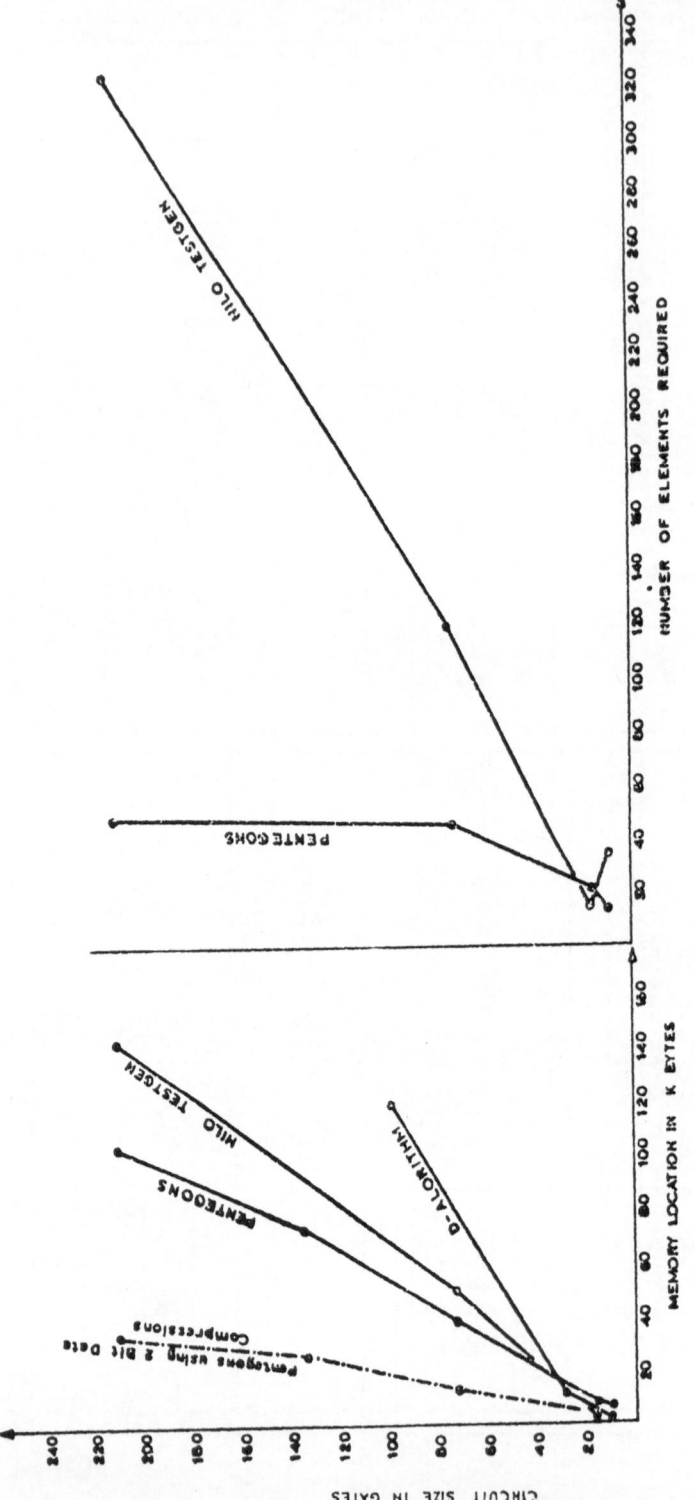

Figure 9

wires such that: one is assigned specifically for logic one, the
second is assigned specifically for logic zero and the third is
assigned specifically for logic feedback activities.
4.   Since any CPN model must have a PP, by definition then the modelled
     devices must be connected to this PP, by means of Petri Net wires via
     Primary Inputs (PI), Primary Outputs (PO), Pseudo Inputs (PsI), and
     Pseudo Output (PsO), as shown in Figure 8.
5.   The Circuit Description Matrix (CDM) is then defined according to the
     numbering set up shown in Figure 8. This is an extensive sparse
     matrix.

By the application of any initial marking the net will be fired giving a
sensitised path and at the same time ensuring logic stabilisation, in other
words consistency path. Because the nets are safe and live other markings
are reached giving another set of vectors which accelerate the findings of
test sets.

The PENTEGONS was run on several circuits of different structures with
varying sizes, and was compared with two known testing procedures the D-
algorithm and the HILO TESTGEN [15] which applies Critical Path
Sensitisation Algorithm. The results are plotted in Figure 9 which shows
that PENTEGONS is really more suited to VLSI era. The reason being due to
the size of the circuit description matrix, which grows linearly with the
number of devices in the CUT while similar description matrices for the
other procedures grow quadratically. Further optimisation is possible for
the PENTEGONS technique by fuller utilisation of all bits of a computer
word. Just the simple 2 bit data compression gives the improvement shown
in Figure 9.

In this paper we presented a new testing technique called PENTEGONS based
on transforming the circuit under test from its circuit logic description
into its equivalent Cubical Petri Net domain. The transformation process
was an automated procedure.

Our work with PENTEGONS has revealed a number of features of interest to
circuit testing. Most importantly:
1.   PENTEGONS generates sensitised paths and also provides stabilisation
     to other paths
2.   Running PENTEGONS on an initial state always provides the next
     desirable state at the end of the run
3.   PENTEGONS does not suffer from contradiction states
4.   The size of PENTEGONS' circuit description matrix means that
     PENTEGONS uses less memory resources than other methods.

Thus, PENTEGONS appears to provide a promising method of coping with VLSI
testing problems.

REFERENCES

1.   Schorr H. Computer-Aided Digital Systems Design and Analysis Using a
     Register Transfer Language, IEEE Trans. Electron Comp. EC13 1964
2.   Durley J.R. Dietmeyer D.L. A Digital System Design Language (DDL),
     IEEE Trans. Computers C17 1968
3.   Flake P.L. Musgrave G. Shorland M. The HILO Logic Simulation
     Language, Proc. CHDL 75, New York 1975
4.   Iverson K.E. A Programming Language, Pub. John Wiley, New York 1962

5. McNaughton R. Yamada H. Regular Expressions and State Graphs for Automata, IEE Trans. Elec. Comp. EC9 1960

6. Holt A.W. Introduction to Occurrence System, Associative Information Techniques, (Ed. EL Jacks) Elsevier 1971

7. Patel S.S. (University of Utah) Lecture Notes - Advance Course on General Net Theory of Processes and Systems, Hamburg 1979

8. Rose C.W. A system of Representation for General Purpose Digital Computer Systems, Ph.D. Thesis, Case Western Reserve University, Cleveland, Ohio, September 1970

9. Clare C.R. Designing Logic Machines Using State Machines, McGraw-Hill, New York 1973

10. Bell and Newell, Computer Structures: Readings and Examples, McGraw-Hill 1971

11. Petri C.A. Communication with Automata Vol 1 Supplement 1. RADC TR65-377, Applied Data Research, Princeton N.J. Contract AF 30(602)3324 January 1966

12. Reisig W. Petri Nets, An Introduction, Springer-Verlog 1982 ISBN 3-540-13723-8

13. Alukaidey T. Petri Net Test-Generation on Systems, Ph.D Thesis, Brunel University 1983

14. Alukaidey T. Musgrave G. Petri-Net Test Generation on Systems, IEE, EDA 1984, Warwick University, pp 72-78

15. HILO Manual 1982, Cirrus Computers Limited, Uxbridge, Middx. UK

FUNCTIONAL TEST OF ASICS AND BOARDS

G. Saucier, M. Crastes de Paulet, F. Tiar

Circuits and Systems Laboratory
46 Av. F. Viallet
38031 GRENOBLE FRANCE

INTRODUCTION

The debugging, the end of manufacturing test as well as the maintenance test of complex ASICs, boards and systems have to be performed through their global functional activations to reach efficiency and high coverage quality especially with respect to dynamic functioning. This is not a contradictory with local test targets but needs some efforts in order to correctly handle the global problem.

Global functional specification by a set of functional timed activations may not be available for testing but are of course mandatory. For complex ASICs and boards, each block is usually described at a functional or structural level. The global system is viewed as an interconnection of these components. The obtention of the global activations needed for test generation is painful and requires a learning procedure or designer's help. This knowledge acquisition may be compared with reverse engineering of functional specifications and may be avoided if test specifications are defined together with design specifications. The adequate language to handle these global descriptions plays a strategic role.

The test generation problem is addressed by creating an adequate environment. This environment has to provide an easy generation of a set of test activations according to several criteria. Automatic test pattern generation may be available at the block level and is replaced at the global level by test generation "primitives" or facilities. These test generation facilities are for example : functional coverage estimation based upon simulation, automatic generation of a set of functional activations with a given set of values assigned to some variables, controlability expertise for some blocks, ...

1 - FUNCTIONAL MODELLING

1.1 - Preliminary definitions

This paper is based on the multilevel Hardware Description Language CADOC.LD [1] but the principles are easily adaptable to other High Level HDLs (such as VHDL, ...).
Using CADOC.LD, hardware entities or blocks are instances of Generic

F. Lombardi and M. Sami (eds.), Testing and Diagnosis of VLSI and ULSI, 273–286.
© *1988 by Kluwer Academic Publishers.*

274

parameterized Functional Resources (GFR) ; a GFR model consists of 3 parts :
a heading (IO and external parameters definition), a declarative part and a
functional part. The functional part is given by means of an interpreted timed
state diagram and expresses the behavior of the related component in terms of
a set of pathes going from the initial state (or place) to final ones ; each path is
associated with an activation mode of the circuit. Notice here that it is possible
to define an ordering on these global activations : for example, path 1 in the
following example has to be activated before the other ones in order to correctly
initialize the circuit.

The description of an edge-triggered flip-flop is given in figure 1. The main
features of the CADOC language as well as the timing diagrams
manipulations are explained in [1].

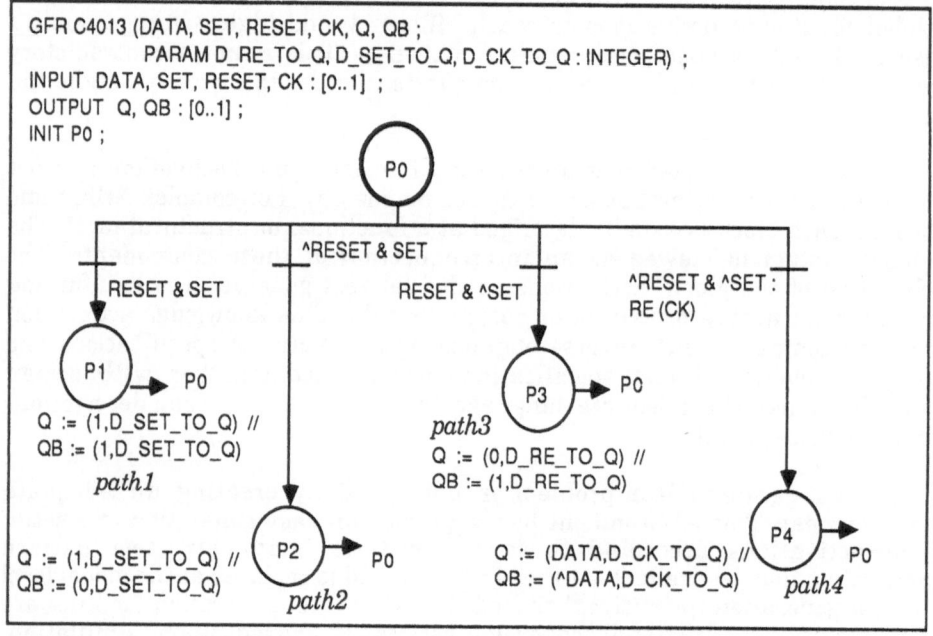

Figure 1 : positive edge triggered flip-flop

Notations
* In the model of C4013, Q := (1, d_set_to_q) means that the variable Q will be
assigned to the value, 1 d_set_to_q time units after the activation of place P1.
* RE (ck) (resp. FE (ck)) indicates a rising edge (resp. a falling edge) on the
signal CK.

A variable appearing on a transition predicate of a resource description
will be referred as **a control variable** of the resource. A control variable is a
global control variable if it is declared as input or bidirectional (**external
variable**) and if it is not modified by the function performed by the resource. In
the C4013 description, SET, RESET and CK are global control variables for this
resource.

1.2 - Dependency graph associated with a resource

A dependency graph is a simplified model expressing the dependencies between the variables of the resource.
Using the example of the C4013, the dependency graph is the following :

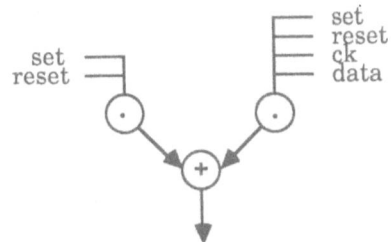

Figure 2 : dependency graph of a resource

1.3 - Composition of resources and global functional model

Let us consider an interconnection of 3 resources : a counter (CPT), a comparator (C15, which compares to 15) and an edge sensitive register (RG). The interconnection is indicated in the figure 3a and the local descriptions of these circuits are given in the figures (3b, 3c and 3d).

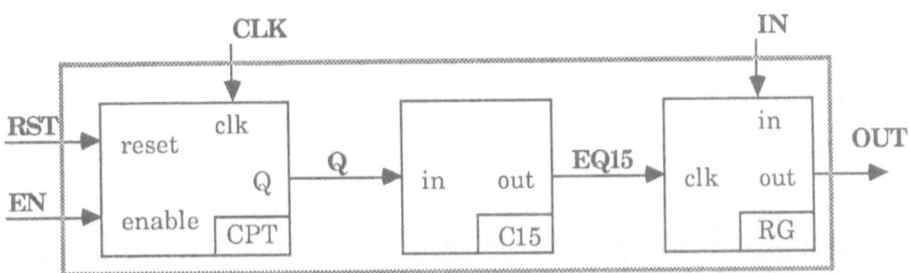

Figure 3a

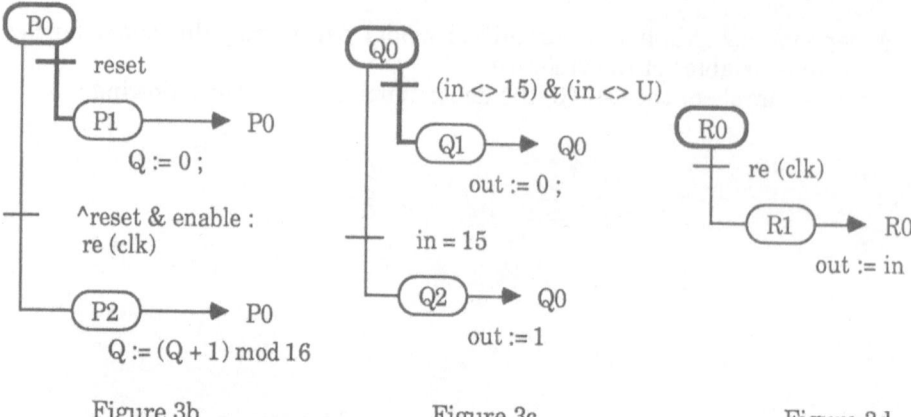

Figure 3b   Figure 3c   Figure 3d

The global description is obtained by a concatenation of the local models according to the connections and followed by a simplification of both the unfirable arcs and the unnecessary places. The initial global model is given in the figure 4a (the simplifiable items are underlined) and the final model is given in the figure 4b (the 2 main global activations are underlined).

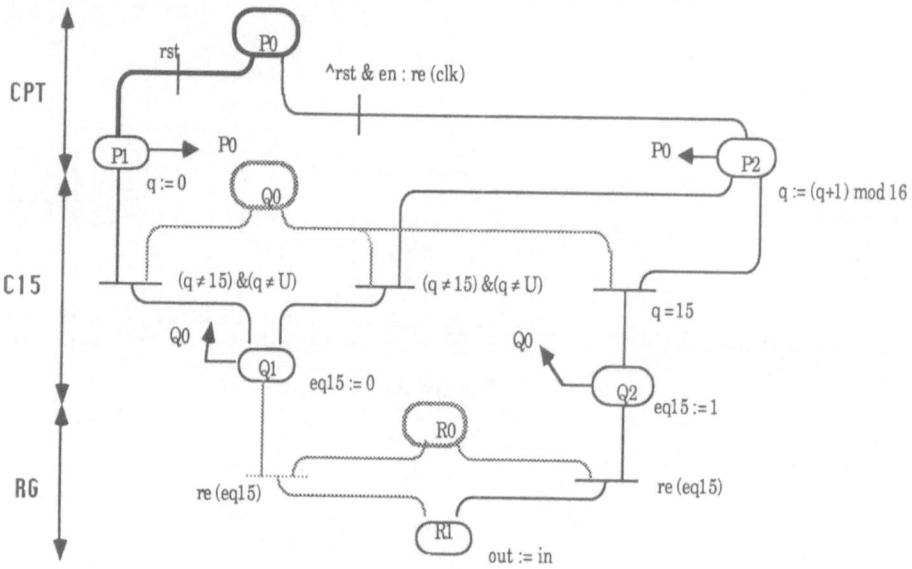

Figure 4a : initial global model

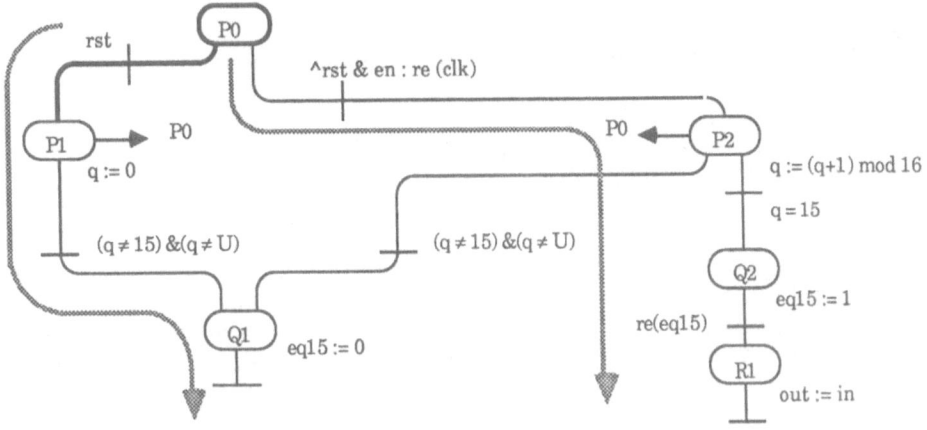

Figure 4b : final global model

The definition of the global functional activations may be helped by analyzing the dependency relations between the variables ; these dependencies for the previous circuit, as well as the 2 extractable pathes are indicated in the figure 5. Note here that this dependency analyzis is facilitated if the considered circuit is data-driven and that such an assistance for finding global specifications in a data driven circuit or a board has been implemented in the SATAN system [2].

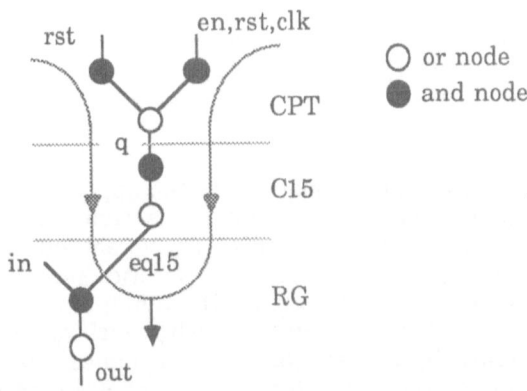

Figure 5 : global dependency graph

1.4 - Rule-based formalisation

Instead of a the representation used in the figure 1, each path of a resource can be described by a rule which summarizes the related activation, the general form of which is :

if <input timing diagrams> **then** <output timing diagrams>

For simple resource this model is immediately deduced from the graphical one. The interest of such a formalism will be detailled in section 4.2.

Exemple 1
The description of C4013 can be given by the 4 following rules :

rule 1 : if (RESET and SET) then (Q := ...) // (QB := ...)
rule 2 : if (not RESET and SET) then (Q := ...) // (QB := ...)
rule 3 : if (RESET and not SET) then (Q := ...) // (QB := ...)
rule 4 : if (not RESET and not SET when re (CK)) then (Q := ...) // (QB := ...)

For more complex resources, this model is more interesting as it summarizes the different global activations of the resource. Such a model is actually given by the designer but may be extracted in the future from the graphical circuit description by means of symbolic simulation techniques [3].

Exemple 2
Consider the graphical model of the figure 4b ; the activations related to the left and the right branchs of the model will be expressed in the rule-based formalisation as follow :

rule 1 : if reset then nil / initialization rule /
rule 2 : if (RST = 0) & (EN = 1) & repeat (16, re (CLK))
    then OUT (rex (CLK,16)):= IN (rex (CLK,16))
Notes
* (rex (CLK,16)) indicates the 16th rising edge on CLK.
* The global activation given by rule 2 is significant only if the circuit has been initialized using rule 1.

2 - TEST OBJECTIVES AND COVERAGE ESTIMATION USING FUNCTIONAL SIMULATION

2.1 - Test objectives

Once the circuit has been modelled at the appropriate level, the next step towards test generation is to define the activations associated with the different functionings of the circuit. These activations will be the test objectives and will guide the global test generation process.
Classical functional test goals rely on the automaton identification [4] ; the goal is to find test sequences covering each transition of a state table and to identify each state by a distinguishing sequence observed at the circuit outputs. For complex resources, test activations are associated with pathes in the functional model : pathes are declared by their activation condition and are associated with a set of values which are the test cases within the path ; pathes conditions are described in the CADOC language and are attributes of the resource. For a given resource a test objective will be enounced either by a set of pathes or by a set of values to be assigned to defined variables to be covered during the test.

Example 1

For the register given in the figure 3d, the test activations are derived from the following general form :

value (IN) when re (CLK)

and, as example, the following test objectives may be associated with the resource :

obj. 1 : IN = 01010101 when re (CLK)
obj. 2 : IN = 10101010 when re (CLK)

Example 2

For the composed circuit, the global model of which is given in the figure 4b, the test objectives will be :

obj. 1 : RESET = 1
obj. 2 :
(RESET = 0) & (EN = 1) &
repeat (16, re (CLK)) & (in = 00000001 when (rex (CLK,16)))
obj. 3 :
(RESET = 0) & (EN = 1) &
repeat (16, re (CLK)) & (in = 00000010 when (rex (CLK,16)))

Note that test objectives 2 and 3 are derived from the same global form (using various values for variable IN) :

(RESET = 0) & (EN = 1) &
repeat (16, re (CLK)) & (value (IN) when (rex (CLK,16)))

Note

The syntax used in this section is only an indication, at the present time, test objectives are indicated on the functional model by means of local variables. As example,a local test variable named LOC_CPT and assigned in the functional model, can be used in order to express activations 2 and 3. Such a representation for test activations facilitates the coverage estimation using the functional simulator (see section 2.2) but introduces possible confusions between the functional model itself and the test informations (test activations). Once a correct syntax will be defined, the test activations will be a separated part of the knowledge associated with the resource.

2.2 - Coverage estimation using functional simulation

Once test targets have been defined for each resource of a complex ASIC or of a board, global input sequences have to be estimated with regard to the local test targets associated to each resource. This coverage estimation is obtained by the use of the CADOC functional simulator ; the global simulation sends a set of functional activations and each time a local test target is reached, the global functional activation can be memorized and a library of test sequences covering the local test cases is then created.

2.3 - Example

Let us consider the square root extractor published in [5] ; the structure of the data-path of the circuit is given in the figure 6 and the controller flowchart is indicated in the figure 7.

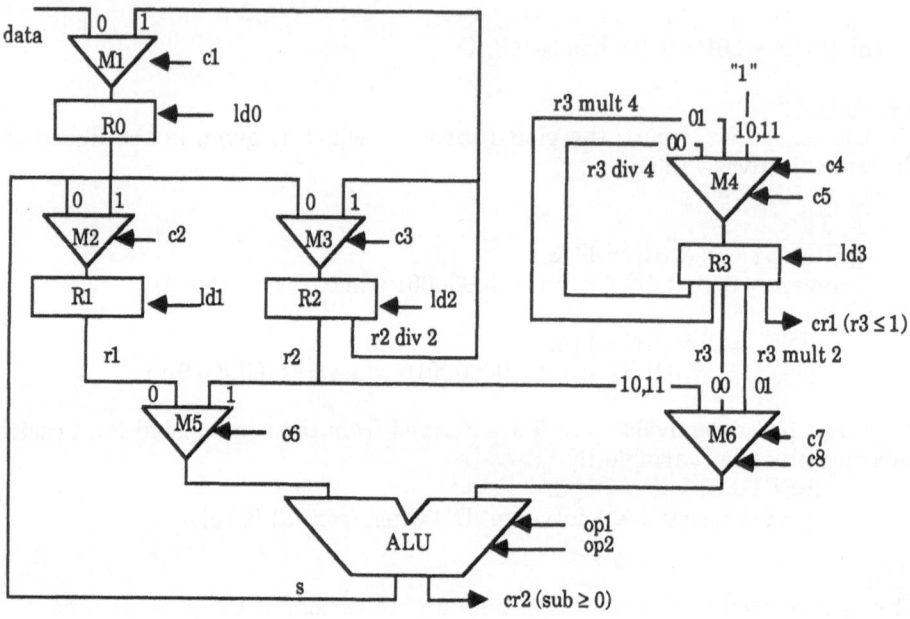

Figure 6 : data path

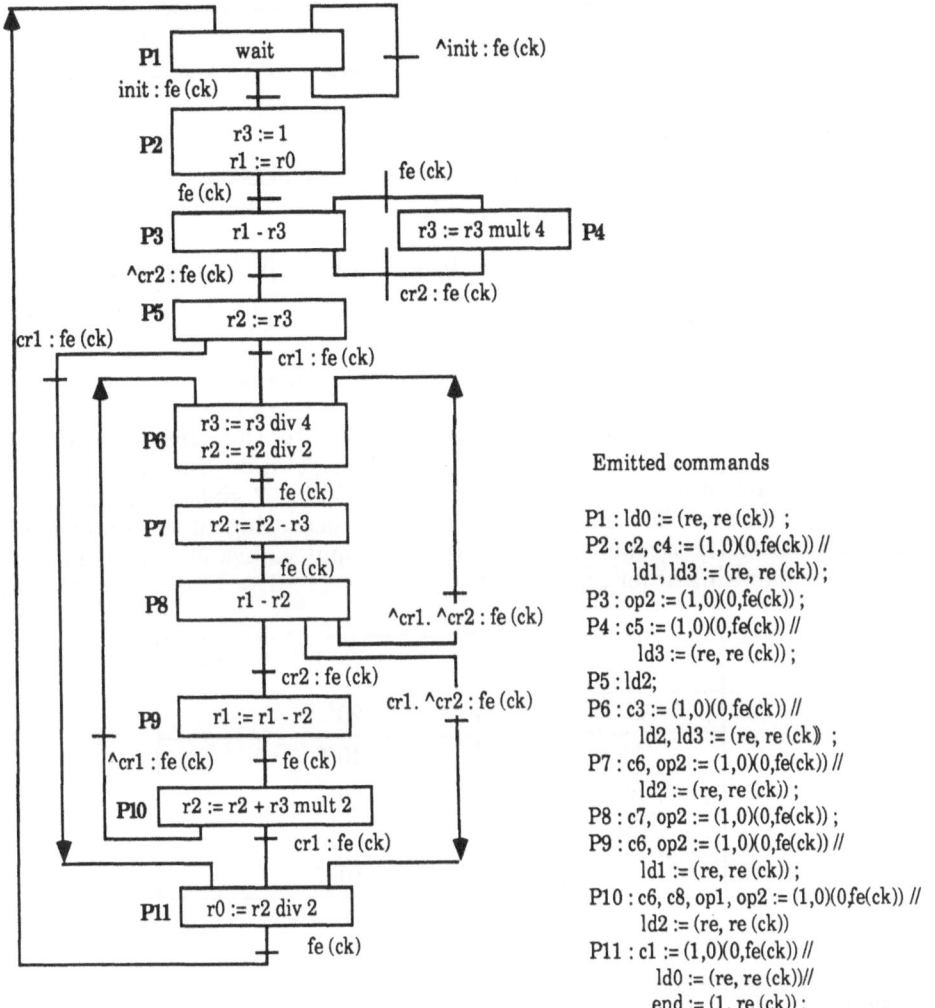

figure 7 : control part

Given a set of local test targets, the test generation assistance system concludes about the coverage reached for a given set of inputs (ratio of test targets reached during simulation).

Let us consider registers R0 to R3 and stuck-at faults : the test activations (which are in this case a simple pair test vector / sampling event) needed to detect these faults are summarized in table 1. Notice that these activations have been chosen in order to be sent to these registers through normal functioning (see discussion on the input domain in the next section).

R0	IN_R0 = {01010101,10101010} when re (LD0)
R1	IN_R1 = {01010101,10101010} when re (LD1)
R2	IN_R2 = {011000000, 000011111, 001110000, 100000000} when re (LD2)
R3	IN_R3 = {000000001, 000000100, 000010000, 001000000,100000000} when re (LD3)

table 1 : test activations for the registers

Suppose we apply (through R0) random values to the circuit and run simulations, the system will provide, after each simulation, the ratio of local test vectors reached for each block of the circuit. Table 2 gives an idea about the coverage reached for the local test set of R3 according to the number of values applied to the circuit and assuming the input values are different and given in the following order : 0,1, ... 255.

number of input values	coverage of R3 test set
1	20%
4	40%
16	60%
64	80%
from 65	100%

table 2

Remarks
* these percentages indicate the worst case, if the first value used is greater than 64, the 100% coverage is reached after one simulation.
* input values are selected randomly but the initialization protocol has to be the same.

3 - CONTROLLABILITY EXPERTISE IN FUNCTIONAL TESTING

This expertise takes place when the informations given by the use of the functional simulator (tracing system) are not satisfactory.

3.1 - Top-down expertise : semi-symbolic simulation

In order to study how resources are controlled in a system, a semi-symbolic simulation tries to evaluate the input domain of the resources.

An input domain indicates the values that can be reached in a normal functioning.

Let us consider for example the previous square root extractor ; symbolic execution has shown that 16 different functional activations exist for this circuit, each of these activations is associated with different values of the input data (data = 0, 1 ≤ data ≤ 3, ...). If we assign the symbolic value #data to the input, the internal resources will receive data function of this value. These data are given in the table 3 and table 4 ; table 3 indicates the input domains of the resources of the data path for the fourth activation (9 ≤ #data < 16) and table 4 indicates the input domain for the add and the substract function of the ALU.

R0	#data, 3
R1	#data, (#data - 4), (#data - 9)
R2	16, 8, 4, 12, 6, 5, 7
R3	1, 4, 16
ALU	(#data,1,-),(#data,4,-),(#data,16,-),(8,4,-),(4,8,+), (6,1,-),(#data,9,-),(5,2,+)

Table 3 : input domains for 9 ≤ #data < 16

path	add	sub
0		(#data,1)
1	(1,2)	(#data,4), (2,1)
2	(4,8)	(#data,16), (8,4), (6,1), (#data,,9)
3	(5,2)	
4	(16,32)	(#data,64), (32,16), (24-4), (#data,36)
5	(9,2)	(10,1), (#data,25)
6	(20,8)	(14,1), (#data,49)
7	(13,2)	
8	(46,128)	(#data,256), (128,64), (69,16), (#data,144), (40,4), (#data,100), (18,1), (#data,81)
9	(17,2)	

Table 4 : input domain of the ALU

To study data dependency for controllability and observability expertise, the system offers different possibilities which allows to cope with the main problems met when dealing with symbolic execution technique : the symbolic expression complexity (greatly increased by the time dimension) and the conditional branchings [6].

i/ Unique absorbant value method

The user defines the clock and the global control signals and set the data

inputs to a symbolic value #S. During simulation, the value of each internal variable depending on #S is replaced by #S ; #S is called absorbant value.
The user may check if, for example, local control signals are absorbed by #S.

ii/ Multiple absorbant values method

This method is a way to carry out a more accurate analyzis of the data flow dependency : the clock and the control signals are still defined by the designer but the data inputs are partitionned into classes and are set to distinct symbolic values #S1, #S2, ... During simulation, these values are absorbant values for variables depending on them. By analyzing values of the variables at the end of the simulation, one can determine possible information exchanges through the circuit blocks.

Both of the previous methods use absorbant value principle and allow to obtain some information - usually obtained through symbolic execution - without handling complex symbolic expressions. One of the drawbacks of this principle is the quasi global contamination of the circuit variables by the absorbant values.

iii/ semi interactive mode

The clock and the global control signals are still defined by the designer and the data inputs are set - as in the multiple absorbant value method - to distinct symbolic values #S1, #S2, ... A classical symbolic execution is run which establishes path conditions associated with the circuit functional pathes.
Because of the complexity arising when handling circuit loops, symbolic execution is performed by a semi interactive system. The user intervention is requested at 2 levels :
  - conditional branchings : he has to choose the block or the functional path within a block to be activated,
  - loops : he has to decide the number of times the loop has to be executed.
In both cases, he has to assign a non symbolic value to one or many of the variables involved in the conditional branching.
Such a semi symbolic process allows :
  - the determination of the data flows : sequence of activated blocks and functional pathes used within the different blocks,
  - the analyzis of the functional activation coverage of reached blocks and pathes. In order to increase this coverage, the user can act either on the initially defined values or on the values defined during simulation.

As a symbolic execution system is very difficult to implement, the following simplified procedures is under implementation on the CADOC system : first the clock and the global control signals are defined interactively by the user ; input variables are set to symbolic values which are absorbant only with respect to constant values (in other cases, these values are processed the same way as in symbolic simulation) : for example, #S1 + 1 will be replaced by #S1. This procedure gives informations about the data flow dependency and is supposed to help the user to perform interactive propagation towards the outputs and consistency towards the inputs.

3.2 - Bottom-up expertise

If we focus on the test pattern generation problem for a given resource, an interactive functional dependency analysis may help to perform the consistency step. Analyzing the dependency graph of the connected resources will help to perform backward and forward chainings to find out a reverse path to the inputs and a path to the outputs, this point has been formalized in [7].

The goal of the forward chaining is to observe the test results at the primary outputs [8]. This is done by following a classical forward chaining procedure similar to the one used to find sensitive path at a structural level :
step 1/ determine all the resources connected to the outputs of the resource under test (RUT),
step 2/ find the rule of these resources the left part of which is satisfied,
step 3/ for all the resources identified in step 2, determine all the resource connected to the outputs and loop on step 2 until reaching primary outputs.
A feedback loop between the pathes identified by this procedure is not a problem but means that this resource is activated several times

The consistency or backward chaining is similar to the consistency procedure at a structural level :
step 1/ determine all the resources connected to the inputs of the resource under test,
step 2/ find the rules of these resources the right part of which satisfied the left part of the RUT ,
step 3/ for all the resources identified in step 2, determine all the resource connected to their inputs and loop on step 2 until reaching primary inputs.

Example
The backward and the forward chaining to observe a given signal is illustrated on signal EQ15 of one of the previous examples (figure 3a) :

CPT/R1 : if RST then Q := 0

CPT/R2 : if ^RST & EN when re (CLK) then Q := (Q+1) mod 16

backward propagation

C15/R1 : if Q ≠ 15 & Q ≠ U then $\boxed{EQ15}$:= 0
C15/R2 : if Q = 15 then $\boxed{EQ15}$:= 1

forward propagation

RG/R1 : if $\boxed{EQ15}$ = re then OUT := IN

CONCLUSION

In this paper we have considered the general problem of the functional test of ASICs and boards (obtention of global functional specifications an generation of test activations). The results of the presented methods will be used for the two following points : test result analysis and diagnosis expertise. Test result analysis has to be planned as carefully as the test generation itself to reach test efficiency. The test result analysis primitives have to perform tasks such as :

- identification of faulty activations after execution of an activation set,
- identification of the suspected block (those which are exercized by an activation giving erroneous results),
- identification of the instance of occurrence of an error,
- definition of the data chronograms at each point and at each coordinate of the circuit for a given activation,
- definition of compacted results at a given point of the circuit,
- assistance to the chronograms comparison.

Taking profit of the previous primitives, the diagnosis expertise will be greatly facilitated. Three steps will be followed in the diagnosis procedure will proceed as follows : first the expert system will propose error hypotheses starting from test results (functional activations and exercized blocks) and secondly the system will generate a test program to corroborate or invalidate an hypothesis ; note here that the system will be able to go from a level of abstraction to a lower one.

REFERENCES

[1] C. Bellon et al., "CADOC system, a tool for multilevel description and test generation for VLSI circuits", 7th. Int.Conf. on CHDL and their applications (CHDL85), pp. 364-380, Tokyo, Japan, Aug. 1985.

[2] C. Robach et al., "Computer analysis testability : evaluation and test generation", Int. Test Conf., pp.338-345, Philadelphia, USA, Oct. 1984.

[3] T. Lin, S.Y.H. Su, "VLSI functional test pattern generation - a design and implementation", Int. Test Conf., pp. 922-929, Philadelphia, USA, Nov. 1985.

[4] Z. Kohavi, "Switching and finite automata theory", McGraw-Hill, Computer Science Series, USA, 1978.

[5] Ph. Basset, G. Saucier, "Top down design and testability for VLIS circuits",19th. DAC, pp. 851-857, Las Vegas, USA, June 1982.

[6] W.C. Carter et al., "Symbolic simulation for correct machine design", 16th. DAC, pp. 280-286, San Diego, USA, June 1979.

[7] J. Rarivomanana, "Système CADOC, génération fonctionnelle de test pour les circuits complexes", PhD Thesis, Circuits & Systems Lab., Grenoble, France, Nov. 1985.

[8] R. Khorram, "Functional test pattern generation for integrated circuits", Int. Test Conf., pp. 246-249, Oct. 1984.

FAULT SIMULATION TECHNIQUES - THEORY AND PRACTICAL EXAMPLES

M. Melgara

CSELT - Centro Studi e Laboratori Telecomunicazioni Torino - Italy.

1. INTRODUCTION

The design of every digital system goes through several steps of specification, synthesis and validation. Different levels of description are
adopted, each one tuned to the level of abstraction under examination,
starting from languages for system specification, going to behavioral,
register transfer, logical and electrical level.

These simulation sessions are not limited to the evaluation of a digital
circuit in normal operation conditions, to verify the correctness of the
design. It is usually worth to analyse the circuit behaviour under abnormal
situations, due to physical failures or to particular environmental conditions.

This different approach of the problem is called "Fault Simulation".
Starting from the description of a fault-free network, the fault simulation
process includes the following steps:
- identify a set of faults according to selected fault hypothesis;
- modify the circuit model by injecting a fault taken from the fault set;
- stimulate the correct and the faulty circuit description with the same
  input pattern;
- compare the outputs looking for differences that can highlight the presence of the fault.

The fault hypotheses can spread from the physical failures (coupling between two adjacent wires, metal or polysilicon line cut), to the wrong behaviour of functional blocks.

The main goals of a fault simulation process are:
- to estimate the capability of an input sequence to detect and locate
  faults within the circuit;
- to evaluate the response of the circuit under the effect of a selected
  set of faults.

The problem normally encountered for fault free simulation, deriving from
the great amount of time required to handle very large circuits, becomes
dramatic when a fault simulation is performed. The computational complexity
of the fault simulation has been evaluated to be proportional to the number
of the net elements to the square power [1]. As the circuit dimension
increases, the time needed to perform fault simulation becomes unacceptable.

VLSI circuits are composed of tens of thousands of gates, hence a logical
level approach, i.e. describing the network with elementary logical functions (AND, OR, inverters) cannot be faced with normal general purpose
mini-computers. One way to obtain results in a reasonable time is to use ad
hoc hardware simulation engines. These machines, structured as multiprocessors, can evaluate in parallel more than one element at a time,
achieving a very high through-put. This solution is not always acceptable
due to the high cost of the hardware, and to the fact that a simulation
engine is normally a hard-wired implementation of a given algorithm. If a
more powerful algorithm is found, it might be very difficult to modify the
hardware in order to adopt the new solution.

F. Lombardi and M. Sami (eds.), Testing and Diagnosis of VLSI and ULSI, 287-310.

A more feasible approach can be to change the level of circuit description, trying to reduce the total number of the primitive elements employed, saving computation time. Functional cells, like registers, multiplexers, ROM, RAM, ALU can be modeled with a single or few lines of code, in spite of hundreds of logical gates. This solution doesn't affect the accuracy of a fault free simulation, but it can create some problems to the fault simulation: few nodes are observable in the circuit description and it is difficult to inject faults into a primitive: particular care must be taken in the fault hypothesis definition.

Normally, a fault simulation procedure is used both to verify if a test pattern is able to detect a given set of faults, and to build a diagnostic dictionary, that allows to identify the fault which caused the observed error. When fault simulation is performed to compute the fault coverage, it is better to drop from the fault list a fault, as soon as it has been detected, in order to reduce the number of faults to be simulated. This approach, called SOFE (Stop On First Error), is very time effective, since the first patterns have a high probability of covering a large amount of faults.

However, SOFE techniques may cause some ambiguity if the final aim of the fault simulation procedure is to provide information about fault location. In this case, all the faults must be simulated with all input patterns. A table is built, putting in the rows the test patterns and in the columns the faults, eventually storing, for each pattern, the faulty primary outputs. If the test patterns are applied to the circuit in a given sequence, it is easy to identify the fault that generated the errors by intersecting the rows of the table corresponding to the input set. Clearly, this technique makes the simulation process very heavy or even unapplicable to VLSI devices, due both to the high device count and to the relevant number of test patterns to be applied to test these circuits.

In order to overcome the problem of fault simulation complexity, several algorithms have been developed [2][3]. Among these, the most popular are: deductive simulation [4], concurrent simulation [5], parallel simulation [6], parallel value list simulation [7], parallel pattern single fault propagation [8]. These methods have been applied to several commercial fault simulation programmes. In the sequel, the simulation algorithms will be described, analysing real implementations of these approaches.

2. FAULT MODELS

The main goal of the fault simulation process is to analyze the behaviour of the circuit containing failures. The choice of the kind of failures to inject is both linked to the adopted description level and the accuracy required.

A first classification of errors tries to distinguish between logical and parametric faults.

A logical fault can change the logical function of an element into another function. Logical faults are usually caused both by design errors (the expected function was synthesized wrongly by the designer) and by physical failures (pin shorts, cracks, open bonds, open connections, bulk shorts, shorts due to scratches, shorts through dielectric and so on).

A parametric fault generally modifies the magnitude of a circuit parameter, such as the delay due to an element, its response speed, its switching thresholds, voltages and currents. Parametric faults are usually due to modifications of the integration process or to environmental conditions, such as temperature, humidity, power supply fluctuations.

Rarely parametric faults can be mapped into logical fault models. Although parametric faults are not considered in logical fault simulation process, they represent a real design problem. It is therefore useful to devote few lines to this subject.

Parametric fault effects can be studied by a fault free simulation, with a careful delay analysis. Parametric faults usually generate changes in element delays and driving capabilities. Using a thorough electrical simulation it is possible to compute the range in which the parameters can fluctuate: for each logical element minimum, typical, maximum delay values are defined. A double delay simulation is performed, in order to verify the overall circuit timing.

We consider for example a NOT gate with minimum and maximum delay dmin and dmax respectively. Suppose that at time t the input signal changes from the logical value 1 to the logical value 0. If single delay simulation is performed, the output will take the logical value 1 at time t+d, where d is the element nominal delay. On the contrary, in double delay simulation, at time t+dmin the output enters an undefined state till t+dmax when it switches to the final logical value.

Using double delay simulation it is possible to verify if all signals can be correctly propagated through the circuit in worst case condition, checking the absence of races, glitches or skew that can compromise the circuit behaviour.

When double delay simulation is performed using classical four level logic, in which values 0, 1, X (undefined) and Z (high impedance) are assumed, some problems may rise. The time interval during which the signal is in an ambiguous state (it is not possible to identify the switching edge) is different from the undefined state X (0 and 1 have the same probability, this is a peculiar condition for not initialised registers). The propagation of X state can create dangerous conditions if they reach the clock control of some register: being the clock undefined, some simulator puts into an undefined state also the register contents!

For those reasons, transition states should be modeled with ad hoc levels, switching to eight, nine or sixteen value simulation, including a subset of possible transition among the logical values (i.e. 0->1, 0->Z, 1->0, 1->Z, etc.). This technique, from a theoretical point of view, can be applied to any simulation level, if we suppose that all basical values are implemented and the delay is concentrated onto the element output.

In the sequel, only logical fault modeling and simulation will be considered, leaving the analysis of parametric fault problems to other bibliography more concerned with simulation topics.

2.1. Structural fault models

Digital circuits are normally described as networks of primitive elements. Each element implements a given function, receiving data from its predecessors and providing the computed outputs to its successors. The function implemented by each element may vary according to the level of description. In the past, the most common level was the logical level, in which basic boolean functions such as bitwise AND, OR, NAND, NOR, NOT, etc., where employed.

The most diffused fault model is derived from the analysis of the interconnection network among the elements; it is based on the fact that a physical fault inhibits the switching capability of a signal, fixing its state either at logical one or at logical zero.

Those faults are called "stuck-at" faults. Given an elementary logical function, it is possible to define both a stuck-at-0 (s-a-0) or a stuck-at-1 (s-a-1) on each input and output of the primitive. The total amount of stuck-at faults for a logical primitive with n input signals and one output signal is 2*(n+1).

If we consider a VLSI circuit, composed of tenth of thousands of logical gates, the number of possible faults connected to each primitive input and output is too high to allow any kind of fault simulation. A partial solution to this problem is to try to reduce the count of distinct faults to be simulated. This procedure, called "fault collapsing", is based on logical properties of faults [9].

Given two distinct faults, f1 and f2, they are said to be "equivalent" if and only if there does not exist a test pattern that can distinguish between them. Let us consider for example a two input OR gate; Single stuck-at-1's on its input wires are equivalent since the only pattern we can apply to detect them is A=0 and B=0. If two faults are equivalent, it is sufficient to consider only one of them, since the test pattern that detects it, surely will detect the equivalent one.

A second rule to collapse fault is based on the concept that a fault f1 may be covered by the test pattern generated for fault f2, although f1 and f2 are not equivalent. Given two faults f1 and f2, f1 is said to "dominate" f2 if the set of tests T which detects f2 also detects f1. This means that it is sufficient to generate tests for f2, the dominated fault, since the dominant fault is detected by them. Note that if also the set of tests for f1 detects f2, it means that f1 and f2 are equivalent.

If we apply the equivalence and dominance rules to the elementary logical gate, we will obtain the following list of faults to be considered:

AND, NAND: each input stuck-at-1, any input stuck-at-0;
OR, NOR: each input stuck-at-0, any input stuck-at-1.

This result may be extended to complex combinational circuits. Given a fanout free combinational circuit, any test set that detects all stuck-at faults on the primary inputs detects also all internal stuck-at faults. If some fanout point is included into the circuit, the test set must also cover the stuck-at faults on the branches of the fanout points.

However, the fault collapsing procedure may distort the fault coverage concept. Suppose, for example, that there are faults f1,..,f10. The collapsing procedure can reduce the faults from f1 to f5. Suppose then there exists a test that covers the faults from f6 to f10. The apparent coverage, defined as the ratio of covered fault versus the total number of faults, is 83.3 %; but f1 represents five faults: the real probability that the test can detect a fault is 50 %. In order to avoid this problem a figure, giving its weight, i.e. the number of faults it represents, must be associated to each fault.

Other classes of faults can be easily modeled on a structural circuit description. It may happen for example that a wire is cut, leaving the input of a gate not controlled, in an high impedance state. According to the device technology, this input can be seen as a logical 0, 1, or better as a high impedance.

Another classical failure condition is the short between two adjacent wire. The circuit behaves as if a WIRED-AND or WIRED-OR gate, according to the device technology, has been applied between the wires.

However, more recent studies have shown that stuck-at, open and short circuit faults are not suitable for MOS VLSI circuits, because the effect of many physical failures in complex MOS gates cannot be represented by the stuck-at fault model. Failures can change the logical function of a gate and also change into sequential, a combinational circuit and vice versa [10-12]. The use of lower level circuit and fault models should be required, to achieve meaningful results from simulation.

This assertion is particularly true for CMOS technology. Let us consider an elementary CMOS NAND gate (figure 1 a,b). The cut of the line creates a dynamic memory node, whose value is related to the previous input sequence. Some proposals have been made [13] to model the CMOS charge storage effect as a consequence of open faults. The output node of the CMOS gate (figure 1c) is substituted by a Set/Reset flip-flop; the NMOS and the PMOS planes are mapped into the equivalent gate level representation, inverting the input signals of the PMOS plane; the NMOS and PMOS network output wires are connected to the reset and to the set inputs of the flip-flop, respectively. The cut fault considered in figure 1b can be modeled as a stuck-at-1 on the NAND gate input.

Although this methodology solves the problem of modeling open faults for CMOS gates in the very simple example shown, a single NAND gate has been substituted by an AND, a NAND and a flip-flop. It is clear that if we apply this procedure to a complete circuit in a flat way, the degree of complexity will increase many times. A fault simulator oriented to this kind of modeling must be able to handle the automatic expansion of a single gate at a time, in order to keep low the total number of elements.

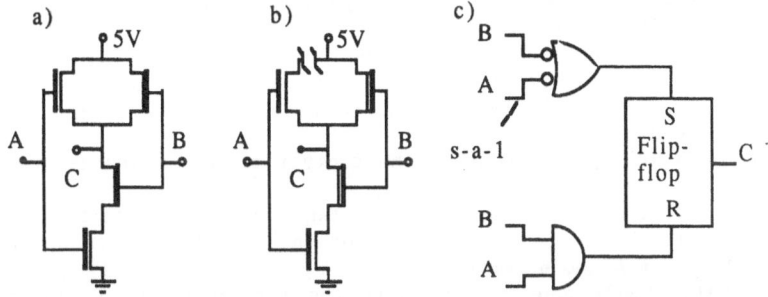

Figure 1. CMOS NAND gate; a) fault free circuit; b) faulty circuit; c) SET-RESET model.

The complexity of a VLSI circuit is too high to perform a complete fault simulation at a low description level in a reasonable time. One way to overcome this problem is to try to rise the circuit description from the gate level to the register transfer level, introducing more complex functional blocks which can represent many logical gates, reducing the number of primitives to be processed.

But, since the functional primitives implement more complex functions, the previously defined stuck-at fault models, already uneffective for logical gates, cannot cope with the problem of describing wrong internal behaviours of the new elements. Hence, the adoption of higher level functional primitives calls for the definition of more powerful fault models.

2.2. Functional fault models

The adoption of an higher level description language, such as Karl III [14], a register transfer (RT) -level simulation system, allows to reduce the number of primitives and to adopt the array notation to represent the typical parallel structures of RT-level elements. An RTL usually includes primitives that implement registers, multiplexers, arithmetic functions, RAM, ROM, decoders, encoders, busses, etc.. Those primitives can substitute tens, hundreds, even thousands (in the case of RAM and ROM) of logical gates.

A drawback in the adoption of RT and functional primitives derives from their inherent complexity. It is very restrictive to limit the fault analysis only to stuck-ats, opens and shorts of the interconnection lines, ignoring all internal faults.

As far as simple combinational circuits are concerned, it can be assumed that the internal faults are modeled by changes of the primitive truth table. On the contrary, if we consider complex and sequential primitives, it would be advisable to modify the procedures that describe their function. Furthermore, if the circuit is described in a nested way, as the interconnection of complex functional blocks, each one composed by elementary primitives, it might be interesting to deal with the faulty behaviour of such large device sub-units.

The reasons for analysing the possible faulty behaviour of complex, but well defined functional blocks are the trial to reduce the problem dimension and the concept that faults can be seen as differences from normal behaviour of the block. In the literature, it can be found a good number of fault models based on behavioural consideration for general structures, composed by registers, busses, RAM, Programmable Logic Arrays [15].

Let us consider, for example a simple structure composed by two registers, A and B, connected to a bus. The behavioural faults associated to the operation "transfer the content of A on the bus" may be:
1) neither A nor B can be transferred on the bus, that is kept in an high impedance state;
2) the content of B, in spite of the content of A, is transferred on the bus;
3) both the contents of A and B are written together on the bus: the final data will be either the logical AND or the logical OR, according to the technology, of A and B contents.

These fault models can be used both for test generation and for fault simulation procedures. However, this kind of fault hypotheses covers only the functionality of the blocks, leaving uncovered many physical failure conditions. Furthermore, it is a very hard job to try to redefine by hand both the truth tables and the functional descriptions of complex units, taking into account all possible faulty behaviours.

A solution to this problem is to use low level circuit and fault models to determine the faulty behaviour of each circuit block. Fault simulation can then be executed at the RT-level by substituting the faulty blocks in the circuit with automatically built functional description of their behaviour in presence of the simulated physical fault. This approach allows one to work at RT-level with transistor level resolution.

Some tools were developed to perform such circuit analysis. Among them we will remember FERT [16], a transistor level fault simulator. This program takes as input the electrical description of a circuit macrocell and a list of physical failures, both extracted from the circuit layout; it provides as output the flow table of each faulty version of the macrocell. FERT can handle dynamic and static memory conditions, as well as oscillations, hazards and races. The faulty truth tables computed by FERT may be used as a functional fault model in a RT-level fault simulator.

3. FAULT SIMULATION ALGORITHMS

The main goal of fault simulation is, given a circuit described at a certain level of abstraction, to verify if a computed input sequence can detect and locate faults belonging to a given set. The simpliest way of performing a fault simulation is:
- simulate the fault free machine;
- for each fault in the fault set:
  - inject the fault;
  - simulate the faulty machine;
  - compare the two results looking for differences.

The inherent complexity of the problem derives from the high count of faults to be modeled, since the simulation process must be repeated for each possible fault. In order to reduce the computational complexity of the problem, several algorithms have been developed. In the following sections, the most common fault simulation techniques will be described; some commercial implementations of those algorithm will be also reported.

3.1. Parallel fault simulation

Parallel fault simulation exploits the computer word parallelism to handle many faults at a time. Let us first consider, for simplicity, two value simulation.

A multi-bit computer word, for example a 32 bits word, whose bits are named from 0 to 31, is given. Suppose one assigns to bit 0 the good machine and to the other bits (1,...,31) 31 faulty machines. At primary input nodes, the input pattern, expanded to all the 32 bits, is assigned to the word. Suppose now this computer word is propagated through the circuit, performing on it all bitwise operations requested by the encountered primitives. When a faulty wire is met, the corresponding bit in the word is forced to the faulty value, using precomputed masks. For example, in order to inject a stuck-at-1 in the i-th bit, an OR operation is performed between the input word and a mask containing all zeros, except the i-th bit at 1; on the contrary, to inject a stuck-at-0, an AND operation is performed with a mask of all ones except the i-th bit at 0. When the activity in the circuit, due to a new input pattern, is expired, the primary output are considered: bits from 1 to 31 are compared against the first one. The faults corresponding to the bits whose value is different from the first one, are marked as detected at the primary outputs.

Let us consider the example in figure 2. A four bit computer word is used; the following assignment is assumed:
- bit 0: the fault free machine,
- bit 1: a1 (wire a stuck-at-1),
- bit 2: d0 (wire d stuck-at-0),
- bit 4: e0 (wire e stuck-at-0).

Suppose to apply the pattern a=0, b=1, c=0, and propagate the four bit word through the circuit. At node a, the fault a1 generates a change in bit 1, since this faults implies a fixed value 1 in that node. This condition is obtained by computing a bitwise OR between the word and "0100": the other bits are not influenced by the fault. At nodes b and c no modification is observed. At node d, being bit 2 already at 0, d0 doesn't effect the value (i.e. ANDing the word with "1101" nothing changes). On the contrary, at node e, bit 3 is forced to 0 by e0. The e0 effect is propagated to node f, but it is masked at node g by the value a=0. At node g, the primary output, the only observable difference from bit 0 is carried by bit 1: only a1 is detected.

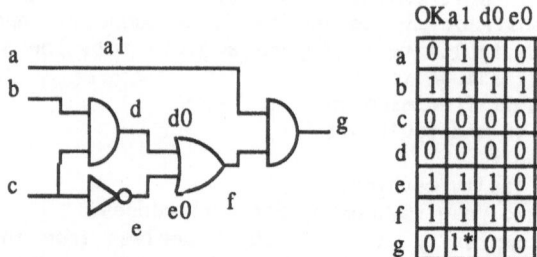

	OK	a1	d0	e0
a	0	1	0	0
b	1	1	1	1
c	0	0	0	0
d	0	0	0	0
e	1	1	1	0
f	1	1	1	0
g	0	1*	0	0

Figure 2. Example of parallel fault simulation.

The estimated computational complexity of parallel fault simulation is proportional to the cube of the number of gates. However, in [2] is referred that, if SOFE technique is applied, the complexity will be reduced to the 2.5 power rule.

Parallel fault simulation allows to handle at the same time many faults. Given an n bits computer word, if we exclude the time needed for mask generation, output comparisons and house keeping procedures, the theoretical net gain of parallel simulation is (n-1), the number of faults simulated during a single step. If m faults are included in the fault set, the procedure must be repeated m/(n-1) times. The number of faults considered simultaneously may be greater than the computer word if powerful string operations can be used. When a given number of faults has been observed, the bit assignment is modified, in order to assign no longer used bits to other faults, reducing string length.

The gain factor can be improved if several "independent" faults are simulated in the same bit position. Two faults f1 and f2, defined on wires a1 and a2, are said to be independent if the set of wires S1, interested by f1 effect, has no intersection with the set of wires S2: there are distinct path from the fault site to the primary outputs, hence f1 and f2 never affect the same signal lines. A reasonable upper limit to the number of independent paths is the number of primary outputs p: the minimum number of steps to be performed is m/((n-1)*p). However, the procedure of identifying independent faults may be so complex that the time saved in simulation is lost during this process.

Parallel simulation can be easily extended to any number of logical values, by considering more than one word at time. The logical values are coded using one bit of the different words, assigning the same bit position to the same faulty machine. In the case of three value logic, two words are used; the first word is identified as X', and the second ones as X''. A possible coding scheme for three logic value is:

```
X' X''   value
0  0    unknown
0  1       1
1  0       0
1  1    unused
```

Suppose, now, to propagate the two coded words through elementary logical gates. The propagation rules to be applied are listed below, where A', A'', B', B'', C' and C'' represent the first and the second word of A, B, C, respectively.

$$\text{AND: } C = A \text{ AND } B$$
$$C' = A' \text{ OR } B'$$
$$C'' = A'' \text{ AND } B''$$
$$\text{OR: } C = A \text{ OR } B$$
$$C' = A' \text{ AND } B'$$
$$C'' = A'' \text{ OR } B''$$
$$\text{NOT: } C = \text{NOT } A$$
$$C' = A''$$
$$C'' = A'$$

If an eight value logic is chosen, three words are needed. In this case, more complex functions must be defined, such as $C'=f(A',A'',A''',B',B'',B''')$.

Parallel fault simulation can be also employed for functional primitives: the function describing the relation among the different words of input and output signals are more complex than those defined for logical gates. If array notation is used, to compact the description dimension, some problems may rise when high level functional primitives are considered. When these functions are encountered, the parallel word must be unpacked, the bits rearranged to create arrays to be evaluated by the primitive; afterwards, the results are packed again in the parallel words.

To cope with fault hypotheses, different from classical stuck-at faults, the functions used for primitive evaluation and for mask construction must be changed. For example, a delay fault may be handled by modifying the scheduling procedure that should update the "faulty bit" when a change in the "fault free bit" is observed. Other faults, like internal wrong behaviour of high level primitives, could require heavy modification of the transfer function.

A commercial implementation of parallel fault simulation can be found in Hilo2 system. Hilo2 [17] is a simulation environment developed by Cirrus Computer Inc. Brunel (GB), sold by GenRad, Milpitas (Ca,USA).

Hilo simulator relies on a multi level description language that allows to describe, in the same context, bi-directional transfer gates, elementary logical gates, buffers, tristates, together with registers and high level behavioral constructs.

These levels provide the user with the facility of defining different description granularities. This approach is used both to reduce the simulation time, expanding only some blocks to be deeply analysed during the actual simulation session, and to perform a top-down design development, proceeding step by step in the refinment process.

An interesting feature of Hilo behavioral language is the concept of "event" that can condition the execution of statements. An event, more than a simple logical variable, is a flag that can be set by a process and tested by another to implement a sort of synchronisation. Using the events it is also possible to describe finite state machines, to implement particular algorithms.

Array notation can be adopted both to group many wires into a bus, and to create blocks of iterated cells (a cell is replicated many times, generating a sort of array structure).

Hilo2 simulation algorithm is based on a four level logic (0, 1, high impedance, undefined). Particular care is given in Hilo2 to delay modeling.

It is possible to specify, for each output wire, absolute rise and fall time, or marginal rise and fall time, defined as the slope of the changing edge, related with the capacitance to be controlled by the output. The input capacitance may also be defined. Those parameters allow the modelling of the effects of big loads, due to long connection lines, big buffers or large number of controlled gates, that heavily modify the delay introduced by a gate. Furthermore, for each parameter, minimum, typical and maximum value can be specified, in order to perform worst case simulations.

Hilo2 simulation system includes a fault simulator and a test pattern generator, both working at logical level.

The test pattern generator is based on a critical path algorithm [9]. It only deals with combinational circuits described as logical gate networks. Any asynchronous feedback loop, included in the circuit, must be cut using ad hoc primitives to avoid signal propagation during a test generation step.

The fault simulator is based on a parallel algorithm. Up to 159 are treated simultaneously. The fault models supported are stuck-at-0, stuck-at-1, wired-and, wired-or faults.

The fault simulator has shown to work very well on complex circuits, too. For example a fault simulation for a chip for speech synthesis, including 7500 logical gates, 600 functional primitives and 1200 sub-circuits, having injected a sampled set of faults (650 faults), applying a test pattern corresponding to 270 microseconds, required 395 minutes on a VAX 11/8650.

However, parallel fault simulation technique, as previously shown, has a complexity that increases as the 2.5 power of the number of gates. Cirrus Computer, in the new version of Hilo, named Hilo3, has implemented a new algorithm, called "Parallel list fault simulation". The algorithm, together with its application in Hilo3 will be described in a following session.

3.2. Deductive fault simulation

Deductive fault simulation was developed by Amstrong [18] and Godoy [19]. Although this technique is not so diffused in commercially available implementations, it is interesting for its theoretical importance. In deductive simulation technique, a fault list is associated with each gate output. The evaluation of a gate is scheduled if there is a change in the fault free machine or in the fault list of the fanin nets. A change in the fault list is called a list event. The content of the fault list, at the gate output, is deduced from the predecessors state and the fault list.

Let $Li = \{f1,..,fk\}$ be the set of faults at the gate input i. If two value logic is considered, the rules for fault list calculation for elementary logical functions are:
- if all inputs are at non-controlling value (0 for OR, 1 for AND): Output fault list = (∪ Li | all gate inputs) ∪ (output stuck-at controlling value);
- if some inputs are at a controlling value (1 for OR, 0 for AND): Output fault list = (∩ Li | controlling inputs) ∩ NOT (∪ Li | non controlling inputs) ∪ (output stuck-at non controlling value);
where (Li | inputs) means for all inputs, and NOT (L) means set of faults not in L. Figure 3 shows an example of fault list calculation.

Suppose a four input AND gate is given. Wires a and b are at controlling value: the second rule applies. The intersection of wire a and b fault sets gives {f1,f2}, while the union of c and d fault set generates {f2,f3,f4,f7}. The intersection between the first and the complement of the second fault set obtained is {f1}, since f1 is in the first, but not in the second set. Finally, the union with the non controlling output value gives {f1,f8}: this is the output fault set of gate E.

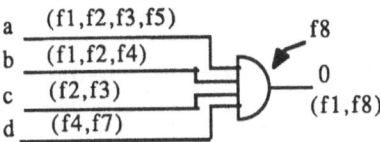

Figure 3. Example of deductive fault simulation.

Let consider now a three value logic. It is more difficult to compute the fault list. A possible solution to this problem is to keep distinct fault lists for the two states different from the good machine: this solution doubles the gate processing time. Another solution is to ignore the faults when the good machines in the X state. If the good machine state is either 0 or 1, and the faulty state is X, let the fault be denoted by f*.

The adoption of f* modifies the definition of set union and intersection (see figure 4).

A	B	A$\bigcup$B	A$\bigcap$B	A$\bigcap\bar{\text{B}}$	B$\bigcap\bar{\text{A}}$
f*	0	f*	0	f*	0
f*	f	f	f*	0	f*
f*	f*	f*	f*	f*	f*

Figure 4. Rules for list operations with faults at X state.

Ignoring faulty machines, when the good machine is X, it can cause some troubles. Some tested faults may not be marked as detected, and glitches like 0-X-0 may occur and set off oscillations in the sequential elements, in the presence of faults. This behaviour increases the simulation time. In order to work around to this problem, it is possible to add to the list a fault even if the good machine enters the X state, from a 0 or 1 previous state. This technique avoids glitch generation when the good machine changes its state. Another way of escaping from this trouble is to use special latch models to manage fault list evaluation in the sequential blocks [18].

When deductive technique is adopted for very large circuit simulation, the amount of memory required, especially for the first test patterns, may grow exaggerately. In that case, the fault set must be split into subsets, to be simulated in following steps.

The deductive fault simulation algorithm has been extended to the functional level of description by Menon [4], defining new fault set propagation rules both for combinational, functional blocks and for memory elements.

The computational complexity of deductive simulation is estimated to be in the magnitude order of the square of the number of gates. Compared against parallel fault simulation, there is an improvement in the performances. However, the problems related to the difficulties of calculating the fault lists and of scheduling events in nominal delay simulation make deductive fault simulation less attractive than other techniques, such as concurrent simulation.

3.3. Concurrent fault simulation

Concurrent fault simulation was introduced by Ulrich and Baker in 1974 [20]. This technique combines the features of parallel and deductive simulations. In parallel simulation, the correct and the faulty circuit are simulated together, computing the responses of all primitives, even if the logical values are equal; in deductive simulation only the fault free machine is simulated, while faulty machine behaviour is derived from the good machine state. In concurrent fault simulation, a fault is simulated only when it causes some wire to get a different value from the corresponding ones in the good machine. The concurrent technique, with respect to the parallel ones, allows the reduction in the total number of faults to be simulated for each input pattern.

Let us consider the NAND gate in figure 5. The input signal A has a stuck-at-0 fault (s-a-0). Since there is a difference in the logical state of good and faulty machines, the gate will be simulated. If A switches to the logical state 0, the fault will have no influence on the network behaviour: its state is identical to the good machine; hence its simulation is stopped.

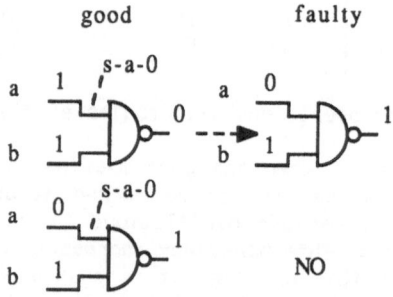

Figure 5. Condition for gate calculation in concurrent fault simulation.

Given a network composed of primitives, a fault list is associated with each element, as in deductive simulation. An entry in the fault list contains the fault index and the input-output value of the gate induced by the fault. A faulty machine is said to be "active" at a given node if the logical state of that node differs from that of the good machine. Gate calculation are scheduled for significant events of the good machines and for active faults. Separate entries in the work stack are kept for good and faulty machine.

Primitive evaluation in a classical concurrent simulator is organised as a three step procedure. We assume that the simulator will be an event driven simulator, in which the computation of a gate successor is conditioned by the presence of a significant event, i.e. by a change in the previous logical value. In the sequel the three step procedure is described, supposing that the calculation of a gate has been scheduled by a previous event.

```
begin
step 1: if the gate was scheduled by an event in the good machine
        then begin
                    evaluate the new good machine output value;
                    if there is a change in the output value
                    then schedule the successors of the gate for next good
                            machine calculation step;
            end;
step 2: if any faulty machine is active in the considered node
        then begin
                    if the fault is in the list
                    then update the state in the list entry for the faulty
                            machine;
                    else add a new entry to the list;
                    compute the output state;
                    schedule the gate successors if the fault is still
                    active;
            end;
step 3: if good machine output changes creating significant events at the
            site of not yet activated faults
        then schedule those faults for calculation at successor sites;
        remove any fault entry no longer active (i.e. with all input and
        output states equal to the good machine ones);
end.
```

Consider the example in figure 6. Gates are identified by capital letters, wires by small ones; a small letter followed by a number (0 or 1)
represents a stuck-at-0 or a stuck-at-1 of the wire. If we apply a=1, b=1,
it follows that c=1: fault a0, b0 and c0 are active at gate A. But the
value of c, under the effect of a0 and b0 is equal to the good machine
ones: a0 and b0 are canceled from the entry list, only c0 is scheduled in

A: a b c	B: c d f	C: e f g
1 1 1	1 1 0	1 0 0
a0 0 1 1	c0 0 1 1	c0 1 1 0
b0 1 0 1	d0 1 0 1	d0 1 1 0
c0 1 1 0	f1 1 1 1	f1 1 1 0
		e0 0 0 1
		g1 1 0 1

A: a b c	B: c d f	C: e f g
0 1 1	1 1 0	0 0 1
a1 1 1 1	b0 0 1 1	b0 0 1 0
b0 0 0 0	c0 0 1 1	c0 0 1 0
c0 0 1 0	d0 1 0 1	d0 0 1 0
	f1 1 1 1	f1 0 1 0
		e1 1 0 0
		g0 0 0 0

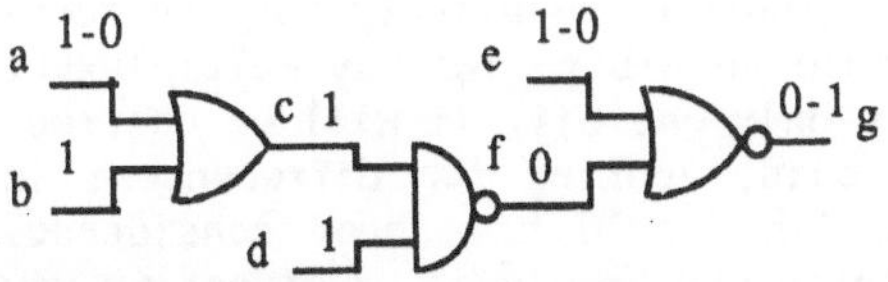

Figure 6. Example of concurrent fault simulation.

gate B list. The output of gate B is f=0, being c=1 and d=1. Faults d0 and f1 are added to the list and scheduled for calculation at gate C. We assume now e=1, which forces g=0. Faults c0, d0 and f1, scheduled for calculation, generate no active machine at gate C output: only faults e0 and g1 are detected at the primary output.

Suppose to switch a and e to 0. Faults a0 and e0 are no loger active: they are dropped. Since c=1 does not change, no activity is scheduled on gate B. Being b0 active, now, at A and B output, it is added to gate B and C entry list. Since e has changed, g is recomputed: its new value cause all faults to be active, the old and the new ones are all observed.

From the example it can be seen that, any time there is a change in the circuit, and a faulty machine is scheduled for calculation, the successor gates fault list must be scanned to verify if the fault is already present.

Concurrent fault technique requires more storage space than deductive, since the fault list in the former simulator contains entries for each active machine, for any circuit node, while deductive simulators preview only entries to the difference of the gate output state. Furthermore, owing to the separation of scheduling between good machine and faulty values, the work lists are longer.

The problem of the large memory requirement can be solved by splitting the fault list and performing many simulation steps with sub-sets of faults. The number of faults to be considered during each step is related to the memory space and of the circuit structure. The limit may be defined as an upper bound of the active fault number, in spite of a fixed number of faults per step, since the memory requirement is directly related with the activity in the circuit. This solution is also preferable because it allows to tune the simulation system to the computer configuration, rather than to each different circuit to be simulated.

Gate evaluation is faster than in deductive simulation. The computational complexity of concurrent simulation is estimated to be proportional to the square of the number of gates in the network. It has been shown that concurrent simulation works better with "short and fat" circuits (i.e. circuits in which many short, possibly distinct paths may be traced between primary inputs and outputs): for such circuits, the entry list is not too long and the searching time in the list is reduced.

Concurrent simulation is maybe the best technique for dealing with multi-valued logic and particular with delay conditions, since the procedures used in the faulty circuit evaluation are the same as those developed for the fault free network.

This characteristic of concurrent simulation has allowed its easy extension to functional models which are more complex than elementary logical gates [5,21]. No general theoretical problem is encountered during this extension. As far as fault lists are concerned, all primitives can be seen as black boxes, since all decisions about the state of a given fault are taken after having performed the primitive evaluation. However, the great advantage of concurrent simulation (the analysis of active faults only) may be partially lost if this technique is applied to relatively complex functional blocks. Let us consider for example an n-bit, two way multiplexer. If the new data to be propagated affects only one bit, it will be uneffective to recompute and check all the n bits, looking for differences: in gate level simulation, only the active line would have been considered. This disadvantage decreases if the simulator can handle in an effective way primitives modeling a high number of equivalent gates and, by keeping the capability of manipulating bit information of gate level primitives, injecting faults only on the active single-bit wires.

Some commercial applications of concurrent simulation technique are CADAT, LASAR, OFSKA. In the sequel OFSKA, a RT-level concurrent simulator, will be analysed in detail.

The CVT Project has been a research plan, partially funded by the European Economic Community, that ran from 1983 to 1986: 28 research centers from Italy, France and Federal Germany cooperated. The RT-language adopted by CVT project was Karl III [14], developed by the University of Kaiserslautern.

The term Karl III is used to identify a simulation environment composed by a compiler, a simulator, an handler of compiled network libraries and various utility programmes.

Karl III simulator works on a four values logic (0, 1, undefined, floating). A network is represented in Karl by a cell , identified by its inputs, its output and the logic it contains. The logic consists of primitives and other lower level cells. There are about sixty primitives, ranging in complexity from simple gates to registers, multiplexers, RAMs, ROMs. The compiler generates the Register Transfer Code (RTC) that is fed to the other tools. The simulator makes use of a Simulation Commands Language (SCIL) that allows to apply stimuli to the network inputs and to observe the logical state of various circuit nodes.

The fault simulator [22], called OFSKA (Olivetti Fault Simulator based on KARl), designed by Olivetti's and CSELT researchers within CVT project, is one the tools for test pattern generation and validation developed around Karl III [23]. The fault simulator is organised as a shell around the Karl III simulator. The concurrent algorithm was chosen since it seemed the only one to meet the requirements of working speed and compatibility to minimise the development time.

OFSKA takes as input the RTC of the fault free network, an RTC-like description of faults to be injected, libraries of fault free and faulty cells. The fault set, generated by a user friendly fault generator, includes the following class of faults:
- stuck-at-0/1, stuck-at-undefined, stuck-at-floating;
- wired-and bridge, wired-or bridge, X-bridge (if values are different, the undefined state is entered);
- truth table changes;
- function changes.

Table and function change models can be automatically derived from the circuit layout, using FERT, the functional fault model generator quoted in 2.2. Fault coverage figures are computed taking into account the real fault weight, that may be different from one, due to collapsing procedures (see 2.1.).

The fault simulator commands have been added to standard SCIL. They allow the definition of primary outputs, fault manipulation and examination. The assignment commands act simultaneously on fully functional network and on the faulty networks, to guarantee that the test pattern applied is identical for all the networks. It is possible to save and restore the whole state of the simulator, stopping the simulation at any time, since a simulation session may last for a considerable time and it might have to be suspended and taken up again later.

All the faults are loaded at the beginning of the session. If no fault is loaded, the fault simulator behaves as a normal RT-simulator. In presence of faults, however, it also simulates the faulty nets, and updates their "state". The state is a variable associated to each fault; its content can

302

be seen at any moment during the simulation: it tells the designer whether the fault has already been observed, controlled or neither. Figure 7 shows the finite state machine graph of fault states. A short description of each state follows.

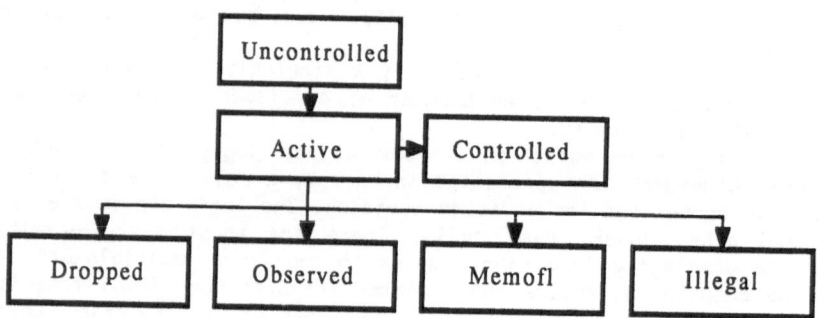

Figure 7. States of faults in OFSKA.

UNCONTROLLED: it is, normally, the initial state of all faults.
ACTIVE: the state of the faulty machine differs from that of the fault free one in at least one node: the machine associated with this fault net is actually simulated.
CONTROLLED: the good and faulty machines have been different in the past, but now they aren't: no simulation performed.
OBSERVED: a terminal state. The fault has been observed at the outputs. The corresponding faulty machine is deleted, and the time of observation and the name(s) of the output(s) on which the fault was observed are saved to be displayed in the final reports.
X-OBSERVED: an unknown state has been observed on an output: the fault is not dropped (hoping that some other pattern can provide a "solid detection') unless explicitly required.
MEMOFL: this is also a terminal state. If all the available memory has been occupied, while more is still needed to simulate all the faults, the "exceeding" ones are put in this state, and henceforth ignored. Those faults shall constitute the input set of a subsequent simulation session.
DROPPED: a terminal state: it is entered if the user decides to stop the simulation of a specific fault; this might be the case for one or more X-OBSERVED faults.
ILLEGAL: a terminal state. It is entered if a fault causes some condition that the simulator cannot handle (typically, an "unknown" value on a control input of a primitive, or a feedback loop with no delay...).

Every time the computing command STEP is given, all the networks (both faulty and fully functional) are simulated. Figure 8 shows a flow chart of STEP commands.

There are three areas of memory on which STEP works continuously: MAIN, OLD, NEW. MAIN is the main work area, in which all the calculations are performed; OLD and NEW contain the initial and final states respectively of the simulation step of the fully functional network, and are of fundamental importance in the operation of reconstruction and comparison of the states of the faulty machines.

```
Procedure STEP;
begin
   save status in OLD;
   compute new status;
   save status in NEW;
   while not end of faults do
   begin
      next fault;
      restore status from OLD;
      inject fault;
      update the status of the nodes following the indications contained in
       the faulty machine description;
      compute new status;
      compare the status with NEW and generate
       new faulty machine description;
   end;
   restore status from NEW;
end STEP.
```

Figure 8. Procedure implementing the STEP command in OFSKA.

At first the fully functional network is simulated, saving its initial and final states in OLD and NEW respectively. Then, one at a time, the all faults in the ACTIVE state and those which, are excited by the preset circuit state, are simulated. Simulating only these faults permits considerable CPU time saving without negatively affecting the result obtained.

The simulation of each fault may be subdivided into three distinct phases: the reconstruction of the associated faulty machine, the actual calculation, and the comparison with the fault free machine. The reconstruction is prompted by consideration of memory economy: since it is likely that the effects of a fault will have repercussions on a reduced number of primitives, it is preferable to record on the faulty machine only the state of those primitives, and to take the value of the remaining ones directly from the good machine. The injection of the fault is a part of the reconstruction phase. At the end of these operations the state of the faulty machine has been built in the main work area exactly as it was at the end of the previous step.

There follows the simulation of the faulty machine, at the end of the which the new state of the faulty machine is in the main work area. Finally the output nodes and the registers of good and faulty machines are compared, to compute the fault state (observed, X-observed, active, controlled).

When the simulation of all faults is finished, the new state of the fully functional machine is reinstalled in the main work area, and control passes to the next command.

The OFSKA fault simulator belongs to the test pattern generation and validation environment built around Karl III and connected to the CVT design data base COSMIC. The other tools (figure 9), described in [23], are:

- OFGKA: fault generator;
- OTAKA: testability analyser;
- FERT: functional fault model generator;
- TIGER: test pattern generator and validator.

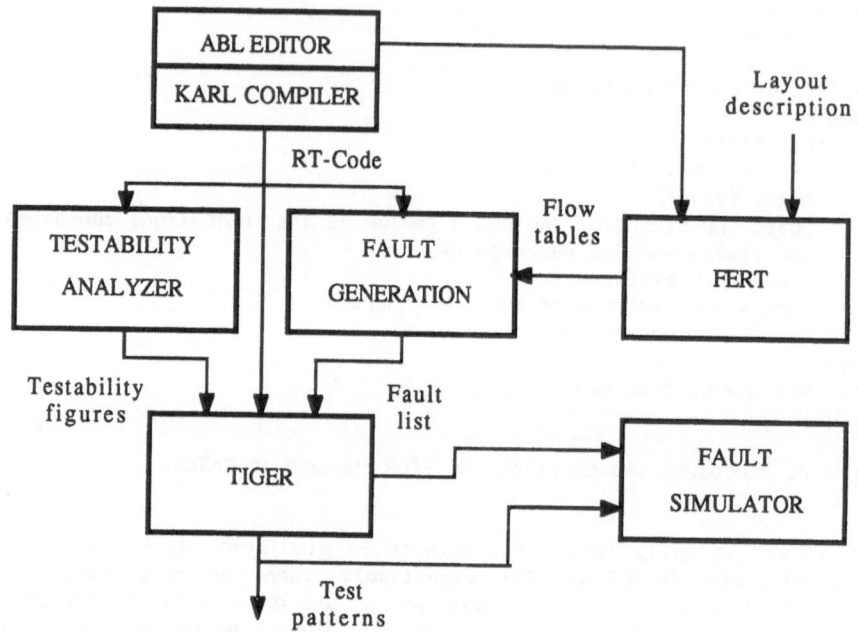

Figure 9. Test generation and validation tools in Karl III environment.

TIGER has been built as a further shell around the OFSKA fault simulator. A designer, after having described and compiled his network using the ABL graphic editor [24] and the Karl III compiler, can enter the user friendly, interactive environment of TIGER, to perform fault free simulation, test pattern generation and fault simulation, using an extended set of SCIL commands. The advantage of this approach is to provide the user with a single environment, allowing him to switch easily from the validation of the fault free design to the analysis of faulty conditions.

3.4. Parallel value list fault simulation

The parallel value (PV) list approach [7] is a combination betweeen the parallel technique and a list structure, similar to the deductive and to the concurrent technique. The faults are gathered into groups of a size equal to the computer word size. Each fault is identified by the group number and the bit position in the word. Each wire has a fault free value and a fault list of equal sized cells, containing the value of a group. Each group is present if at least one bit in it has a value different from the fault free.

The parallel value list technique requires less memory than concurrent technique, since values are coded in parallel words; dynamic creation and deletion of the cells is fast. Parallel fault simulation is more suitable for highly sequential circuits, while concurrent fault simulation works better with "short and fat" circuits. PV list is a trade off between those characteristic; if faults are chosen to collect in a group "nearly equivalent" faults, the faulty activity produced will lead to shorter lists and a higher concentration of the faulty effects in the cell. When faults are detected, they are deleted and the remaining faults are compressed to reduce the groups number.

The PV lists are propagated, following two methods, according to the circuit activity. If the fault free value changes, involving many changes in the faulty values, the evaluation process is similar to the deductive techniques, where set union and intersection operation are applied to the fault lists. If the fault free value is not affected and only a few faulty values change, the evaluation procedure ignores the inactive groups; the fault effects are propagated using an approach similar to the concurrent technique that evaluates and schedules only the active faulty elements.

Let us discuss the example in figure 10, derived from [7]. The given computer word has eight bit (0,...,7). The following notation is adopted to represent groups and values. Given the string a=0-G7(p2=1,p3=X), it means that the fault free value of the wire a is 0; in group 7, bits 2 value is 1 the bit 3 value is X. Suppose now to switch the value of wire a to 1-G1(p7=0)-G3(p3=0). The evaluation of the NAND gate A involves the set union of the two PV lists 1-G1-G3 and 1-G3-G7-G11. The groups 1,7,11 will simply be copied and inverted, while group 3 will be processed. We will obtain the PV list 0 - G1(p7=1) - G3(p0=p2=X,p3=1) - G7(p5=X,p7=1) -G11(p6=1). The list is put into a event cell; wire c switching and wire c PV list assignment are scheduled. The evaluation of gate B involves the intersection of the PV lists 0-G1-G3-G7-G11 and 0-G7: only group 7 will require Boolean equation processing, the other groups being ignored. The result of this evaluation is 1-G7(p5=X). At this point, because the fault free value in node e is not changing, the resultant PV list is converted to a different type of list where there is no fault free value and the cells

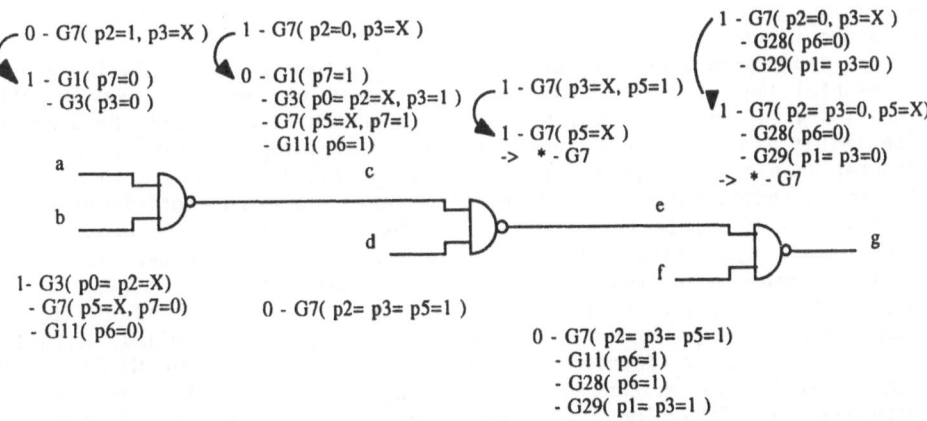

Figure 10. Example of parallel value list simulation.

in the list indicate active groups only. The new PV list scheduled on wire e is identified by *-G7(p5=X). When the value on node e changes, during the evaluation of gate C only the active group 7 will be considered; groups 28 and 29 will be ignored. The result of the evaluation is *-G7(p2=p3=0,p5=X), which will be scheduled on the output wire g and propagated until the activity in the group dies.

The PV list technique has been implemented, by Cirrus Computer in Hitest and in Hilo3.

Hilo3 simulation environment derives directly from Hilo2 one. The same HDL is adopted, to guarantee a full transportability between the two environment. The main differences between the two packages are the number of logical values and the fault simulation algorithm. Hilo3 is a 15 value simulator. The five basic values are: strong 0(0), weak 0(L), high impedance (Z), weak 1 (H), strong 1 (1); furthermore, the following transition are defined: P(1-H), R(H-Z), F(L-Z), N(0-L), T(1-Z), B(0-Z), W(H-L), U(1-L), D(0-H) and X(0-1). All these levels allow an accurate MOS circuit modeling. These values are applied both to fault free and to fault simulation, at switch, gate and RT level of description.

In Hilo2, the fault simulator was based on a parallel technique, while in Hilo3 a PV list technique is implemented. In Hilo3 fault sets are first of all collapsed, looking for fault equivalence not only at the primitive boundary, as Hilo2 did, but also analysing the network. A classical example of simple global fault collapsing may be applied to a chain of n inverters. If the single inverter is considered, its collapsed fault set includes a s-a-0 and a s-a-1; it follows that the total number of faults to be injected into the chain is 2*n. However, if a global fault collapsing is performed, since there are no fan-out points in the chain, the real collapsed fault set includes only one s-a-1 and one s-a-0.

Afterwards, the collapsed fault set is subdivided into groups, and lists of groups are built. It is possible to select the number of faults to be considered during each simulation pass, splitting the complete fault simulation procedure in subsequent sessions. This parameter must be chosen in order to optimise the memory size and the simulation time, according to the circuit dimension and total fault count. If too large a set of faults is considered, the virtual memory size will explode, loosing computer time for page fault handling. On the other hand, a very small fault subset requires many simulation passes, and many initialisation procedures. A reasonable set would include from 500 to 1000 faults.

Other improvements are the possibility to perform fault simulation in an incremental way: the procedure may be stopped at any time, saving the overall status, and restarted later. This approach also prevent the simulation result from dangerous system failures, avoiding the loss of partial information already computed. A percentage of all faults may also be considered to reduce the overall simulation time, obtaining statistical fault coverage. In Hilo3, fault simulation has been extended to the behavioural level. Fault hypotheses added for this abstraction level deal with the "event". A behavioural fault may be seen as an event forcing or inhibition. This approach allows to easily model faults like those presented in [15].

Some benchmarks performed by Cirrus Computer show that Hilo3, with PV list simulation, is from two to seven times faster than Hilo2, with parallel fault simulations. However, there are some unlucky circuits for which the two simulation times are almost the same. If we consider the example of the speech synthesizer, described in the session 3.1., the adoption of Hilo3 fault simulator has provided a 25% time reduction, always on a VAX 11/8650.

PV list simulation technique, implemented in Hilo3, represents a theoretical enhancement, with respect to the parallel technique, as for simulation time, and to the concurrent ones, as far as memory requirements are concerned, although PV list still requires some fine tuning, maybe, in group building procedure.

3.5. Parallel pattern single fault propagation

All the previously described algorithms can be applied to any kind of logical circuit, providing different throughput according to the network structure. Quite recently, some new algorithms have been proposed, able to deal only with Level Sensitive Scan Design structures (LSSD) [1], but very powerful in exploiting computer calculation capabilities. These techniques, known as "Parallel Pattern Single Fault Propagation" (PPSFP) have been introduced by J. A. Waicukauski et al. [8] from IBM.

The new procedures were introduced to calculate fault detection probabilities for LSSD network tested with random patterns. The method can be applied to combinational structures that satisfy to the following constraints:
1) No unknow state can may be generated at any logical gate with the primary inputs fixed at given states;
2) Any occurrence of the high impedance state can be controlled such that it behaves as either a logical "1" or "0".

These conditions will allow to perform an exact two-value, zero delay simulation and the simulation of a new pattern will require no storage of the circuit previous state, saving memory space and computer time. The adoption of the technique of propagating parallel patterns, together with such time saving allow to perform an extremely efficient fault simulation. The simulation algorithm consists in the following steps.
1) A fault free, 2 value simulation is performed for 256 patterns in parallel: a string of 256 bits with the defined sequence of 1's and 0's is placed at each primary input. These pattern are then propagated through the network, and all nodes values are computed using simple logical bitwise string operations (AND, OR, NOT).
2) For each single stuck fault still being considered, the faulty values are propagated forward, starting at the fault site and stopping the calculation when all differences from the good machine disappear. If a fault become detectable at an observation point, it is dropped from the next simulations.
3) Repeat steps 1 and 2 until all faults have been detected or the maximum number of pattern have been simulated.

PPSFP efficiency is due to the ability of exploiting the parallelism, without the waste of operations of parallel fault simulation, where many faults are simulated in parallel with one single pattern. In this last case the potential effect of each fault is computed for all gates, even if most of gates cannot be affected by most of faults. Furthermore, in parallel fault simulation one bit is always reserved to the good machine, and when the number of faults simulated in one pass is less than the maximum number allowed, the remaining bits are wasted.

In [8] an interesting comparison between PPSFP and parallel fault simulation is performed, assuming the same simulation procedure and the same parallelism factor of 256 for the two techniques. Let be:
G = number of gates of the structure
P = number of patterns to be simulated

F = number of faults to be simulated (assume F = 2G)

k = average percent of total fault to be simulated

For PPSFP, the number of simulation steps is P/256. During each fault free simulation pass G calculations are required. If we assume that the average number of operations for each remaining fault is 20, we get:

$$\text{No. calc.} = (P/256)(G+20)kF = (PG/256)(1+40k)$$

In parallel fault simulation, each pattern must be repeated kF/255 times, and each gate must be recalculated at each pass. Hence the number of calculation required is:

$$\text{No. calc.} = P(kF/255)G = 2kPG^2$$

Given a circuit of a total of 10,000 gates and an average of 10% of faults to be simulated per pattern, it is expected that PPSFP is 400 times faster than parallel fault simulation. According to the referenced paper's authors, PPSFP will also win against concurrent and deductive fault simulation techniques for a ratio approaching the parallelism factor (256), since all expensive list treatment procedures, peculiar to the aforementioned methods, are avoided.

In conclusion the PPSFP technique represents a very good approach that, if it possible to assume that the average number of faults to be simulated for each pattern is a constant, the complexity increases linearly with the circuit size. If also "k" increases with the circuit density, the simulation time will rise at a faster rate. The main drawback of PPSFP is that it can only be applied to combinational circuits that satisfy to the aforementioned restrictive conditions.

4. CONCLUSIONS

The problems and the reasons related to the fault simulation of digital integrated system have been discussed in this chapter. The fault simulation is a fundamental step during the design of a VLSI device, since it allows to analyze the circuit behaviour in critical and faulty conditions, and to evaluate the coverage offered by a test pattern. If the first reason may acquire importance for particular devices for which a fault secureness is compulsory, as it has in the aerospace, telecommunication and military applications, the second reason involves commercial considerations.

The cost of an integrated device is related also to the quality degree assured by the supplier: the higher the coverage of the test sequence applied during the device validation, the lower the residual defect level of a given stock, the higher the probability that the boards, built using these components, will work.

However, the fault simulation procedures, due to their inherent complexity and to the amount of computation time required, are rarely performed, especially in the case of VLSI devices. This class of circuits, composed of hundreds of thousands of transistor, are hardly handled by fault free simulators.

Some alternative techniques have been proposed to estimate the fault coverage without performing fault simulation. Among these we would remember the attempt of compute the detection probability using random pattern testability [25] and the program STAFAN [26] that compute a statistical fault coverage only looking at the fault free circuit activity.

Another encouraging answer to the problem of fault simulation complexity was given in [27] by V.D. Agrawal. By applying the Monte Carlo method he showed the statistical validity of coverage figures derived considering a reasonable sample set of the universe of faults: the closer to 100% (or to 0%) is the coverage, the smaller is the confidence interval. Suppose to consider a sample set of 1,000 to 2,000 faults (note that the set dimension is independent from the circuit size); if we obtain a fault coverage of 95%, the confidence interval is about 2%. The importance of this result derives both from the fact it is applicable to any fault simulation method and that it does not require to tune the fault sample set to the circuit complexity.

Another reasonable way of dealing with VLSI fault simulation is to perform their validation at an high description level, like register transfer level, reducing the number of elements to be considered. Some algorithms have been presented, together with their commercial implementation. In our opinion, concurrent and parallel value list techniques are the most suitable for functional fault simulation.

However, as previously underlined, an abstraction level that is too high may provoke a discrepancy between the description and the physical circuit, although the behaviour is the same. This separation requires a careful selection of the fault models to be considered. New functional models, that take into account topological information, transferring it to the higher level of description, represent the only reasonable link that can assure both acceptable fault simulation time and reliable simulation results.

REFERENCES

1. T.W. Williams, Design for testability, in: P. Antognetti, D.O. Pederson, H. De Man, Computer design aids for VLSI circuits (Sijthoff & Noordhoff, Alphen aan den Rijn, N, 1981).
2. P.S. Buttorff, Computer aids to testing - An overview, in: P. Antognetti, D.O. Pederson, H. De Man, Computer design aids for VLSI circuits (Sijthoff & Noordhoff, Alphen aan den Rijn, N, 1981).
3. Y.H. Levendel, P.R. Menon, Fault simulation methods - Extension and comparison, The Bell System technical Journal, vol.60, n. 9 (1981) 2235-2258.
4. P.R. Menon, S.G. Chappell, Deductive fault simulation with functional blocks, IEEE Trans. Comput., vol. c-37, n. 8 (1978) 689-695.
5. L.P.Henckelels, K.M. Brown, C. Lo, Functional level, concurrent fault simulation, 1980 IEEE Test Conf., (1980) 479-485.
6. S. Seshu, On an improved diagnosis program, IEEE Trans. Electron. Comput., vol. EC-14 (1965) 76-79.
7. P.R. Moorby, Fault simulation using parallel value list, ICCAD-83, (1983) 101-102.
8. J.A. Waicukaski, E.B. Eichelberger, D.O. Forlenza, E. Lindbloom, T. McCarthy: "A statistical calculation of fault detection probabilities by fast fault simulation", IEEE International Test Conference 1985, Digest of papers, 779-784.
9. M.A. Breuer, A.D. Friedman, Diagnosis & reliable design of digital systems, (Computer Science Press, Inc., Woodland Hills, Ca-USA, 1976).
10. J. Galiay, Y. Crouzet, and M. Vergniault, "Physical versus logical fault models MOS LSI circuits: impact on their testability", IEEE Trans. Comput., vol. C-29, No. 6, June 1980, 527-531.
11. B. Courtois, "Failure mechanisms, fault hypotheses and analytical testing of LSI-NMOS (HMOS) circuits", VLSI 81, (1981) 341-350.
12. P. Banerjee and J. A. Abraham, "Generating tests for physical failures in MOS logic circuits", IEEE International Test Conference 1983, Digest of Papers, October 1983, 554-559.

13. S.M. Reddy, V.D. Agrawal, S.K. Jain, A gate level model for CMOS combinational logic with application of fault detection, 21 Design Automation Conference, (1984) 504-509.
14. K. Lemmert et al., Karl III reference manual, CVT report, University of Kaiserslautern (1984).
15. S.M. Thatte: "Test generation for microprocessor", report R-842, University Of Illinois, May 1979.
16. M. Melgara, M. Paolini, R. Roncella, S. Morpurgo, CVT-FERT: Automatic Generator of Analytical faults at RT-level from Electrical and Topological Descriptions, 1984 IEEE Test Conf., (1984) 250-256.
17. Hilo2 user-manual, (Cirrus Computer, Gen-Rad, Milpitas, Ca-USA, 1984).
18. D.B. Amstrong, A deductive method of simulating faults in logic circuits, IEEE Trans. Comput., vol. C-21, (1972) 464-471.
19. H.C. Godoy, R.E. Vogelsburg, Single pass error effect determination (SPEED), IBM Tech. Journal. vol. 13, (1971) 3343-3344.
20. E.G. Ulrich, T. Baker, Concurrent simulation of nearly identical digital networks, Computer, vol. 7 (1974) 39-44.
21. M. Abramovici, M.A.Breuer, K. Kumar, Concurrent fault simulation and functional level modeling, 14th Design Autom. Conf., (1977) 128-137.
22. S. Morpurgo, A. Hunger, M. Melgara, C. Segre, RTL test generation and validation for VLSI: an integrated set of tools for Karl, CHDL Conf., (1985) 261-271.
23. I. Stamelos, M. Melgara, M. Paolini, S. Morpurgo, C. Segre, A multi-level test pattern generation and validation environment, International Test Conference, Washington DC, 1986, pp. 90-96.
24. G. Girardi, R. Hartenstein, U. Welters, ABLED: a RT-level schematic editor and simulator interface, Euromicro '85, (1985) 193-200.
25. J. Savir, G.S. Dilton, P.H. Bardell, Random Pattern Testability, IEEE Trans. Comput., vol c-33, pp. 79-90, January 1984.
26. S.K. Jain, V.D. Agrawal, Statistical fault analysis, IEEE Design & Test, vol. 2, pp. 38-44, February 1985.
27. V.D. Agrawal, Sampling techniques for determining fault coverage in LSI circuits, Journal of Digital Systems, vol. V, pp. 189-202, 1981.

THRESHOLD-VALUE SIMULATION AND TEST GENERATION

Vishwani D. Agrawal
AT&T Bell Laboratories
Murray Hill, NJ 07974-2070

Kwang-Ting Cheng
University of California
Berkeley, CA 94720

A simulation approach to test generation is developed. Advantages of this approach are that backtracking is completely avoided and circuit delays are taken into account. The method is based on a directed search in the vector space. In order to compute the cost function that is minimized during the search a new form of logic simulation, called the threshold-value simulation, is developed. Logic gates are modeled as threshold functions with the standard definition of threshold functions modified to propagate signal controllability information. The method allows accurate three-value (0, 1, X) simulation of logic circuits and the controllability information is used for test generation and fault simulation. Thus, the threshold-value model provides an unified framework for verification and test.

1. INTRODUCTION

The automatic test generation problem is considered by some people as "completely solved". The reality is that adequate methods are available only for combinational circuits and for test generation to be automatic one must use scan design [1].

While scan design is a viable technique, a substantial number of VLSI circuits are still designed without scan. The only recourse the designer of these circuits has is to generate tests manually. Although several sequential circuit test generators have been developed, their performance is questionable. They have one or more of the following limitations: 1) circuit must be synchronous, 2) limited number of flip-flops, 3) limited number of gates, 4) limited number of vectors per fault, and 5) circuit delays must be neglected.

Primarily, most sequential circuit test generators have been devised on the basis of the fundamental combinational algorithms like D-algorithm [2] and PODEM [3]. Muth [4] showed that if a nine-value algebra was used then tests can be generated for synchronous circuits. However, the method does not guarantee success for asynchronous circuits. Even for synchronous circuits, the implementation of the algorithm is very complex and may be inefficient for large circuits.

In the present work we recognize that conventional test generation methods have two

F. Lombardi and M. Sami (eds.), Testing and Diagnosis of VLSI and ULSI, 311–323.

problems: 1) they rely on backtracking, and 2) they ignore circuit delays. For sequential circuits, backtracking must be done both in space and time and could easily become unmanageable. Neglecting delays can often make a test impossible or produce a test that causes race in the circuit. Our motivation is, therefore, to research an entirely different approach.

Our approach is based on simulation. Simulation involves no backtracking and the event-driven algorithms deal with circuit delays in a very natural way.

2. Problems with Conventional Methods

We will give three examples to illustrate the situations that are responsible for the high complexity of the commonly used *path-sensitization* algorithms. The first example will serve as a reference.

Consider the combinational circuit shown in Fig. 1. Test generation for the fault "line c stuck-at-1" is carried out in the following steps:

1. Set line c to 0. Its value, shown as 0/1, means 0 in good circuit and 1 in faulty circuit.
2. Justify step 1 by setting line a to 0.
3. Propagate the value of line c to the output e by setting line d to 1.

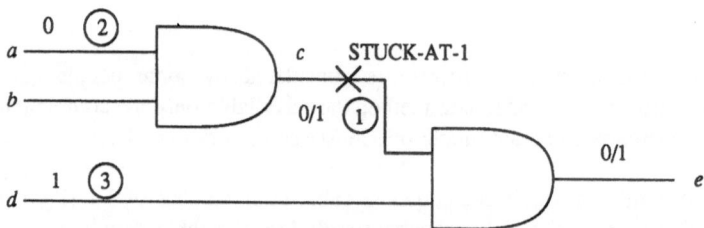

Circled numbers show the sequence of test generation steps.

Fig. 1 A simple test generation example.

The test $0X1$ is found in three steps.

2.1 Backtracking

Consider the circuit of Fig. 2. Even though this circuit may seem functionally trivial, the fault "line c stuck-at-1" is detectable; faulty circuit output function is $e = a$ while the good circuit realizes the function $e = a \cdot b$

Test generation steps for this example are:

1. Set line c to 0. Its state is denoted by 0/1 as before.

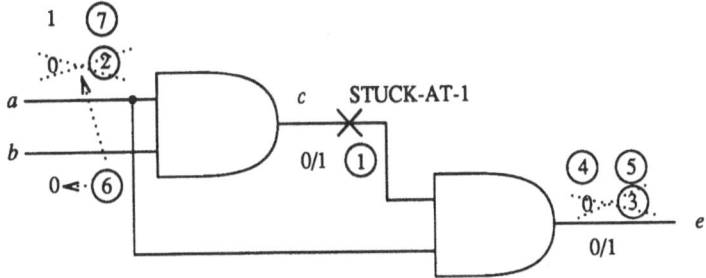

Circled numbers show the sequence of test generation steps.

Fig. 2 Test generation with reconvergent fanout.

2. Justify $c=0$ by setting line a to 0.
3. Carry out forward implication of $a=0$, i.e., set e to 0.
4. Propagate the state of c forward. It is impossible. Backtrack until a new choice is available.
5. Backtrack to step 3 and undo it.
6. Backtrack to step 2, use alternate choice: set $b=0$ and leave a unspecified.
7. Propagate state of c forward by setting $a=1$.

Test 10 is found in seven steps. Although this circuit has about the same size as that in the last example, the test generation took more than twice as many steps. The added complexity is due to the reconvergent fanout of a which made backtracking necessary. In general, large logic circuits contain numerous reconvergent fanouts and the test generation complexity increases rapidly with the size of circuit.

2.2 Sequential Circuit

Our third example illustrates conventional procedures for sequential circuits. Consider the latch in Fig. 3. Let us assume the gates have zero delay. The result of applying the combinational test generation approach is shown in the figure. The process stops with inputs at 10 and the output at 1. Since the output is the same in both good and faulty circuits, test is not found.

Figure 3 also shows that the faulty function, when "line c is stuck-at-1", is $z = \bar{b}$. The fault destroys the feedback and the circuit no longer has the storing capability. The fault can be easily detected by first applying a 10 input and then following it up by 11. The first pattern will produce a 1 output irrespective of the fault. The second pattern simply stores the state of latch. In the good circuit, the output will remain as 1 while it will change to 0 in the faulty circuit.

Our combinational test generator was not able to solve this problem due to the zero-delay assumption. If we consider finite delays of gates, it is possible to apply the value of $z=1$ to d and then change b to 1 to sensitize the path for the fault. A common practice is to cut the feedback path. This is shown in Fig. 4. A copy of the circuit is attached to

314

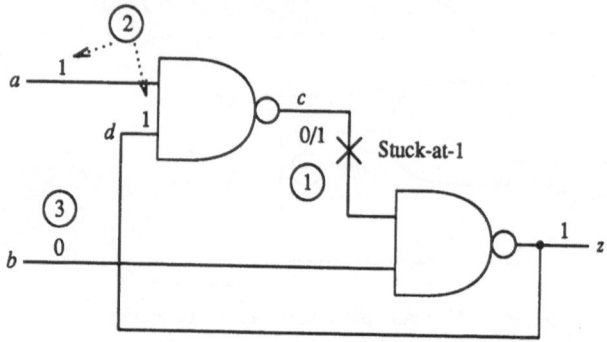

Circled numbers show the sequence of test generation steps.

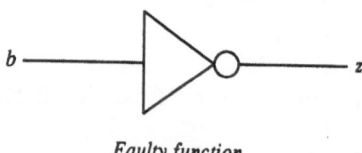

Faulty function

Fig. 3 A NAND latch example.

generate the feedback signal d. The test generation, as shown in the figure, begins from the copy shown as *current time frame*. In general, any number of time frames (previous or future) can be added on either side of the current time frame. However, the complexity of circuit increases. Another problem occurs due to the signals left unspecified by the test generator. For example, the test in Fig. 4 is a 11 pattern preceded by $X0$. If we set X to 1, we get the desired test. But if X was set to 0, the test will cause a *race* in the fault-free circuit. Thus sequential circuit tests generated by such procedures require special process-

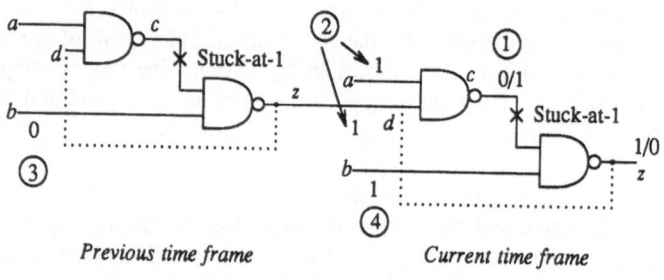

Previous time frame   *Current time frame*

Circled numbers show the sequence of test generation steps.

Fig. 4 Time frame extension of NAND latch.

ing to avoid timing problems.

3. Outline of the New Method

The purpose of examples in the last section was to illustrate three difficulties, namely, 1) backtracking, 2) time frame processing, and 3) generation of race-free tests. Our objective in the new method is to avoid these problems.

Our problem is to find a test for a given fault. We conduct a directed search in the space of all input vectors. The search is directed by a *cost function*. The cost function $C(V_i)$ of input vector V_i only depends on the result of simulation of the good and the faulty (with single fault) circuits. Also $C(V_j) \leq C_0$, for some C_0 to be defined later, if and only if V_j detects the fault. Starting at any vector V_i, cost is computed for all vectors that are at unit Hamming distance from V_i and we adopt the vector with minimum cost. Successive moves will either lead to a test or the search will terminate at a local cost minimum. In the latter case, the search process can be restarted with a new initial vector.

Since the cost function entirely depends on the results of simulation, circuit delays, and timing problems (race, etc.) will be taken into account by the simulator. Also, all decisions during the search are made on the basis of inputs and outputs and no explicit consideration of the internal structure (reconvergent fanouts) is required.

For combinational circuits, the final vector at which the cost drops below C_0 is the test. For sequential circuits, all vectors from the initial to the final vector form the test sequence.

3.1 Cost Function

The cost function of a signal in the circuit is computed for an input vector and a specific fault. We define it as follows:

$$C(V) = \frac{1}{|F(V) - F'(V)|}$$

where $F(V)$ is the function of the input vector V and $F'(V)$ is the faulty function. For binary values, the cost will be 1.0 if the fault is detected and will be infinity if it is not detected. If we have two vectors, V_i and V_j, such that none of them detect the fault under consideration, then the cost for each vector will be infinity. Thus the cost as defined above will not allow us to discriminate between these vectors. However, to conduct a directed search for a test we would like to know which vector is better. In other words, we would like to discriminate between logically identical signal values.

Consider the two AND gates shown in Fig. 5. In both cases the logical output is identical. However, in gate (a) the output can be changed by changing just one input while in (b) we must change all three inputs for changing the output to 1. We might express this difference by saying that the output signal in (a) is *somewhat closer* to 1. The following model of Boolean gates represents this phenomenon numerically.

316

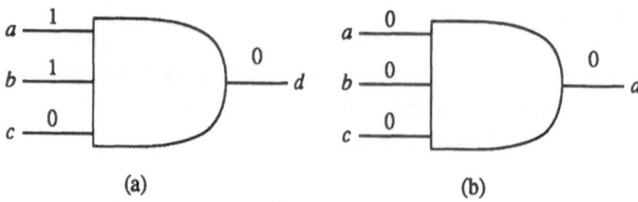

(a)   (b)

Fig. 5 Discriminating between identical signals.

4. Threshold-Value Model

We must point out that our use of threshold model is different from the conventional threshold logic [5]. We define two *signal thresholds*, s_0 and s_1, as shown in Fig. 6. Signal values in the range $[0,s_0]$ are equivalent to logic 0 and those in the range $[s_1,1]$ are equivalent to logic 1. We compute the average input value for a gate as

$$v = \frac{1}{n}\sum_{i=1}^{n}v_i$$

where the gate has n inputs and the signal value of the *ith* input is v_i. Next, we define a *gate threshold* t_G as the value of v at which the logical output of gate G switches. In Fig.

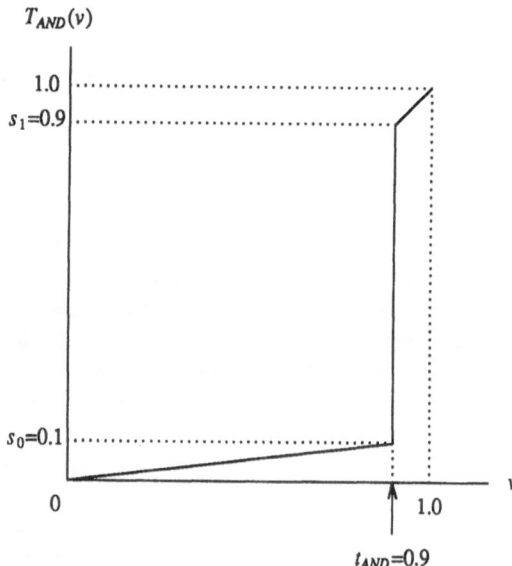

Fig. 6 Threshold function model for AND gate.

6, we have constructed a threshold function $T_{AND}(v)$ for an AND gate. In order to avoid ambiguity in the output, signal thresholds and the gate threshold should satisfy the following condition:

$$1 - \frac{1-s_0}{n_{max}} < t_{AND} \leq s_1$$

where n_{max} is the maximum fanin in the circuit. The function in Fig. 6 is for $n_{max}=8$. If we evaluate the outputs of gates in Fig. 5 using the threshold-value model, then the outputs in the two cases will be 0.074 and 0.0, respectively. According to signal thresholds both outputs will be interpreted as logical 0. However, the output in (a) is slightly higher indicating that it will be comparatively easier to change to 1.

Threshold function for OR gates with $n_{max}=8$ is shown in Fig. 7. The condition for thresholds here is,

$$s_0 \leq t_{OR} < \frac{s_1}{n_{max}}$$

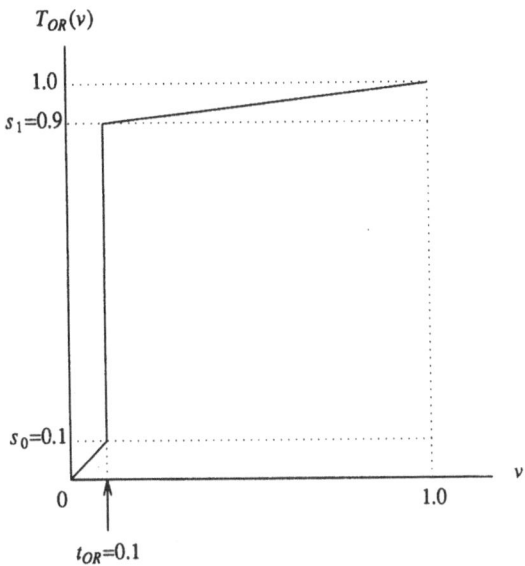

Fig. 7 Threshold function model for OR gate.

For an inverter, the threshold model is $T_{INV}(v) = 1-v$. For other gates or Boolean functions, threshold function models can be derived easily [6].

5. A Test Generation Example

First consider combinational circuits. Suppose we take any input vector and perform threshold-value simulation to compute the outputs. We then repeat the simulation after introducing the fault for which a test is desired. This is done simply by fixing the faulty line to the value (0.0 or 1.0) to which it is stuck. Next, we compute the cost at each output as the reciprocal of the absolute difference between the good and the faulty values. We take the minimum (over all outputs) as the cost of this vector. Notice that when

$$Cost \leq \frac{1}{|s_1 - s_0|} \ or \ 1.25$$

the input vector will be a test. If the cost is higher than 1.25, we compute the cost for all vectors that are at unit Hamming distance from the input vector and select one that has the lowest cost. We continue to lower the cost by repeating this process until a test is found.

Fig. 8 gives a simple illustration. The starting vector is 100. The cost, as computed in (a), is 500. A one bit change in the input vector leads to 000 and a cost of infinity as shown in (b). Next, in (c), the cost for 110 is found as 27.02. In this example we adopted a *greedy* heuristic and accepted this vector. However, the cost is still higher than 1.25 needed for a test. Another one bit change in the input vector provides a test (111). This is shown in (d).

This simple example was given to illustrate the principle of our method.

6. Implementation of Test Generator

We will discuss a preliminary implementation of a test generator and some results. The test generator was developed for general circuits: combinational and synchronous or asynchronous sequential. It begins with an initial vector which can be supplied by the user or, as default, the all 0 vector is used.

6.1 A Multipass Process

The test generator works in multiple passes. We will first describe the combinational test generator and then discuss the differences required in handling sequential circuits.

First Pass. All faults under consideration are simulated through a concurrent fault simulator. Cost of the initial vector is computed for each fault. The fault with the lowest cost is selected for test generation. A directed search for test is conducted. If test is found, all faults are simulated with this vector and the detected faults are removed from consideration. As a result of this fault simulation, costs of this test vector is also obtained for each undetected fault. Again, the fault with the lowest cost is used for generation of the next test.

If the search for test for a fault terminates without the fault being detected, then that fault

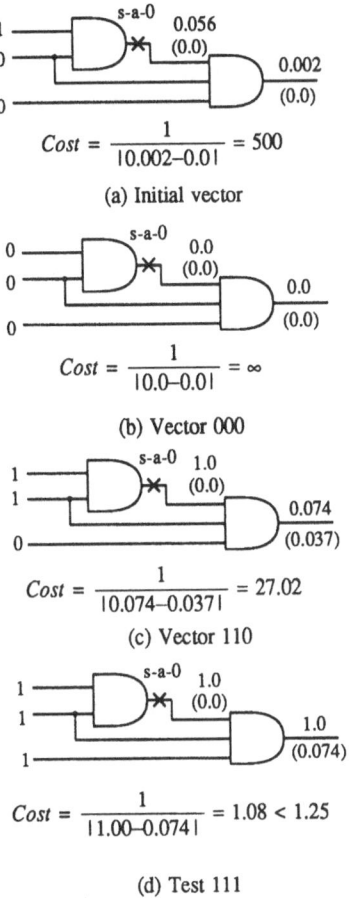

$$Cost = \frac{1}{|0.002-0.01|} = 500$$

(a) Initial vector

$$Cost = \frac{1}{|0.0-0.01|} = \infty$$

(b) Vector 000

$$Cost = \frac{1}{|0.074-0.0371|} = 27.02$$

(c) Vector 110

$$Cost = \frac{1}{|1.00-0.0741|} = 1.08 < 1.25$$

(d) Test 111

Fig. 8 Test generation example (faulty values are shown in parenthesis).

is classified as *hard to detect*. This fault is not classified as detected or redundant but it is not used for test generation again in this pass. It can, however, be detected by the fault simulator.

Subsequent Passes. There are two possible reasons for a fault not being detected during a pass: 1) the fault is redundant, or 2) the search terminated at a *local* minimum in the search space. In the latter case, the often used procedure is to start with a different initial vector. This is exactly what the subsequent passes do. Any number of passes can be made at user's option. Each pass can be initiated either with a new random vector or with a vector that is at a large Hamming distance from the initial vectors of all previous passes.

6.2 Combinational Circuits

To verify our test generator's capability of correctly handling the situation where a multipath sensitization is essential, we ran it on the eight-gate circuit example of Schneider

[7]. All detectable faults were detected in the first pass. Our next example is a four-bit ALU circuit with 104 gates and 268 (including 4 redundant) faults. The first pass took 7 seconds on a VAX11/780 computer to generate 33 vectors to cover 97.4 percent faults. A second pass required 2 seconds, produced 3 new vectors and left only the four redundant faults. For an eight-bit ALU circuits (210 gates, 524 faults, 8 redundant faults), again two passes were needed to cover all detectable faults with 65 vectors that were generated in 36 seconds. Another circuit with 469 gates had 942 faults. There were no redundant faults. The first pass produced 116 vectors for 99.2 percent fault coverage. A second pass improved the coverage to 99.7 percent and increased the vector count to 121. This last circuit is one of the benchmarks (C880) that have been used for evaluation of test generators [8].

Several other circuits of varying sizes were also used. We made an interesting observation: The average number of trial vectors simulated during each search was always of the same order as the number of primary inputs in the circuit. If we assume that simulation complexity per vector is on the order of the number of gates in the circuit, the the time for each test search should be proportional to *gates* × *primary−inputs*. We should, however, remember that some of these searches do not result in a test but get terminated at local cost minima.

6.3 Sequential Circuits

Test generation for sequential circuits required several additional features. These are described below.

Unknown Signals. It is necessary to simulate sequential circuits with at least three signal states, 0, 1, and X (unknown). We assigned a range [0.45,0.55] to the unknown state. Since threshold-value model discriminates between only two states, the evaluation of a gate with three input states is done in two steps. For example, an AND gate with five inputs 10XX1. We split it into two AND gates. The first gate has three inputs with values 0XX. The output of this gate is computed from a three-input threshold-value model. The resulting output is applied to a second three input AND gate whose other inputs are 11. Thus, by a dynamic two-level splitting, any gate can be evaluated.

Race Analysis. We used a logic model proposed by Ulrich and Baker [9] for automatic race analysis. This method requires one additional gate per latch. Whenever an input sequence, that would cause a race, is applied to the latch, the output is automatically set to the unknown state.

Feedback Loops. In case of feedbacks, even when the logic values are stable, because of continuous values used in threshold-value model, it is necessary to evaluate the gates repeatedly around the feedback loop. It is, however, possible to identify loops before simulation and derive closed form expressions for feedback signals to avoid iteration.

Cost Computation. In a sequential circuit, memory elements are considered as pseudo-observation points for cost computation. For this purpose, all flip-flops are levelized. Primary outputs are considered level 0. Flip-flops directly (or through combinational logic) connected to primary outputs are level 1. The flip-flops that can only feed into level 1 flip-flops and not into primary outputs are level 2. And so on. Cost for an input vector is computed for each level separately. That is, the minimum of costs at the inputs of all level n flip-flops will be the cost at level n. The total cost is a weighted sum of the costs at various levels. If the fault is not detected at the level 0 (primary output) then the weighted cost is used to direct the search.

Initialization Test generation can be started from all unknown (both good and faulty) states. In this case, the test generator automatically initializes the circuit. Alternatively, a user-supplied initialization sequence can be used. All faults are simulated with this sequence to find the lowest cost fault as the target for search. This fault simulation yields the good and the faulty states that are used at the beginning of the search. After a test sequence is generated, subsequent fault simulation, in a similar manner, gives the states for the next target fault.

Synchronous Circuits. In synchronous circuits, the clock signals are identified and their sequences are specified to the test generator. The non-clock (or data) signals are manipulated by the test generator in a similar manner as in a combinational circuit to minimize the weighted cost. After all data signals have been changed, only the final vector is used as test. This vector is then followed by the prespecified clock sequence. Cost is again computed. The search uses combinational mode and clock sequence, repeatedly, until a test is found.

Asynchronous circuits. When no clock signal is identified all signals are treated alike during the search. However, the test generation for asynchronous sequential circuits differs from that of combinational circuits in two ways. First, memory elements are identified to compute the weighted cost. Second, each single bit change that is accepted by the test generator (because it lowers the cost) produces a test vector in the test sequence.

A Sequential Circuit Example. The present version of sequential circuit test generator has been run on several circuits. An example circuit is shown in Fig. 9 [10]. Test generation was done in the asynchronous mode. Results are given in Table 1.

Test generation was started without initialization. During the test generation, *potentially detectable* faults were also considered. These are the faults that produce an X signal at a primary output while a 0 or a 1 is produced in the good circuit. Even though the search for a test was terminated when the target fault became potentially detectable, the fault was saved for the second pass. After two passes, all 67 faults were detected by a sequence of 35 vectors.

7. Conclusion

We have presented a new methodology for test generation. Primary idea is to use simulator for test generation. The concepts of cost computation and directed search are introduced in the context of test generation. Advantages of this approach are, 1) Backtracking is avoided, and 2) circuit timing is taken into account. Preliminary results show the feasibility of the method for combinational and sequential (synchronous and asynchronous) circuits.

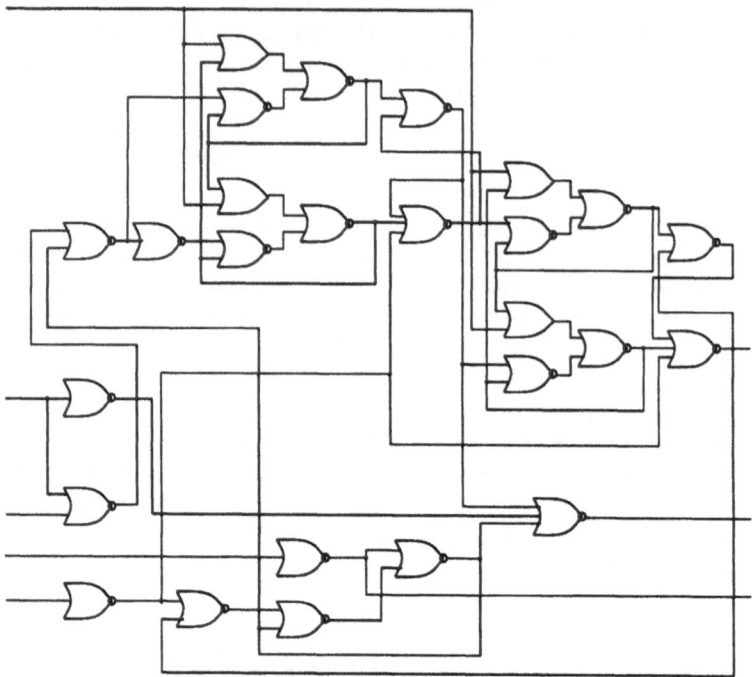

Fig. 9 A sequential circuit example.

Table 1 - Test Generation for Fig. 9 Circuit		
Pass No.	Vector Number	Faults Detected
1	2	3
	5	21
	7	31
	10	45
	11	45
	12	47
	16	53
2	17	53
	18	54
	19	57
	25	62
	29	63
	30	64
	35	67

REFERENCES

[1] V. D. Agrawal, S. K. Jain, and D. M. Singer, "Automation in Design for Testability," *Proc. Custom Integrated Circuits Conf.*, Rochester, NY, May 1984, pp. 159-163.

[2] J. P. Roth, W. G. Bouricius, and P. R. Schneider, "Programmed Algorithms to Compute Tests to Detect and Distinguish Between Failures in Logic Circuits,"*IEEE Trans. on Computers*, Vol. EC-16, pp. 567-580, October 1967.

[3] P. Goel, "An Implicit Enumeration Algorithm to Generate Tests for Combinational Logic Circuits,"*IEEE Trans. on Computers*, Vol. C-30, pp. 215-222, March 1981.

[4] P. Muth, "A Nine-Valued Circuit Model for Test Generation,"*IEEE Trans. on Computers*, Vol. C-25, pp. 630-636, June 1976.

[5] S. Muroga, *Threshold Logic and its Applications*, Wiley, New York, 1971.

[6] K. T. Cheng and V. D. Agrawal, "A Simulation-Based Directed-Search Method for Test Generation," *Proc. Int. Conf. Computer Design (ICCD-87)*, Port Chester, NY, October 1987.

[7] P. R. Schneider, "On the Necessity to Examine D-Chains in Diagnostic Test Generation,"*IBM J. Res. and Dev.*, Vol. 11, p. 114, January 1967.

[8] "Recent Algorithms for Gate-Level ATPG with Fault Simulation and their Performance Assessment (Special Session)," *Proc. IEEE Int. Symp. Circuits and Systems (ISCAS)*, Kyoto, Japan, June 1985, pp. 663-698.

[9] E. G. Ulrich and T. Baker, "Concurrent Simulation of Nearly Identical Digital Networks,"*Computer*, Vol. 7, pp. 39-44, April 1974.

[10] *AIDSSIM - Fault Simulation and Verification Simulation System*, Gateway Design Automation Corp., Westford, MA, 1984.

BEHAVIORAL TESTING OF PROGRAMMABLE SYSTEMS

F. Distante
C.S.I.S.E.I. (CNR)
Department of Electronics
Politecnico di Milano (ITALY)

1. INTRODUCTION

While testing of digital systems has been a subject of study for at least thirty years [1], introduction first of LSI and then of VLSI and ULSI has given new emphasis to the problem and, consequently, has made it necessary to study it under a new point of view.

Conventional approaches, based upon detailed structural information and fairly simple fault models, are hardly suited to the very complex structures of devices such as microprocessors or even of the more complex structures afforded by VLSI and ULSI: test patterns based upon these approaches and guaranteeing high coverage would easily involve unacceptable power requirements. Moreover, the number of observability and controllability points is very limited with respect to on-chip functional complexity, so that fault identification or even fault coverage may become impossible; on the other hand, addition of test points is practically impossible when dealing with ICs, at least as far as the user is concerned.

All above considerations have led, on the one hand, to development of "design for testability" techniques (see [2], [3], [4]) and, on the other hand, to study different testing techniques based upon reduced information but still providing good coverage figures and allowing to create tools for computer-assisted generation of test patterns (ATPG) [5]. In particular, so called functional testing has been largely advocated and is at present subject of a number of studies.

Several definitions of functional testing are adopted by different authors (see [6]), and the most important will be here recalled. In general, it can be said that functional testing aims at verifying correct operation of a digital system (of varying complexity) with respect to its functional specifications, as opposed to tracking of signal values through a circuit structure (as in structural testing). Even within the same functional testing approach, two major philosophies may be stretched out.

The first (more throughly discussed in the next subsection) is based upon description of a complex systems in terms of suitably interconnected functional blocks. This approach is often referred to as the "real" functional testing approach . In this case, testing aims at verifying whether such functional blocks and their interconnections work properly. Fault models adopted relate to functional operations of the blocks and are thus at a higher level than traditional stuck-at models; they allow also to take into account fault models typical of VLSI/ULSI implementation (technology-dependent models, pattern sensitivity, etc.). This kind of approach still requires a certain degree of detailed information on system's implementation, being therefore hardly suited for user's test. Moreover, test patterns are strongly dependent on the particular architecture of functional blocks, thus implying a complete re-design of test procedures even whenever a different implementation of the same device is considered.

325

F. Lombardi and M. Sami (eds.), Testing and Diagnosis of VLSI and ULSI, 325–353.
© 1988 by Kluwer Academic Publishers.

The second approach is often (and more properly) referred to as behavioral testing: typically it has been considered for programmable devices. In this case, information adopted to create any test sequence is strictly user-available and sufficient to describe the external behavior of the device independently of its internal structure. Behavior is described in terms of "operators" applied to the device together with a set of "data", and producing a set of results. Since each operator may be seen as activating an "operational block" inside the device, faults (even if detected as erroneous result produce by an operator) may be ascribed to such blocks. Even so, it can be seen that a measure of structural information is still required to provide an acceptable fault model.

Behavioral test patterns can then be created even by the user and adapted to externally equivalent devices produced bu different firms and with different internal structure. Still, as we shall see in later sections, coverage evaluation and fault model definition constitute two most critical points of, such approach.

1.1 Simulation Based Methods

Conventional approaches to structural testing of digital systems require description of the structure at gate - or transistor - level and subsequent tracking of faults effects through the circuit. In a similar way, an approach to functional testing is still based upon a description of the physical architecture (even though at an higher abstraction level) and a simulation of its operation.

To this end, Computer Description Languages (CDL) are often used, allowing both to describe a digital architecture and to simulate its operation ([7], [8], [9]). In general, all these languages allow reference to functions of different abstraction levels (from complex functional units down to registers or gates) and to their interconnections; simulation involves not only logic operation, but also transient behavior and timing.

The first problem to be solved is then identification of the primitives and of the fault model suited to a functional description and consistent with the device's physical behavior. It should be noticed that not all CDLs allow efficient simulation in presence of faults (fault injection) and that even fewer of them are suitable for automatic generation of test patterns starting from the description of a given architecture. We will now point out approaches specifically based upon ad-hoc hardware description languages.

Main problems to be afforded when dealing with functional testing are definition of "fault model" (i.e. of the set of faults whose identification is the final aim of the procedure) and definition of an optimum algorithm for their identification. In the functional approach errors, rather than faults, are considered and they are best modeled by their influence on the truth table of the block itself.

A relevant proposal in this sense was made by Breuer and Friedman [10]; they introduced the concept of "functional level primitives" such as adders, counters or shift registers, and considered error propagation in a network of corresponding functional blocks. The algorithm proposed to create ATPGs at functional level is based upon Roth's D-algorithm [11]: while classic D-algorithm refers to gate-level description of an architecture and to corresponding fault assumptions, subsequent modifications at functional level substitute functional blocks (described by a CDL) and fault propagation among them.

Levendel and Menon [12], [13] further extended this approach to create a version of the D-algorithm applicable to functions described in a computer description language. The language aspect is specifically examined in [13], where suitability of the algorithm both to procedural and non-procedural languages is considered, and the problem of automatic test generation is examined. As fault modeling is concerned, the authors consider both control faults and functional faults, defined upon the functional description of a block. In a later paper [14], the same authors have presented an alternative approach based upon a similar extension of a different gate-level test generation procedure, (the "critical trace algorithm") allowing simultaneous detection of multiple faults. Complexity of test generation is reduced in comparison with the previous method, although some loss of information is suffered.

A completely different point of view, adopted also for design of digital systems, is that based upon "binary decision diagrams" [15]. These can be used for generating testing experiments for digital devices, starting from their logical description; since binary decision diagrams allow to partition the system into subsequently more refined subsystems, they allow also to identify parameters such as testability for such subsystems or even to define test procedures.

Logic simulation is the basis for other approaches such as [16] and [17]. In [17] a hierarchical description of a complex subsystem is advocated, making use of a set of primitives or "devices" comprised in a library and going from elementary gates to complex functional blocks. The hierarchical approach allows to obtain - at high level - a very compact description of an architecture without incurring into any actual loss of information. A problem not completely dealt with is creation of satisfactory fault models corresponding to the device library.

The same question of obtaining at the same time modeling and fault simulation for complex circuits is discussed in [16]. Major features of the simulation program there proposed (FANSIM3) are, again, a hierarchical approach to system description, a functional and timing specification language for modeling primitives at the functional level, and the extension of concurrent fault simulation to functions, timing faults and functional faults.

Possibility of a functional description at high abstraction level allows also to represent failures across multiple implementations of a system. An example of a sophisticated simulator is discussed in [18], where a spectrum of faults ranging from hard to transient, from stuck-at and bridged to operation and functional ones, with different occurrence distributions, is allowed. A fault injected at high abstraction level can provide coverage of multiple faults at lower level, then affording better efficiency of the simulation program.

Availability of a suitable circuit description and of a corresponding fault model definition allows Kwok-Woon Lai [6] to propose a complete functional testing methodology which generates test directly from the functional specification. In this case, digital system specification language allows descriptions in a wide range of abstraction levels. An interesting factor is that such an approach can be adopted by the designer himself, so to create a test pattern concurrently with development of system design.

Another approach to functional test of digital systems described by means of a RT-level CDL is that of Su [19]: particular attention is there dedicated to techniques of fault injection into the RTL specification. Test generation is achieved making use of two different methodologies. The fist one is still based upon the D-algorithms, here applied on data graph representation of the system under test. The second ("symbolic simulation") implies the creation of two symbolic copies of the system under test: one is the good ones, and the other one is the faulty one. The test is then performed making the two machines "execute" upon a set of suitable symbolic data "variables".

2. BEHAVIORAL TESTING

Behavioral testing has been defined as a testing approach based upon command of a device's externally controllable functions and observation of corresponding results at the external output pins. Given the complexity of the system and the reduced number of control and observation points this kind of test action is far from a detailed, RT-level test of the functional units and of related information transfers.

An approach to behavioral testing that in some way recalls more classical functional testing (making use of hardware description languages and simulation) was presented in [20]. The authors describe a device at high level via function tables (identifying the behavior of the functional blocks making up the device) and of action and condition tables further detailing its operation; functional tables comprise a "fault" section in which functional faults are defined and that ultimately allows fault injection and simulation. Further developments in this field were introduced by the same authors in [21], where the concept of functional tables is expanded so as to model circuits as interconnected functional blocks. Reachability matrices, representing interconnections among such blocks, allow to cut down path searching time.

A basic philosophy for behavioral testing of microprocessors has bee introduced in [22], [23], [24]; they suggest adoption of a graph theoretic model for microcomputer architectures, allowing to use the microprocessor organization and instruction set as guiding parameters for test generation procedures. An oriented graph representing information flow inside the microprocessor is derived from the instruction set; functional fault models are defined for the basic functions and the graph model is used as a guide for generating test actions covering such faults. This approach has been widely adopted in later literature; while definition of functional errors obviously does not prevent introduction of fault models extracted from structural information, still the overall approach allows the user to create test patterns from information available to him and for devices characterized by the same external behavior (even if with different internal structure).

An extension of this last approach can be found in [25]. Here the authors develop a comprehensive model allowing to easily detect faults in instruction execution process; test generation process does not require a detailed knowledge of instruction execution process, yet it is capable to detect a greater number of faults while having relatively low complexity. The generated test sequences mat be executed by the microprocessor in a self-test mode, thus dispensing with the need of complex testers.

Graph theory is also the basis of the approach presented in [26], where in fact the dual aspects of "design for testability" and if test-pattern definition are afforded at the same time. A multi-level approach to description of complex structures is chosen, and at any level a graph representing at once dynamic and static characteristics of the device is used.

Most papers dealing with behavioral testing advocate the "start-small" philosophy; the underlying justification is that staring test operation on simple instructions, involving a limited number of internal registers and functional units, allows to eliminate a first subset of possible error areas and to acquire information useful for subsequent test of complex instructions. Such point of view is adopted in [27], [28].

A general approach to behavioral testing is the one described in [29]; the method involves three subsequent steps: a "consistency test" allowing to verify correct execution of instructions independently of any error assumption, a "scanning test" identifying correct behavior of the various functional blocks and a "control signal test" where the system's behavior is examined under different configurations of control signals. In a later paper [30] the same authors claim total independence of the method from any structural information or fault model; actually, the claim is only partly valid as physical characteristics may lead to serious modifications on how the methodology is applied.

As a general definition, we may state that a behavioral test procedure for a programmable device will consist of:

- instructions (whose correct execution is in fact tested)

- data application

- control signals application

A problem too often overlooked when behavioral testing is considered is definition of a figure of merit and a methodology allowing to identify an "optimum" test procedure. In the next sections we will discuss the meaning of "optimality" and will introduce some techniques which allow to identify the optimum test procedure. On the theoretical basis discussed in the next section, design of an ATPG for programmable devices will be described.

3. MULTIPLE LEVEL BEHAVIORAL TESTING

We consider here behavioral testing as related to all those VLSI devices that can be described by:

- a set of operators $\{o_j\}$, each defined by input and output operands and by the functions it performs;

- a set of sequencing informations describing for each operator the sequence of phases externally visible;

- a set of external control and observation points (typically, external pins).

Behavioral testing then aims at verifying correct execution of all operators, ideally for a set of input operands (or "data") making visible all possible errors.

This in turn can be realized by defining for each operator a "test sequence", i.e. a sequence of operators allowing to check correct execution of the operator under test when applied to suitable operands.

The sequence is repeated upon different sets of data, so as to exhaust (ideally) all error possibilities in a given error model. The set of all test sequences constitutes a "test procedure" checking correct behavior of the device.

Errors rather than faults are defined, being associated with operator execution. In principle, no structural information is available, so that no structural error model can be adopted and used for subsequent coverage evaluation. This makes it necessary to take into account such possibilities as error masking, indeterminate error location, etc.; as a consequence, additional figures of merit characterising test sequences with respect to such instances must be defined. To this purpose, in [31] the concept of "ambiguity" was introduced: whenever a test sequence comprises other un-tested operators besides the operator under test, possibility of either error masking or of incorrect error attribution arises, thus generating a measure of "ambiguity". Ambiguity is thus related to error coverage: optimization of a behavioral test procedure involves both figures of merit.

When more complex devices are considered – such as, e.g., very advanced CPU's with a high degree of silicon software – the behavior can actually be described at a number of different abstraction levels.

Whatever the level of abstraction at which a complex device is described, each operator may be defined as a suitable sequence of **phases**, each phase being in turn made up of a number of parallel functionalities. A given functionality may well appear in a number of different operators. Complexity (or "cardinality") of an operator, seen as the number of subsequent phases involved in its execution, may then be evaluated purely from the device's behavior; as proved in [31], ordering of the operators based on increasing cardinality is necessary to achieve optimum test procedures. Since usually the set of operators does not allow to completely avoid ambiguity in test procedures, it is necessary to identify minim-ambiguity ones.

Considering now a complex device described at multiple levels of abstraction, the problem arises of identifying a connection between behavioral test procedures defined at different abstraction levels: it is obviously very important to find which relationships may be proved to exist between such optimum test procedures, and in particular to state whether a direct connection exists between coverage figures evaluated at different abstraction levels. The approach here presented allows to separate generation of test procedures (consisting of operators' sequences and independent of the particular fault model adopted) from definition of data sets to be applied (which are strongly dependent on the fault model). Actual coverage obviously depends also on data sets applied; analysis of the pure test procedures provides a "limit" value to be achieved by using optimum data sets.

In the next subsection it will be seen that, assuming a top-down device description approach, identification of optimum test procedures at high abstraction level is useful also when greater detail is meant to be achieved, since optimum test procedures at lower abstraction levels can be found simply inside the expansion of the previously determined one. This allows to proceed through subsequent refinements in definition of behavioral test procedures.

Then, the problem of relating error coverage figures concerning the different abstraction levels will be afforded. It will be seen that behavioral test coverage figures at high abstraction level may be accepted as actually significant if the device's design satisfies some basic guidelines; moreover, definite relationships between behavioral coverage figures corresponding to increasing detail will be introduced.

3.1. Multiple abstraction levels and multiple test levels

A complex VLSI device can usually be described at multiple levels of abstraction going from the highest behavioral level down to - possibly - RTL level. When a top-down design methodology is afforded, actually, even the behavioral description may proceed through various levels of abstraction; initially, global functional specifications only are available, then increasing detail is reached as design information is added by further identification of the various functionalities. Thus, for instance, a VLSI device comprising a high measure of silicon software will be defined at first by a set of high-level primitives implementing a given virtual machine; then, an actual architecture will be detailed and implementation of primitives will be outlined, etc. until the register-transfer and microorder level is reached.

At any level l, behavior of the device can be described by the set of operators of level l and by their definition. Description of operators as consisting of subsequent phases allows to identify their relative complexity and therefore to perform an ordering following increasing cardinality, as outlined in the previous section.

We assume that going from level l to level $l+1$ will not increase the number of control and observation points (an assumption consistent with an approach dealing with VLSI devices, since the number of external pins remains fixed) but that granularity of control and observation may increase, as regards both detail of information and time phases. It may be safely deduced that at high abstraction level a limited set of high-level operators will be available, and that anyway relationships between such operators will be rather simple. For example, if the high-level primitives are operating-system ones, few and straightforward relationships will be identified among them, while at lower level operators dealing directly with registers and behavioral units have more extensive and complex interrelationships.

Therefore, at high level, identification of optimum test procedures will be relatively simple, having to deal with few alternatives and with a tractable problem description. On the other hand, since no information on error models is available, definition of significant data sets to guarantee error coverage might lead to excessively large data sets. Obviously, increasing architectural detail leads to increase also error model information (and thence to focus on the most significant data sets): it is important to find whether a connection exists between optimum test sequences at different abstraction levels. To this end, we will first deal with the relationships between test sequences and then (in the next section) with the relationships between coverage figures.

Passing from level l to level $l+1$, each operator o_i^l expands into a set of operators at level $l+1$: it is
$$o_i^l \to \{o_{ij}^{l+1}\}$$

332

Given a test sequence at level l, all operators constituting it expand to level $l+1$ independently of each other. In [32] some properties concerning such expansions have been proved, allowing to conclude that it is possible to go from optimum test sequences at level l to other – still optimum – test sequences at level $l+1$.

When going to abstraction level $l+1$, information units and functions are defined with finer detail but – obviously – none of them will be deleted; thus, testing requirements do not decrease. Moreover, as already seen, increasing detail will not introduce new control points and any new observability will involve only information units and functions that were considered "transient" at level l. Therefore, if a cycle[(*)] existed at level l it will also be present in the level $l+1$ expansion (in general, it may not involve all operators in the expansion) thus localizing ambiguity.

As a consequence, optimality with respect to ambiguity is defined essentially as the number of cycles; if two test sequences $A_1(o_1^l)$ $A_2(o_1^l)$ have the same number of cycles, cycle length and cardinality of operators involved will be taken into account. In fact, importance of cardinality appears when expansion to level $l+1$ is performed, since a higher-cardinality operator will expand into a longer sequence of operators at level $l+1$.

The main result that can be deduced from the above is possibility of defining a top-down approach for generation of behavioral test sequences through multiple abstraction levels. The following procedure is justified:

1. Identify an optimum test procedure TP_l at a high abstraction level l.

2. Going to level $l+1$ (lower abstraction), an optimum test procedure TP_{l+1} may be found simply inside the expansion of TP_l, i.e. without considering all possible alternatives at level $l+1$. And so on, recursively.

Test procedures defined by the previous properties guarantee that – given suitable data sets upon which each test sequence must operate – maximum coverage and minimum ambiguity may be reached. Actually, it becomes necessary to state a relationship between behavioral test coverage and architecture-related "functional test coverage". The next subsection will deal with this problem.

3.2. Behavioral test coverage: evaluation and credibility

The classical definition of fault coverage C for a given test procedure is

$$C = \frac{E_d}{E}$$

where E_d is the set of faults detected by the test procedure and E is the total set of faults corresponding to a given fault model.

It becomes now necessary to define a comparable "behavioral error coverage" figure. As already said, errors are defined and detected as related to a device's behavior, observed in terms of operator execution and

(*) The situation in which operator o_1^l requires in its test sequence operator o_2^l which, in turn, makes use of o_1^l in its own test sequence (thus implying ambiguity in testing since to overcome this deadlock one of the two operators will be considered as already tested [33]) is defined as a cycle.

corresponding results.

The complete set of functionalities appearing in the definition of o_i^l constitutes the set of "<u>proper</u>" functionalities of o_i^l.

Each functionality involves at least one information unit (e.g. a register, an external observation/control point, etc.). Execution of a functionality will in general modify the state of associated information units, or, more concisely, the "state of the functionality".

We introduce two different sets of possible errors related to execution of any given o_i^l:

- "Combinatorial" or "proper" errors; they appear as an incorrect state of the operator's proper functionalities. For example, an incorrect state of the carry bit after execution of an ADD instruction would be a combinatorial error.

- "Sequential" or "improper" errors; they appear as the modification of the state of a functionality which is proper neither of o_i nor of any operator in the test sequence. Phenomena such as pattern sensitivity or bridging may lead to errors of this second class. Actually, only such improper errors that affect at least one of the <u>device's</u> proper functionalities will be considered. In fact, should - e.g. - a bridging phenomenon create a parasitic capacitance acting as a storage element whose contents are in no way reflected by any proper information unit, the error itself would not be considered in a behavioral approach.

Notice that the above immediately denotes a difference between behavioral and structural testing. While in structural testing all faults related to the chosen error model are taken into account to evaluate the total set of faults E, in behavioral testing only errors that in some way affect the device's behavior - and that can therefore be observed through application of suitable test sequences and data sets - are considered. Behavioral testing is then particularly suited to user's testing of VLSI devices, where no additional observation and control points can be provided besides the device's pins.

From the above consideration it can be deduced at once that behavioral testing must be related to the abstraction level l at which it operates. If by E_k^l we denote the set of errors related to o_k^l and by E_{dk} the number of errors detected through the test sequence and the data set applied, we can give a "theoretical" figure of behavioral coverage as

$$C_{b\,k}^l = \frac{E_{dk}}{E_k^l}$$

Given now the multiple-level approach to behavioral testing, it becomes necessary to find the relationship (if any) between C_b^l and C_b^{l+1}; this would allow to state whether a coverage figure obtained at high abstraction level, with no actual information on the device's architecture and structure, is effectively meaningful. Since, as said previously, the approach can be iterated until the "functional block" level is reached, this would also ultimately provide a link between behavioral and functional testing.

In particular, should the coverage figure evaluated at a level l be <u>higher</u> than that evaluated at level $l+k$, this would imply presence of a severe error masking at high abstraction level and it would in fact make the corresponding coverage figure actually useless, if not even misleading.

The relationships defined in the sequel may constitute a design guide not simply in definition of test procedures, but also in architecture design, providing an insight into the device testability.

To this aim, we introduce first some restrictions to the class of architectures we consider.

- Any functionality is controlled through the "data" applied ("data" has here an extended acception; for instance, an interrupt request is part of the "data" for the interrupt control functionality) and it is made visible through its results. We assume that <u>all functionalities — even improper ones</u> — may be observed and controlled, possibly in indirect way (i.e. through other functionalities) and with a measure of ambiguity.

- There is no "behavioral redundancy", i.e. for any given functionality f_i^l at level l there is only one architectural block (or set of architectural blocks) that implement it. This implies that expansion of f_i^l when going to level $l+1$ is unique, i.e. there is only one set of functionalities $\{f_j^{l+1}\}$ (and of related operations such as information transfers) performing it. As a consequence, expansion of f_i^l is independent of the operator o_k making use of it.

We further make two assumptions concerning the error model adopted:

1. Define "data" to be correct if their value coincides with the expected one, incorrect otherwise (i.e. if they have been affected by error of a functionality). Then, a correctly working proper functionality operating upon incorrect data will produce incorrect results.

2. An improper functionality (i.e. affected by a sequential error) by definition gives incorrect results, since its information units should not have been previously modified in any way.

Any functionality f_i^l may be associated with:

- a <u>proper</u> behavioral area B_k^l upon which it acts;
- an <u>improper</u> behavioral area S_k^l affected only as a consequence of a sequential error.

Obviously, any B_k^k and S_l^k will relate to a number of functionalities. It can be noticed that, while B_k^k is technology-independent and relates only to the architecture, S_l^k is technology-dependent.

At the highest abstraction level, no information concerning improper areas is available; to guarantee complete technology-independent coverage, in principle, an exhaustive analysis of all possible error areas should then be performed in each test sequence. It is self-evident that such an approach is totally impractical: on the other hand, two considerations are in order, namely:

- only improper areas involving a "memory" of the error need be considered; related errors affect the device's behavior only if they persist in the system long enough to be used as incorrect data.

- a test sequence — unless it consists of the operator under test, o_k^l, only — may usually make evident also a subset of S_k^l. This is due to the fact that, actually, a test sequence may be seen as:

1. presetting a number of functionality states

2. affecting a subset of the above via the operator under test, o_k^l

3. making visible a set of functionality states comprising the ones proper to o_k^l but possibly other ones previously preset.

Thus for any given o_k^l and for the adopted test sequence, the theoretical limit to the errors detected may be defined as $E_d(B_k^l)$ U $E'_d(S_k^l)$, where $E'_d(S_k^l)$ is the set of improper errors identified by the test sequence.

The _theoretical_ coverage figure stems from an assumption that considers only proper error areas; it can be defined as:

$$C_b = \frac{E_k(B_d^l)_k}{E(B_k^l)_k} \quad (1)$$

C_b can be seen as the limit value of behavioral coverage when technology and design are such that only proper functionalities are affected by an operator's error. We introduce then the "real coverage" figure, taking into account also improper error areas; for the same test procedure for which C_b has been evaluated, (that therefore adopts only "theoretical" test sequences, as previously defined) it is:

$$C_r = \frac{(E_k(B_d^l)_k \; U \; E'(S_d^l)_k)}{(E(B_k^l)_k \; U \; E(S_k^l)_k)} \quad (2)$$

Notice that in this case the value of the denominator is actually unknown. In general it will be $_k E'(S_k^l) <_k (E(S_k^l)$, therefore $C_r < C_b$. This is clearly a relevant point since it means that the theoretical coverage figure may be excessively optimistic by comparison with the actual coverage obtained.

We aim to see how C_b and C_r vary depending on the abstraction level adopted.

Should error coverage appear to _decrease_ with increasing level of detail, this would mean that high-level test procedure gives overly optimistic results, and that in fact a number of errors actually affecting the device's performance have been overlooked. Therefore, if a behavioral approach is to be acceptable, it must be $C_b^l \leq C_b^{l+1}$.

Such relation is satisfied [34] if the "relative" coverage concerning only those functionalities that are transient al level l and visible at level l+1 must be at least as good as the coverage achieved at level l. While on the one hand if test procedures are not yet fully defined, this last condition may be a guide for their best choice, on the other hand if we relate to optimum test sequences, whenever the condition is not satisfied this means that the device's design does not lend itself to satisfactory behavioral testing.

Consider now the "real" coverage figure: it is

$$
C_r^l = \frac{E_d(B^l) + E_d(S'^l)}{E(B^l) + E(S^l)}
$$

Consider now the expansion of total improper error areas from level l to level l+1. Complete knowledge about improper error areas can be obtained only if very detailed structural and technological information is available; otherwise, as already hinted, only a very rough evaluation of C_r^l could be made. By adopting worst-case criteria, each operator could be considered as affecting - as a consequence of an error - all memories defined at the same level, either through proper functionalities or through improper ones. If this pessimistic point of view is adopted, it can be safely assumed that increasing information will lead to finer specification of improper error areas;

Relationship $C_r^l < C_r^{l+1}$ is more easily satisfied if r>t, i.e. if improper error areas become more and more restricted while definition of the device's structure increases. Better still, the higher is restriction of improper error areas for any pair l,l+1 the more easily we achieve a limit value of C_t^{l+1} nearing C_r^{l+1}. This goes to confirm an intuitive evaluation; in fact increasing detail brings the behavioral description of the device nearer and nearer to its structural description. Under the previous assumptions, non-decreasing observability both of "architectural" errors (corresponding to proper functionalities) and of "technology-dependent" errors (corresponding to improper functionalities) is guaranteed. This condition concerning observability can be assumed as a "design rule" in the design of a complex device.

4. TEST PROCEDURE DESIGN AND OPTIMIZATION

Design and optimization methodologies for test procedures will now be introduced. All of the following is independent from the abstraction level adopted, even if, for simplicity sake, an "instruction level" will be assumed.

Due to the abstraction level assumed, information considered to create a test procedure is the conventional user-available information, that is:

- the set of internal registers and functional units (e.g. control unit, ALU, etc.);

- the set of instructions and of control signals (interrupt requests/acknowledge, DMA, etc.);

- general information derived from device's timing charts.

Instructions, depending from their behavior as seen at the external pins, are classified as:

- <u>directly controllable</u> instructions: these will receive input signals (data) directly from the external pins of the device;

- <u>non-directly controllable</u> instructions: functional units involved in the execution of such instructions cannot be directly reached from the external pins (i.e. input data must be provided by executing a sequence of other instructions);

- <u>directly observable</u> instructions: providing results of computation directly to the external pins;

- <u>non-directly observable</u> instructions: results of execution of such instructions directly observable at the external pins.

If every instruction belonging to the instruction set was directly controllable and observable, the problem of testing would reduce to identification of optimum data sets for each instruction (optimality being defined with reference to coverage for a given fault model). Such assumption is though quite unrealistic; a <u>test action</u> consists of a sequence of instructions allowing control (i.e. data loading into input registers) and observation (i.e. transmission of results to external pins) of the instruction under test. Obviously, different test actions may be defined for each instruction since data loading and output retrieving may be generally performed in different ways. A set of test actions allowing to achieve the complete test of all functionalities of the device is called here a <u>test procedure.</u>

When designing a test action for a given instruction I^*, only already tested instructions should be used. In fact if even one instruction besides I^* has not been already tested, error masking or erroneous error attribution may occur: this implies presence of <u>ambiguity</u> in the test action and, then, in the whole test procedure. Again, ambiguity arises whenever a "cycle" links two different test actions, e.g. test action for I' involves I'' and test action for I'' involves I'. Thus, to obtain the optimum test procedure it is necessary to choose the set of test actions minimizing total ambiguity.

4.1. Modeling of test actions

Test action (I^*) for an instruction I^* is a partially ordered set having the following general structure:

$$(I^*) = \{I_1 I_2 \ldots I_k I^* I_{k+1} \ldots I_{n-1} I_n\}$$

Three different subsequences may be identified:

- the <u>"set-up"</u> subsequence $(I^*) = \{I_1 I_2 \ldots I_k\}$, i.e. the set of "predecessors" providing input data to the instruction under test;

- the instruction I^* itself;

- the <u>"observation"</u> subsequence $(I^*) = \{I_{k+1} \ldots I_{n-1} I_n\}$, (set of "successors") allowing external observation of the results of execution of the instruction under test.

To identify the set of test actions constituting the optimum test procedure, we require a model capable to represent both the set of instructions and the error spaces (through the register and/or functional blocks involved).

Graph models provide a compact representation of test actions; in particular, Petri nets [33] have proved very useful by providing a tool that also represents the ordering and the "dynamics" of test actions.

Functional units related to an instruction I^* may be outlined as:

a. a set of input registers $R_i(I^*)$ accessed by I^* to receive data;

b. a set of output registers $R_o(I^*)$ accessed by I^* to store results of its operation.

External pins are considered as registers, and intersection of any pair of sets may be non-void.

Registers belonging to such two sets are represented in the Petri net model by means of places connected as input to the transition representing I^*.

Composition rules adopted to build the Petri net representing a test action are:

a. any direct predecessor is connected to the places representing shared registers in $R_i(I^*)$ and so on recursively;

b. any direct successor is connected to the places representing shared registers in $R_o(I^*)$ and so on recursively.

These composition rules will allow firing of transition I^* only when all transitions representing predecessors and successors have already fired, thus loading all input places. It is then immediate to associate the "firing" of a transition to the achievement of test for an instruction. Ambiguity is revealed by the presence of cycles in the net. Transitions (i.e. instructions) involved in cycles are not live, thus forbidding propagation of tokens. Since tokens must propagate through the whole net to assure complete test of all functionalities, it will then be necessary to open the cycles, forcing tokens in the appropriate places. Each "cut" that has to be performed to obtain a live net, increases the degree of ambiguity of the obtained test procedure.

Graphs related to every possible test action of all instructions may be merged to create a "supergraph" representing the set of all possible test procedures. A test procedure may be identified in the supergraph by choosing for each instruction a set of predecessors and successors among the ones available.

To achieve a test procedure which minimizes ambiguity it would appear reasonable to perform the following steps:

1. build all possible test actions for every instruction;

2. for each instruction, choose the lowest ambiguity test action;

3. merge such test actions to obtain a test procedure.

Anyhow, global optimum would not be assured since optimization is performed upon single test actions without considering relationships among different test actions.

A Short-sighted Greedy algorithm presented in [35] implements the above steps. That algorithm has proved to be a powerful tool for ambiguity evaluation of test actions, while providing only good (generally not the best) test procedure.

Optimum could obviously be reached by adopting an exhaustive approach; all possible combination of choices may be performed and ambiguity can be evaluated for each of them: only combinations with minimum ambiguity will be accepted, thus generating the optimum test procedure.

Nevertheless, this approach is very time-consuming since the number of possible choices is very high. Thus, to generate a test procedure with lower effort, an heuristic approach has to be adopted.

4.2. A hill-climbing heuristic for optimum test procedure identification

Detection of the best test actions may be misleading to identify the optimum test procedure, since an algorithm which considers all instructions and their relationships has to be adopted. A guided heuristic algorithm based upon the branch-and-bound technique was proposed in [36]: even if such algorithm is not exhaustive, two major problems must be solved to limit the growth of the B&B tree (ultimately of the computation time): initial test procedure P_o (the root of the tree) must be as close as possible to the optimum solution, and a lower bound to ambiguity must be carefully defined to assure fast convergence of the algorithm.

An alternative to such classical optimization method is based upon simulated annealing technique [37], which allows detection of optimum test procedure in a reasonable number of steps and with low storage requirements.

Simulated annealing is a randomized algorithm which has been proposed for finding globally optimum least-cost configurations in large NP-complete problems, involving many degrees of freedom, with cost functions having local minima [38]. Its most interesting feature consist in the possibility of exploring the configuration space of optimization problem allowing "hill climbing" moves, i.e. at each step new configurations with increased cost may be accepted. These moves are controlled by a parameter, in analogy with the temperature in the annealing process, which decreases towards the end of the process. In this way, provided a suitable tuning of particular variables of the algorithm, it was been proved [39] that it is possible to reach the global minimum also when other local minima are present.

For the simulated annealing algorithm, the combinatorial optimization problem may be specified by a finite set of configurations, called states S, and by a cost function defined on all the states. In our case, such entities have been mapped upon appropriate figures of the supergraph: the state is a collection of test actions defining a test procedure, while the cost function is the ambiguity introduced by test actions selected in that particular procedure.

The algorithm is characterized by a rule to generate a new state with a given probability, and by an acceptance rule according which the new configuration is accepted or rejected.

At each step a new solution S' is randomly generated from the last accepted solution S: the new test procedure may be easily obtained from the last procedure considered by substituting a of the test action with one of its possible alternatives. Two states which differ only by a single test action are defined as neighbors.

The new test procedure is surely accepted if its ambiguity is lower than the ambiguity of the previous one. In the other case, a random rule is used: the acceptance rule is controlled through the current value of a parameter T, defined as the temperature of the annealing process. The new solution, if worse than the previous one, will anyhow be accepted if a function f(S,S',T) (ranging from [0,1]) is greater than a random number (with uniform probability distribution in the interval [0,1]). Function f(S,S',T) is defined as:

$$f(S,S',T) = \min \left(1, \exp\left(- \frac{A(S')-A(S)}{T}\right)\right)$$

where A(.) is the ambiguity evaluated for a given state (test procedure).

The control parameter T decreases in time behaving as the temperature in the annealing process. Thus, probability of accepting worse solutions decreases in time since function f will have higher probability of being smaller than the random number which controls acceptance or rejection.

Properties of the simulated annealing algorithm can be studied using the Markov chains as theoretical modelling tool. Theoretical analysis states that the algorithms generates (with probability 1) the global optimum, when given conditions on the number of iterations at each value of the parameter T or on the updating rule for T are satisfied.

Theoretical behavior which should be followed by temperature T to assure convergence of the algorithm has been proved [39] to be:

$$T = \frac{}{\lg (n + n_o + 1)}$$

where n is the current number of iteration and n_o is a parameter satisfying $1 \leq n_o \leq \infty$. If $n_o \geq r * L$ (where r is the radius of the graph underlying the Markov chain and L is the Lipschitz constant of the cost function), the Markov chain is strongly ergodic. This fact implies that any state probability vector converges component-wise to the constant vector of the optimum solution, i.e. the vector in which all elements are zero except those corresponding to the global least-cost configuration.

It can be easily seen that exact evaluation of parameters such as r and L may be performed through theoretical analysis, but a detailed knowledge of the graph underlying the Markov chain is required and their actual computation implies a non trivial effort. Therefore, a careful estimation of parameters is required to preserve algorithm's behavior, i.e. preserving strong ergodicity, without introducing great computational overhead.

We choose the worse-case value for the radius r: if we suppose there exists a path connecting two states S_1 and state S_2 through all the other states of the graph, r is given by $|S|-1$, where $|S|$ is the cardinality of the set of states. We have then obtained $r_{apx} \geq r$.

To give an approximate value L_{apx} to the Lipschitz constant, it has been noticed that L is the maximum variation of cost function (i.e. of ambiguity) between two neighboring states: thus, in the worst case, it is surely less then the maximum ambiguity introduced by opening one of the possible cycles, i.e. $L_{apx} \geq L$.
Therefore, we assume $n_o = r_{apx} * L_{apx}$ assuring strong ergodicity.

Using such approximate parameters the convergence is assured, but this is only an asymptotical behavior. As long as rate of convergence is concerned, better results are obtained by allowing more than one iteration at the same temperature T and by changing the annealing schedule.

Experimental results state that a faster convergence is assured by starting with a quite large value of the parameter T, whereby virtually all new states are accepted. Besides, the annealing schedule previously presented is substituted by the following one:

$$T_k = _k * T_{k-1}$$

where k is number of changes of temperature and   is tuning factor affected by the difference of ambiguity between the last two accepted states [40]. The best suited values for   are comprised between 0.8 and 0.95: in particular,   must be the largest as possible (approximately 0.95) when the cost function changes rapidly, while it is possible to choose a value near to 0.8 if the cost function is quite steady.

To obtain the best results, the system has to be allowed to achieve "equilibrium" at each value of the parameter T of the annealing process. In other words, a sufficient number of iteration must be performed at each temperature T, so that the state probability distribution is close to the stationary probability distribution of the considered Markov chain. The required number of iterations is chosen accordingly with the values assumed by the cost function before the two previous temperature reductions. If there is a notable difference between these values, then it is necessary to bring the system near to a steady-state by allowing a suitable number of iterations at the same temperature: experimental results show that about 30 iterations are enough for high ambiguity differences. Otherwise, a lower number of iterations may be performed: three or four iterations allow to achieve a satisfying behavior of the algorithm.

Due to asymptotical convergence of the algorithm, it is necessary to define a stopping criterion. To consider an exact stopping criterion, the difference between the next-to-least-cost solution and the least-cost solution has to be known. But this again requires too much computational effort: a reasonable stopping criterion (validated through extensive experiments) is to assume that the global optimum has been reached when the last three accepted solutions at different temperatures are equal.

Branch-and-bound and simulated annealing may be considered as complementary algorithms. In fact while B&B requires a great amount of storage to trace the open solutions, simulated annealing considers only the current solution and the last accepted one. On the other hand, simulated annealing converges to the optimum solution generally through a great number of iterations, while the former algorithm may achieve the optimum test procedure with a minor number of steps, provided an accurate choice of the function adopted to bound the growth of B&B tree.

5. BAT: OPTIMIZATION ALGORITHMS AND OVERALL DESIGN OF A BEHAVIORAL AUTOMATIC TESTER

BAT's design may be subdivided into four distinct phases: the first step that has to be afforded is to define a functional description language for a programmable device. Elements of such language must be both user-available and cover completely all device's functionalities. Moreover, all possible relationships among functional units of the considered device must be pointed out with no great effort by the user.

Possible test actions for a given instruction are identified by examining relationships intercurring among different functionalities, thus definition of a data structure emphasizing such mutual bindings must be afforded. Figures of merit upon which the data structure is designed are flexibility, low storage requirement and fast data retrieve.

Fast data retrieve is instrumental to the third step, i.e. the optimization algorithm. Due to NP-completeness of the problem, it will be shown in the next subsections that an heuristic algorithm has to be adopted. It is well known that convergence of heuristic algorithms is a delicate issue, so it becomes essential to speed-up as much as possible each iteration.

As result of the previous step, a Petri net model of the optimum test procedure is created. Petri nets are used since parameters upon which evaluation of test procedure is performed, i.e. degree of ambiguity and error coverage, are well evidenced [33]. A graphic representation is thus required to obtain general information on the test procedure and to be able to inspect with deeper detail relationships among different instructions and/or test actions.

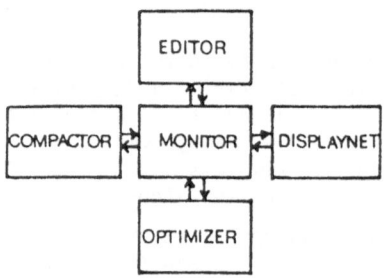

Fig. **1.** BAT's architecture

Due to this logical partitioning of the problem, four separate environments have been defined and implemented:

1. BAT-editor, which provides an easy data input phase (i.e. description of the device) to the user;

2. BAT-compactor first allows reductive modifications to the input file, and then stores description data into an optimized structure;

3. BAT-optimizer builds the supergraph and then identifies the optimum test procedure;

4. BAT-displaynet displays the Petri net associated to the optimum test procedure and its details.

Though these phases are logically piped, it may be required to run only given parts of the whole package, so a monitor has been designed to manage the interactions among different environments.

5.1. Editor and data compaction

To our purpose, all the information required to describe the behavior of the device may be found within the microprocessor instruction set. Additional information which has to be added consists in instruction's cardinality value and the class (branch, manipulation, transfer) to which the instruction belongs.

Since an instruction interacts with other instructions reading or writing in commonly shared registers, to complete the description of an instruction we will have to add the set of registers used by the instruction itself.

An instruction may be completely described by the following structure:

```
{class} {cardinality} {mnemonic code}
{input register set}
{output register set}.
```

Fig. 2.

The description of the whole device will consist of a collection of such structures. For example, Z-80 instruction LD A,x will be stored as in figure 3.

```
T 2 LD A,x
I DB
O A

(DB = Data Buffer)
```

Fig. 3.

The user is aided in this data entry phase by the editor provided by BAT.

The editor may either update an already existing microprocessor description file or create a new one. Operations provided by the editor are the usual utilities such as search, insert, replace, delete, etc. During the editing session, the screen is divided into three windows: the Menu, the Dialog and the Display windows. The Menu window displays all options available to the user in the current mode: options will be different if the user is in insert mode or -e.g.- in search mode. The Dialog window is where prompt messages are presented and where data or commands are entered by the user. Results of such operations are shown in the Display window.

Moreover, BAT-editor checks the correctness of input data. Possible errors are classified as fatal errors and warnings. While a fatal error (e.g. missing or erroneous class code, missing cardinality, missing mnemonic code) requires correction before enabling any further activity, correction of a warning (e.g. multiple insertion of the same instruction or register) may be delayed or omitted by the user. At the end the detected procedure is stored on an external file.

Final output of BAT-editor is a sequential ASCII text-file containing all the instructions of the device under test, each described as in figure 3.

Though very compact, such organization of the device description is not well suited to identify relationships among different instructions when building the supergraph; a more flexible data structure is then required to optimize the execution time of the set of procedures involved in the supergraph design and optimum test procedure identification.

Before entering into detail of the proposed data structure, it is interesting to notice how the initial quantity of data may be reduced.

Given an instruction I^*, choice of a possible predecessor I_p (successor I_s) is performed considering only cardinality value and the set of registers affected by the execution of $I_p(I_s)$; this information is sufficient since cardinality value allows us to build minimum-ambiguity actions [31] while the set of registers shows the possible relationships intercurring between $I_p(I_s)$ and I^*; functionally <u>equivalent</u> instructions are then defined as instructions having equal cardinality and affecting the same input and output registers. When identifying predecessors and successors, only one instance of equivalent instructions has to be considered.

Moreover, a reduced test may be required: as an example the user may like to "force" only particular cardinality instructions as predecessors (successors), or evaluate the importance of a set of instructions in test actions design. This may be performed by excluding the chosen set of instructions from the microprocessor description file and then compare the two different test procedures obtained.

While in the first case (compaction) no information is lost on the behavior of the device (i.e. optimum test procedure can be obtained), in the latter (exclusion) there is an actual loss of information. Exclusion of instructions should be used only for statistical evaluations or for experiments in test procedure design.

Both these facilities are provided to the users by means of two different menus.

Information is thus stored in a suited data structure according to the options [36] the user has chosen in the compaction-exclusion phases. In general, information on instructions and related registers are organized as in fig. 4.

The basic criteria followed in designing of this structure are:

- dynamic storage allocation, which avoids an "aprioristic" determination of the maximum number of instructions and relationships among them;

- modularity of storage representation, which allows to achieve an efficient storage management;

- efficiency for data retrieve, in particular to allow a fast identification of the relationships among instructions.

Instructions of the device under test are grouped accordingly with their cardinality to guarantee an easy querying during the construction of the supergraph. A cardinality value is represented by an element of a dynamic list, which is the pointer to the set of instruction with the same cardinality. The elements of such list are ordered by increasing cardinality values. Each set of instructions with the same cardinality is stored as a dynamic list of records representing the instructions

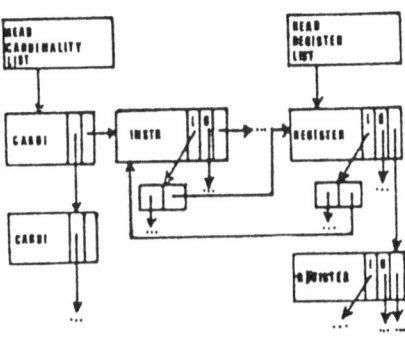

Fig. 4.

themselves.

An instruction is described by the following fields:

- mnemonic code, which uniquely identifies the instruction;

- cardinality, containing the cardinality value of the instruction itself;

- a pointer to the list of input registers used by the instruction;

- a pointer to the list of output registers used by the instruction;

- a pointer to the next instruction having the same cardinality.

For a fast identification of predecessors and successors of each instruction, another data structure is introduced to store information about registers (which are the interconnection points among instructions). This data structure is similar to the previously presented one; a register is described by the following fields:

- mnemonic code, identifying the register or the external pins;

- type of register, specifying if the record represents a register or an external pin;

- a the pointer to the list of instructions using this register (or pin) as input;

- a pointer to the list of instructions using this register (or pin) as output.

- a pointer to the next record.

The lists of instructions using a register or a pin as input or output are linked lists of elements containing the following fields:

- a pointer to the record describing the instruction (see above);

- a pointer to the next elements in the list associated to the considered register.

Similarly, the lists of registers or pins used by an instruction are linked lists of elements with the following fields:

- a pointer to the record describing the register or the pin (see above);

- a pointer to the next elements in the list associated to the considered instruction.

In this way a quite complex graph, holding all information about instructions, registers and relationships among them, is created. Though, due to such structural complexity, it is possible (fig. 4) to retrieve easily all data required to build the supergraph.

5.2. A guided heuristic for optimum test procedure identification

Before running the optimization algorithm, it is necessary to create a suitable representation of the supergraph, on which the optimum test procedure will be identified. The data structure adopted for the supergraph is shown in fig. 5.

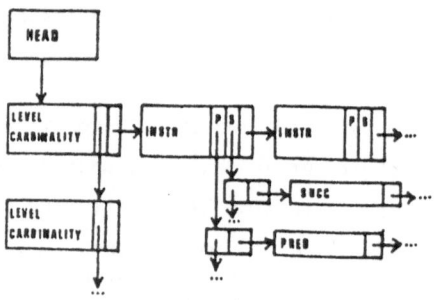

Fig. 5.

Instructions are organized onto levels of increasing cardinality value. Each instruction is represented by a record containing the following informations:

- the code of the instruction;

- its cardinality;

- its predecessors (i.e. the possible subsets of instructions whose outputs load all the inputs of the considered instruction);

- its successors (i.e. the possible subsets of instructions whose inputs use all the outputs of the considered instruction);

- a pointer to the next instruction with the same cardinality.

Each list of predecessors or successors is constituted by an header and a set of elements representing the instructions. The header holds the following data:

- a pointer to the first instruction of the list of predecessors or successors;

- a pointer to the next list of predecessors or successors.

Each element of these lists has the following fields:

- a pointer to the record representing an instruction in the supergraph (see above);

- a pointer to the next element of the list.

Theorems presented in [31], [33] state that to introduce minimum ambiguity predecessors and successors of an instruction I^* have to be chosen among instructions with cardinality values not greater than the cardinality of I^*.

Thus the supergraph may be built by means of an incremental method using the partitioning of instructions into levels of cardinality, as follows. Levels are examined for increasing values of cardinality: for an instruction of level k, all possible set of instructions of lower levels covering its input and output registers are identified and the proper places and transitions are inserted in the supergraph, i.e. the proper lists are inserted into the data structure representing the supergraph. This is iterated for each level upon every instruction starting from the lowest cardinality level.

It is now possible to use the supergraph to identify the optimum (minimum ambiguity - maximum error coverage) test procedure for the considered device.

A test procedure which minimizes ambiguity reasonably appears to be achieved by performing the following steps:

1. build all possible test actions for every instruction;

2. for each instruction, choose the lowest ambiguity test action;

3. merge such test actions to obtain a test procedure.

An algorithm presented in [35], belonging to the class of the Short-sighted Greedy algorithms, implements such steps; generally only a good test procedure is obtained, not the best one. In fact, global optimum would not be assured since this kind of optimization is performed upon single test actions without considering relationships among different test actions. While performing a non satisfactory optimization, that algorithm has proved to be a powerful tool for cycle detection and ambiguity evaluation of test actions.

The optimization problem could be obviously solved by adopting an exhaustive approach; all possible combination of choices may be performed and ambiguity can be evaluated for each of them: only combinations with minimum ambiguity will be accepted, thus generating the optimum test procedure. Nevertheless, this approach is very time-consuming since the number of possible choices is very high.

A possible alternative may consist in adopting a branch-and-bound technique [41], [42] to scan the set of possible test procedure: even if such guided heuristic algorithm is not exhaustive, complexity may be not trivial unless a suitable bounding function is defined.

The general structure of the algorithm for identification of the optimum test procedure consists of the following steps:

- <u>initialization</u> - The supergraph is built by collecting all possible test actions of every instruction; a test procedure P_o is selected among those contained in the supergraph. It is assumed as the <u>reference test procedure</u> P^* (i.e. as the reference solution), while its <u>ambiguity</u> $A_o = A(P_o)$ is the <u>reference ambiguity</u> (i.e. the reference value). The branch-and-bound algorithm creates a tree of subproblems: each node of such tree is related to a particular subproblem in which test actions have been chosen and fixed for a subset of instructions. Initialization creates the root and associates it with the supergraph, in which none of the alternative test actions for any instruction is fixed.

- <u>selection</u> - An open node, i.e. a node not yet explored, is selected. To bound the growth of the branch-and-bound tree, the node it is chosen among the ones opened in the last iteration of the algorithm; in particular, to achieve early a better solution it is useful to consider the node with minimum value of bounding function.
  For such node, an instruction without fixed test action is chosen. Also this choice may influence the dimensions of the branch-and-bound tree (i.e. the amount of storage required); to prevent its growth, two rules are identified for instruction selection:

  1. the instructions are considered by increasing cardinality value,

  2. within the same cardinality, the instructions are examined by increasing number of alternative test actions.

- <u>branch</u> - All possible alternatives to test this instruction are identified: for each of them a new branch is created and a new node is introduced in the branch-and-bound tree. To each new node is associated the set of fixed test actions obtained by adding the corresponding test action to the set of fixed test actions of the "father" node.

- <u>bound</u> - The bounding function is evaluated for each new node: it gives a lower bound to the ambiguity of every test procedure that may be generated in such subtree.
  If the bound is greater than the reference ambiguity, the node is closed because no test procedure will be found having an ambiguity lower than P^*.
  If every instruction has been considered, a complete test procedure is produced; it is compared with P^*: if its ambiguity is lower than A^*, it is assumed as the new reference test procedure and the bound comparison with new reference ambiguity has to be performed for opened nodes. The algorithm is iterated from the step "selection" until no more opened nodes are present. When the algorithm stops, the current reference test procedure is the optimum one.

The most critical points are the initialization and the definition of the bounding function for ambiguity: therefore they have to be carefully designed in order to obtain good performances from the algorithm. Lower computational effort will be spent if the initial solution is quite near to the optimum test procedure. Besides, if the lower bound is very restrictive, a lot of nodes will be closed soon, thus limiting the number of developed subtrees.

Initialization is the most complex step: information about instructions and their relationships has to be stored in order to allow fast retrieve. Instructions are then organized onto levels, corresponding to cardinality values. When the instructions of one level are examined, all possible test actions for the instructions of lower cardinality levels are supposed to be already identified.

To reduce iterations of the branch-and bound algorithm, the initial test procedure P_o must be as close as possible to the optimum solution. An initial test procedure may be identified by means of the greedy algorithm presented above.

The second problem, which has to be solved to assure fast convergence of the branch-and-bound algorithm, is the definition of a lower bound for the ambiguity in each node.

Ambiguity cannot decrease whenever an alternative test action is fixed in a node of the branch-and-bound tree; thus, the lower bound of ambiguity in each node will be given by the sum of ambiguities introduced by selected test actions along the path from the root to the considered node. If such value is greater than the reference ambiguity A^*, none of the test procedures which could be generated in the subtree may be better than the reference test procedure P^*; thus, they may be "a priori" discarded.

5.3. Displaynet

The output of the optimizer consists of two ASCII text file containing the description of the optimum test procedure for the considered programmable device.

The first file holds the sequence of test actions that have to be performed to test the whole device. An example of an element of this sequence is shown in fig. 6.

```
LD A,(DE)
ambiguity = 3
length = 6
LD E,DB
LD D,DB
LD A,(DE)
LD DB,A
```

Fig. 6.

The second file contains the Petri net representing the optimum test procedure, i.e. it gives the order following which the instructions have to be tested. Such file consists in two parts: the first one is the set of labels associated to places and transitions, e.g. the name of the instructions, registers or pins; the second one is the incidence matrix of the Petri net. (see [43]). An example of this description is shown in fig. 7.

This last file is accepted as input by Displaynet, a tool which provides a graphic representation of a Petri net on a large set of graphic output devices. A matrix description of a Petri net in fact is not very easily

```
parameters
HEADER "LD A,(DE)"
ROWS = 7
COLUMNS = 17
LABEL = 2
  0 0 0 0 0 0 0 0 1 0 0 0 0 0-1-1
  0 0 0 0 0 0 0 1 0 0 0 0 0-1-1 0 0
  0 0 0 0 0 1 0 0 0 0-1-1 0 0 0 0
  0 0 0 1 0 0 0 0-1-1 0 0 0 0 0 1
  0 1 0 0 0 0 0-1-1 0 0 0 0 1 0 0 0
  1 0 0 0 0-1-1 0 0 0 0 1 0 0 0 0 0
  -1-1-1-1-1 0 0 0 0 0 0 0 0 0 0 0

"D" "E" "DB" "AB" "A" "DB" "D" "DB"
"E" "A" "DB" "D" "DB" "E" "DB" "DB"
"A"
"LD A,DB" "LD DB,E" "LD DB,D" "LD DB,A"
"LD E,DB" "LD D,DB" "LD A,(DE)"
```

Fig. 7.

understandable by the users, while a graphic representation allows to detect immediately the presence of cycles in the Petri net, i.e. the presence of sources of ambiguity for the test procedure.

Displaynet draws places and transitions upon a virtual plane without boundaries to obtain a nice representation of the Petri net. To enhance understandability of relationships among instructions, places and transitions are drawn minimizing the intersections among arcs connecting the nodes of the Petri net.

When this step is completed, the picture is scaled down to display it onto the graphic device. However, the resolution of the device reduces the readability of the Petri net when the number of nodes is high.

To overcome this problem two features are available to the user:

- zooming a part of the picture to magnify details;

- drawing a node of the Petri net and the nodes connected to it; predecessors and successors of an instruction are then easily identified and possible cycles may be graphically detected. This can be iterated allowing to draw step by step the whole net.

Design and implementation of the graphic user interface of BAT is based upon MFB, a graphic library developed at the University of California, Berkeley: this library allows to generate outputs for a great number of graphic devices, as raster-scan displays, plotters and dot-matrix printers.

6. REFERENCES

[1]   M.A. Breuer, A.D. Friedman, Diagnosis and reliable design of digital systems, Pitman Publishing Limited (1976)

[2]   E.J. McCluskey, S. Bozorgui-Nesbat, "Design for autonomous test", Digest of Papers, 1980 Test Conference (November 1980)

[3]   E.H. Porter, "Testability considerations in a VLSI design automation system", Digest of Papers, 1980 Test Conference (November 1980)

[4] C.H. Chen, "VLSI design for testability", Digest of Papers, 1979 Test Conference (October 1979)

[5] F. Distante, V. Piuri, G. Galvani, A. Maderna, M. Minotti, "BAT: optimization algorithms and overall desogn of an automatic behavioral tester", Proc. EUROMICRO '86, Venice (Italy) (Sept. 1986)

[6] Kwok-Woon Lai, "Functional testing of digital systems", PhD Thesis, Carnegie-Mellon University, Pittsburgh (December, 1981)

[7] Computer, special issue on hardware description languages (Dec. 74)

[8] Computer, special issue on hardware description languages application (June 77)

[9] "VHDL: The VHSIC Hardware Description Language", in Design & Test, ed. IEEE (April 1986)

[10] M.A. Breuer, A.D. Friedman, "Functional level primitives in test generation", IEEE Trans. Computers, vol. C-29, n. 3, pp. 223-235 (March 1980)

[11] J.P. Roth, "Diagnosis of automata failures: a calculus and a method", IBM J. Res. Develop., vol. 10, pp. 278-281 (1966)

[12] Y.H. Levendel, P.R. Menon, "Test generation algorithms for nonprocedural CHDL", Proc. FTCS-11, pp. 200-205 (June 1981)

[13] Y.H. Levendel, P.R. Menon, "Test generation algorithms for computer hardware description language", IEEE Trans. on Comp., pp. 577-588 (July 1982)

[14] Y.H. Levendel, P.R. Menon, "The *-algorithm: critical traces for functions and CHDL constructs", Proc. FTCS-13, pp. 90-97, Milano (June 1983)

[15] S.B. Akers, "Functional testing with binary decision diagrams", Proc. FTCS-8, pp. 82-92, Toulouse (June 1978)

[16] D.M. Schuler, T.E.Baker, R.S.Fisher, R.V.Bosslet, S.S.Hirschhorn, M.B.Hommel, "A Program for the Simulation and the Concurrent fault Simulation of digital circuits described with gate and functional models", Test Conference, pp. 203-207 (1979)

[17] K.P. Wacks, F.J. Hill, M. Masud, P. DeBruyn Kops, "An integrated system for LSI device modeling", Digest of Papers, pp. 473-478, 1980 Test Conference

[18] M.R. Barbacci, "Instruction set processor specifications (ISPS): the notation and its applications", IEEE Trans. on Comp., vol. C30, n. 1 (Jan. 1981)

[19] S.J.H. Su, Yu-I Hsieh, "Testing functional faults in digital systems described by register-transfer language", Journal of Digital Systems, pp. 161-183 (1981)

[20] K. Son, J.Y.O. Fomg, "Automatic behavioral test generation", Digest of Papers, pp. 161-165, 1982 Test Conference

[21] K. Son, J.Y.O. Fong, "Table driven behavioral test generation", Proc. ISCAS 84, vol. 2, pp. 718-722, Montreal (May 1984)

[22] S.M. Thatte, J.A. Abraham, "Methodology for functional level testing of microprocessors", Proc. FTCS-8, pp. 90-95, Toulose (1978)

[23] S.M. Thatte, J.A. Abraham, "Test generation for general microprocessors architectures", Proc. FTCS-9, pp. 203-210, Madison (Wisconsin) (1979)

[24] S.M. Thatte, J.A. Abraham, "Test generation for microprocessors", IEEE Trans. on Comp., vol. C-29, n. 6, pp. 429-441 (June 1980)

[25] D. Brahme, J.A. Abraham, "Functional testing of microprocessors", IEEE Trans. on Comp., vol. C-33, n. 6, pp. 475-485 (June 1984)

[26] J.M. Gobbi, "Test et autotest de circuits complexes", PhD Thesis, Institut National Polytechnique de Grenoble (1981)

[27] G. Saucier, R. Velazco, "Microprocessor functional testing using deterministic test patterns", Proc. EUROMICRO 7, pp. 221-231, Paris (1981)

[28] C. Robach, G. Saucier, "Application oriented microprocessor test method", Proc. FTCS-10, pp. 121-125, Kyoto (Japan) (1980)

[29] R. Velazco, "Test comportamental de microprocesseurs", PhD Thesis, Institut National Polytechnique de Grenoble (1982)

[30] C. Bellon, E. Kolokithas, R. Velazco, "Generation automatique de programmes de test pour microprocesseurs", Proc. Journees d'Electronique 1983, pp. 161-170, Lausanne (October 1983)

[31] M. Annaratone, M.G. Sami, "An approach to functional testing of microprocessors", Proc. FTCS-12, Santa Monica (1982)

[32] M.G. Sami, M. Bedina, F. Distante, "A formal approach to computer-assisted generation of functional test patterns for VLSI devices", Proc. ISCAS 84, vol. 1, pp. 19-23, Montreal (May 1984)

[33] M. Bedina, F. Distante, M.G. Sami, "A Petri-net model for microprocessor functional Test", Proc. Journees d'Electronique 1983, pp. 149-160, Lausanne (October 1983)

[34] M. Bedina, F. Distante, M.G. Sami, "Some properties of behavioral testing and of related coverage", in Rap. int. 85-8, ed. Dipartimento di Elettronica, Politecnico di Milano

[35] F. Distante, "A Petri net matrix approach in VLSI functional test", EUROMICRO '85, Bruxelles (Belgium) (Sept. 1985)

[36] F. Distante, A. Maderna, L. Galvani, M. Minotti, "A Software Tool for Microprocessor Functional Test", in Microprocessing and Microprogramming, ed. North Holland, vol. 16, n. 2-3, pp. 107-112 (September 1985)

[37] F. Distante, V. Piuri, "Optimum behavioral test procedure for VLSI devices: a simulated annealing approach", ICCD'86, Port Chester, NY (Oct. 1986)

[38] S. Kirkpatrick, C. Gelatt, M. Vecchi, "Optimization by simulated annealing", Int. Rep., IBM Comp. Sc./Eng. Tech. Watson Res. Center, Yorktown Heights, NY (1982)

[39] F. Romeo, A. Sangiovanni-Vincentelli, Probabilistic hill-climbing algorithms: Properties and Applications, ERL, Memo, University of California, Berkeley (1984)

[40] S. White, "Concepts in simulated annealing", Proc. ICCD 84 (1984)

[41] Abadie, _Integer and Non-linear programming_, pp. 437-450, North-Holland, Amsterdam (1970)

[42] Allen J. Scott, _Combinatorial Programming, Spatial Analysis and Planning_, Methuen & Co. (1971)

[43] J.L. Peterson, _Petri net theory and the modelling of systems_, ed. Prentice Hall, New York (1981)

TESTING OF PROCESSING ARRAYS

F. Distante, M.G. Sami, R.Stefanelli
Department of Electronics
Politecnico di Milano (ITALY)

1. INTRODUCTION

Processing arrays have been in recent years the subject of many research activities, with particular reference to their implementation by means of VLSI or WSI devices; in fact, while their regular structure, characterized by high locality of interconnections, makes them well suited to high integration, relevant speed performances and the possibility of mapping upon them advanced signal-processing algorithms creates an obvious interest for such demanding applications as real-time radar imaging (and, more in general, signal processing) etc.

Testing of processing arrays actually is made more difficult by one of the characteristics that enhance their performances as regards operation speed and their suitability for VLSI implementation. In fact, the mode of operation of all such devices is defined as *computation intensive* rather than *I/O intensive*, and the architecture has a very limited number od I/O points with respect to the amount of internal information transfers (and, obviously, also of functional complexity): while this is of immediate relevance for device implementation (pin-out problems are reduced) and for overall processing speed (I/O transfers, in particular when involving external memories, constitute a bottle-neck with reference to global system speed) on the other hand it translates into a very low ratio of observation/control points with respect to functional complexity and therefore in an unfavourable situation as far as testing is concerned. Moreover, device complexity and application demands led to introduction of techniques for fault-tolerance through reconfiguration that - by requiring an amount of redundancy - may generate further testing problems.

Before discussing the various techniques proposed for testing, it is necessary to define the *fault models* adopted for this particular class of devices. Let us refer to a general structure of processing array, as consisting of:

- processing elements (PEs) or *cells*, whose complexity may vary with the application class envisioned (from simple, fixed-point arithmetic units to full microcomputers);

- interconnection network, consisting of data paths (buses);

F. Lombardi and M. Sami (eds.), Testing and Diagnosis of VLSI and ULSI, 355–381.

- interconnection-controlling circuits: these are introduced to provide the array with reconfiguration capacity, both as regards fault-tolerance and (in some instances) as regards topology (see e.g. [Sny82]) and consist of devices such as switches or multiplexers and (if on-chip self-reconfiguration is envisioned) of the circuitry necessary to generate their control signals.

Three different fault assumptions have been considered in the literature, namely:

1) Faults are assumed to be located only in the processing elements, not in the interconnection network or in the related control circuits. This assumption is usually justified by the consideration that PEs are much more complex than the interconnection system and - therefore - much more likely to incur into failures. The assumption is particularly valid for run-time testing: when this fault model is used, testing involves only the processing elements.

2) Besides PEs, also interconnection-controlling devices (e.g. switches) are assumed to be possibly faulty.

3) Interconnection links are considered as possible fault locations, as any other component of the array. Actually, few authors consider this instance outside of end-of production testing (see [And85]); sometimes, testing approaches lead to detecting faults in interconnection links *as if* they were faults in one of the connected PEs.

We shall not deal here with approaches for end-of production testing that in practice do not introduce relevant differences from general techniques (e.g. use of electron-beam testing or of laser testing): rather, we shall concentrate on *specific* approaches, tailored on the particular characteristics of processing arrays (such as their regularity or the mode of operation for systolic arrays). In the sequel, we shall first analyse some proposals aiming at definition of *testability conditions* for processing array; then, some main solutions oriented to *self-testing* will be discussed. Many such solutions have been designed to support subsequent fault-tolerance through reconfiguration, and actually exploit the redundancy introduced to this end in order to obtain self-testing capacities. Last, a technique for *host-driven* testing will be discussed in detail: this also is designed for a class of reconfigurable arrays, and it fully exploits both the augmented interconnection network provided to that purpose and the reconfiguration philosphy so as to grant concurrent run-time testing with the highest amount of parallelism.

2. TESTABILITY CRITERIA FOR PROCESSING ARRAYS.

As an example of main lines of research, we will examine some different testability criteria of processing arrays.

In [Kau67] an ILA (Iterative Logic Array) is said to be *testable* if it is possible to detect presence of any faulty cell in it, independently of the array size. An ILA is said to be *C-Testable* [Fri73] if it is testable with a constant number of tests, again independently of its size.

Kautz used a general fault model and an exhaustive test pattern for the individual combinatorial cell: i.e. the cell is tested for all its possible faults by applying all possible cell inputs to that cell. The following general conditions for testability were given:

1) cell inputs must be accessable independently of the position of the cell within the array, in order to be able to apply a complete set of tests for detecting all faults in the cell;

2) it must be possible to propagate the effect of such tests to an observable output of the array.

Kautz gave also the necessary and sufficient conditions for optimal testability, i.e. one granting maximum fault-coverage.

In [Fri73] the author considered C-testability of a class of mono-dimensional ILAs in which only the horizontal outputs of the rightmost cell are observable. In a later paper ([Dia76]), Dias studied ILA's in which a number of outputs lines were observable. In order to verify the truth table of such arrays, he gave a procedure to generate a test set whose size is independent of the array size, suggesting also the kind of modification that can be made to any given basic cell for this purpose.

In [Elh85] test of two dimensional arrays of combinatorial cells is considered, based on results presented in [Par81] concerning testing of ILA structures. The authors present an approach based on the idea of simultaneously testing cells at a given horizontal and vertical distance from each other. Two problems have to be faced: the difficulty of applying proper inputs to cells in an array of arbitrary dimensions and the need to propagate the effect of a fault to at least one observable output.

To achieve such goal, test sequences must be capable of regenerating at given horizontal and vertical intervals. Regeneration of test sequences means that while propagating through the array such sequences must periodically reconfigure themselves to their initial state. In this way also cells internal to the array may receive *external* data.

A cell in the array is characterized by a truth table (called in [Elh85] a *flow table*) which has a row for each x horizontal input and a column for each y vertical input. Each pair x, y is called a *state*. It can be demonstrated that if the flow table of a cell has m row and n columns, to test an $N \times M$ array we will need $M \times N \times m \times n(m \times n - 1)$ tests [Fri71], [Kau67]. Obviously, the fact of getting complexity of testing dependent on array dimensions is negative, the more so when very large devices are considered. To avoid this problem, the concept of *C-testability* has been introduced, defined as possibility to test the array by a constant number of test vectors, independently of the number of cells. In [Elh85] the authors prove that in order to achieve C-testability is sufficient to add four rows and four columns (at most) to the state table of the basic cell of the array, so that given constraints on the transitions of an arbitrary cell are satisfied. Such constraints make sure that every pair of states is distinguishable, thus allowing a possible fault to propagate while passing through succeeding cells. They also show how, provided the flow table is modified accordingly to the above, a $N \times M$ array of combinational cells whose basic cell flow table has m rows and n columns can be tested in at most $(3m + 1)(3n + 1)(m + 4)(n + 4)$ steps. This is due to the fact that x-inputs and y-inputs are repeated every $(3m + 1)$ and $(3n + 1)$ cells at most and that the size of the modified table is at most $(m + 4)(n + 4)$.

Another testability criterion that has been pointed out with regards to processing arrays is that of the *I-Testable* ILA's: an ILA is said to be I-testable if test responses from each cell can be made identical. An ILA is *CI-Testable* if it is I-testable with a constant number of tests, independent of its size.

An ILA can be I-testable and C-testable but this does not imply that it is CI-testable without any structural modification. Even if the cell structure is modified to suit CI-testability, problems then arise in designing a test generator usable in the BIT (Built-In-Test) environment. In fact, tests derived for CI-testable arrays may have different loop test length for each of the test vectors (see [Dia76], [Sri79]).

A last testability criterion is derived from the following considerations.

Suppose an ILA is partitioned into cell blocks of equal length, each block forming a *super cell*. If the basic cell is C-testable, but not CI-testable, it is possible to find a super-cell (the smallest feasable block size) which is CI-testable. This is the definition of *PI-testability*. An ILA is PI-testable (partitioned I-testable) if it can be partitioned into blocks of equal length such that all blocks have the same test response (note that cells within blocks may have different responses), and such that the number of tests is independent of the ILA size. In [Abo84] it is shown that all C-testable arrays are also PI-testable, without any structural change of the cells.

3. SELF-TESTING APPROACHES FOR PROCESSING ARRAYS.

Self-testing has been advocated for many authors, and at least two main justifications can be recalled:

a. complexity of the devices is such that self-testing capacities are mandatory: this is particularly true when wafer-scale arrays are considered;

b. the testing phase is a preliminary step for subsequent self-reconfiguration: in that case, and unless we can accept to suspend nominal operations for periodical externally-driven test phases, self-testing is again a basic requirement.

While some approaches are specifically designed for concurrent self-testing, other foresee self-testing phases periodically alternating with nominal operation ones. Moreover, as it will be seen in the sequel, different assumptions on computing power and internal structure of the individual processing arrays are made in the different instances.

We can identify three main classes of approaches for array-level self-testing. A first class is based on a concept that can be defined as *time redundancy*, deriving from the *Repeated operation with shifted operands* that was introduced for linear iterative structures such as the ones implementing some arithmetic functions (see [Pat83]).

Basically, *time redundancy* implies that, in a modular structure such as a processing array, each processing step be repeated a number of times in such a way that the individual operations making up the complete array-level processing step be performed, at each phase, by different modules. Thus, for example, in a linear array we can denote the physical position of the various modules as $P(1), P(2), \cdots, P(N)$, and the individual function (corresponding to a data quantum) as $F(1), F(2), \cdots, F(N)$. If a time-redundancy technique is adopted, during phase 1 $F(i)$ will be executed by processor $P(i)$ $(1 \leq i \leq N)$; during phase 2 a rotation of the operands will be effected so that $F(i)$ be executed by $P(i-1)$ for $2 \leq i \leq N$, and $F(1)$ by $P(N)$, etc. If results obtained in the various phases are stored and subsequently compared (provided suitable fault assumptions are acceptable) detection of a fault and even location of the faulty module is possible. There will be, of course, a measure of structure redundancy (due to the comparators, the memories and the added interconnection links supporting *operand rotation* during the various phases), but if the module are of even moderate complexity such redundancy will be much lower than that required by conventional replication and comparison: on the other hand, of course, array operation speed sharply decreases.

An extension of this basic philosophy to the case of rectangular arrays (and, by

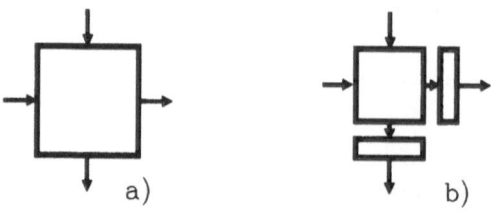

Fig. 1: Fault models: a faulty block affects all
the outputs of the block itself

immediate extrapolation, of any regular array with n-neighbor meshes) was presented in
[Sam84]: since keeping interconnection locality was considered as instrumental for VLSI
implementation, *shifted operands* rather than *rotated operands* have to be considered
(the latter case would involve very long interconnection links between extreme rows and
columns) and as a consequence a number of redundant cells must be added in order to
perform the complete operation during each phase.

Faults are assumed to be located in PEs only and are considered as *functional* ones
- i.e., a fault is detected when a PE produces incorrect results; further assumptions
concern the failure mode of PEs, that are initially summarized as follows:

1. there is at most one undetected fault in the array;

2. a faulty PE always produces incorrect results on all its output lines, whatever the
   inputs it may receive; in the case of a rectangular array, this leads to assuming the
   PE model of fig. 1.a - i.e., the PE is a *monolithic* device and a fault affects all its
   functions;

3. whenever a fault-free PE receives inputs different from the expected ones (as a
   consequence of a fault in a PE that produces information transmitted to it) the
   results it outputs are also different from the expected ones.

Basically, a $N \times N$ rectangular array operating in a *wavefront computation* way is
considered (i..e., information flows in one direction along each axis); a spare column
in position $N + 1$ is added to support the repeated operation with shifted operands.
The testing technique is best described by referring to the example in fig. 2. Denote
as *physical indices* the coordinates of a PE in the array, and as *logical indices* the
coordinates of the *elementary function* in the array-level processing step performed by
the individual PE. Two phases are involved:

Phase a: logical and physical indices of all cells coincide - i.e. cell with coordinates (i, j)
performs elementary functions identified by logical indices $(i', j') = (i, j)$. A

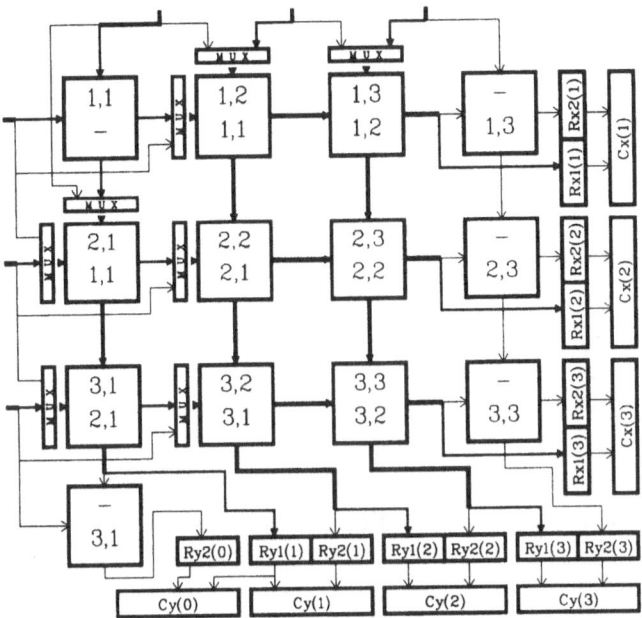

Fig. 2: Testing of a 3x3 nominal array through time redundancy (two phases). R_x and R_y are registers and C_x and C_y are comparators. Thick line interconnections are active in both phases. Medium line interconnections are active in the first phase. Thin line interconnections are active in the second phase. Upper logical indices: first phase. Lower logical indices: second phase.

register $R_{x1}(i)$ at the end of each row stores results produced by cell (i, N), while a register $R_{y1}(j)$ at the bottom of each column stores results produced by cell (N, j);

Phase b: logical indices are re-assigned as $i' = i$, $j' = j - 1$. The same operands used in phase a are fed again to the logical array; the shift grants that no cell processes twice the same inputs. Results of each cell $(i, N + 1)$ are stored in a register $R_{x2}(i)$, and results of each cell (N, j) in a register $R_{y2}(j - 1)$.

Comparisons between each pair of registers with the same row (column) index will - under the given fault assumptions - give agreement whenever both cells feeding the registers (and performing the same elementary function, respectively, in the first and in the second phase) receive correct inputs and give correct results - while disagreement appears as soon as one of the pair is faulty or when one of them receives incorrect inputs. Thus, a very simple algorithm allows to locate the faulty cell (by checking the first row and the first column in which disagreement is detected): actually, even failures in the interconnect links can be detected, but the algorithm attributes them to the receiving cell. Adding a spare cell in tha first colums provides a solution in case of a fault in the first or second column.

Structure redundancy is limited to $N+1$ spare cells, $4 \times N+1$ registers, $2 \times N+1$ comparators and to the augmented interconnection network (links and multiplexers). Operation speed - if diagnostic phases are introduced at each processing phase, so as to minimize error latency - is obviously halved, and worst-case error latency is in that case proportional to N.

On the other hand, the fault model corresponding to fig. 1.a is quite restrictive. A more satisfactory one is obtained if we accept that a faulty cell can produce an unexpected value either on both outputs or on one output only - an assumption adequate to relatively complex PEs with separate output sections on the two output lines (see fig. 1.b). Let us assume moreover that, whenever a cell receives unexpected data on at least one input line, its results will both be unexpected. The previous technique would lead to erroneous fault location whenever one output line only is faulty: a three-phase testing action provides the solution, but it requires both a spare row and a spare column and mapping of logical indices onto physical indices in the three phases such that no single cell will communicate with the same physical neighbors in all three phases corresponding to each single processing step.

Figures 3.a, b, c show, respectively, for each of the three phases, interconnections activated among working cells and logical indices associated with such cells. While presence of a fault is again diagnosed very easily, location becomes more complex than in the previous case - depending as it is on actual type of fault in the cell (global, horizontal output, vertical output) and on the indices themselves: table I allows to pinpoint the faulty device and even the kind of fault associated with it. Redundancy has increased sharply with respect to the previous technique, both as regards structure (there are $2 \times N$ spare cells, $6 \times N$ registers. For every three registers there is a pair of comparators (C_{13} and C_{23}) which compare, respectively, results of the 1st and 3rd phase and of the 2nd and 3rd phase; moreover, the interconnection network also is quite more complex) and as regards time (each processing step now requiring three phases): error latency has not been modified.

If the array, besides being self-testing, is also self-reconfiguring, both techniques allow detection of *multiple* faults (up to the limit for which reconfiguration is possible) provided faults appear singly in the time interval between two subsequent testing phases. Failure of multiple cells in such time interval could easily lead to erroneous identification of a *dummy* faulty cells (and, even in the optimum case, only one fault would be detected).

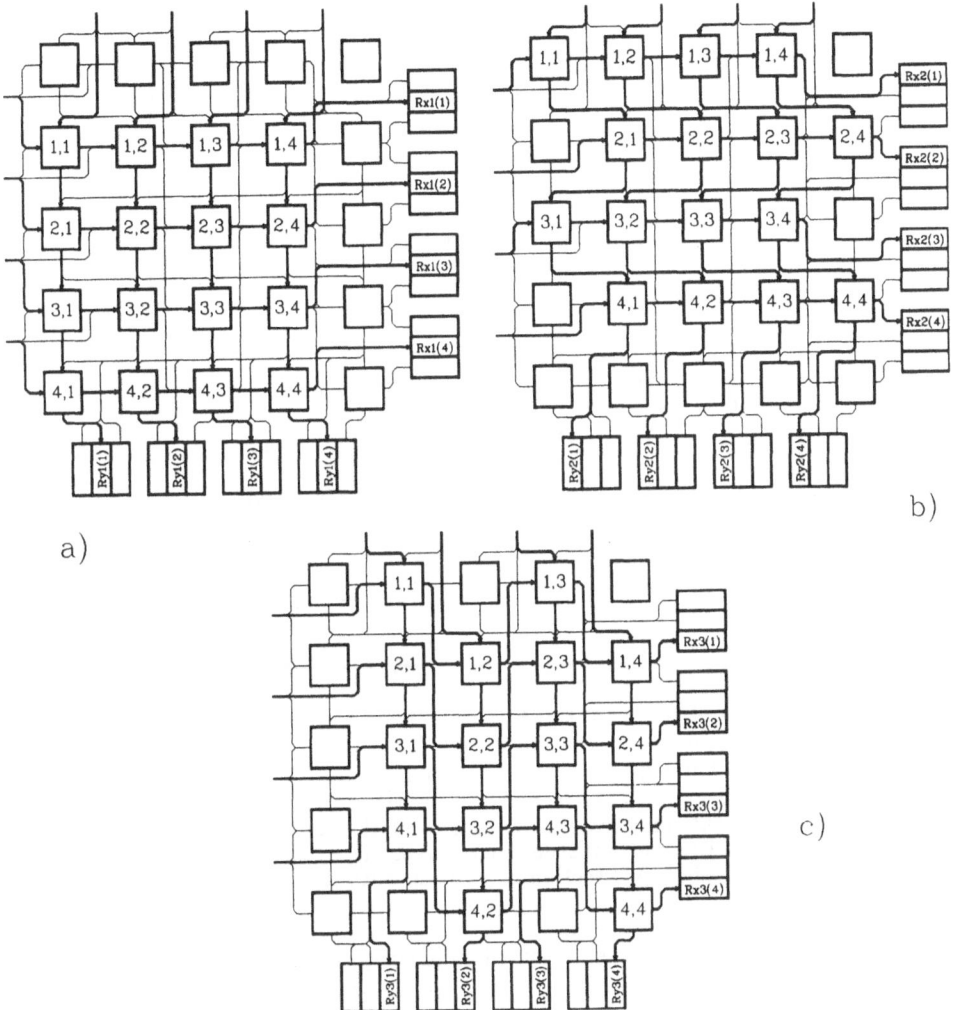

Fig. 3: Testing of a 4×4 nominal array through time redundancy (fault model of fig. 1.b). Fig. 3a,b,c correspond respectively to the first, second and third phase.

A second class of approaches still aims at concurrency (even though only partial) of testing with nominal operations, but here the speed decrease required by time redundancy is avoided: this is an aim well justified by the stringent real-time requirements often present in applications of processing arrays. Once more, availability of a large number of identical processing units is exploited to avoid massive replication: the technique is related to an approach used in general multiprocessor systems and often defined as the *roving monitor* one, by which a spare unit is used to check operations of different nominal units, monitoring each of them in sequence.

Cx23	CY13	n	m	i	j	fault type
D	A	E	E	n+1	m+1	both
D	A	E	O	n+1	m	hor
D	A	O	E	n	m+1	vert
A	A	O	O	n+1	m+1	both
A	A	O	E	n+1	m	hor
A	A	E	O	n	m+1	vert
D	D	O	E	n+1	m+1	both
D	D	O	O	n+1	m	hor
D	D	E	E	n	m+1	vert
A	D	O	O	n+1	m+1	both
A	D	O	E	n+1	m	hor
A	D	E	O	n	m+1	vert

Table 1. Symbols used in Table 1 have the following meaning: **A, D** (agreement and disagreement); **E, O** (even and odd); n is the lowest index associated with a horizontal comparator Cx23 or Cx13 giving disagreement; m is the corresponding index for vertical comparators; i,j indices of the recognized faulty cell; **hor, ver** and **both** denote respectively a fault on horizontal or vertical output section or on both output sections.

An approach of this type is presented in [Cho84], with reference to linear arrays (extension to rectangular arrays is simple). If N cells are required to perform the *nominal functions* of the array, K cells are added to allow testing operations: K may be as low as 1 or as high as N, depending on the degree of error latency that can be accepted for each individual PE. In fact, the K spare cells duplicate operations of K *nominal* ones: a comparator on each pair of cells checks correctness of results (note that, here again, a *functional* fault model is assumed for the cells). Fault assumptions are now less restrictive than previously: in fact, they are:

1) a faulty PE will always produce faulty results;

2) two faulty devices, given the same input data, will produce different results;

3) faults are located in PEs only;

4) there may be multiple faults, but there is never a fault-free PE located between

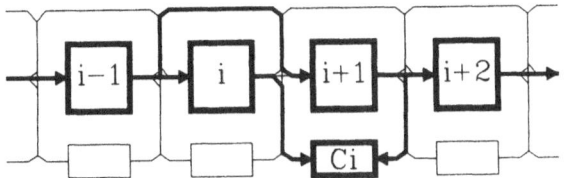

Fig. 4: Testing of a linear systolic array through structural redundancy. Modules i and $i + 1$ receive the same input data; circuit C_i compares the outputs of the two modules.

two faulty ones;

Assume for simplicity $K = 1$; testing is performed as follows (refer to fig. 4):

- set i to 1; testing is performed on PE (i); PE $(i + 1)$ acts now as the roving monitor, therefore a bypass link is activated such that $(i + 1)$ will not perform nominal operations (its functions are performed by $(i + 2)$, and functions on the whole chain are shifted one position right), $(i + 1)$ is fed the same nominal data fed to (i) and it performs the same function. At the end of the processing step, comparison on results is performed by a local comparator;

- if the comparison gives agreement, both (i) and $(i + 1)$ are assumed to be fault-free (we use again the assumption that a faulty device will always give unexpected results), index i is incremented by one so that the PE that was a monitor in the previous step becomes now a *nominal* cell and is in turn a subject of testing; if there was a disagreement, since a two-way comparison would not allow to pinpoint the faulty unit, again testing is shifted one position right and the following phases will allow to detect which of the units was faulty.

Dynamic assignment of monitor functions to different cells is effected through a *Token-passing* mechanism: a test token (or, more in general, K test tokens) are passed along the array together with the data stream, activating bypass links and modifying functions as they move along the chain. This requires addition to each PE of a suitable control circuitry capable of recognizing the token and of acting correspondingly. Error latency is related to the level of redundancy accepted, i.e. to the number of monitors K present in the array. If a reconfigurable system is envisioned (the links added for testing can be exploited for reconfiguration) we can foresee a soft degradation in the level of concurrency, with spare PEs substituting faulty ones and a corresponding decrease in the number of available monitors. There is no speed degradation (it can be safely assumed that time required by comparisons is negligible with respect to nominal operations performed by the cells). The technique just discussed has been extended

366

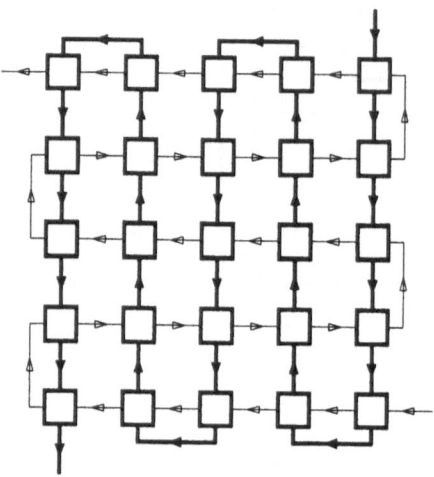

Fig. 5: Orthogonal linear arrays configured for
two-stages testing token algorithm.

in [Cho85] for off-line (possibly end-of production) testing of wafer-scale arrays. The concept of a token-passing organization is exploited to select the PEs subject to the testing procedure, while complete test vectors are fed from the outside in order to reach acceptable fault coverage. Test tokens are now more complex than in the previous case, since end of an individual test vector must also be identified; moreover, an extension of the algorithm has to be introduced because for end-of production testing no patterns of faulty PEs can excluded. To this end, it is suggested that multiple passes of the algorithm be repeated, involving pairs of cells at increasing distance (a solution whose simplicity is offset by the increasing amount of added interconnection links and the decrease in interconnection locality) or alternatively that the linear algorithm described be repeated along two dimensions; the interconnection network would still be augmented, but locality would be kept (see fig. 5).

A third class of approaches is again derived from classical techniques proposed for general multiprocessor systems, namely from the *system-level* techniques. Probably the best system-level solution is the Preparata-Metze-Chien (PMC) one, [Pre67] with the subsequent Hakimi-Amin [Hak74] extension, also known as *t-testability*. In the PMC criterion, in a system consisting of $N > 2 \times t$ units each unit is tested by t neighbors, and no preliminary assumption on fault status of these unit is necessary. In fact, test of each unit leads to creation of a table, and only final analysis of the complete test results will allow to identify faulty and fault-free units. The only restriction is that no more than t units can be faulty at any given time. While quite useful for general multiprocessor systems with a high degree of connectivity, this basic technique would be less than satisfactory for processing arrays such as systolic arrays, given their low connectivity: for the rectangular array commonly envisioned, the limit would be of four faulty PEs in the whole array, too stringent given the chip dimensions and the fault-tolerance requirements.

A generalization of the PMC technique to the case of rectangular systolic arrays was introduced by Somani and Agarwal in [Som84]: there, testing (obviously, performed off-line) involves execution of a suitable systolic algorithm made of two phases, one for detection, the second one for fault location.

Detection follows the classical t-testability pattern: each PE is tested by its t direct neighbors, and each PE i testing a PE j generates a binary result stating whether it declares j to be fault-free ($A_{ij} = 0$) or faulty ($A_{ij} = 1$). At the end of this phase, if at least one A_{kh} is 1 it can be declared that there is at least one faulty PE in the array: a complete table of variables $A_{kh} = 0$ could be misleading only if we accepted the alternative that *all* PEs in the array were faulty at the same time - and that moreover all agreed on the same test result (a possibility that can be very reasonably discarded).

If presence of faults is detected, the phase dedicated to fault location is begun: the restriction introduced for allowable fault distributions is far less stringent then in the Hakimi-Amin solution, and it can be stated as follows:

- let an *Uniform open system* centered on PE i be a subarray consisting of i, of the t neighbors at distance 1 (i.e. directly connected to it) and of the t neighbors at distance 2 such that the total number of external edges of the subgraph is $t \times (t-1)$. The Somani-Agarwal algorithm performs correctly whenever the fault distribution is such that no uniform open system in the array will contain more than t faulty units.

Assuming an acceptable fault distribution, the location phase starts from the set of A_{ij} created in the detection phase; majority votes on all A_{ij} are computed by each PE, thus generating a set of *confidence level* values that identify whether a given PE i is *likely good* or *likely faulty*. All such values are communicated by each PE to all its neighbors, so that in a third pass each PE will combine preliminary test results with confidence levels and revised confidence levels will be generated. Iteration of this operation as many times as the maximum distance between units in the systolic array leads to a syndrome locating all faulty units (under the previous fault assumption). Performances of this algorithm are immediately assessed by noticing that up to 44 faults can be located (if the fault distribution is favorable), against the 4 that would be granted by the classical t-testability technique.

Here no structure redundancy is required, and performance degradation can only be defined in terms of the periodicity with which test actions are repeated. On the other hand, previous technique did not require that the individual PE be capable of sophisticated processing actions, besides the ones required by the particular application: even the *test-token* approach only involved a simple control unit (possibly external) processing an elementary protocol. Here, on the contrary, PEs must be capable of feeding test vectors to neighbors, checking results, voting, updating confidence levels etc. Thus, the approach is oriented only to those processing arrays in which individual cells are full computers (even though reasonably simple ones).

4. HOST-DRIVEN TESTING.

In many instances, systolic arrays are in a way *coprocessors* controlled by general-purpose host machines or at least by external controllers: such is the case, typically, in signal-processing applications. It is then quite reasonable to assume that the same host machine will drive the testing actions, be they periodical or concurrent: this philosophy, obviously, allows to take into account far reduced redundancy than in the previous cases, since the host machine provides the support that would otherwise be required from spare units inside the array.

Actually, even end-of production test procedures usually belong to this class: we will here consider only some sample approaches that can also be adopted for *concurrent* or *semi-concurrent* testing, and therefore become a viable alternative to self-testing.

Evans and McWhirter [Eva86] have introduced a *mixed* technique, whose purpose is to achieve a complete test of the whole array including PEs, interconnection network and reconfiguration-controlling circuits. They suggested to adopt individual self-testing for the PEs (a technique not discussed here since it is strongly dependent on PE functions) and to rely on host-driven actions for the interconnection network (in particular, for the switches) and for the reconfiguration-controlling circuits. The approach is explicitly related to the reconfiguration philosophy defined as the "WINNER" approach [Eva85]: still, the main criteria can be considered fairly general and therefore applicable also in a different context.

Partitioning of test actions, as outlined above, leads to a hierarchical testing strategy. In order to avoid bottlenecks, even the circuits used for PE self-testing are subject to further testing: since a signature-analysis strategy is adopted, the host-driven phase of testing will also involve the comparators included in the signature analyzers so as to grant that no hard-core is required there.

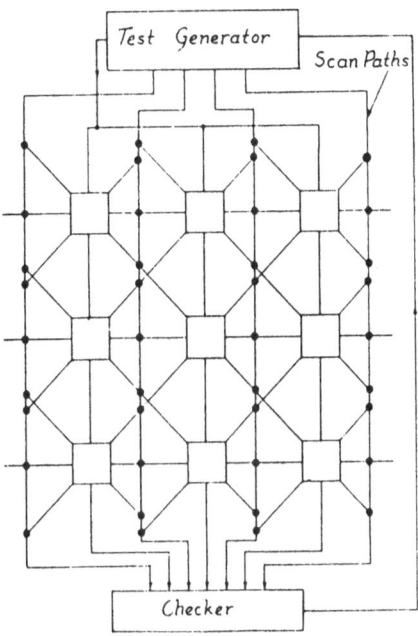

Fig. 6: Schematic of the scan path testing approach

External testing is performed upon the whole interconnection network and its related reconfiguration-controlling circuitry by using a scan-path technique: creation of the scan-path inevitably involves further redundancy, but this is reasonably limited (see fig. 6). Testing of the signature comparators is also performed by means of the scan-path technique.

The obvious criticism could be that - finally - the scan-path themselves would become the ultimate hard-core section of the system: although simple, they involve additional area and thus can lead once more to yield (or reliability) decrease. In [Eva86] it is suggested to duplicate them to avoid such bottleneck: alternatively, a *fail-safe* philosophy could be adopted, by a proper design of test vectors.

The above solution (that can be considered as a *semi-concurrent* approach) leads to a full test of the array, including interconnections; other host-driven techniques are designed with the same scope of achieving also a satisfactory test of interconnection network and of related circuits. Since many techniques for array fault-tolerance adopt

Fig. 7: Triple-bus interconnection network supporting reconfiguration algorithms in rectangular arrays

an interconnection network consisting of switched buses running inside routing channels, such structures have been discussed in detail - e.g., in [Cho86]. Although the solution there described refers to a specific structure, it is easily extended to a large class of instances. The set of faults considered in [Cho86] includes switches, buses and interconnection links: although faults are listed as if belonging to the *stuck-at* type, actually switch faults are seen as functional, since only faults that cause the switch to malfunction are taken into account. No specific restriction is introduced as regards fault distribution: it is simply assumed that a high number of switches and connection lines will be active in a wafer (otherwise, even with reconfiguration, harvesting would be so poor as to make the wafer practically useless). The authors avoid augmenting the interconnection network to grant testing possibility (actually, a functional redundancy exists, since PEs are capable of switching functions so as to allow testing of communication links between cells and switches). Relatively simple test actions (*straight-line tests*) allow to identify regions inside which faulty switches are confined: iterations of these tests leads to restrict such regions until fault location is practically reached. The complete algorithm is relatively complex (for a $N \times N$ array, $O(N)$ comparisons are required to find a fault-free routing path for test data to be applied); on the other hand, it is *fault-secure*, so that presence of any fault will certainly be detected.

A fully host-driven approach is presented in [Dis86]: there, no self-testing capacity is required from the PEs - on the other hand, a measure of redundancy is again required, even though limited to the interconnection network. The solution is presented with relation to a class of reconfigurable structures, all making use of a switched-bus interconnection network (see fig. 7): again, it can be immediately extended to any structure based upon the same type of interconnection networks. No extension need be made to the interconnection network used for reconfiguration: it must only be assumed that all buses are externally accessible. Since interconnection network will provide the actual medium to *reach* (i.e. to control and observe) PEs internal to the array, a first mandatory step consists in testing the interconnection network (with particular care to switches) itself.

To such aim, the following assumptions are made:

- functional faults are considered: a switch may be open, shorted or *stuck-at state* (i.e. it will not react to setting commands);

- no two inputs can merge into a single output;

- no single input can split into two outputs.

With these limitations, the test procedure outlined in fig. 8 allows to test for faults in all buses and switches, by connecting (through all possible paths) input pins of the arrray to the output pins. It will then be possible to apply suitable test vectors to the input pins and verify actual correspondence of outputs. Test vector definition will strongly depend on fault models adopted for buses and (most important) for switches.

It has though to be noted that test procedure of fig. 8 does not allow a *complete* test of the interconnection network: in fig. 9 thick lines denote bus segments and paths between pairs of switches that actually undergo testing. Yet, it is still incapable of identifying faults in links connecting PEs to adjacent buses - faults that will then subsequently be detected, during the second phase, *as if* the interconnected PEs were faulty. While giving incorrect fault location, for reconfiguration purposes it can be noticed that a PE incapacitated from interconnecting itself with the adjacent buses is effectively isolated from array operation - so that considering it as inactive is quite correct.

Refer now to testing of PEs: here, aims of concurrent testing can be summarized as follows:

Fig. 8: Test patterns for bus and switch testing

- reasonable fault coverage must be reached;

- error latency should be kept as reduced as possible - this, in turn, leads to requiring that the largest possible subset of PEs be concurrently subject to testing at any given time.

An off-line testing procedure can be created, by which the reconfiguration buses are used - through a suitable switch setting - to make I/O pins of a subset of PEs externally accessible and to apply test vectors to such PEs (buses and switches have been previously tested). Given a $N \times N$ array with X buses in each horizontal channel and Y buses in

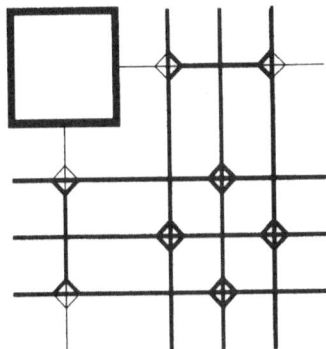

Fig. 9: Thick lines denote interconnections which are tested by
the procedure outlined in fig. 8

each vertical channel, if $Z = min(X,Y)$ it is possible to test concurrently $Z \times N$ PEs. In fig. 10 patterns of PEs that are simultaneously tested are shown: the number inside the PE represent the step during which the PE is tested.

We consider now in detail a technique that can be defined as a *semi-concurrent* one, in the sense that interconnection network and switches are tested by periodical actions, off-line, while PEs are tested concurrently with nominal operation.

Such testing approach can be described as follows:

1) At any array processing step, a specified set of PEs is considered to be *under test*;

2) *nominal* inputs are fed to these PEs, and their results are forwarded, as required by array operation, to their *logical* neighbors: i.e., they continue their nominal operation: at the same time, both their input data and their outputs are monitored by the host. The host compares such outputs with expected results; in the assumption of a reasonably simple PE structure, such comparison can be completed by simple means such as access to ROM-stored information;

3) whenever results are found to be in contrast with the expected values, the offending PE is declared to be faulty and it is excluded from operation; a reconfiguration step is then immediately initiated.

To perform this type of concurrent testing, the interconnection network must be augmented so as to allow monitoring of inputs and outputs of PEs under test; two slightly different interconnection structures may be considered, each supporting different testing strategies.

374

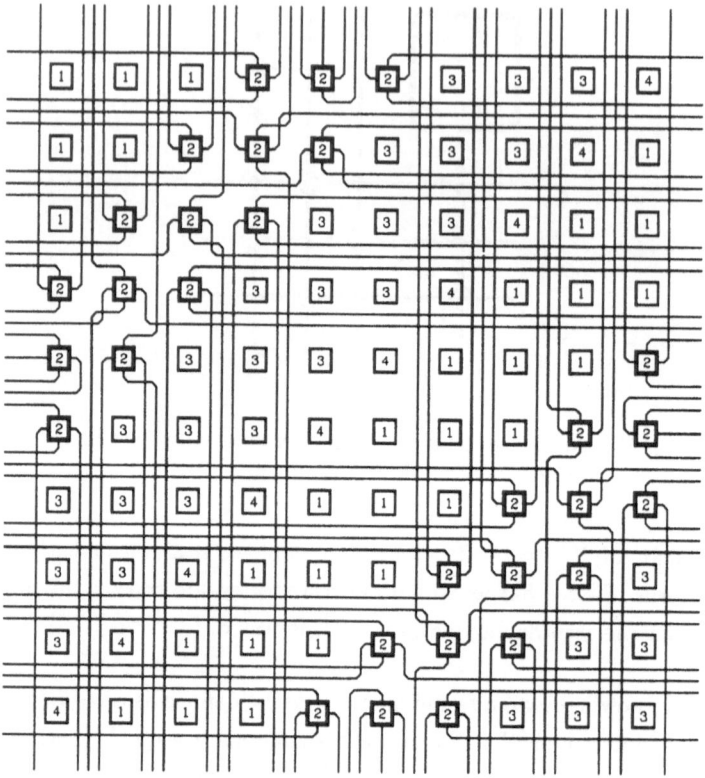

Fig. 10: Switch setting for off-line parallel PE testing

Starting from the nominal structure of fig. 7, observation and control capacity can be reached simply through addition of two diagnostic buses (one vertical and one horizontal). Switches on diagnostic buses are slightly different from the ones on reconfiguration-supporting buses, since a larger number of possible switch settings becomes necessary in order to route test information to external pins. Obviously, these do not allow to test *simultaneously* all PEs in the array, but rather to test any PE singly and - in general - fairly large subsets of PEs in parallel.

It should moreover be noticed that in order to test correct operation of a PE we need to *observe* the values of its inputs and outputs, not necessarily to pre-determine the value of inputs and observe the values of corresponding outputs. The set of input values that can be obtained without direct external control has though to be large enough to cover the given fault assumptions (at run-time, less detailed fault coverage and higher ambiguity than for initial testing may be accepted as a trade-off against concurrency).

As long as there are no faults in the array, physical neighbors are also logical neighbors, so that the segment of diagnostic bus between two neighbors will allow to monitor at the same time the output of one PE and the input of the other one. In this case, then the degree of parallelism that can be reached in testing is evaluated as follows:

- let the array dimensions be $N \times N$ (comprising also spare rows and columns): pins available are $4 \times (N+1) + 4N = 8N + 4$ (given the limitations of switch settings on all buses but diagnostic ones, only pins directly connected to border PEs and ends of diagnostic buses are freely available for testing purposes);

- then, since each PE can be controlled and observed by using four independent segments of diagnostic buses: given possible switch settings, each diagnostic bus can be used as *split* into two independent segments so that the four segments required for a PE do not block complete use of four distinct buses;

- given the above, examine PEs on a diagonal of the square array: if inputs to these PEs are routed consistently in one direction (e.g. upwards and leftwards, respectively) and outputs in the opposite one, no conflict will arise on the diagnostic network. Consider in fact two PEs on a diagonal, with physical and logical indices (i, j) and $(i-1, j+1)$: the previous strategy guarantees that the vertical diagnostic bus between them will be used in the upper segment by $(i-1, j+1)$, in the lower segment by (i, j), and no attempt to sharing any bus segment will be made. (The same holds for horizontal buses).

Obviously, this strategy allows to test simultaneously the largest possible set of PEs only when the main diagonal is used; otherwise, a pair of diagonals at opposite extremes of the array will have to be suitably chosen and adopted simultaneously. The distance between the two diagonals can be evaluated as follows:

- Let a diagonal with extreme points $(1, i)$ and $(i, 1)$ be chosen initially (obviously, $i < N$). The *extreme* diagnostic buses used for its test are, in order, the vertical bus between columns i and $i+1$, to test the horizontal output of $(1, i)$, and the horizontal one between rows i and $i+1$, to test the vertical output of $(i, 1)$.

- as a consequence of the conditions ruling use of buses for observation and control, the two above buses (and all the ones with lower indices) cannot be used for the testing procedure of any PE (j, k) with $j, k \leq i+1$: thus, the longest diagonal that can be tested simultaneously with the previous one is that from $(N, i+2)$ to $(i+2, N)$. (In this way, actually, $N-1$ PEs are tested at each step).

The major drawback of this approach consists in the following: assume that some PEs in the array have failed, and that reconfiguration has been performed: in this case, in general, *logical* and *physical* indices associated with PEs will be different. As a consequence, considering two active PEs whose logical row indices are k', $k'+s$, the physical row indices l, m upon which they map in order will not necessarily respect relation $m > l$ (the same, obviously, holds also for column indices).

Consider then, in particular, two cells on a *logical diagonal*, (h', k') and $(h'-1, k'+1)$; rules previously described for observing their inputs and outputs, would not any more guarantee absence of conflicts for access to diagnostic bus segments.

An alternative strategy must then be adopted. The simplest one could be to test one single PE at a time: this is obviously always possible, but it may lead to relevant error latency. Alternatively, the largest possible set of PEs that can be tested in parallel with a given PE could be identified from time to time (and depending from present reconfiguration) by the host. To this end, the following conditions must be met for two PEs to be testable in parallel (we keep the criterion of *observing* inputs to PEs under test on uppermost and leftmost segments of diagnostic buses, while outputs are observed respectively from right and lower ends):

- define as *row predecessor* of (i', j') the PE having logical indices $(i', j'-1)$ (and conversely for a column predecessor)

- no two PEs onto the same physical row (column) can be tested in parallel;

- no two PEs whose row (column) predecessors are mapped onto the same physical column (row) can be tested in parallel;

- whenever a PE and its row (column) predecessor are mapped onto the same physical column (row), the column (row) physical index of the predecessor must be lower than that of the PE under test.

Still, identification of *largest testable sets* is totally reconfiguration-dependent and it may be quite complex, so as to make the strategy just described rather debatable (in particular, it would require a relevant computation effort on the part of the host). A second structure of diagnostic network is then considered, that - while slightly more complex - allows higher parallelism in testing, independently of presence or even distribution of

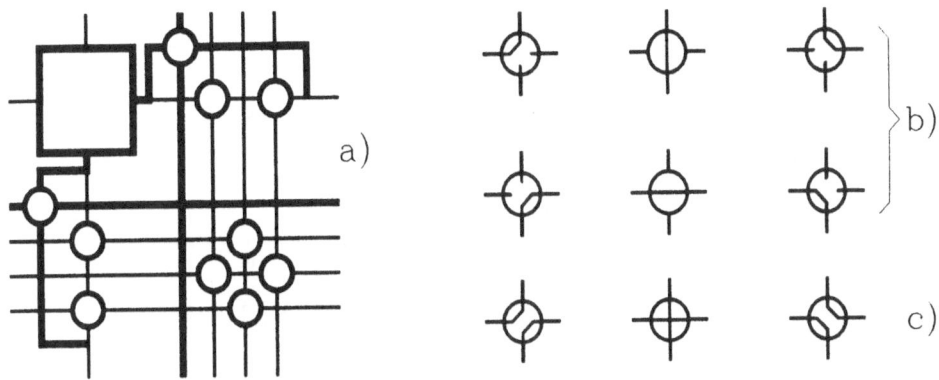

Fig. 11: Triple-bus structure with addition of diagnostic buses; a) array structure; b) possible setting of diagnostic switches; c) possible setting of reconfiguration switches

faults, and thus ultimately leads to lower error latency.

Consider the diagnostic network represented in fig. 11 it has been modified, with respect to the previous one, by addition of a link (with the related switch) between direct horizontal (vertical) input of each PE and the nearest vertical (horizontal) diagnostic bus. The added switch has the possible settings shown in fig. 11.b. In this way, immediate independent access to all four terminals of each PE can be granted via the four diagnostic buses enclosing it, whatever the configuration actually present in the array. The testing procedure is outlined in fig. 12; the number inside the PE represents the step during which the PE is tested.

Due to dimensions of PEs with respect to buses and switches, no actual area increase is introduced by adopting a suitable layout, as long as the *edge* of a PE is longer than the area occupied by switches; switch control complexity is only slightly increased.

In absence of faults, no modifications to the testing strategy are introduced with respect to the previous structure; even in presence of reconfiguration, anyway, the identical criterion can still be applied (now, reference can be made directly to the four diagnostic buses enclosing the single PE, without need of identifying the logical neighbors and their related diagnostic buses), so that again at each processing step N PEs can be monitored in parallel. Therefore, total fault latency decreases due to increased test parallelism: obviously, this is paid in terms of increased complexity of interconnection network testing.

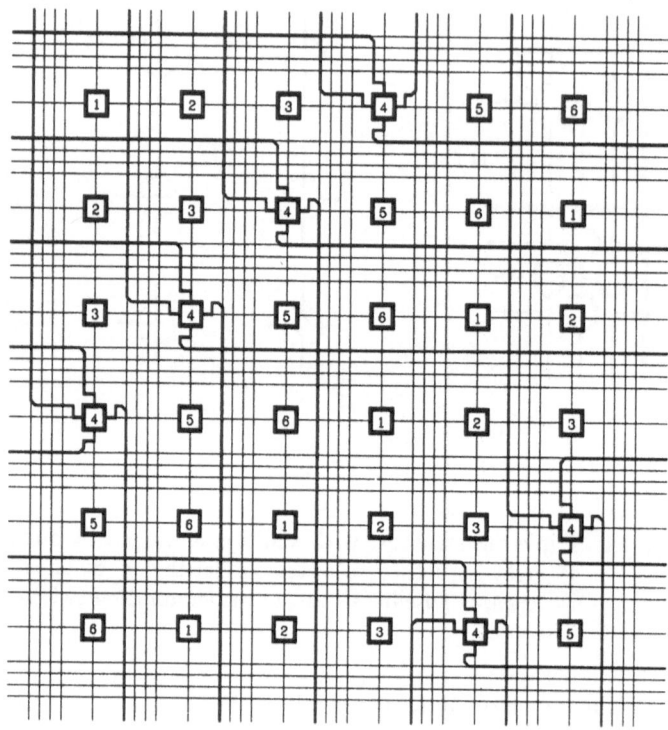

Fig. 12: Diagnostic switch setting for concurrent monitoring of
PEs

While in the previous case only non-faulty PEs underwent testing, here no assumption on state of a PE are necessary. This enables to test also PEs that have previously been declared faulty, provided the reconfiguration control system is such that even faulty PEs are fed nominal inputs (their outputs, obviously, are blocked by the interconnection network and routed only on the diagnostic buses). Thus, it is possible to account for transient faults: if outputs of a *faulty* PE are found to be correct for a predetermined number of subsequent operation steps, the fault can be considered as transient.

Considerations on fault latency strongly depend on the fault model assumed for the PEs; following the assumption (chosen by several authors) that a faulty PE will always produce faulty results, whatever the inputs, error latency will depend only on test strategy: if, on the contrary, presence of inputs capable of masking some faults is allowed, this probability also will have to be taken into account.

Let T be the time interval between two subsequent test actions upon the same set of PEs. If at each operation step the subset of PEs under test is changed, when full diagonals of an $N \times N$ array are tested at each step this time interval is proportional to N (actually, it is $N^2/(N-1)$). This holds for both diagnostic structures as long as there are no faults, and for the second one independently of the fault distribution; for the first diagnostic structure, in presence of faults the time interval will vary with reconfiguration - if variable subsets are defined - going towards N^2 for increasingly hard fault distributions (and, obviously, it will go with N^2 if the simpler strategy of testing one single PE at a time is chosen).

Should the simpler fault assumption be adopted, then, error latency would coincide with the above time interval T. Refer now to the second, more realistic fault model.

Let $p(t)$ be the probability that data input to a given PE (k,l) at time t will mask the errors assumed as possible for the PE itself. Then, $1 - p(t)$ is obviously the probability of detecting faults in PE (k,l) at time t by means of our concurrent test strategy. If no information is available on the distribution of input data, it is impossible to state whether probability of detection will increase in the subsequent test cycle for the same PE: thus, in the worst case error latency could increase indefinitely. Anyway, since the host keeps observing the input data fed to PE (k,l) - in order to check results of operation - it can be assumed that:

1) the host knows what input patterns mask possible faults; this information is automatically desumed by the error model and the PE structure;

2) whenever such pattern is observed at the inputs of at least one PE in the subset under test, the subset itself is kept under test until all PEs have received at least once data allowing fault detection.

Obviously, such strategy increases error detection probability for the single PE, but at the same time it increases error latency at the array level. Still, processing speed of a systolic array, as compared to run-time failure rate in a VLSI device, is such as to make a possible increase in error latency less critical that a decrease in credibility of the test action itself.

REFERENCES

bo84: E.M. Aboulhamid, E. Cerny: "Built-in Testing of One-Dimensional Unilateral Iterative Arrays", *IEEE Trans. on Computers*, Vol. C33, pp. 560-564, June 1984

nd85: H A.Anderson: "Computer-aided design and testing for RVLSI", *International*

380

Workshop on WSI, Southampton, 1985

Ann83: M. Annaratone, R.Stefanelli: "A multiplier with multiple error correction capabilities", *Proc. 6th Symposium on Computer Arithmetics*, 1983

Cha86: A.Chatterjee, J.A.Abraham: "C-Testability for generalized tree structures with applications to Wallace Trees and other circuits", *ICCAD-86*, pp 288-291

Cho84: Y.W.Choi, S.H.Han, M.Malek: "Fault diagnosis of reconfigurable systolic arrays", *ICCD-84*, 451-455

Cho85: Y.H.Choi, D.S.Fussell, M.Malek: "Token-triggered systolic diagnosis of Wafer-scale arrays", *Int'l workshop on WSI*, Southampton, 1985

Cho86: Y.H.Choi, D.S.Fussell, M.Malek: "Fault diagnosis of switches in Wafer-Scale arrays", *ICCAD-86*, 292-297

Dia76: F.J.O. Dias: "Truth-table verification of an iterative logic array", *IEEE Trans. on Computer*, vol. C25, pp. 605-613, June 1976

Dis86: F. Distante, M.G. Sami, R. Stefanelli: "Testing techniques for Complex VLSI/WSI Processing Arrays" S. Kartashev, S. Kartashev eds., Van Nostrand Reinhold Publisher

Elh85: H. Elhuni, A. Vergis, L. Kinney: "C-testability of two-dimensional arrays of combinatorial cells", *Proc. ICCAD 85* pp. 74-76

Eva85: R.A.Evans, J.V.McCanny, K.W.Wood: "Wafer scale integration based on self organisation", *Int'l Wowkshop on WSI*, Southampton, 1985

Eva86: R.Evans, J.McWhirter: "A testing strategy for self-organising fault-tolerant arrays", *Oxford Workshop on systolic arrays*, Oxford, July 1986

Fri71: A.D. Friedman, P.R. Menon: "Fault detection in Digital Circuits", Englewood Cliffs, NJ: Prentice-Hall, 1971

Fri73: A.D. Friedman: "Easily testable iterative systems", *IEEE Trans. on Computer*, vol. C22, pp. 1061-1064, Dec 1973

Hak74: S.L. Hakimi, A.T. Amin: "Characterization of connection assignment of diagnosable systems", *IEEE Trans Computers*, Vol. C-23, pp86-88, Jan 1974

Kau67: W.H. Kautz: "Testing for faults in combinatorial cellular logic arrays", *Proc. 8th Symp. Switching Automata Theory*, 1967, pp. 161-174

Mangir: T.Mangir: "Sources of failures andyield imporvement for VLSI and restructurable interconnect for R-VLSI and WSI:

Part I, Sources of failure and yield improvement for VLSI", *Proc. IEEE*, June 1984, 690-708

Part II, Restructurable interconnects for RVLSI and WSI", *Proc IEEE*, Dec 1984, 1687-1694.

Par81: R. Parthasarathy, S.M. Reddy: "A Testable design for Iterative Logic", *IEEE Trans. on Computer*, Vol. C30, pp. 883-841, 1981

Pat83: S.Laha, J.H.Patel: "Error correction in arithmetic operations using time redundancy", *FTCS-13*, 298-305

Pre67: F.P.Preparata, G.Metze, R.T.Chien: "On the connection assignment problem of diagnosable systems", *IEEE Trans. Electronic Computers*, Vol. EC-16, N.6, Dec 1967, p.848

Sam84: M.G.Sami, R.Stefanelli: "Self-testing array structures", *ICCD-84*, pp. 677-682

Sny82: L. Snyder: "Introduction to the configurable highly parallel computer", *IEEE Computer Magazine*, vol. 15, n. 1, Jan 82, pp 47-56

Sca86: J.T.Scanlon,W.K.Fuchs: "A testing strategy for bit-serial arrays", *ICCAD-86*, pp. 284-287

Sri79: T. Sridhar, J.P. Hayes: "Testing bit-sliced Microprocessors", *Proc. FTCS 9*, 1979, pp. 211-218

OLD AND NEW APPROACHES FOR THE REPAIR OF REDUNDANT MEMORIES*

Fabrizio Lombardi
Department of Electrical and Computer Engineering
University of Colorado, Campus Box 425
Boulder, CO 80309, USA

ABSTRACT.

This paper describes some approaches for repairing memories. Repair is implemented by deletion of either rows and/or columns on which faulty cells lie. These devices are commonly referred as redundant memories, because columns and rows are added as spares. Existing repair approaches are reviewed. New repair techniques and algorithms are proposed. The first algorithm is based on a fault counting technique with a reduced covering approach. The innovative feature is that reduced covering permits an heuristic, but efficient criterion to be included in the selection of the rows and/or columns to be deleted for repair of a memory die. This feature retains the independence of the repair process on the distribution of faulty cells in memory, while it allows a good repairability/unrepairability detection. The second approach, namely the Faulty Line Covering Technique, is a refinement of the Fault-Driven approach. This approach generates all possible repair-solutions within a smaller number of iterations than the Fault-Driven algorithm. It is proved that the Faulty Line Covering technique will execute faster under most fault distributions. This approach is perfect. The third approach exploits a heuristic criterion in the generation of the repair-solution. The criterion is based on the calculation of efficient coefficients for the rows and columns of the memory. Two techniques for coefficient selection are proposed. It is proved that these techniques require very little additional testing hardware while still providing fast (and generally good) repair-solutions. Comparative analysis with existing approaches is presented. Early-abort techniques are discussed. An adaptive repair technique which can be run in parallel with memory testing, is also analyzed. Illustrative examples and simulation results are provided to substantiate the validity of the proposed repair techniques.

1. INTRODUCTION.

Large density memories have become a reality [1]. This has been possible by integration techniques such as Very Large Scale Integration (VLSI) and Wafer Scale Integration (WSI). *Redundancy* has been extensively used for manufacturing memory chips and to provide repair of these devices in the presence of faulty cells. This type of memory has been referred as *Redundant Random Access Memory* (RRAM). Redundancy consists of spare cells arranged into spare rows and columns. These rows and columns are used to replace those rows and columns in which faulty bits lie. This process is commonly referred as memory *repair,* and it has increased considerably the yield of these devices and hence decreased their cost [2,3,7,8]. Redundancy is not free of penalties. The

---

*This research supported in part by grants from AT&T and NATO.

F. Lombardi and M. Sami (eds.), Testing and Diagnosis of VLSI and ULSI, 383–427.
© 1988 by Kluwer Academic Publishers.

insertion of spare rows and columns contributes to an overall larger chip area than in the original irredundant design. Degradation of performance (such as a longer access time [3]) and yield can also occur [6] if no appropriate precautions are taken into account.

Many algorithms for memory repair by row/column deletion have appeared in the literature [3,4,5]. Recently, it has been proved that row/column deletion is an NP complete problem [22] and it reaches an asymptotic value in repairing large memory arrays [17]. It has been demonstrated that the repair strategy of row/column deletion is effective for the current generation of memories because of the small expected number of faulty bits.

This paper is organized as follows. Section 2 presents the preliminaries (definitions and notation) which are used. Section 3 reviews existing techniques for memory repair. The basic operations of existing schemes are discussed in Section 4. An approach which achieves repair and repairability, is discussed briefly in Section 5. The notion of fault counting is introduced in Section 6. This is used in Section 7 to propose reduced covering as an initial and basic criterion for repair. The repair process is analyzed in detail in Section 8 (to generate the repair-set), Section 9 (the refinement criteria for repairability), Section 10 (the procedure to generate the repair-solution). Section 11 analyzes the implications of fault distribution on memory repair by reduced covering. The issues involved in an implementation of a defect analysis system are discussed in Section 12. Section 13 introduces Fault Line Covering as a perfect approach (but with an exponential complexity) to repair a memory die. A heuristic criterion which is referred as the effective coefficient, is proposed in Section 14. Two algorithms are also presented. Early-abort and user-defined criteria are outlined for an efficient execution of the repair process. Simulation results and comparison between the proposed approaches are presented in Section 16. Section 17 establishes a new technique for memory repair. This technique adaptively considers the two processes of testing and repair in parallel. Conclusions are addressed in the last section.

2. PRELIMINARIES.

This paper deals with two dimensional memory arrays. An array is made of n bits on a single die. The number of columns is given by N, while the number of rows is given by M. Figure 1 shows a 16×16 array. The numbers of spare rows and columns are given by R_A and C_A respectively. A *fault counter* is associated with each row and column of the array. The content of the fault counter of row i (column j) is denoted as $C(R_i)$ ($C(C_j)$) for $i = 1, \ldots, M$ ($j = 1, \ldots, N$). The *total fault count* is denoted as T_F, where

$$T_F = \sum_{i=1}^{M} C(R_i) = \sum_{j=1}^{N} C(C_j) \tag{01}$$

and rows are numbered from 1 to M and columns are numbered from 1 to N. The tested memory die is represented by an $M \times N$ diagnostic matrix $A = (a_{ij})$. Element a_{ij} in the ith row and jth column, is 0(1) if the (i,j) cell is fault free (faulty). It is important to define the *leading element* a_{ij} of A, as the first non-zero element of A, such that the ith row is the first non-zero row and the (i,j)-

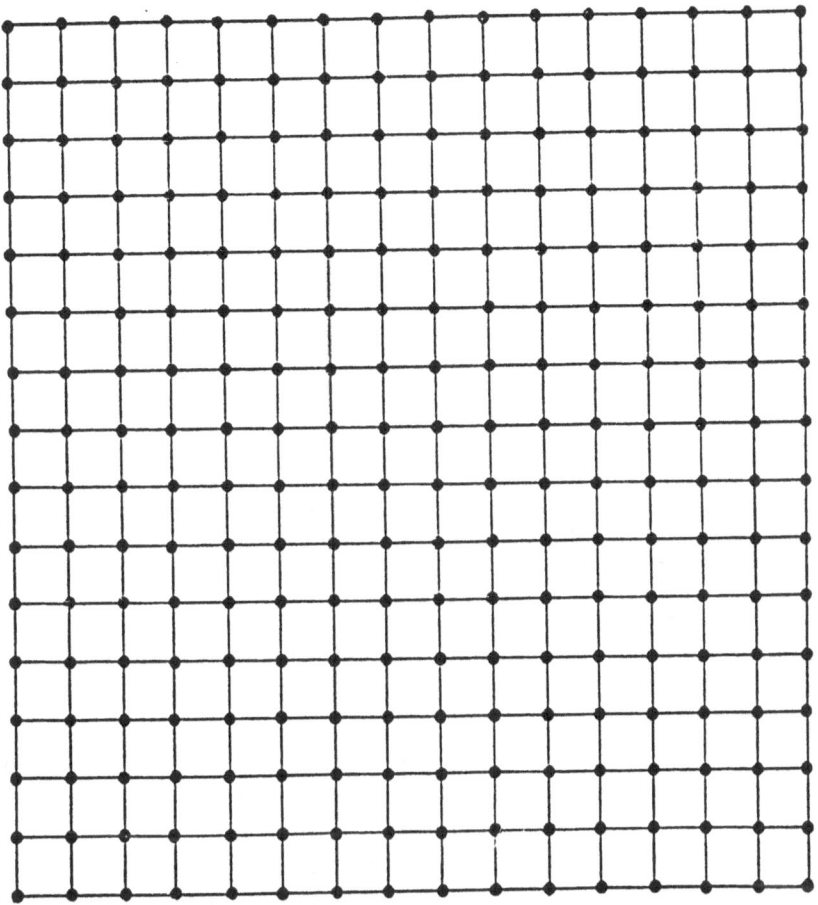

Figure 1: 16x16 array

entry is the first non-zero element of row i.

The following characteristics are assumed in this paper.

I. Redundancy is applied locally to each die (subarray). This does not imply a large redundancy overhead for large memories, but deletion of a row/column does not affect the whole memory array [17].

II. Only faulty bits affected by hard defects are considered [17].

III. Faulty bits appear in the die according to different distributions. This depends on the technology [6,9]. Fault distributions are analyzed in detail in Section 11.

The following definitions are used in the analysis.

DEFINITION 1.

A *fault line* is either a row or a column on which one or more faulty bits exist.

DEFINITION 2.

A fault line is said to be *covered* if all faulty bits on the line are repaired by using spare rows and/or columns.

DEFINITION 3.

A faulty bit which does not share any rows or columns with any other faulty bit is referred as a *single faulty bit*. The number of single faulty bits in memory is denoted as S_F.

DEFINITION 4.

The row (column) *complement bits* of a faulty bit (i,j) are all faulty bits in R_i (C_j), except for faulty bit (i,j) itself.

DEFINITION 5.

A *leading faulty bit* (i,j) is the first faulty bit of the memory die such that row i is the first faulty line and the (i,j) entry is the first faulty bit on row i.

3. REVIEW.

Repair of memory devices has been directed toward minimization of repair-time. Initially, the memory die undergoes a testing phase. Sites of faulty bits are logged into a diagnostic table. *Row-fault-counters* and *column-fault-counters* are used. These counters keep track of defective bits for each row and column of the memory. A *total-fault-counter* records the overall number of faulty bits.

A repair algorithm is executed to generate a *repair-solution* (whenever one exists) using the provided redundancy. The repair-solution relies on the information supplied by the counters and the diagnostic table. An ideal repair algorithm is not limited to provide a repair-solution, but it also satisfies the following *benchmark criteria* [4,6].

A. Repair-time should be contained within a reasonable execution.

B. Algorithm *execution* should be *aborted* at the earliest detection of unrepairability. Early-abort conditions should be used to promptly identify unrepairable arrays.

C. Efficient repair at polynomial time complexity for the worst case analysis.

D. Applicability to an ATE implementation.

E. *User-friendly* to permit a human selection for types of repair-solution that have been generated by the repair algorithm.

F. The number of rows and columns to be deleted should be minimized. An optimal repair-solution is desirable.

G. The repair technique should be compatible with hardware implementations in a defect analysis system.

Manufacturers have used different algorithms to compute repair-solutions for redundant devices. These algorithms are based on techniques like the Broadside approach [6], Repair-Most approach [6], Fault-Driven approach [4] and the Branch and Bound approach [22].

The *Broadside approach* employs a crude technique to locate each faulty bit and immediately repair it. No optimization is used. Spares are allocated in a very inefficient fashion, because no overall distribution of the faulty bits is considered. This results in failure to identify a potentially repairable die. Hence, yield enhancement is possible when the repair-solution is easily generated. The Broadside approach has a very fast execution and it does not require a complicated defect analysis system. It is very efficient when the number of faulty bits and the redundancy are small.

A limited usage of optimization techniques can be found in *Repair-Most* [6]. In this technique, row-/column-fault counters are employed to determine spare allocations. Repair-Most is implemented in a two-stage algorithm: Must-Repair and Final-Repair. *Must-Repair* determines either the row, or the column that must be replaced by a fault free spare to repair the maximum number of faulty bits. This process is iteratively repeated until no more faulty bits are left uncovered in the memory array by using the spares. This corresponds to a *maximization* criterion in the row/column selection and allocation. A minimization in the allocation of spares can be accomplished by an initial covering of faulty bits. This information is supplied to *Final-Repair* to find a balanced time allocation for the desired repair-solution. This is accomplished by considering processing time, laser time and spare utilization [8]. Although this approach gives better results than the Broadside approach, undesirable features like inability to provide repair-solutions for certain devices and no provision for user-defined preferences, are still present. An example of the inability of Repair-Most to repair a device is shown in Figure 2 [6,8]. Circled vertices identify faulty bits. It has been proven that this fault pattern is unrepairable using Repair-Most. The main disadvantage of Repair-Most is the inflexibility in the generation of the repair-solution. This is caused by the restrictive selection provided by the contents of the counters. Also, the maximization criterion does not necessarily guarantee an efficient repair.

The *Fault-Driven* approach [4] partially avoids the drawbacks of [6]. In the Fault-Driven approach, repair-solutions are generated according to user-defined preferences. Repair is implemented using a two-stage analysis: Forced-Repair and Sparse-Repair. Fault-counters are still employed. *Forced-Repair* determines specific rows or columns that must be replaced by redundant copies.

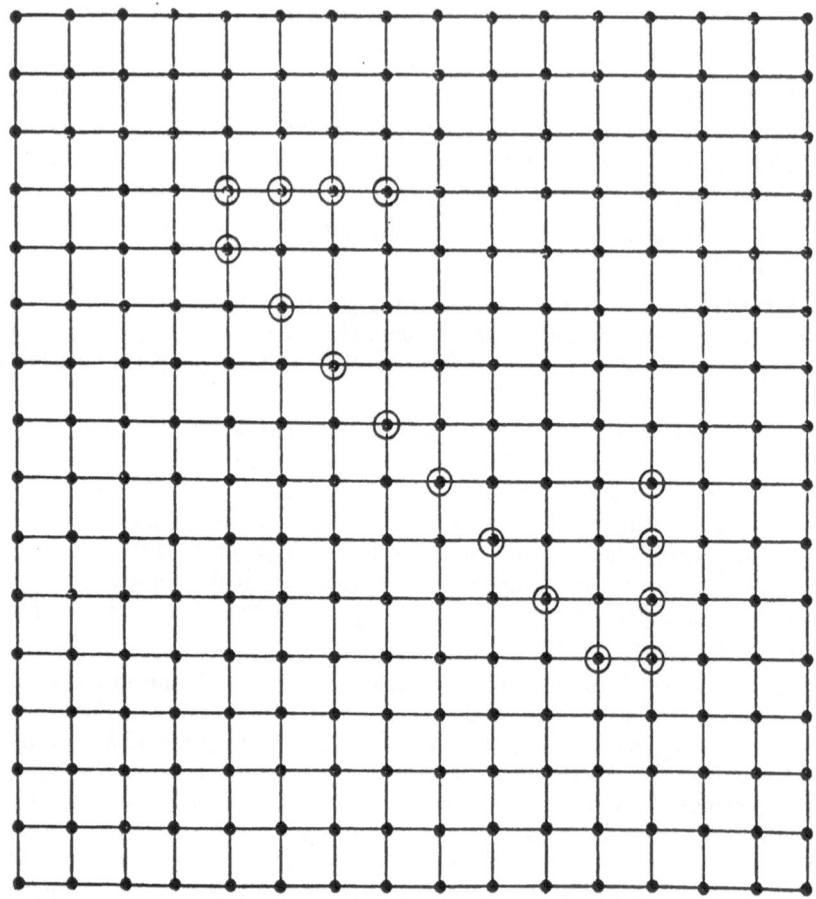

Figure 2.

Sparse-Repair determines repair-solutions for all remaining faulty bits at completion of Forced-Repair. A solution exists in Forced-Repair, if and only if each of the fault-counters exceeds the spare availability, i.e. a column (row)-fault counter is greater than C_A (R_A). This corresponds to a condition in which repair is not possible, if no spare column (row) is utilized to reduce the number of unrepaired faulty bits. A number of iterative passes over row- and column-fault counters is required for completion of Forced-Repair.

If Forced-Repair is not successful in repairing all faulty bits, *Sparse-Repair* is evoked. A solution database is generated to update and expand the records of all possible repair-solutions. Sparse-Repair is based on invalidating the records generated by Forced-Repair. A record is invalid if either the repair algorithm attempts to use more than the available number of spares, and/or it is determined that a particular spare is not required. Invalidation is implemented by a solution optimization routine. This routine is executed if at least one existing solution record already repairs a faulty bit [4]. The Fault-Driven approach operates as a greedy algorithm. All possible repair-solutions are generated. Selection of a repair-solution depends on the desired objectives set by the provided redundancy.

The most significant advantage of the Fault-Driven approach lies in user selection of repair-solutions. This allows an operator to select a repair-solution. This usually results in a significant saving of repair-time, since no efficient automated program is available. Optimality is maintained, because only solutions to repairable devices, that have successfully passed Sparse-Repair analysis, can be judged. This relieves the user of unnecessary information. However, the large number of records which are generated, requires a large computational overhead which if either the size of the array, or the redundancy is large, can result in unacceptable costs.

Two new algorithms for the repair of RRAMs have been proposed in [22]. The first algorithm uses a Branch-and-Bound approach with early screening in the repair process. A bipartite graph approach is used to obtain the least required number of spare rows and columns to repair the memory array. This is given by the minimum vertex cover [12] and it can be solved in polynomial time. Final analysis is based on a Branch-and-Bound algorithm. This algorithm is rather easy to implement and it generates an optimal solution whenever one exists. A lower number of records than the algorithm of [4] is also accomplished using a similar cost function. The second algorithm of [22] uses a heuristic criterion. The proposed approach is based on an approximation of vertex cover for general graphs [2,3]. First a row (or a column) with a single faulty cell, whose corresponding column (row) has the lowest cost, is selected. The corresponding spare is allocated. If none exists, a row or a column with minimum value of cost divided by the number of faulty cells, is allocated. Costs are readjusted and the process continues until all faulty cells are repaired. The execution time of this approach is $O((R_A + C_A)V))$, where V is the sum of the vertices in the bipartite graph. Although optimal solutions cannot be guaranteed, the results are generally good. A significant reduction in repair-time is achieved for large memory arrays. The main disadvantage of this algorithm is the still rather high polynomial time complexity (V can be a rather large number).

4. EXISTING SCHEMES.

One of the most commonly used technique for repairing RRAMs consists of the detection of *unrepairability* [5]. Initially, a test is performed on the bits along the *major diagonal* of the memory. This provides the initial information to establish if spare availability is sufficient for repair. No bit along the major diagonal shares either a row, or a column with any other remaining bits in memory. Each of these faulty bits requires the allocation of a separate spare. The die is found to be unrepairable [6,7] if the number of diagonal faulty bits exceed the number of spares. A further detection technique must be used to supplement the conditions found by diagonal testing and to detect the unrepairability of a device. This detection process utilizes *region-fault-totalizers* [6.7]. Region fault totalizers are counters, that record the number of faulty bits within regions of a memory die. The memory is fully tested and the content of the totalizer is appropriately adjusted. The *total number of faulty bits* that can occur in a die prior to repair, can be compared for the worse case analysis with the content of a totalizer. Repair can take place if this limit is not exceeded.

If a diagonal test and the totalizer yield a repairability condition, then repair can continue. Suppose that the numbers of unused spare rows and columns are less than the provided redundancy. Let the number of faulty bits left unrepaired be either less than, or equal to the maximum number of faulty bits in a die. Then the die may be repairable. If this condition is not satisfied, then the die is definitely unrepairable. Further repair techniques such as Sparse-Repair, are required. This is a time consuming process in the repair of a RRAM, because the relatively sparse pattern of faulty bits must be analyzed within a usually very large memory die. Determination of repairability (i.e. the conditions by which the faulty bits in a die can be repaired using the repair-solution) can also create more overhead.

Consider the fault pattern shown in Figure 3, $C_A = R_A = 3$. The number of spare rows and columns is insufficient to repair this die [8]. This array is unrepairable. However, no information on repairability can be derived by using diagonal testing [6], because no faulty bit appears on the major diagonal. Detection of unrepairability would also fail by using a fault totalizer. The total number of faults in Figure 3 is 14 and this does not exceed the theoretical limit of 87 faulty bits that is theoretically possible by utilizing a totalizer. If Forced-Repair is executed, no improvements for unrepairability detection would be accomplished. Spares are still unutilized in Forced-Repair, because the maximum number of faulty bits exceeds the number of faulty bits detected in the fault pattern. Sparse-Repair must be executed to efficiently terminate repair. This process has exponential complexity because all possible combinations in the use of spares must be generated. Only after a large computational overhead, the unrepairability of the memory die is found. This overhead is very severe, because if $R_A = C_A = 0$, Sparse-Repair [6] takes only linear time with respect to the number of bits.

5. REPAIR AND REPAIRABILITY.

A new approach to memory repair has been proposed in [5]. This approach consists of establishing the repairability/unrepairability conditions of a memory

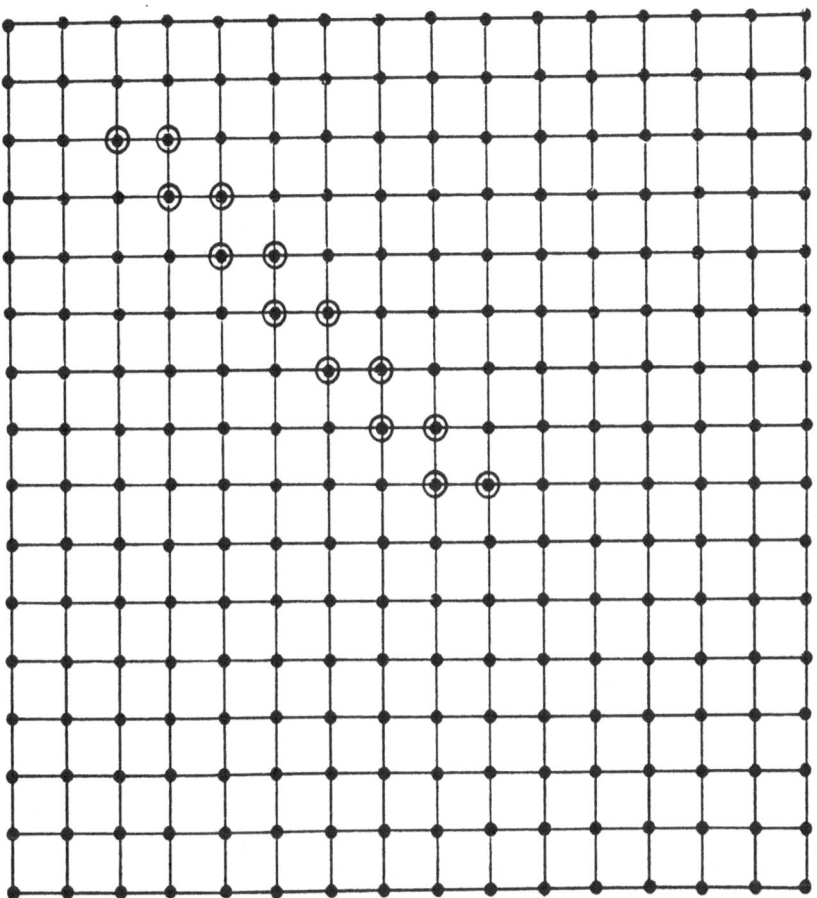

Figure 3.

die prior to finding its repair-solution. *Repairability* (unrepairability) is defined as the probability of *correctly diagnosing* a memory die as repairable (unrepairable) *prior to* executioning the repair algorithm. A brief description of the approach of [5] will follow as applicable to the proposed algorithms.

The first stage of the repair algorithm of [5] is similar to Repair-Most, because it establishes repairability detection using fault counters. Repairability detection consists of calculating the total fault count [7]. The *maximum number of faulty bits* that permit repair can be compared with the total fault count. If this limit is not exceeded, repair can take place. Detection of unrepairability by a fault count can be extended to repairability. This should not be mislead with repair, because the content of a counter does not account for the locations of the faulty bits. An alternative solution for establishing repairability is to execute one of the algorithms of [6]. It is evident [5] that unrepairability should be *determined together* with the repair-solution if a low repair-time is required. Uniqueness conditions in the allocation of spares for memory repair have been presented in [5]. These conditions apply to those faulty cells that satisfy two selection criteria. The first criterion is that no faulty bit shares either a row, or a column with any other faulty bit. The second criterion is that the number of these faulty bits should not exceed the number of spare rows and columns.

It has been proved that the occurrence of faulty bits in a memory array can determine a repair process that can be uniquely described in most cases by the above conditions [5]. This characterization is very important, because repair is independent of the locations of faulty bits. This can be used to define an unrepairable memory as a memory in which there exists a number of faulty bits that do not share either a column and/or a row with any other faulty bit and their number exceeds the number of spare rows and columns. Equivalently, a memory is R_A/C_A-repairable if there exists a set of rows and columns (referred to as *repair-solution*), such that all nonzero elements of the diagnostic matrix are covered. This implies that repair in [5] is a *covering process* [12]. This characteristic can be used to derive the repair-solution of the memory die. However, an approximation to the repair-solution can be initially found at low overhead. This is required not to incur in a large and rather wasteful repair-time if the memory is unrepairable. This approximation is given by the *repair-set*. The repair-set is denoted by S and it can be constructed as follows. Initially, the leading element of A is selected. All entries along the row and column of the leading element are replaced with zero entries. This process is repeated until the diagnostic matrix is a zero matrix.

Selection in the covering of a leading element is based on spare availability. For a_{ij} if either the row or column redundancy has been fully utilized, the leading element is defaulted in the repair-set to either a_{0j}, or a_{i0} (the index corresponding to the unavailable spare is given by 0). The repair-set is constituted by S_R and S_C, i.e. the sets of rows and columns which cover the leading elements. R_A/C_A-repairability has been established in [5] as function of the cardinality of the repair-set and the number of spare rows and columns.

The advantage of this approach [5] is that it establishes the repairability of a memory die by only checking the *cardinality* of the repair-set. This condition is *independent of the distribution of the faulty bits*. The benefit of this approach is that unrepairability can be established in parallel with the repair-set and the repair-solution. This is advantageous, because computational overhead is reduced in most circumstances [5]. If the condition of unrepairability is met, testing and repair processes can be aborted, yielding a low repair-time. The repair-solution is derived by refinement of the repair-set using *complements*. A column complement for a_{ij} is defined as a nonzero element, say a_{ik}, such that the kth column $C_k \notin S_C$. Similarly, a nonzero element a_{lj} is a row complement of a_{ij}, if the lth row $\notin S_R$. The set of rows and columns required to cover all faulty bits (i.e. the repair-solution) can be found by using the repair-set found previously and complementing all of its elements. Those rows and/or columns whose covered bits are repaired using other elements of S, can be deleted from the repair-set. The repair-solution denoted as S''_{FC}, is therefore given by a refinement of S. This approach is referred as *repair by full covering* due to the covering characteristics of repair and the generation of the repair-set. It has been proved that the approach of [5] provides good repairability/unrepairability detection at a rather modest algorithmic complexity.

6. FAULT COUNTING.

The approach of [5] can be extended to fault counting, i.e. using fault counters to determine the elements of the repair-solution from the repair-set. Fault counters are commonly used in current defect analysis systems. Fault counting is an attractive technique, because it has a fast execution time. This can provide good repair-solutions if the contents of the counters are appropriately manipulated.

If fault counting [6] is used, repairability using the approach of [5] is characterized by the following Theorem.

THEOREM 1.

Let $\max\{R_A, C_A\} < |S| \le R_A + C_A$. Let each element (i,j) of S have either $C(R_i)$ and/or $C(C_j)$ be greater than 1. The memory is R_A/C_A repairable if

$$r' = |R'| = |\{(i,j) \in S \mid C(R_i) > 1\}| \qquad (02)$$

and/or

$$c' = |C'| = |\{(i,j) \in S \mid C(C_j) > 1\}| \qquad (03)$$

where $C_j \in S_C$ and $R_i \in S_R$ and

$$r' \le R_A \qquad (04)$$

and

$$c' \le C_A \qquad (05)$$

PROOF:

In the repair-solution, those rows that correspond to elements of R' and those columns that correspond to elements of C', are selected by construction of the repair-set S. All the nonzero elements are covered. r' and c' do not exceed R_A and C_A. The die is R_A/C_A-repairable by definition. $\square$

Theorem (1) can be used either in the Algorithms of [5] to establish the repairability of a memory die. This new algorithm is similar to the repair Algorithm of [5] and will not be explicitly given. The only difference is the inclusion of a repairability condition to satisfy Theorem (1).

Consider the fault pattern shown in Figure 4, $S = \{(1,1),(2,5),(3,7),(5,2),(7,3)\}$. The row-fault counts are: $C(R_1) = 4$, $C(R_2) = 3$, $C(R_3) = 1$, $C(R_5) = 1$ and $C(R_7) = 3$. The column-fault-counts are: $C(C_1) = 4$, $C(C_5) = 1$, $C(C_7) = 3$, $C(C_2) = 3$ and $C(C_3) = 1$. This implies that $r' = 3$ and $c' = 3$; $r' = R_A$ and $c' = C_A$. The die is repairable. The repair-solution using full covering with fault counting is given by $R''_{FC} = \{R_1,R_2,R_7\}$ and $C''_{FC} = \{C_1,C_2,C_7\}$.

7. REDUCED COVERING.

Spare availability is not fully utilized by the repair algorithm of [5]. If either row or column redundancy, or no overall redundancy is available, a row and a column are added to the repair-set for each leading element. This is independent of the covering property of the leading element, i.e. for a_{ij} either R_i, or C_j can cover this faulty bit. This selection is refined when the repair-solution is derived by using complements. A suitable use of fault counters can avoid this characteristic of the elements of the repair-set.

The new selection criterion for element membership in the repair-set is given by the content of the fault counters of the row and column of each leading element. For a_{ij} either a row and/or a column are added to the respective row and column sets in the repair-set if and only if $C(R_i)>1$ and/or $C(C_j)>1$. This criterion has the following implications on the generation of the repair-set.

I. An element in the repair-set (row or column) can cover more than a single faulty bit. This is initially similar to Repair-Most.

II. If either one or both of the contents of the fault counters are 1, then the faulty bit is covered by only one spare (either a row or a column). Selection of a spare in this case is governed by the same criteria as in [5].

This criterion defines a new covering approach to repair redundant memories. This approach is referred as *reduced covering*, because leading elements in the diagnostic matrix are not fully covered as in [5].

The repair-set which is found using reduced covering, is denoted as S'^r. This repair-set must still preserve the repairability conditions set by [5]. Let R'_A and C'_A denote the number of unutilized spare rows and columns available for selection by the repair algorithm and S'^r_R and S'^r_C be the row and column set in S'^r i.e. $|S'^r| = \max\{|S'^r_R|,|S'^r_C|\}$. The following conditions establish the covering selection in the repair-set.

A) If R_A, $C'_A > 0$ and $C(R_i) > 1$ $C(C_j) > 1$, $R_i \epsilon S'^r_R$ and $C_j \epsilon S'^r_C$.

B) If R'_A, $C'_A > 0$ and $C(R_i) > 1$ $(C(R_i) \leq 1)$ $C(C_j) \leq 1$ $(C(C_j) > 1)$, $R_i \epsilon S'^r_R (C_j \epsilon S'^r_C)$.

C) If R'_A, $C'_A > 0$ and $C(C_j) \leq 1$ $C(R_i) \leq 1$, $C_j \epsilon S'^r_C (R_i \epsilon S'^r_R)$ provided that $C'_A > R'_A (R'_A > C'_A)$. If $C'_A = R'_A$, $C_j \epsilon S'^r_C (R_i \epsilon S'^r_R)$, provided that $|S'^r_C| > |S'^r_R| (|S'^r_C| < |S'^r_R|)$. If $C'_A = R'_A$ and $|S'^r_C| = |S'^r_R|$, $C_j \epsilon S'^r_C$ and

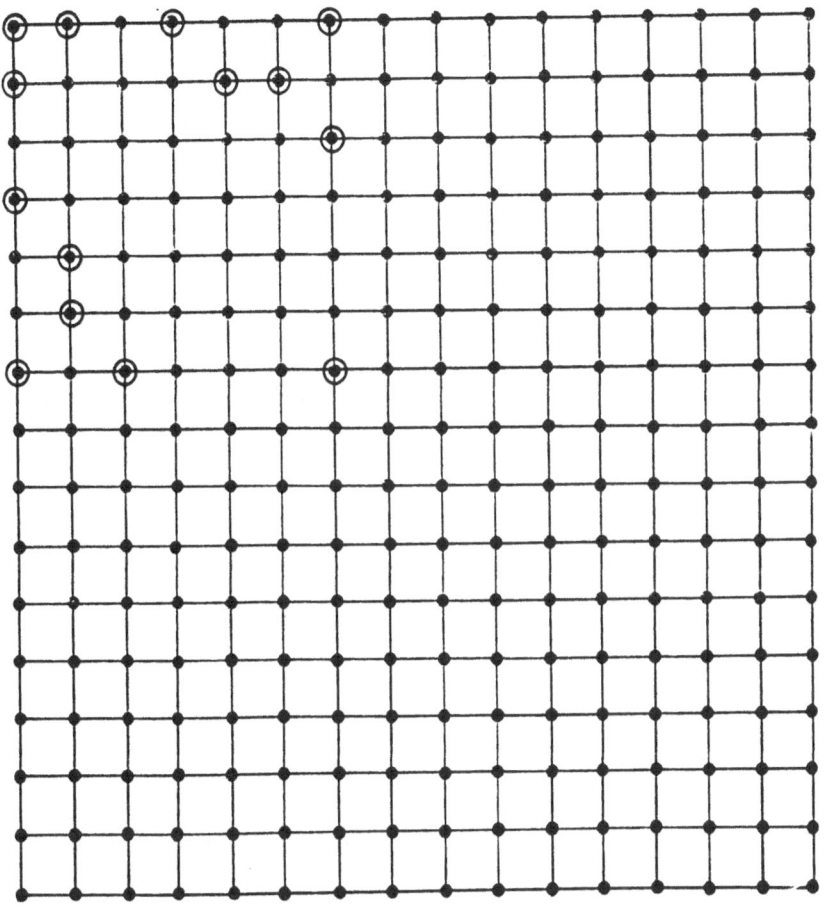

Figure 4.

$R_i \in S_R^{'r}$.

D) If $R_A' > 0$ and $C_A' = 0$, $R_i \in S_R^{'r}$.

E) If $C_A' > 0$ and $R_A' = 0$, $C_j \in S_C^{'r}$.

F) If $R_A' = C_A' = 0$, $R_i \in S_R^{'r}$ and $C_j \in S_C^{'r}$.

The above conditions rely on the contents of the fault counters of the leading elements and the availability of spare resources for the selection of rows and columns in $S_R^{'r}$ and $S_C^{'r}$. Hence, this is a *fault counting* technique. The following extreme scenarios are possible.

I. $R_A = C_A$. Hence, initially $R_A' = C_A'$ by definition. For totally random faulty cells, each leading element a_{ij} has much likely $C(R_i) = C(C_j) = 1$. C_j and R_i are added to the repair-set by rule (C).

II. If spares are exhausted, the repair-set is generated as in [5] by rule (F).

III. If the number of leading elements in the diagnostic matrix A is denoted by $|L|$, $|L| = \sqrt{NM}$, by rules (A)-(F) in a repairable memory

$$|L| \le |S_R^{'r}| + |S_C^{'r}| \le 2|L| . \tag{06}$$

If $R_A = C_A$, in a repairable memory $|L| \le R_A$.

Consider the fault pattern shown in Figure 5. The first leading element is faulty bit (2,3). For this cell $C(R_2) = 2$ and $C(C_3) = 1$. If $R_A = C_A = 3$, then $R_2 \in S_R^{'r}$, but $C_3 \notin S_C^{'r}$.

8. REPAIR-SET GENERATION.

Reduced covering follows the same rules as the approach of [5] to generate the repair-solution from the repair-set. A reduction in repair-time can be accomplished by generating initially a repair-set [5]and establishing repairability/unrepairability. The algorithm for generating the repair-set by faulty counting is as follows.

PROCEDURE 1: REPAIR-SET GENERATION BY FAULTY COUNTING.

Step 1: Initialize, $S_R^{'r} = \{\}$; $S_C^{'r} = \{\}$; $flag = 0$; $var = 0$; $R_A' = R_A$; $C_A' = C_A$. If $R_A' = 0$, $flag = 1$; if $C_A' = 0$, $flag = 2$.

Step 2: Find the leading element of the diagnostic matrix A, say a_{ij}.

Step 3: Establish row/column selection using conditions (A) through (F) of Section 7.

Step 4: If only R_i has been selected and $flag \ne 1$, $var = 1$. If only C_j has been selected and $flag \ne 2$, $var = 2$. If both R_i and C_j have been selected, $var = 3$.

Step 5: If $var \ne 1$, go to Step (6). Otherwise, $S_R^{'r} = S_R^{'r} \cup \{R_i\}$. Reset all 1-entries along R_i. Adjust all column fault counters. Go to Step (8).

Step 6: If $var \ne 2$, go to Step (7). Otherwise, $S_C^{'r} = S_C^{'r} \cup \{C_j\}$. Reset all 1-entries along C_j. Adjust all row fault counters. Go to Step (9).

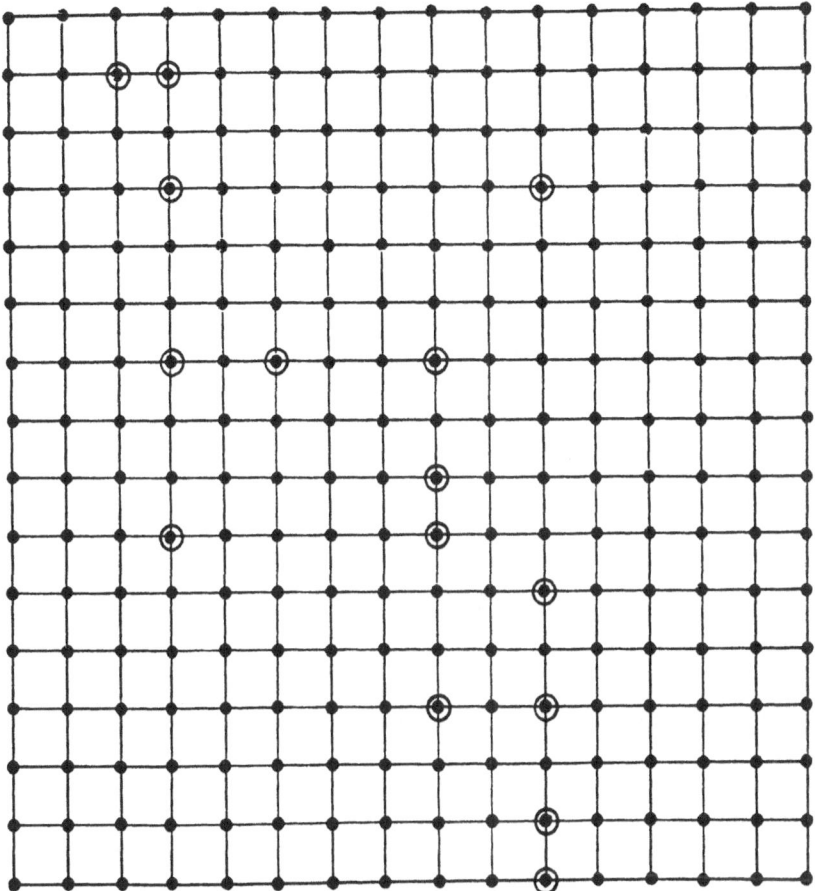

Figure 5.

Step 7: $S_R^{'r} = S_R^{'r} \bigcup \{R_i\}$, $S_C^{'r} = S_C^{'r} \bigcup \{C_j\}$. Reset all 1-entries along R_i and C_j. Adjust all counters; $R_A' = R_A' - 1$ and $C_A' = C_A' - 1$. Go to Step (10).

Step 8: $R_A' = R_A' - 1$. Go to Step (10).

Step 9: $C_A' = C_A' - 1$.

Step 10: If $R_A' = 0$, $flag = 1$.

Step 11: If $C_A' = 0$, $flag = 2$.

Step 12: If $C_A' = R_A' = 0$, $flag = 3$.

Step 13: If the contents of all fault counters are not zero, go to Step (2).

Step 14: The repair-set by fault counting has been generated.

Consider the fault pattern shown in Figure 6. If $R_A = C_A = 4$, $S_R^{'r} = \{R_2, R_5, R_8, R_{10}, R_{16}\}$, $S_C^{'r} = \{C_3, C_7, C_8, C_{12}\}$. For the pattern shown in Figure 5, if $R_A = C_A = 3$, $S_R^{'r} = \{R_2, R_4, R_7\}$ and $S_C^{'r} = \{C_4, C_9, C_{11}\}$.

The repair-set found by Procedure (1) is characterized by the following Theorem.

THEOREM 2.

The repair-set found by Procedure (1) has lower or equal cardinality than the repair-set found by the Algorithm of [5], i.e. $|S_R^{'r}| \leq |S_R|$ and $|S_C^{'r}| \leq |S_C|$.

PROOF:

This Theorem will be proved by contradiction. Assume that $|S_R^{'r}| > |S_R|$ and $|S_C^{'r}| > |S_C|$. This implies that a row and a column are always selected to cover a leading element a_{ij}. This is the covering condition of [5]; hence $|S_C^{'r}| = |S_C|$ and $|S_R^{'r}| = |S_C|$. This is the worst case condition, because if $C_A' \neq 0$ and $R_A' \neq 0$ the covering algorithm of [5] will assign for a leading element a_{ij} both the row and the column to the repair-set independently of the contents of the fault counters, i.e. $R_i \epsilon S_R$ and $C_j \epsilon S_C$. This is not generally the case with Procedure (1): only one element will be added to the repair-set if a fault counter of the leading element is 1. Hence, $|S_R^{'r}| < |S_R|$ and/or $|S_C^{'r}| < |S_C|$. The Theorem is proved. $\square$

Consider the fault pattern of Figure 5, $S_R = \{R_2, R_4, R_7, R_9, R_{11}\}$ and $S_C = \{C_3, C_4, C_6, C_9, C_{11}\}$. Hence, $|S_R| = 5 > |S_R^{'r}| = 3$ and $|S_C| = 5 > |S_C^{'r}| = 3$. Theorem (2) is satisfied.

9. REFINEMENT CRITERIA.

The repair-solution in [5] is found by refining the repair-set through complementing all its elements and finding those columns and rows which are essential in covering all faulty cells. It has been proved that this process is very effective in determining the repair-solution, but it has quadratic execution time [5].

The elements of the repair-set can be distinguished according to three criteria.

399

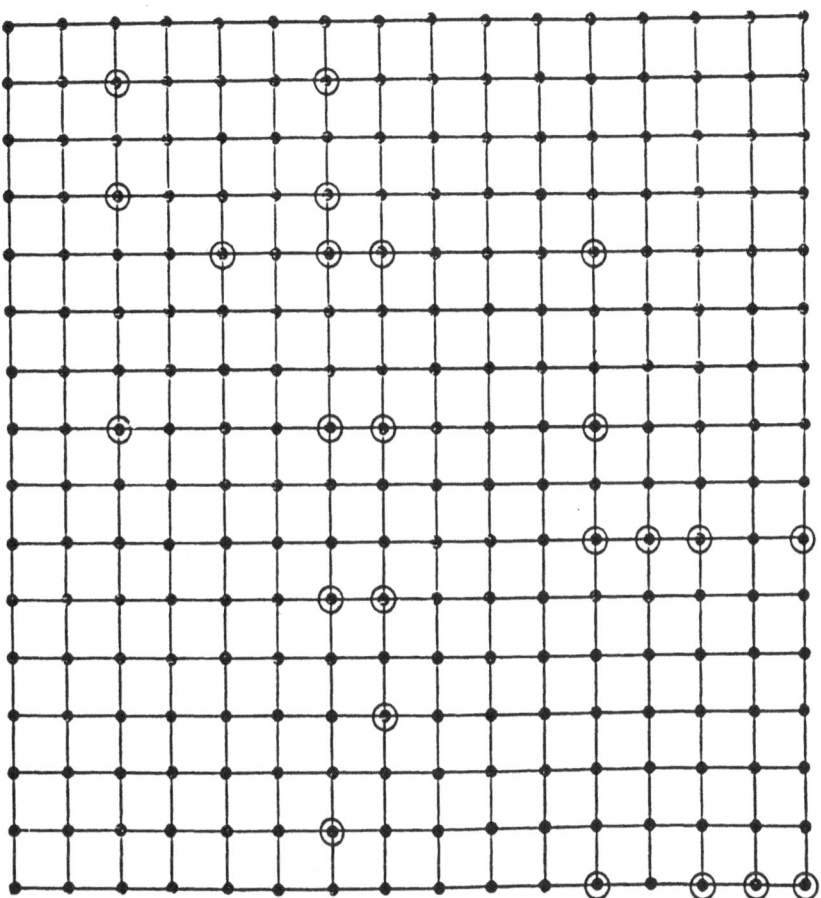

Figure 6.

I. An element of either $S_R'^r$ or $S_C'^r$ is said to be *redundant*, if all faulty bits covered by this element can be also covered by other elements of the other repair-set.

II. An element of either $S_R'^r$ or $S_C'^r$ covers at least a faulty bit that no other element in the repair-set can cover. This element is referred as a *primary element*.

III. An element of either $S_R'^r$ or $S_C'^r$ can cover some (but not all) faulty bits which are also covered by other elements. This element is referred as a *complementary element*.

Consider the pattern of Figure 5. R_4 is a redundant element, because bits (4,4) and (4,11) can be covered by C_4 and C_{11} also; hence, C_4, $C_{11} \in S_C'^r$. R_2 is a primary element, because bit (2,3) cannot be covered by C_3, $C_3 \notin S_C'^r$.

All primary elements must be in the repair-solution because of (II) and the covering properties of repair [5]. If $R_i(C_j)$ is a primary element, R_i must belong to the row set in the repair-solution (C_j must belong to the column set of the reconfiguration-solution).

Complementary elements can be selected to belong to the repair-solution by finding the *repair-factor*. Let $R_i \in S_R'^r$ and $S(R_i)$ denote the set of columns that belongs to $S_C'^r$ such that all faulty bits along R_i can be repaired using these columns. $S(R_i)$ is referred as the *complement set* of R_i. The repair-factor is denoted as

$$\gamma(R_i) = N_R - N_{SR} \tag{07}$$

where N_R is the number of distinct rows (except R_i) which appear in all complement sets $S(C_h)$ for every C_h in $S(R_i)$. N_{SR} is the number of times the elements of the row set in the reconfiguration-solution appear in all complement sets $S(C_h)$ for every C_h in $S(R_i)$. Equivalently, for $C_j \in S_C'^r$

$$\gamma(C_j) = N_C - N_{SC} \tag{08}$$

where N_C and N_{SC} are defined in a manner similar to N_R and N_{SR} by changing the appropriate quantities. The repair-factor introduces a heuristic criterion in the selection of the elements of the repair-solution. Redundant elements can be immediately deleted from the repair-set if and only if the faulty bits which are covered by these elements, can be also covered by primary elements. If by deleting certain rows and/or columns some faulty bits are uniquely covered by complementary elements, these elements become automatically primary by definition.

Consider Figure 7. The repair-set is given by $S_R'^r = \{R_2, R_4, R_5, R_6, R_7\}$ and $S_C'^r = \{C_1, C_2, C_5, C_7, C_9\}$. The primary elements are R_5 and C_9. Hence, $R_5 \in S_R''$ and $C_9 \in S_C''$. The repair-factor of R_2 is $\gamma(R_2) = 4-1 = 3$, where $S(C_3) = \{R_4\}$ and $S(C_5) = \{R_2, R_5, R_8\}$.

The repair-set found by Procedure (1) still satisfies the repairability/unrepairability conditions given in [5]. However, new conditions can be found by inspecting the contents of the fault counters as relating to a fault counting technique. These conditions are applicable after the execution of

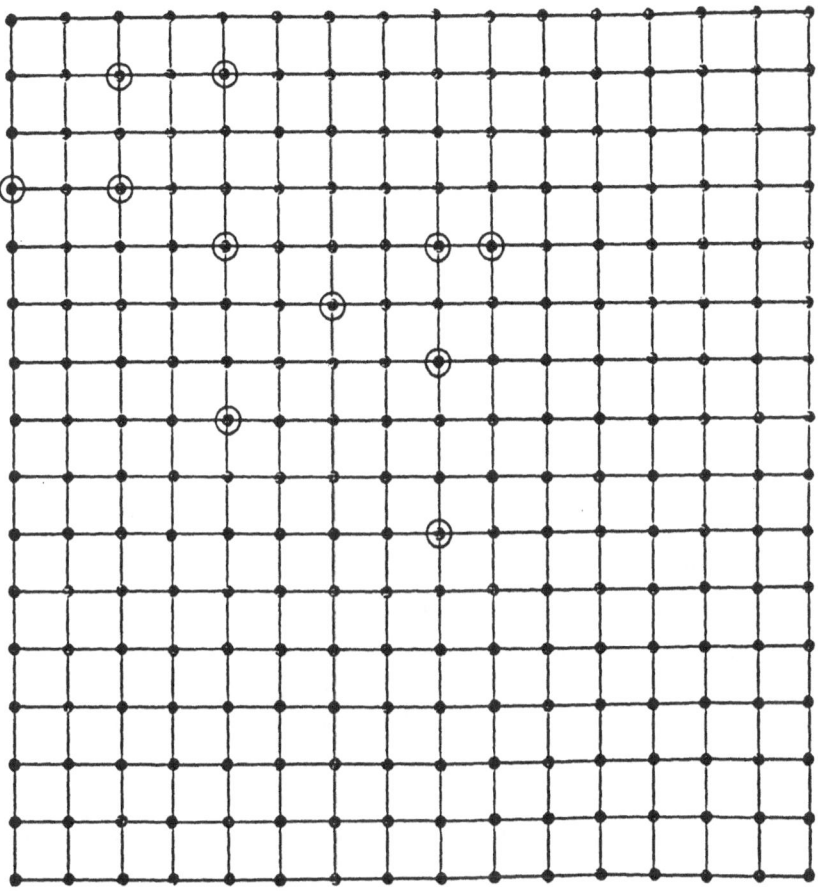

Figure 7.

Procedure (1). They relate the contents of the fault counters to the elements of the repair-set. The conditions of repairability/unrepairability in a fault counting technique use only the contents of the fault counters, not the locations of the faulty bits. Hence, these conditions are very conservative and they are not strictly applicable if the locations of faulty bits were directly known. This however results in a process with low complexity for detecting repairability/unrepairability, because only the contents of the fault counters (whose number is related to the dimensions of the die) are considered.

Let the row (column) fault counters of all $R_i \in S_R'^r (C_j \in S_C'^r)$ be sorted in a list $L_R (L_C)$ of ascending order, i.e. $|L_R| = |S_R'^r|$ and $|L_C| = |S_C'^r|$. The following Lemma establishes unrepairability.

LEMMA 1.

Let C^z denote the number of columns in $S_C'^r$ for which $C(C_j) > |S_R'^r| \geq R_A$ for every $C_j \in S_C'^r$. A memory die is not repairable, if

$$C^z > C_A \tag{09}$$

PROOF:

The condition

$$C(C_j) > |S_R'^r| \geq R_A \tag{10}$$

implies that if C_j is not selected in the repair-solution, more than R_A spare rows are required to cover the faulty bits which lie along C_j. If the number of deleted columns is in excess of the provided column redundancy, the memory is not repairable by definition [5]. Hence the Lemma is proved. □

THEOREM 3.

Let $|S_R'^r| > R_A$, $|S_C'^r| \leq C_A$ and $|S_R'^r| > |S_C'^r|$. A memory die is unrepairable if

$$C(R_i) > |S_C'^r| \tag{11}$$

where R_i is the row corresponding to the (R_A+1)th entry in L_R.

PROOF:

If $|S_C'^r|$ is no more than the provided column redundancy, all columns in $S_C'^r$ must be in the repair-solution. However, these columns must be used to cover faulty bits which are already covered by some of the rows in $S_R'^r$. This is required to reduce the cardinality of $S_R'^r$ to be at least equal to the provided row redundancy [5]. The first R_A row entries L_R can be used for covering. The remaining entries of L_R must be covered using the spare columns in $S_C'^r$. If (11) is satisfied, this implies that the number of columns in the repair-solution is not enough to cover all faulty bits along R_i. There exists a faulty bit a_{ij} in which $C_j \notin S_C'^r$. The memory is unrepairable by definition. □

THEOREM 4.

Let $|S_R'^r| > R_A$ and $|S_C'^r| < C_A$. A memory is repairable if

$$\sum_i C(R_i) + |S_C'| \leq C_A \tag{12}$$

for all R_i corresponding to the last $(|S_R'| - R_A)$ entries in L_R and

$$|S_R'| \leq R_A \qquad (13)$$

where S_R' is now given by only all those rows R_j corresponding to the first R_A entries in the list L_R.

PROOF:

The hypothesis implicitly assumes that some columns in the provided redundancy are not utilized. The unutilized columns can be used to cover faulty bits which are covered by some of the elements in the row set of the repair-set. These columns can be used to reduce $|S_R''|$ to meet the repairability condition of [5]. Obviously, the first R_A entries of L_C can be utilized by using the provided R_A spare rows. The remaining entries in L_R can be deleted using the unutilized spare columns. (12) is proved, because no distinction with respect to the elements of those rows left in L_R can be made by looking at the contents of the fault counters. The overall number of ultimately utilized columns in a repairable memory die cannot exceed the provided column redundancy [5]. The Theorem is proved. □

Lemma (1) and Theorems (4) and (3) are also applicable to column redundancy by interchanging the appropriate quantities. Note that complexity of unrepairability/repairability detection using fault counting is $O(\log|S'r|)$, because sorting is required to generate L_R and L_C and upgrading of the fault counters is $O(1)$ (by assumption).

Consider Figure 7. If $R_A = C_A = 3$, $L_R = \{C(R_5),\ C(R_2),\ C(R_4),\ C(R_6),\ C(R_7)\}$ and $L_C = \{C'(C_9),\ C'(C_5),\ C'(C_3),\ C'(C_7),\ C'(C_1)\}$. $C^z = 2$, hence $C^z < C_A = 3$. The memory die is not unrepairable.

10. THE REPAIR-SOLUTION.

Let S_{RC}'' denote the repair-solution. S_{RC}'' consists of the sum of two sets, S_R'' and S_C''. Procedure (1) postulates a new procedure for generating the repair-solution.

PROCEDURE 2: REPAIR-SOLUTION BY FAULT COUNTING.

Step 1:   *flagc* = *flagr* = 0. For each $R_i \in S_R''$, find $S(R_i)$. If at least a faulty cell cannot be covered (R_i is a primary element), $S(R_i) = \{\}$; $S_R'' = S_R'' \cup R_i$, delete R_i from S_R''.

Step 2:   For each $C_j \in S_C''$, find $S(C_j)$. If at least a faulty cell cannot be covered (C_j is a primary element), $S(C_j) = \{\}$; $S_C'' = S_C'' \cup C_j$, delete C_j from S_C''.

Step 3:   For every $R_i \in S_R''$, determine those $S(R_i)$ that can be covered by only those C_j, such that $|S(C_j)| = 0$. If so, R_i is redundant, delete R_i from S_R', $|S_R'| = |S_R'| - 1$. Adjust all those $S(C_j)$ in which R_i is a member, to zero cardinality. Set *flagr*.

Step 5:   For every $C_j \in S_C''$, determine those $S(C_j)$ that can be covered by only those R_i such that $|S(R_i)| = 0$. If so C_j is redundant, delete C_j from S_C', $|S_C'| = |S_C'| - 1$. Adjust all those $S(R_i)$ in which C_j is a member, to zero cardinality. Set *flagc*.

Step 6: If all $|S(C_j)| = 0$ and all $|S(R_i)| = 0$, go to Step (8). If either $|S_R''| = 0$ and/or $|S_C''| = 0$, go to Step (11).

Step 7: If *flagc* = *flagr* = 0, go to step (8). Else, go to Step (1).

Step 8: Calculate $\gamma(R_i)$ for every R_i in S_R''. Calculate $\gamma(C_j)$ for every C_j in S_C''.

Step 9: Sort all $\gamma(R_i)$ and $\gamma(C_j)$ in a single list by ascending order. If the largest entry is unique delete that element from the appropriate set. If C_j is selected, $|S_C''| = |S_C''| - 1$ and $S_C'' = S_C'' \cup C_j$. If R_i is selected, $|S_R''| = |S_R''| - 1$ and $S_R'' = S_R'' \cup R_i$. If not unique, calculate $\Delta_R = R_A - |S_R''|$ and $\Delta_C = C_A - |S_C''|$. If $\Delta_R > \Delta_C$ select a row and upgrade S_R'' and S_R''. If $\Delta_C > \Delta_R$ select a column and upgrade S_C'' and S_C''. If $\Delta_R = \Delta_C$, selection is based on comparing R_A with C_A.

Step 10: If after this selection, faulty cells are still left uncovered, go to Step (8).

Step 11: The repair-solution S_{RC}'' has been found.

Procedure (2) can be divided into two parts.

I. Steps (1)-(7) find redundant and primary elements of S_R'' and S_C''. Primary elements are added to the repair-solution.

II. Steps (8)-(10) use the heuristic criterion of a repair-factor to select the repair-solution out of the remaining elements of S_C'' and S_R''.

Consider the fault pattern shown in Figure 6. The faulty cells covered by deleting R_8, can also be covered by deleting C_3, C_7, C_8 and C_{12} also. These columns belong to S_C''. R_8 can be deleted from S_R'', because this is a redundant row. Hence, the rows in the repair-solution are given by $S_R'' = \{R_2, R_5, R_{10}, R_{16}\}$ through executing Steps (1)-(7).

The new Algorithm for repairing a memory by fault counting with reduced covering is as follows.

ALGORITHM 1: REPAIR BY FAULT COUNTING.

Step 1: Add the contents of either all row or column fault counters to determine the total fault counter τ.

Step 2: If τ is greater than the number of spare cells, go to Step (8).

Step 3: Execute Procedure (1) to generate the reconfiguration-set. Using the repairability/unrepairability conditions of Section 7, establish the memory status. If the memory is unrepairable go to Step (8). Else, continue.

Step 4: Execute Procedure (2) to find the reconfiguration-solution.

Step 5: If $S_C'' > C_A$, go to Step (8).

Step 6: If $S_R'' > R_A$, go to Step (8).

Step 7: The memory is repairable using the elements of S_R'' and S_C''; go to Step (9).

Step 8: The memory is unrepairable.

Step 9: Repair by Fault Counting is complete.

The memory whose fault pattern is shown in Figure 6, is repairable. This is possible due to Procedure (2) and the deletion of R_8 from S_R''. Consider the fault pattern shown in Figure 8. The repair-solution is given by $S_R'' = \{R_2, R_4, R_5\}$ and $S_C'' = \{S_5, S_7, S_9\}$. The primary elements found by Algorithm (1) are R_5 and C_9. If $R_A = C_A' = 3$ this memory is repairable. Note that this array is wrongly considered unrepairable by using the algorithm of [4].

The complexity of Algorithm (1) is $O(\min\{\sqrt{NM}, (R_A C_A)\})$, because the reconfiguration-solution is generated by using only the reconfiguration-set, not the complement generation. This complexity figure is given by either the maximum number of leading elements in the diagnostic matrix in a repairable die or the provided redundancy. Note that in the complexity analysis it has been assumed that the fault counters can be upgraded automatically (i.e. $O(1)$). Procedure (2) accounts for the largest computational overhead, because the elements of the repair-set must be selected according to their covering properties.

Algorithm (1) can be compared with the repair algorithm of [5]. This comparison is characterized by the following Lemma and Theorem.

LEMMA 2.

Let a memory be repairable by using Algorithm (1) and the repair algorithm of [5]. Then,

$$|S_{RC}''| \leq |S_{FC}''| \tag{14}$$

PROOF:

This Lemma is proved by using Theorem (1). Since $|S'_R| \leq |S_R|$ and/or $|S'_C| \leq |S_C|$, (14) is proved by construction and definition of the repair-solution. □

THEOREM 5.

If a memory is repairable using the covering algorithm of [5], it is also repairable using Algorithm (1).

PROOF:

Assumes that the reverse is true. There exists a cell a_{ij} such that $R_i \notin S_R''$ and $C_j \notin S_C''$. This contradicts the definition of reduced covering, because S' must cover all faulty elements. Hence, $R_i \in S_R''$ and/or $C_j \in S_C''$ while Algorithm (1) preserves the covering properties of repair. A repairable memory using reduced covering does not have a repair-solution of greater cardinality than the repair-solution of [5] by Lemma (2), i.e. (14) is applicable. The Theorem is proved. □

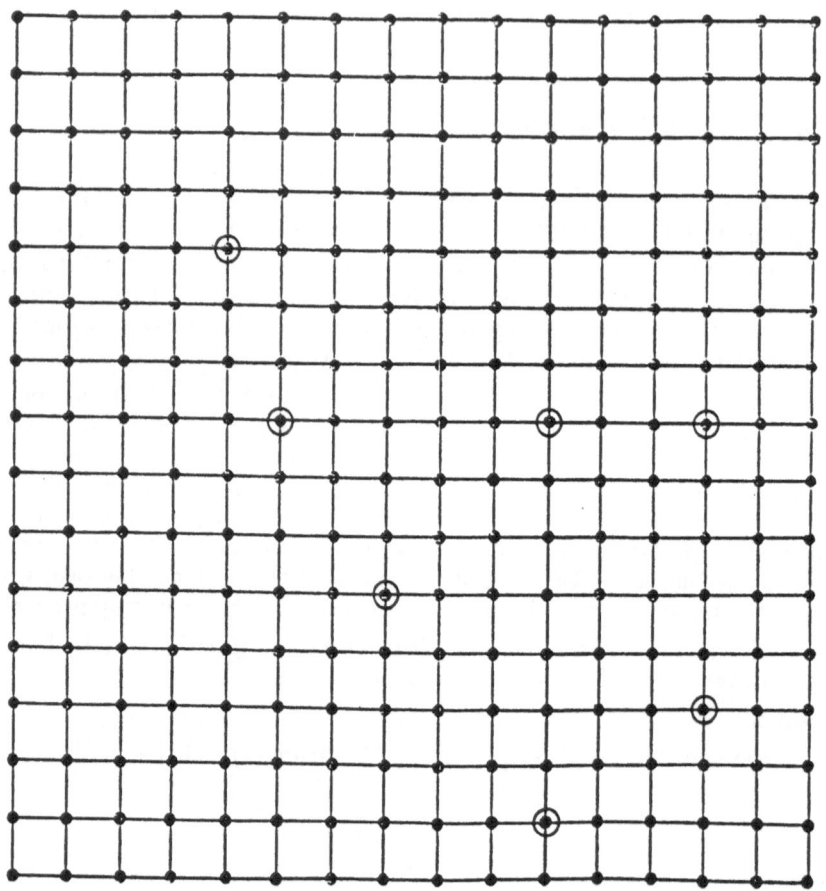

Figure 8.

11. FAULT DISTRIBUTIONS.

Distribution of faulty bits in a memory die follows different models according to the manufacturing technology [9,11,20]. A Poisson distribution (or variations of it [10,18]) is commonly assumed for VLSI devices [9]. This implies that the faulty bits are randomly distributed and failure independence is upheld. This distribution has the following implications on repair by fault counting.

I. The number of primary elements is rather high, because many isolated faulty bits are present in the memory array.

II. Fault counters do not show large variance in the counts.

The repair-set found by Procedure (1) closely resembles the repair-solution for VLSI memory dies. Reduced covering is effective provided that complementary elements can be found using repair-factors. Redundant elements can be identified by the small fault counts. Consider Figure 8, $R_A = 2$ and $C_A = 3$. Algorithm (1) finds the repair-set as $S_R^{'r} = \{R_5, R_8, R_{12}, R_{13}, R_{15}\}$ and $S_C^{'r} = \{C_5, C_8, C_{10}, C_{11}, C_{14}\}$. The primary elements are R_8 and C_8, $R_8 \in S_R^{'r}$ and $C_8 \in S_C^{'r}$. No redundant element is present. The following elements are selected using repair-factors: R_5, C_{11} and C_{14}. Hence $S_R^{''} = \{R_5, R_8\}$ and $S_C^{''} = \{C_8, C_{11}, C_{14}\}$. This memory is repairable.

The Poisson distribution is not applicable to wafer scale integration (WSI). WSI memories present failure modes in which fault dependency is introduced by the wafer [20]. Two scenarios are possible in these circumstances.

I. Presence of clusters of faulty bits within a region of the memory die. This is a contaminated region.

II. Defects in manufacturing cause faulty bits to lie along common rows and/or columns.

Reduced covering is effective for WSI, because it tends to optimize selection of rows and columns by considering initially the fault counts. Those counters that identify the fault scenarios of (I) and (II), are selected in advance, because they are primary elements in the repair-solution. In the presence of arbitrarily shaped clusters, reduced covering repairs the memory die along the major dimension of the cluster. An example is shown in Figure 9. Assume $R_A = C_A = 4$. The repair-solution is given by $S_C^{''} = \{C_5, C_7, C_8, C_9\}$ and $S_R^{''} = \{R_6, R_8, R_9, R_{10}\}$.

12. DEFECT ANALYSIS SYSTEM IMPLEMENTATION.

The repair algorithm by reduced covering can be analyzed for implementation in a defect analysis system [8,13]. The characteristics of reduced covering can be added to existing defect analysis systems, because repair relies mainly on the contents of fault counters, not necessarily on the location of faulty bits. The repairability/unrepairability conditions of Section 9 can be found by inspecting L_R, L_C and the fault counters.

An important feature for implementing the proposed approach to a defect analysis system is given by the calculation of a repair-factor. A repair-factor $\gamma(R_i)$ can be calculated as follows. $|S(R_i)|$ is equal by definition to the content of $C(R_i)$ after having added all primary elements in the repair-solution. By the same argument, every complement sets $S(C_h)$ for a C_h in $S(R_i)$ can be found. Let

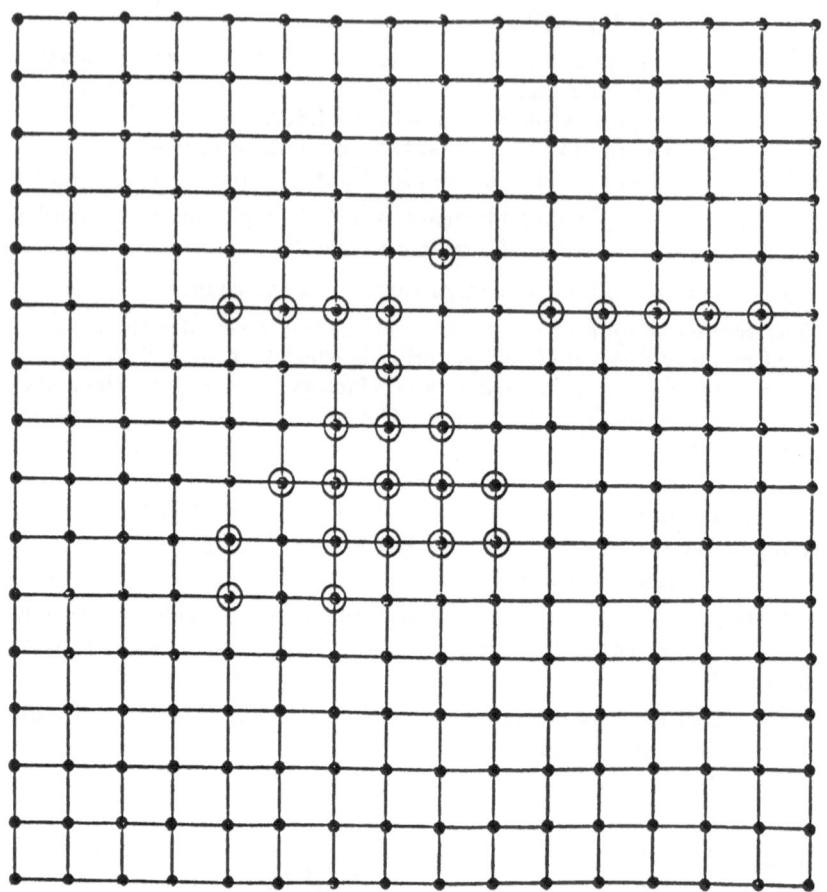

Figure 9.

each complement set $S(R_i)$ $(S(C_h))$ given by a linear boolean list of length N (M). The jth entry in a complement set $S(R_i)$ $(S(C_h))$ is 1 if $a_{ij}(a_{jh})$ is faulty; 0, otherwise. N_R is given by ORing all complement sets $S(C_h)$ and counting the number of 1 entries in the final linear list. N_{SR} can be calculated by counting the number of individual 1 entries in all complement sets $S(C_h)$ which correspond to the row elements of the reconfiguration-solution.

Calculation of the repair-factors can be accomplished at low hardware cost, because no major processing is involved. Current defect analysis systems make extensive use of fault counters [8] for memory repair. The modifications required for the proposed fault counting technique can be easily accommodated in current defect analysis systems.

13. THE FAULTY LINE COVERING APPROACH.

A new repair approach is proposed in this section. The Faulty Line Covering Approach (namely FLCA) is based on a particular covering property of a faulty line. This can be formally characterized by the following Lemma.

LEMMA 3.

A faulty line with k faulty bits can only be covered by a spare row (or column), or by k spare columns (or rows) without redundantly using spare rows (or columns).

PROOF:

If the faulty line has been covered by a spare row (or column), any other columns (or rows) which cover the same faulty bits on this line, are redundant. Equivalently, the inverse is also true. Hence the Lemma is proved. □

Lemma (3) represents an intuitive, but very important characteristic of the repair process. This has the following implications.

I. Repair by line covering can be implemented by making only two choices. This consists of deleting a faulty line either with a similar spare, or with a number of different type of spares (the number of spares is dependent on the content of the fault counter of the fault line).

II. Selection of covering is dependent only on the type of spares (row and column), not on the exhaustive combination in the use of faulty bits in the repair-solution.

FLCA uses a repair algorithm made of three parts.

I. Forced-Repair is used to cover those rows and columns in which the number of faulty bits is greater than C_A or R_A. This is the same as the Fault-Driven Algorithm.

II. Faulty-Line-Covering covers faulty lines either by a spare row (column), or by multiple columns (rows).

III. Single-Fault-Covering covers the remaining single faulty bits.

If the maximum value of the contents of the fault counters is denoted as M_F, the following Algorithm achieves repair by FLCA.

ALGORITHM 2: REPAIR BY FLCA

Step 1: Use Forced-Repair repeatedly to cover those rows and columns on which the number of faulty bits exceeds C_A or R_A. Store current values of C_A, R_A, T_F, S_F and the covered row and column numbers on a record. Use the record as an original parent.

Step 2: If $T_F > 2C_AR_A$, go to Step (7).

Step 3: For all recent parents DO

   3.1: If $T_F = S_F$, set single fault parent flag and finish treating this parent.

   3.2: Count the faulty bits in every row and column of the array. Sort all fault counters to find M_F. Select the line whose fault count is M_F.

   3.3: If $R_A \geq 1$ (or $C_A \geq 1$), cover the selected row (or column) with a spare row (or column). Add the covered row (or column) number to the parent record to generate its descendants. If $C_A \geq M_F$ (or $R_A \geq M_F$) cover the selected row (or column) with M_F columns (or rows). Add the covered columns (or rows) to the parent record to generate another descendant. The descendants will be parents for the next iteration. Record T_F, S_F, C_A and R_A for the descendants.

Step 4: If one or more descendants have been generated in Step (3), let the descendants be parents and go to Step (3).

Step 5: Select all records in which $R_A + C_A > S_F$ from the records which have the single faulty parent flag set. If no record is selected, go to Step (7).

Step 6: Using user-selection criteria, select one record from the records selected in Step (5) and cover the remaining single faulty bits (if any). The repair-solution has been found.

Step 7: The memory is unrepairable. Abort execution.

In the above algorithm, Step (1) is Forced-Repair, Step (3) is Faulty-Line-Covering and Step (6) is Single-Fault-Covering. Early-abort criteria are used in Algorithm (2): in Step (2), $T_F > 2 C_AR_A$; in Step (5), $R_A + C_A > S_F$. User-selection criteria are introduced in Step (6). These are analyzed in depth in Section 15.

Some examples are discussed to illustrate Algorithm (2). Consider the pattern shown in Figure 1, $R_A = C_A = 4$. It has been proved that this memory die is not repairable if Repair-Most is used [6]. There are 16 faulty bits in this pattern. The Fault-Driven algorithm will take a very large number of iterations to find a repair-solution. Only two iterations are required by using FLCA. This is shown in Figures 10A and 10B. * denotes the flag for single faulty parent. Note that parents B and C in Figure 10A have only one descendant each, because $M_F > R_A$ and $C_A = 0$. Parent D in Figure 10B is a single fault parent, but it cannot be used because $S_F > R_A + C_A$. Hence, the repair-solution is given by parent E in Figure 10B.

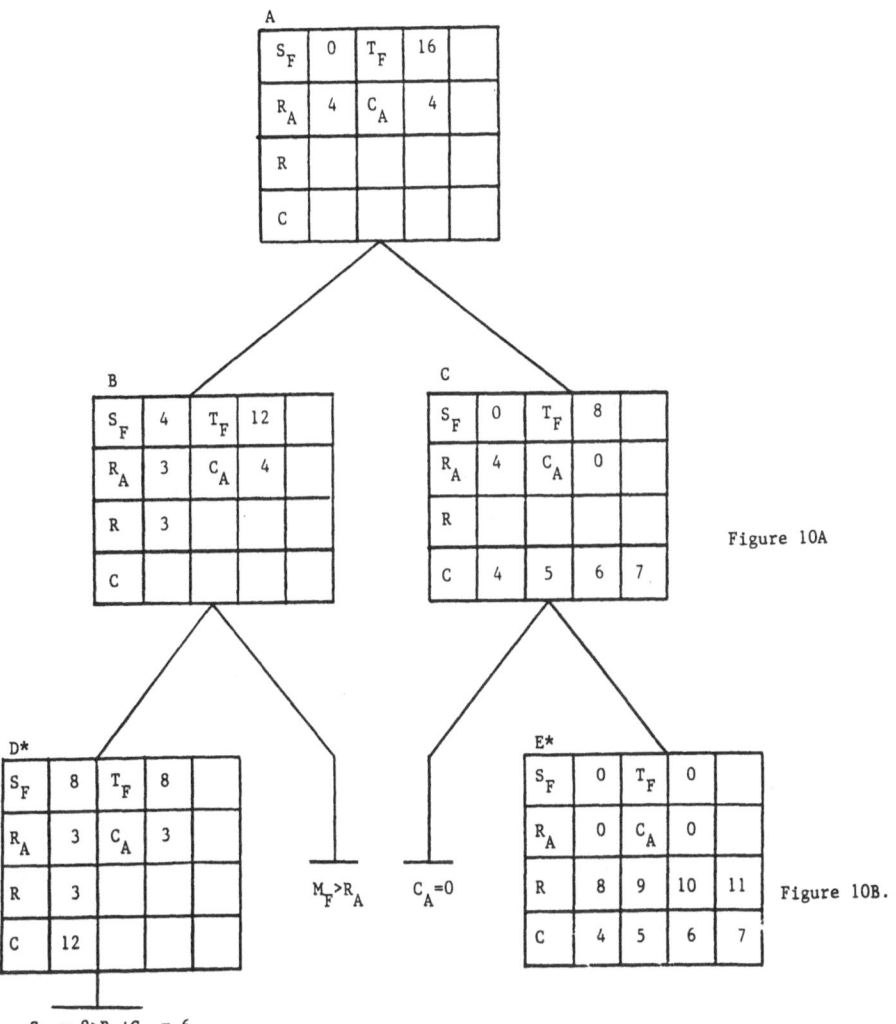

Figure 10A

Figure 10B.

Figure 10.

While the previous example has illustrated the powerful repair capability of FLCA compared with the Fault-Driven algorithm [4], a second example will illustrate that similar improvements can be expected with respect to the number of records required to find the repair-solution. Consider the fault pattern shown in Figure 11, $R_A = C_A = 3$. This pattern is the same as considered in [4]. Figures 12A, 12B and 12C show the iterations required by the repair-algorithm using FLCA. Figure 12A shows the results at completion of Forced-Repair. Two descendants are formed by the execution of the first iteration. This is shown in Figure 12B. Descendant B uses a spare row to cover R_2, while descendant C uses two spare columns to cover C_2 and C_7. Both these repair actions cover all faulty cells on R_2. Descendants B and C are used as parents in the third iteration (Figure 12C). There are two faulty cells for parent B on R_8, or equivalently one faulty cell on C_4 and another faulty cell on C_7. R_8 is selected for covering faulty bits, because it has the maximum count. Descendant D of parent B uses a spare row to cover the faulty bits on R_8, while descendant E of parent B uses two spare columns to cover the faulty bits on R_8. Single faulty parent flags are set for parent C, because $S_F = T_F = 1$. Single faulty parent flags are also set for parents D and E, because $T_F = S_F = 0$. Faulty-Line-Covering is terminated. A number of repair-solutions are possible for selection, because descendants D, E and F are all correct repair-solutions. Note that D is the solution that minimizes the number of utilized spares. An important characteristic is that only 6 records have been used in the repair process. The Fault-Driven algorithm requires 13 records plus the numerous optimization records [4].

Optimization is not required using FLCA, because this aspect is already taken care by Lemma (3). Let min denote the smallest fault counter of the memory. This can be used in connection with Lemma (3) to analyze the complexity of FLCA. This feature can be formally proved by the following Theorems.

THEOREM 6.

Let every line have at least min faulty bits except the S_F single faulty bits. At most $2^{\frac{T_F - S_F}{min}+1} - 1$ records are generated during the iterative execution of the repair algorithm FLCA.

PROOF:

Two choices are possible in FLCA for each faulty line (by Lemma (3)). Hence, the number of descendants per parent is 2. All records make up a binary tree. There are at most $(T_F - S_F)/min$ lines to be covered if no faulty line is covered during the Forced-Repair phase. The last records in the tree have no descendant. This tree has $(T_F - S_F)/min+1$ levels (including the root), so the overall number of records is $(2^{\frac{T_F - S_F}{min}+1} - 1)$. The Theorem is proved. □

Note that the number of records generated by the Fault-Driven algorithm is $2^{T_F+1} - 1$.

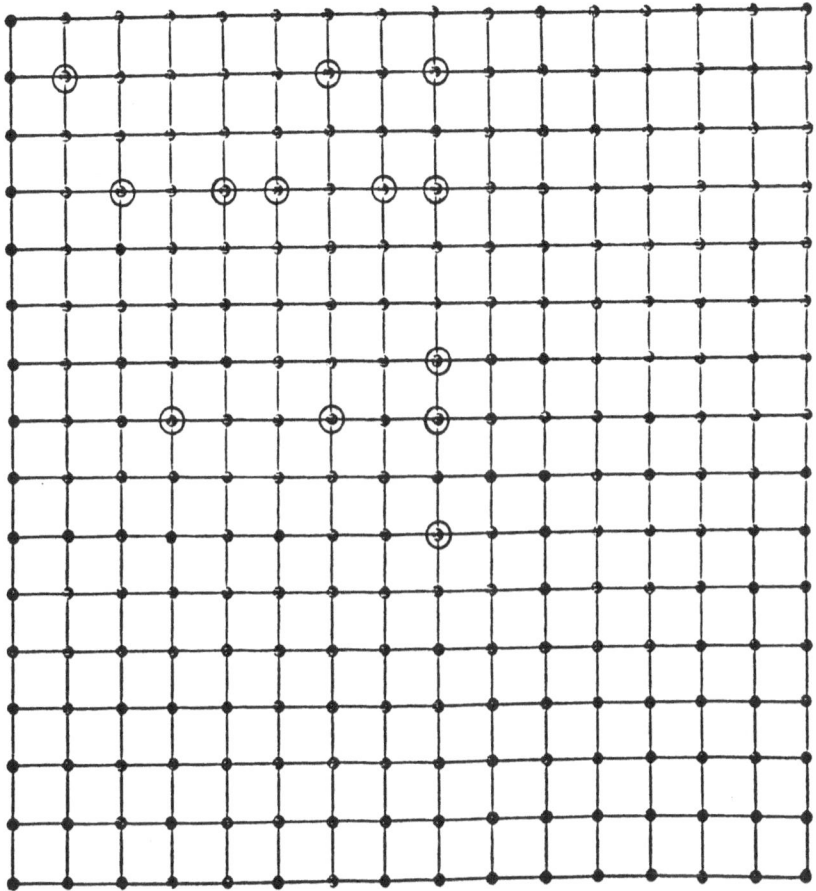

Figure 11.

414

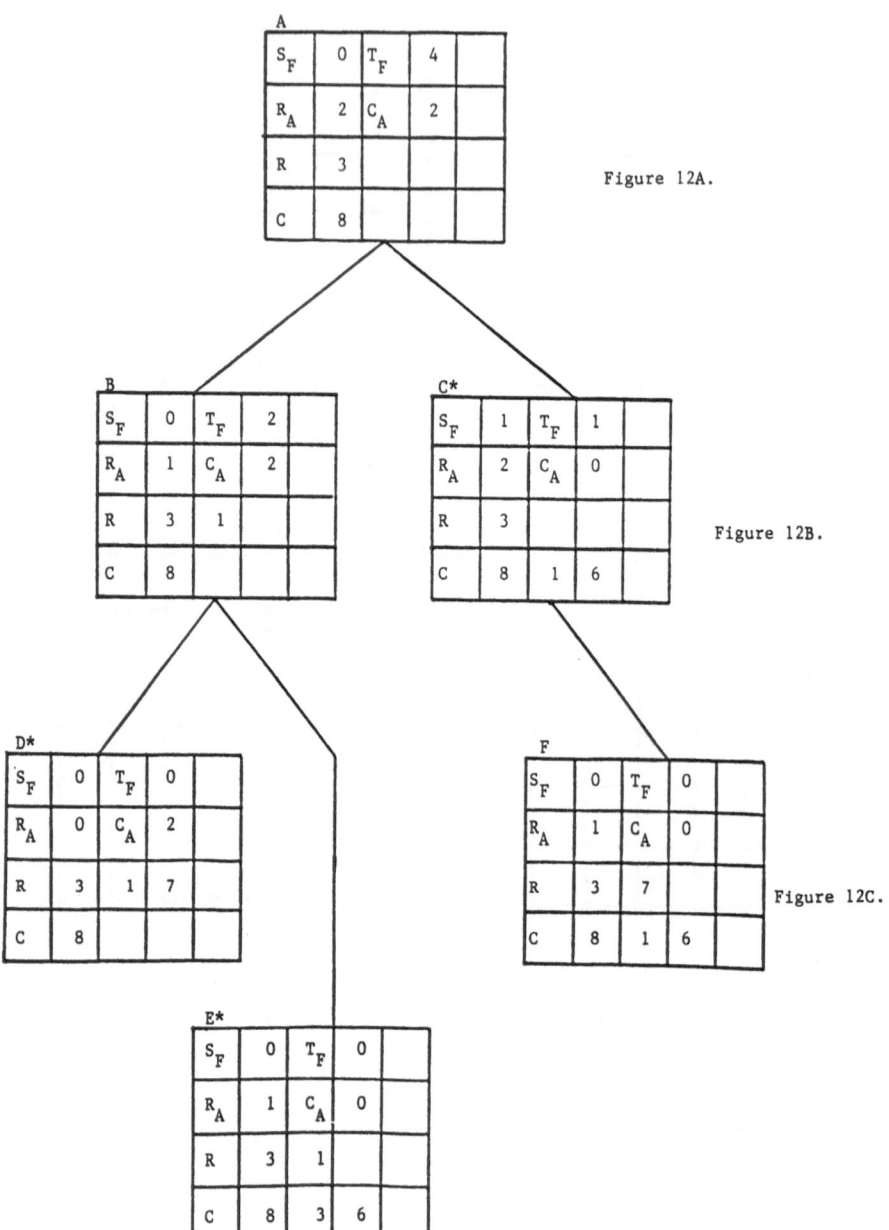

Figure 12A.

Figure 12B.

Figure 12C.

Figure 12.

THEOREM 7.

The repair algorithm using FLCA generates as good a repair-solution as the Fault-Driven algorithm.

PROOF:

This Theorem can be proved by stating that the excess records generated by the Fault-Driven algorithm are made of solutions which use redundant spares to cover faulty bits. The solutions which do not use redundant spares, are only those which are specified by Lemma (3) for the Faulty-Line-Covering phase. Repair using FLCA relies on Lemma (3) to avoid the use of any redundant spares in the covering process. Use of redundant spares in the Fault-Driven algorithm is only established, when all records have been generated. FLCA filters out these solutions prior to generating records. Repair using FLCA and the Fault-Driven algorithm use Forced-Repair. The only difference between these two approaches is the deletion of records which use redundant spares. Hence, the Theorem is proved. □

14. THE EFFECTIVE COEFFICIENT APPROACH.

In the previous section, FLCA has been proposed to generate a lower bound in the number of records to find the repair-solution. However, FLCA has a reduced, but still exponential complexity. New criteria must be formulated to reduce the computational complexity of repairing RRAMs. The proposed heuristic criterion is based on the definition of a complement. This definition relates a faulty bit to all the other faulty bits along its coordinates. The implication of this definition is that a figure of merit can be assigned for each row and column of the array. This figure of merit is referred as the *effective coefficient*, because it establishes a relationship between a faulty line and all other lines to repair faulty cells. The coefficient is referred as effective, because it does not only consider the fault counter of the faulty line, but also the complements. The effective coefficient of R_i (C_j) is denoted as $F(R_i)$ $(F(C_j))$. $F(R_i)$ is given as

$$F(R_i) = N_{F0} + N_{F1} - N_{F2} \tag{15}$$

N_{F0} is the number of faulty bits on R_i. N_{F1} is the number of faulty bits on R_i which have no column complement faulty bit. N_{F2} is the number of faulty bits on R_i which have only a single column complement for each bit and the column complements have no row faulty bit. $F(C_j)$ can be defined equivalently by interchanging row with column (and vice versa) in the definitions of N_{F0}, N_{F1} and N_{F2}. Note that in (15) $N_{F0} = C(R_i)$. Some examples will clarify the definition of effective coefficient. Consider the die shown in Figure 13. For C_5, $N_{F0} = 4$ (there are four faulty bits); $N_{F1} = 2$, because (3,5) and (5,5) have no row complement faulty bit; $N_{F2} = 1$, because (6,4) has only one row complement (6,3) and (6,3) has no column complement. Therefore, $F(C_5) = 4+2-1 = 5$.

Two procedures can be envisaged for repairing RRAMs using effective coefficient.

I. Largest Effective Coefficient Algorithm (LECA).

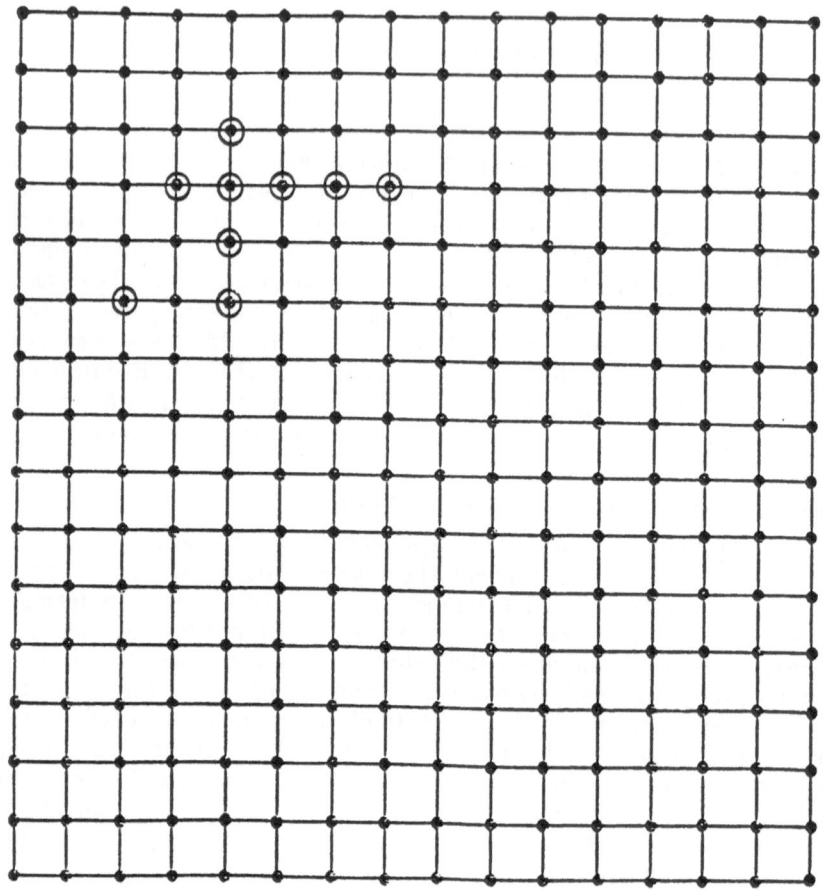

Figure 13.

II. Binary Effective Coefficient Algorithm (BECA).

LECA makes use of a maximal approximation to select either the row (column), or the columns (rows) to be deleted. All effective coefficients are utilized. It is assumed that all $F(R_i)$s and $F(C_j)$s are stored in hardware registers. Their update is automatically (i.e. $O(1)$) performed by the testing hardware [8,13]. LECA is as follows.

ALGORITHM 3: LECA.

Step 1:   Calculate the effective coefficient for every row and column of the memory. Sort and find the maximum value of the effective coefficients. Let it be F_{max}.

Step 2:   Count the number of rows (denoted as N_R) and the number of columns (denoted as N_C), which have F_{max} as effective coefficient. If $N_C > N_R$ go to Step (3). If $N_R > N_C$ go to Step (4). If $N_R = N_C$, compare R_A with C_A. If $R_A > C_A$, go to Step (4). If $C_A > R_A$, go to Step (3). Else, select arbitrarily.

Step 3:   Select one column from the columns which have F_{max} as effective coefficient and cover it with a spare column. Let $C_A = C_A - 1$. Go to Step (5).

Step 4:   Select one row from the rows which have F_{max} as effective coefficient and cover it with a spare row. Let $R_A = R_A - 1$.

Step 5:   Calculate the number of rows and columns to be repaired. Let them be R_R and C_R. If $R_R = C_R = 0$, repair is complete. Else, if $R_A = 0$, go to Step (6). Else, if $C_A = 0$, go to Step (7). Else, go to Step (1).

Step 6:   If $C_A \geq C_R$, cover the remaining columns and repair is successful. Else, the memory die is unrepairable.

Step 7:   If $R_A \geq R_R$, cover the remaining rows. Repair is successful. Else, the memory die is unrepairable.

As an example, consider the fault pattern shown in Figure 14, $R_A = C_A = 3$. Repair using LECA consists of the following steps. The first deletion is given by C_9, because $F_{max} = F(C_9) = 5$. After this deletion, $F(C_5) = F_{max} = 4$ is selected. The third selection is based on the different cardinalities of R_A and C_A, because $F_{max} = F(C_3) = F(C_7) = F(C_{10}) = F(R_4) = F(R_5) = F(R_6) = 2$, but $R_A = 3$ and $C_A = 1$. Hence, R_4 is deleted. The next deletions are R_2, R_6 and C_{10} to cover $(2,3)$, $(6,7)$ and $(5,10)$.

An interesting improvement to LECA can be accomplished if a single faulty bit counter and a total fault counter are used. In this case the selection process can be stopped using effective coefficients after all remaining faulty bits are single faulty bits. This will result in a considerable saving of time, because the calculation of effective coefficient is no longer required (as in the last three deletions of the faulty bits in Figure 14).

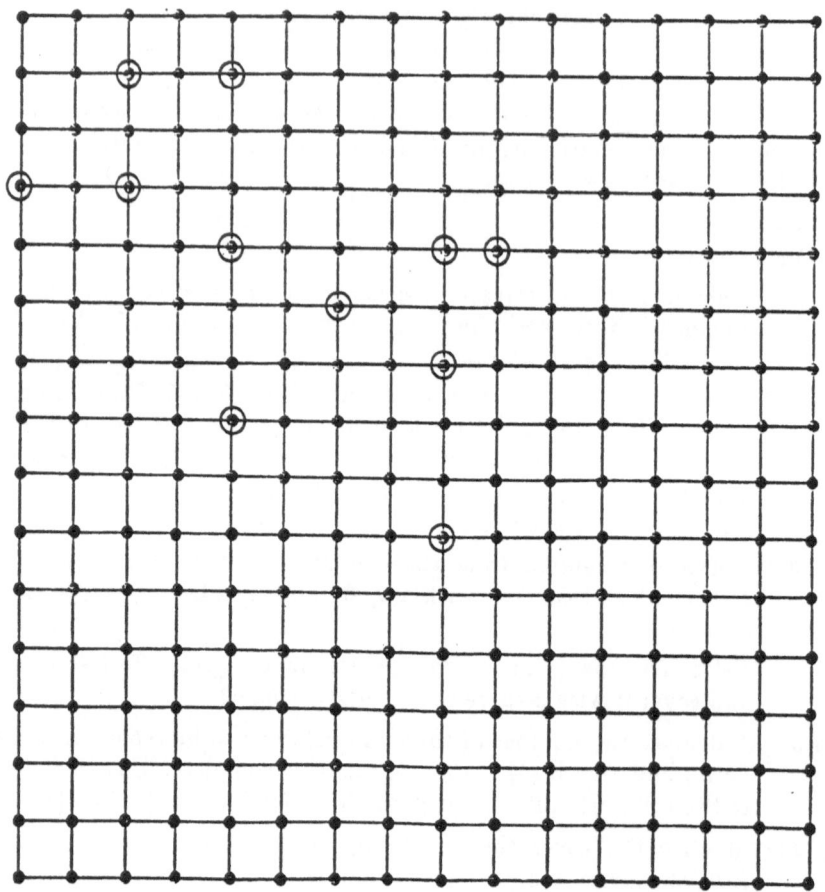

Figure 14.

The complexity of LECA can be established as follows. Sorting to find F_{max} is $O(\log(M+N))$. If the condition of repair is satisfied [5], the loop between Steps (1) - (5) in Algorithm (3) will be executed $(R_A + C_A)$ times. Hence, LECA is $O(\max\{R_A, C_A\}\log(\max\{M,N\}))$.

LECA tends to maximize the criterion of selection for effective coefficient. This procedure is effective under the following fault distributions.

I. *Cluster distribution.* LECA is effective, because fault counters will show large differences in their counts. Complements will vary considerably. These considerations imply that the overhead in finding F_{max} can be retributed considerably by an efficient memory repair.

II. *Connectivity distribution.* This is the fault distribution under which a faulty bit will cause the failure of either all or most bits along a coordinate (either a row, or column) [15]. This is applicable to VLSI, where this failure mode is caused by manufacturing defects (such as failure of a metal line in CMOS devices) [9]. In this case the same considerations as (I) are applicable.

In a more general context of a stable manufacturing technology such as VLSI-MOS [9], a random fault pattern is probably more applicable within a fixed-size die. The interested reader should refer to [10] for a detailed discussion of fault distributions. The calculation of F_{max} could be rather wasteful for random faulty bits, because effective coefficients have no significant count variance. A different approach must be considered.

This approach is based on a binary decision for effective coefficient selection, namely BECA. In this procedure only a total fault count and two effective coefficients are used at a time. BECA is given as follows.

ALGORITHM 4: BECA.

Step 1:   If $T_F \neq 0$ select the first leading faulty bit in the array, say (i,j). Else, repair is successful.

Step 2:   Calculate the effective coefficient for R_i and C_j. If $F(R_i) > F(C_j)$ cover the row with a spare row. If $F(C_j) > F(R_i)$ cover the column with a spare column. If $F(R_i) = F(C_j)$, compare R_A with C_A. If $R_A > C_A$, cover the row with a spare row. If $C_A > R_A$ cover the column with a spare column. If $R_A = C_A$ select arbitrarily. If a row is covered, $R_A = R_A - 1$. If a column is covered, $C_A = C_A - 1$.

Step 3:   If $R_A \neq 0$ and $C_A \neq 0$, update the total fault count and go to Step (1). If $R_A = 0$, go to Step (4). Else if $C_A = 0$, go to Step (5).

Step 4:   Count the number of remaining columns to be covered, say N_C. If $C_A \geq N_C$ cover them and repair is successful. Else, the memory die is unrepairable.

Step 5:   Count the number of remaining rows to be covered, say N_R. If $R_A \geq N_R$ cover them and repair is successful. Else, the memory die is unrepairable.

Consider the fault pattern shown in Figure 15, $R_A = C_A = 3$. Figure 15 is different from Figure 14 by faulty bit (8,6). Repair is as follows. The first leading faulty bit is (2,3). $F(C_3) < F(R_2)$, hence R_2 is selected. The second leading element is (4,1) and again R_4 is selected, because $F(R_4) > F(C_1)$. The third selection is leading element (5,5). R_5 is selected, because $F(R_5) = 4 > F(C_5) = 2$. The remaining selections are C_5, C_7 and C_9 to cover (7,5), (6,7), (7,9) and (10,9).

The complexity of repair using BECA is now reduced to $O(\max\{R_A, C_A\})$, because no sorting is required to find $F_{\max}$ and the number of executions depends on the repairability conditions of [5] i.e. $(R_A + C_A)$.

It should be noted that the fault pattern of Figure 9 can be repaired using LECA and BECA also. R_3 and C_{12} are not selected, because their effective coefficients are zero.

The effective coefficient procedures can be compared with the covering algorithm of [5]. The obvious advantage is the lower order of complexity. The probability of repair achieved by either LECA or BECA, can be stated by the following Theorem.

THEOREM 8.

A repairable memory found by the covering algorithm is also found repairable by an effective coefficient procedure.

PROOF:

The covering algorithm of [5] generates a repair-set as an approximation to the repair-solution. This is achieved by covering each leading faulty bit using a row and a column. Refinement of the repair-set to find the repair-solution is accomplished in a second stage by deleting all redundant spares. An effective coefficient procedure (either LECA, or BECA) still relies on covering using the leading faulty bits. Covering however is implemented by using only a spare /leading faulty bit (either column or row). Presence of complements is taken into account by the definition of effective coefficient in (02). Hence, use of redundant spares is minimized, while covering is in progress. This procedure is equivalent to the approach of [5], because it still utilizes leading faulty bits in the covering process, but it optimizes selection by Lemma (3). Hence, the Theorem is proved. □

15. EARLY-ABORT AND USER-DEFINED CRITERIA.

Many techniques have been used to reduce the time to repair RRAMs. These techniques employ either early-abort conditions (derived from the fault pattern) and/or preferences (derived from user-defined criteria). These criteria are used to terminate the repair-algorithm if it is evident that the repair is going to be unsuccessful. This process has been extended in [5] by defining repairability (unrepairability) as the probability of correctly diagnosis a memory as repairable (unrepairable) prior to the execution of the repair-algorithm. The condition of repairability can be integrated with the early-abort condition of Forced Repair (i.e. $T_F > 2R_A C_A$) to stop executing FLCA if the memory is not repairable. The condition of early-abort of Forced Repair can be modified if

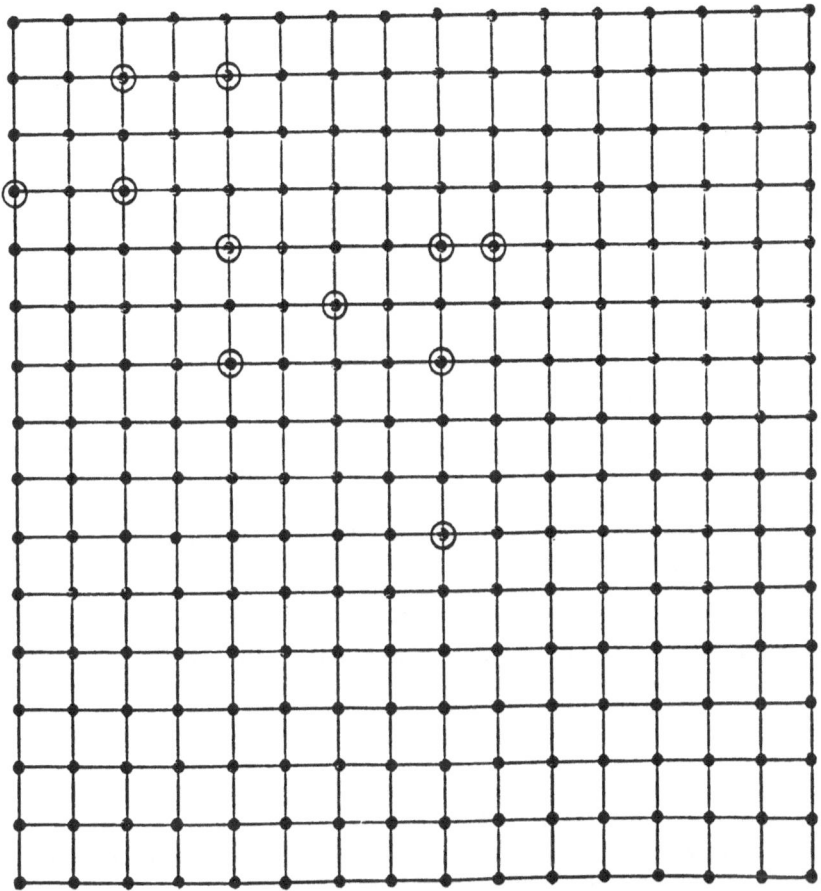

Figure 15.

$C_A \neq 0$ and $R_A \neq 0$. This condition is now given as

$$T_F > 2R_A C_A - S_F \min\{R_A, C_A\} - S_F \qquad (16)$$

by considering that if $R_A + C_A < S_F$, Fault-Line-Covering must be stopped. (16) can be used as early-abort criterion after each iteration of FLCA.

The repair algorithms of Section 14 were analyzed without using early-abort techniques. These techniques were not added for convenience and simplicity of the explanation itself. Criteria for early-abort are as follows.

$$T_F > (MR_A + NC_A) - R_A C_A \qquad (17)$$

and

$$S_F > R_A + C_A \qquad (18)$$

These can be used in LECA and BECA.

A modification however is required for the criteria to find the repair-solution. In FLCA, these criteria are the minimal numbers of utilized spare rows and columns. This number is directly related to the minimal repair cost. If the cost to repair a row (column) is given by $C^R(C^C)$, the total cost of repair C_T is given by

$$C_T = C^R(R_A - R_{AR}) + C^C(C_A - C_{AR}) \qquad (19)$$

where R_{AR} and C_{AR} are the remaining number of unutilized spare rows and columns. (19) can be integrated in FLCA when selecting a repair-solution. (19) introduces a new criterion that relates repair to the optimality of the repair-solution.

16. SIMULATION.

Algorithms (1), (2), (3) and (4) have been evaluated by simulation. A sample of 1000 dies with an equal number of faulty bits has been used. Faulty bits have been injected using a random number generator.

The proposed repair approaches can be compared with respect to their efficiency in finding a repair-solution whenever a repairable memory is analyzed. The figure of merit in this case is given by the *efficiency factor*. This is defined by

$$E_i = \frac{N_i}{N_z} \qquad (20)$$

where N_z is the number of repairable memories out of a sample of z devices and N_i is the number of repairable memories found by Algorithm (i), $i = 1, \ldots, 4$. By Theorem (7),

$$N_2 = N_z \qquad (21)$$

and

$$E_2 = 1 . \qquad (22)$$

Let E_0 denote the efficiency factor of the repair algorithm of [5]. Table (1) shows the efficiency factors for the different algorithms.

A second feature which has been evaluated by simulation, is the *average number of records* generated by Algorithm (2) (and denoted by R_2) versus the Fault-Driven Algorithm (denoted by R_{FD}) for all samples.

Table 1.

$N = M$	$R_A = C_A$	E_4	E_3	E_1	E_0
9	3	0·919	0·952	0·915	0·910
9	4	0·904	0·950	0·918	0·912
16	3	0·944	0·986	0·937	0·925
16	4	0·948	0·986	0·940	0·927
16	5	0·949	0·988	0·940	0·936
25	3	0·960	0·993	0·954	0·938
25	4	0·960	0·994	0·955	0·940
25	5	0·961	0·995	0·956	0·942

Table (2) shows the average number of records per repair-solution.

Table 2.

R_2	R_{FD}	$N = M$	$R_A = C_A$
4	50	9	3
15	160	16	4
26	500	25	5

The following features may be pointed by analyzing the data in Tables (1) and (2).

I. The proposed algorithms are very efficient for repairing memories. The heuristic nature of Algorithms (3) and (4) provides a very good solution to a proved NP complete problem as memory repair [14,22].

II. The significant decrease in the number of records required by Algorithm (2) with respect to a Fault-Driven approach enhances its practical feasibility due to a considerable saving in repair-time.

A further feature that has been evaluated by simulation is repair in presence of clusters, as applicable to memory manufactured using WSI [11]. Table (3) shows the efficiency factors for the algorithms. Clusters have been simulated as oval shaped defective regions having axis not greater than the provided redundancy.

Table 3.

$N = M$	$R_A = C_A$	E_4	E_3	E_1	E_0
9	3	0·911	0·932	0·909	0·901
9	4	0·907	0·933	0·909	0·903
16	3	0·932	0·956	0·932	0·917
16	4	0·937	0·957	0·933	0·922
16	5	0·943	0·961	0·931	0·931
25	3	0·949	0·963	0·947	0·934
25	4	0·952	0·974	0·947	0·938
25	5	0·954	0·976	0·949	0·939

17. ADAPTIVE REPAIR.

The proposed repair algorithms inherently assume that a diagnostic matrix A is available for analysis. The A matrix is directly derived from testing the memory die. This provides full testing information to the repair process. However, in theory $O(NM)$ testing operations are required to test all bits in the memory die. This can create considerable overhead if no special ATE is available to expedite this process. A different approach is proposed in this section. Testing and repair are not considered as two separate processes, but they are executed in parallel, one as a direct result of the other.

The diagnostic matrix is now represented by a *test matrix* $T = \{t_{ij}\}$. t_{ij} has a ternary value: 0 if bit (i,j) is fault free, 1 if bit (i,j) is faulty, 2 if bit (i,j) has not been tested. It is evident that T is equal to A if no entry in T has value 2.

The objective of the proposed approach is to adaptively test the bits in the memory die as a direct consequence of repair and the expected repair-solution. This approach relies on detecting repairability/unrepairability of the memory die. Testing is carried out to satisfy certain conditions for membership in the repair-solution. These conditions have been partially analyzed in [5] by using the bits along the major diagonals as the first criterion for repair by full covering. This information can be used as a starting point for adaptive testing.

Consider the following scenario: bits (i,i) and (j,j) $(i < j)$ are tested and found to be faulty. All bits which lie along R_i, R_j, C_i and C_j, are inherently covered if these rows and columns are members of the repair-solution. Equivalently, (i,i) and (j,j) can be covered by reduced covering using only two spares (either two columns, or two rows, or a column and a row). This condition can be collapsed into full covering if either (i,j), or (j,i) is faulty. Hence, this bit must be tested.

Adaptive testing is based on a *collapsing technique* between full and reduced covering (and viceversa). The collapsing technique is based on an iterative execution of an algorithm to find the bit which could cover the maximum number of already tested (and faulty) bits. This is the inverse approach of [5].

Adaptive testing is carried out for the bits along the largest (and still untested) diagonals which are left in the die. This process is directed in two parallel directions.

I. The detection of unrepairability of the memory die.

II. The optimization of the repair-solution within the available test results.

(I) can be used as an early-abort criterion to further reduce the overhead associatedd with memory repair. (II) can be used to reduce testing, because those untested cells which lie along the elements of the repair-solution are not required to be tested. These cells must be tested if the row (or the column) on which they lie is not part of the new repair-solution. This characteristic is particularly desirable in the following circumstances.

I. The memory die has a significant amount of redundancy, because the probability of having a defective bit is rather high. This occurs when WSI is used [11].

II. The expected yield of the memory die is low and detection of unrepairability can result in a significant saving of repair-time.

Testing is continued not only with respect to the collapsing technique but also for those bits along the remaining (and yet untested) diagonals. This process tends to discharge all the unrepairable memory arrays prior to generating the repair-solution. Hence the results of adaptive testing can be incorporated as an early-abort criterion.

18. CONCLUSIONS.

Repair of redundant memories has been analyzed. New approaches have been presented. The first approach is based on a reduced covering criterion in the selection of spares. This criterion is dependent on a fault counting technique which can be easily incorporated into a defect analysis system. This approach is a direct refinement of a previous approach [5]. Two further new approaches are analyzed. These other two approaches use different criteria (namely Faulty Line Covering and Effective Coefficient) to provide fast, yet efficient repair. It has been proved that using Faulty-Line-Covering a reduction in the number of repair-solution records of the Fault-Driven algorithm [4] is possible. It is proved that using the heuristic criterion of effective coefficient, repair is accomplished at linear cost with respect to the dimension of the array. This is a net improvement over the quadratic execution time of the covering algorithm of [5]. It is also proved that the above improvements are accomplished at no significant expense of the repair-solution. Worst case analysis demonstrates that the proposed approaches are far better than the algorithms of [5]. The lower execution complexity makes these algorithms very attractive for a practical implementation. A new approach based on iteratively finding the repair-solution of a partially test memory die, has also been proposed.

19. REFERENCES.

1   Denker, R.P., Clemons, D.G., Huber, W.R., Petrizii, J.B., Procyk, F.J. and G. M. Trout, "A Fault-Tolerant 64k Dynamic RAM', *IEEE Trans. on Electron Devices*, Vol. ED26, No. 6, 1979.

2   Fitzgerald, B.F. and E. P. Thoma, "Circuit Implementation of Fusible Redundant Addresses of RAMs for Productivity Enhancement", *IBM J. Res. Develop.*, Vol. 24, pp. 291-298, 1980.

3   Stapper, C.H., McLaren, A.N. and M. Dreckmann, "Yield Model for Productivity Optimization of VLSI Memory Chips with Redundancy and Partially Good Product," *IBM J. Res. Develop.*, Vol. 24, pp. 398-409, 1980.

4   Day, J.R., "A Fault-Driven Comprehensive Redundancy Algorithm for Repair of Dynamic RAMS," *IEEE Design and Test of Computers*, Vol. 2, No. 3, pp. 33-44, 1985.

5   Wey, C-L. and F. Lombardi, "On the Repair of Redundant RAMS," to appear in *IEEE Trans. on CAD of ICAS*, Vol. CAD6, No. 2, pp. 222-231, 1987.

6   IEEE ISSCC Digest of Technical papers, section on Memories and Redundancy Techniques, pp. 80-87, 1981.

7   Evans, R.C., "Testing Repairable RAMs and Mostly Good Memories," *Proc. Int. Test Conf.*, pp. 49-55, 1981.

8   Tarr, M., Boudreau, D. and R. Murphy, "Defect Analysis System Speeds Test and Repair of Redundant Memories," *Electronics*, pp. 175-179, Jan 12, 1984.

9   McCluskey, E.J., "Comparing Causes of System Failure," *Proc. Euromicro 86*, pp. 11-22, 1986.

10 Stapper, C.H., "On a Composite Model to the IC Yield Problem," *IEEE J. of Solid State Circuits*, Vol. SC10, pp. 537-539, 1975.

11 Negrini, R., Sami, M.G. and R. Stefanelli, "Fault Tolerance Approaches for VLSI/WSI Arrays," *Proc. Phoenix Conf. on Comp. and Comm.*, pp. 460-468, 1985.

12 Garey, M.R. and D.S. Johnson, "Computers and Intractability," *Freeman*, San Francisco, CA, 1979.

13 Haysaka, Y., Shimotori, K. and K. Okada, "A Testing System for Redundant Memory," *Proc. Int. Test. Conf.*, pp. 240-244, 1982.

14 Dahbura, A., private communication.

15 Schuster, S.E., "Multiple Word/Bit Line Redundancy for Semiconductor Memories", *IEEE J. Solid-State Circuits*, Vol. SC-13, pp. 698-703, 1978.

16 Heckelman, R.W. and D. K. Bhausar, "Self-testing VLSI" *Proc. IEEE Solid State Circuits Conf.*, pp. 174-175, 1981.

17 Fuja, T. and C. Heegard, "Row/Column Replacement for the Control of Hard Defects in Semiconductor RAM's," *IEEE Trans. on Comp.*, Vol. C35, No. 11, pp. 996-1000, 1986.

18 Warner, R.M. Jr., "Applying a composite model to the IC yield problem", *IEEE J. Solid-State Circuits*, SC-9, pp. 86-95, 1974.

19 Sud, R. and K.C. Hardee, "16k Static RAM takes New Route to High Speed," *Electronics*, pp. 117-123, September, 1980.

20 Leighton, T. and C.E. Leiserson, "Wafer-Scale Integration of Systolic Arrays," *IEEE Trans. on Comp.*, Vol. C34, No. 5, pp. 448-461, 1985.

21 Stapper, C.H., "On Yield, Fault Distributions and Clustering of Particles," *IBM Journal of Research and Development*, Vol. 30, No. 3, pp. 326-338, 1986.

22 Kuo, S-Y. and W.K. Fuchs, "Efficient Spare Allocation in Reconfigurable Array," *IEEE Design and Test*, Vol. 4, No. 1, pp. 24-31, 1987.

23 Clarkson, D., "A Modification of the Greedy Algorithm for Vertex Cover," *Info. Proc. Letters*, Vol. 16, pp. 23-25, 1986.

RECONFIGURATION OF ORTHOGONAL ARRAYS BY FRONT DELETION*

Fabrizio Lombardi
Department of Electrical and Computer Engineering
University of Colorado Campus Box 425, Boulder, CO 80309, USA

C-L Wey
Department of Electrical Engineering
Michigan State University, East Lansing, MI 48824, USA

ABSTRACT.
   This paper describes various algorithms for reconfiguration of VLSI arrays. These algorithms are applicable to arrays in which reconfiguration requires the logical deletion of the whole front of computation for algorithmic reduction (either rows, or columns). This family of reconfigurable array is referred to as bussed arrays. A new approach is proposed. The innovative feature of this approach is independence of reconfiguration on distribution of faulty cells in an array. The proposed technique achieves a good detection of reconfigurability/unreconfigurability.

   Algorithms that provide unreconfigurability/reconfigurability detection and the reconfiguration-solution are presented. New and simple switching circuits to implement the proposed reconfiguration approaches are described.

   The main benefit of these algorithms is the reduction in execution-time to determine physical reconfiguration and reclaim operational arrays for processing. Illustrative examples and simulation results are presented.

1. INTRODUCTION.

   Reconfiguration is an important feature of *Very Large Scale Integration* (VLSI) systems [1]. Complexity of these systems has enhanced the necessity of reconfiguration for yield improvement at production-time and fault tolerance at run (application)-time. Due to increasing demand on chips with regular interconnection patterns such as memories, PLAs and cellular architectures [2,3,4,5,10,12], a structured approach to reconfiguration [1] has become of primary importance.

   At production-time as the chip density increases, geometry of the chip shrinks and the probability of finding defective cells increases. *Rejection of a chip* can be caused by a small percentage of faults [2]. Semiconductor manufacturers have extensively employed *redundancy techniques* [7] to attain significant reliability improvements. Redundancy is used to reclaim operational all those chips in which reconfiguration is possible [6]. Fault tolerance at run-time also requires redundancy: as faults occur, fault free cells are reconfigured for continued system operation. Techniques, such as token-diagnosis [18] have been proposed to enhance on-line performance of VLSI devices at run-time.

---

*This research supported in part by grants from AT&T, NATO and by the Engineering Foundation.

F. Lombardi and M. Sami (eds.), Testing and Diagnosis of VLSI and ULSI, 429–467.
© *1988 by Kluwer Academic Publishers.*

Reconfiguration must abide by functional specifications [16]. This makes reconfiguration at run-time a more complex process than redundancy allocation for sparing at production-time. Homogeneous systems are an optimal choice, because they provide modularity in reconfiguration. This is an important feature for fault tolerant computing.

Many reconfiguration techniques have appeared in the literature [8,9,11]. These techniques have been proved to be effective in either *circumventing* faulty components at production-time [9] and/or enhancing system reliability and *fault tolerance* [19,20] of arrays. Redundancy is embedded in homogeneous systems such as memory and VLSI arrays by adding extra rows and columns of cells (and associated interconnection circuitry) [7].

Redundancy in memory chips is used for repair at development-time. Redundancy is not free of penalties. Insertion of spare rows and columns to the original design contributes to an overall larger chip area.

A spare-efficient, but time-consuming approach for reconfiguration of VLSI is proposed in [8]. This approach, referred as the Diogenes approach, is more appropriate for off-line reconfiguration for production and yield improvement.

In [19,20], reconfiguration is accomplished by a *deterministic process* of inspecting neighbouring cells. This technique has a low execution-time. Efficiency of reconfiguration rapidly decrease as either the number of simultaneous faults increases, or certain patterns of faulty cells occur within rows or columns of the array.

A different approach is proposed in [17]. This is applicable to run-time. When a faulty cell is detected, either the row, or column in which the faulty cell is located, is deleted from the array. An algorithm is proposed for logical reconfiguration. Even though the approach of [17] is elegant, reconfiguration is restricted to algorithm reduction.

For on-line reconfiguration, different criteria must be used. Several measures [11] such as spare efficiency and silicon overhead, can be used to assess the effectiveness of redundancy for on-line reconfiguration. To discriminate between these measures, reconfiguration can be analyzed with respect to the reconfigured array and the algorithm it is supposed to execute. This type of reconfiguration is referred as *physical reconfiguration*. This is different from the logical reconfiguration for algorithmic reduction of [17]. Physical reconfiguration assumes that the algorithm can be imbedded into an array by algorithm reduction at the expense of a dynamic, but at a fast rearrangement of the array.

One of the measures to assess physical reconfiguration is the time required to either detect unreconfigurability, and/or determine a reconfiguration-solution. This is commonly referred as *execution-time* [7,21]. Recent research [16,21] has been directed toward reducing execution-time

Several algorithms have been proposed to determine the reconfiguration of VLSI arrays [17,19] by using redundancy. However none of them establishes the reconfigurability. Reconfigurability is defined as the probability of correctly diagnosing a reconfigurable array prior to the execution of the reconfiguration algorithm.

The aim of this paper is to present new approaches, that can be implemented to determine reconfigurability/unreconfigurability and reconfiguration-solutions of VLSI arrays. Reconfiguration is implemented by either row, or column deletion [17]. In the next section, a brief review of reconfiguration algorithms is presented. Advantages and disadvantages of existing algorithms are analyzed. Preliminaries are introduced in Section 3. Criteria for reconfigurability and various algorithms for row/column reconfiguration are presented in Section 4. The analysis is applicable to two dimensional orthogonal arrays. Illustrative examples are also presented. In Section 5, complexity of these algorithms is discussed. Two new approaches based on array partition are proposed. It is proved that reconfiguration by partition (either partial or full) reduces algorithmic complexity. Simulation results are given in Section 6 to demonstrate the effectiveness of the proposed approach to orthogonal arrays. The implications of redundancy on array dimensions are discussed in Section 7. The design of switching circuits to implement the proposed reconfiguration techniques are described in Section 8. Conclusions are addressed in the last section.

2. REVIEW.

Research in reconfiguration of VLSI array has been directed toward reducing execution-time for reconfiguration.

Run-time and production-time reconfiguration techniques employ different approaches, however they share many commonalities. Reconfiguration uses test results to maximize the number of those fault free cells that can be kept operational in the reconfigured array. Sites of faults are logged into a fault-map. This map is used to generate a *reconfiguration-solution* (whenever one exists) by utilizing redundancy.

An ideal reconfiguration algorithm should provide a reconfiguration-solution, but it should also satisfy the following *benchmark criteria*.

A. Execution-time should be contained within a reasonable bound.

B. At the earliest detection of no reconfiguration, *algorithm execution* should be *aborted;* graceful degradation (if applicable) must be initiated.

C. *Adaptability* to different interconnection array configurations.

D. Within a given execution-time, number of fault free cells that can be used for continued operation after reconfiguration, must be maximized.

Different algorithms to compute reconfiguration-solutions have appeared in the literature [8,9,17,19]. A particularly interesting approach to reconfiguration by redundancy is proposed in [20]. Each cell in the array is identified by a pair of *physical* indices. The whole array is represented by a physical *index matrix,* whose entries are the physical coordinates, and a *logical index matrix* that identifies the function of each cell at operation-time. In absence of faults, the two matrices are equal on an entry to entry basis. After reconfiguration, entries of the logical index matrix may differ from the corresponding entries of the physical index matrix. A new logical matrix must be generated. Transformation of the old logical matrix is accomplished by using a reconfiguration operator. The reconfiguration algorithm consists of a two-step procedure.

A) In the first step, a methodology is identified. This considers complexity of the interconnection network, switch organization and routing information for fault free cells.

B) In the second step, the specific reconfiguration-solution is found, i.e. a new logical matrix is generated from the old one.

It has been proved [19,20] that this technique provides a good locality with a modest hardware overhead for routing. Disadvantages of this approach are: decreasing probability of successful reconfiguration in the presence of a large number of simultaneous faults and restricted number of reconfiguration possibilities.

A new approach is proposed in [17]. Reconfiguration consists of removing either rows or columns with faulty cells. Deletion of a row or a column is considered for process partitioning by execution in fixed-size VLSI array architectures [17,21]. The novelty of this approach results from the fact that it exploits the characteristics of both the algorithm and array architecture.

However, physical reconfiguration cannot be completely independent of processing. The following features must be considered in reconfiguration.

I. Processing in arrays is usually synchronous [13,14].

II. Processing proceeds along fixed directions. This permits data exchanges [13,22] between neighbouring cells.

Reconfiguration of synchronous arrays is constrained by many limitations. These are related to both the physical and the logical nature of the processing. These limitations are.

I. Data (inputs/outputs of the array) are usually correlated [13].

II. Delay due to reconfiguration should be minimized for a synchronized flow of processing to avoid uneven delay. Either wait-states for unreconfigured fault free cells, or complex switching circuits are required to avoid uneven delay [13,14]. In the first case cell utilization is low. In the second case hardware overhead can be significant [20].

III. Processing, especially of a systolic nature [5,14], shows significant dependency on the architectural/physical embedding of the algorithm, [13,22]. In a synchronous array, this is a very stringent requirement to achieve high throughput in fixed-size architectures [21].

A new approach to array reconfiguration is proposed to overcome the above limitations. This approach, referred as *front reconfiguration,* exploits the commonalities between processing in adjacent cells (found in a systolic algorithm) and the physical organization of the array. Front reconfiguration consists of the logical deletion of all those fault-free cells that shows an algorithmic dependency with at least a faulty cell. As the word implies, a whole front which has been defined by the algorithm [17,21], is logically deleted from the array. This reconfiguration approach offers the following advantages.

I. Delay is uniform along the whole front of computation [21]. This considerably reduces hardware overhead. Switching circuits are simple and easily controllable [17].

II. Algorithmic reduction and graceful degradation can be accomplished for many known problems [17].

III. Reconfiguration can be dynamically implemented.

3. PRELIMINARIES.

This paper deals with two dimensional arrays with orthogonal interconnections (along rows and columns). This type of array is commonly referred as an orthogonal array (Figure 1).

The array is made of n cells. Cells are arranged in a two-dimensional fashion. The number of columns is given by N. The number of rows is given by M. Therefore,

$$n = NM \tag{01}$$

The array is made of two cell banks: the operational bank and the redundant bank. The numbers of spare rows and columns are respectively denoted as R_A and C_A. The number of cells in the operational bank is given by

$$n_0 = (N - C_A)(M - R_A) \tag{02}$$

The number of cells (spares) in the redundant bank is

$$n_a = n - n_0 =$$

$$= NM - (N - C_A)(M - R_A) = \tag{03}$$

$$= NR_A + MC_A - R_A C_A$$

It is assumed that

$$n_0 \geq n_a \tag{04}$$

i.e.

$$n_a \leq \frac{n}{2} \tag{05}$$

In the analysis the following assumptions are made.

I. Faults are associated with cells. Reconfiguration-controlling and switching circuits are assumed to be either self-testing [11,19] or self-checking [20]. These circuits are not hard core components. A more detailed analysis of switching circuits is presented in Section 7. Interconnection links are assumed to be fault free.

II. Spare cells are identified by all those processing elements in excess to the ones required for a particular operational phase.

III. The algorithms for reconfigurability and reconfiguration are executed by a host computer that has full control over the processor array.

IV. Faults occur randomly in the array, i.e. failure independence of cells is upheld.

V. Even though, the number of simultaneous faulty cells could be small, faults can accumulate over a long period of time. Ultimately a growing number of

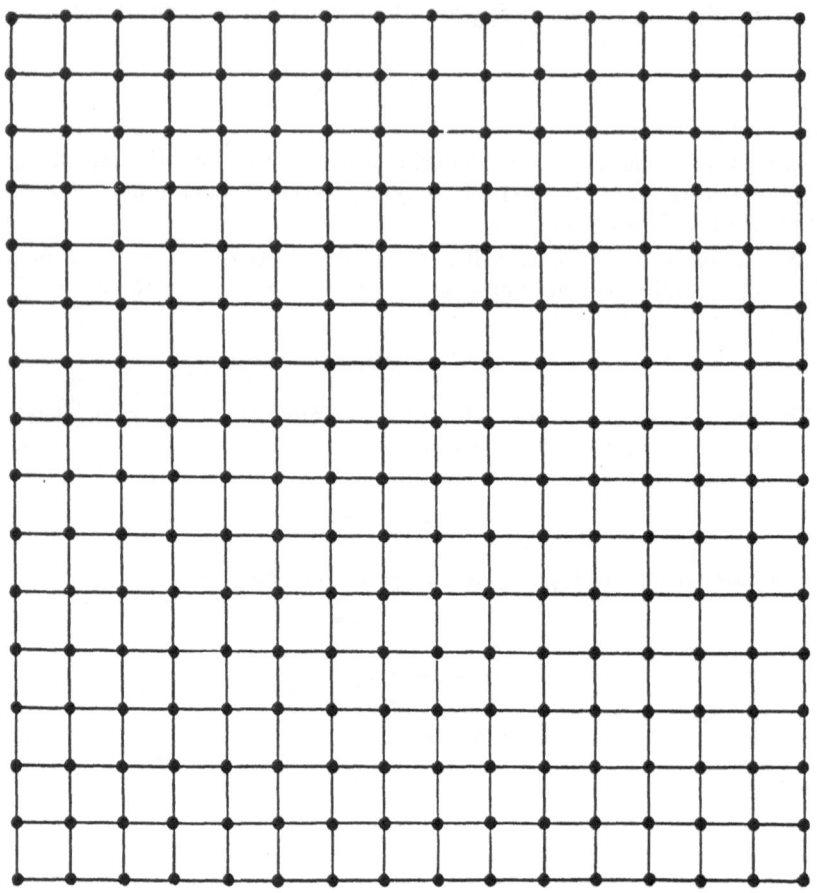

Figure 1: 16x16 array

faulty cells can be present in the array. This implies that dynamic reconfiguration strategies must be applied. Dynamic reconfiguration offers significant advantages compared with static reconfiguration [20].

VI. Processing proceeds synchronously along cell interconnections in a lock-step manner.

If reconfiguration is implemented as a row/column deletion, it is appropriate to refer to this type of array as a *bussed array* (Figure 2). The array can be thought of as crossed by orthogonal busses. A bus can be logically placed between either two rows and/or two columns. This bus is accessed by all cells of a column (row). Failure of at least one of these cells implies corruption of the common bus. The bus and all cells attached to it must be discarded to recover from this fault. Equivalence of a bussed array to row/column reconfiguration of arrays is obvious, since both systems retain the same connectivity, as depicted in Figures 1 and 2.

4. RECONFIGURABILITY AND RECONFIGURATION.

The objective of physical reconfiguration is to reconfigure at least an $(M-R_A)\times(N-C_A)$ array (referred as a *target*) in which cells are connected in the same fashion as in the original $M\times N$ array. This objective should be accomplished using a moderately complex reconfiguration technique and fairly simple switching circuits.

It will be initially assumed the array undergoes a fast test [4,18]. The testing information can be used to establish if spare availability is sufficient and the array can be reclaimed operational. This process is referred as *reconfigurability detection*. Detection consists of inspecting *region-fault-totalizers* [7]. A region fault totalizer is a counter, that records the number of faulty cells within a region of the array. The *maximum number of faulty cells* that permit reconfiguration, is given by

$$F_t = (NR_A + MC_A) - R_A C_A \leq n_a \qquad (06)$$

F_t can be compared with the content of a region totalizer. If this limit is not exceeded, reconfiguration can take place. Inspection of region-fault-totalizers provides the initial condition for reconfigurability detection. Another alternative is to execute one of the algorithms of [7]. The *exponential complexity* [19] of this process requires the inspection of all combinations in the use of spares. Unreconfigurability/reconfigurability of an array should be *determined together* with its reconfiguration-solution, because execution-time can be reduced by an earliest detection of unreconfigurability.

It is important to generalize to the whole bussed array some uniqueness conditions in the allocation of spares to reduce the execution-time. These conditions apply to those faulty cells that satisfy the following selection criteria.

I. No faulty cell shares either a row or column with any other faulty cells.

II. The number of faulty cells, that satisfy (I), exceeds the number of spare rows and columns.

These criteria are necessary and sufficient conditions for unreconfigurability of those cells that lie along the major diagonal of the array. Occurrence of faults

436

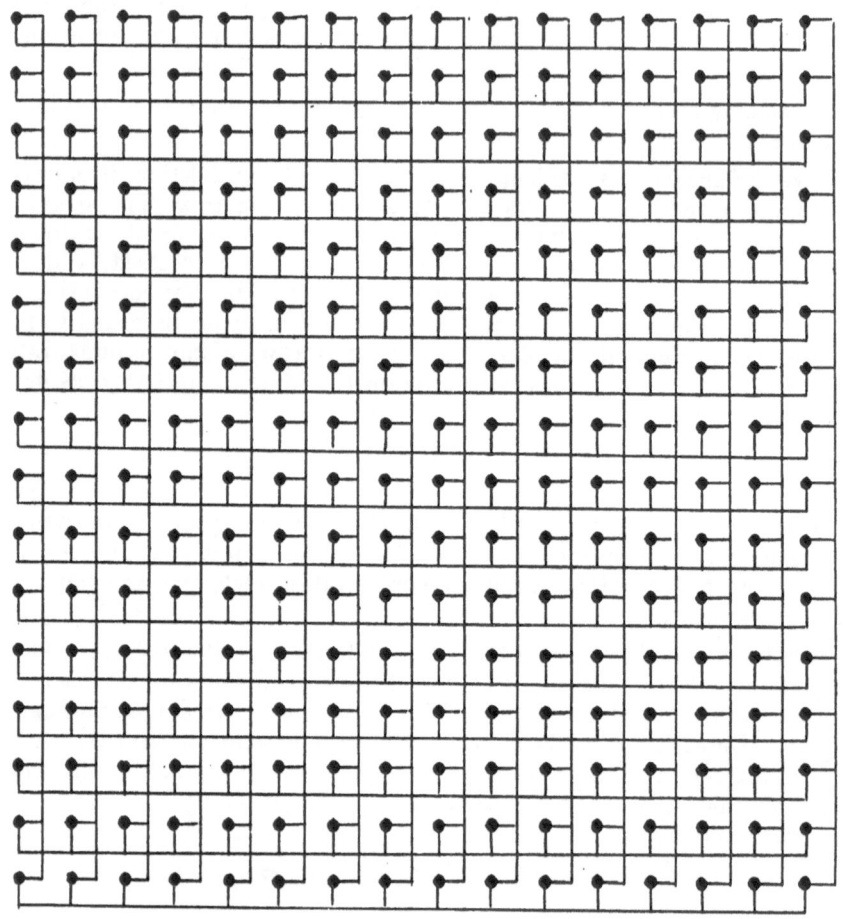

Figure 2: 16x16 Bussed Array

can determine a pattern that can be uniquely described by the above criteria. This characterization is very important, because array reconfiguration in this case is made independent of the particular locations of the faulty cells. Those cells that satisfy (I) and (II) can be used to establish an appropriate reconfiguration strategy.

Some examples illustrate the usefulness of the above criteria. For the fault pattern of Figure 3, it is verified that condition (II) applies if $R_A = C_A = 3$. This yields an impossible reconfiguration. Consider the fault pattern shown in Figure 4, where faulty cells that do not share a column or a row are represented by large black dots. In Figure 4, the number of these cells is 7. This number is in excess of the number of spare rows and columns ($R_A = C_A = 3$). The bussed array is promptly identified to be unreconfigurable.

It is possible to deduce that if there exists a number of faulty cells in a bussed array that do not share either a column and/or a row with any other faulty cell and whose number exceeds the number of spare rows and columns, the array is not reconfigurable.

However, these faulty cells must be promptly identified for reconfiguration to start. A tested array is represented by an $M \times N$ diagnostic matrix $A = (a_{ij})$; element a_{ij} in the ith row and in the jth column, is 0(1) if the (i,j) cell is fault free (faulty). A bussed array is R_A/C_A-reconfigurable, if there exists a set of R rows and C columns in A (referred as *reconfiguration-solution*), such that all nonzero elements are covered by $R \leq R_A$ and $C \leq C_A$. This *covering property* [15] can be used to derive the reconfiguration-solution of the bussed array.

An approximation of the reconfiguration-solution is initially given by the reconfiguration-set. To construct the reconfiguration-set, it is important to define the *leading element a_{ij}* of A. The leading element is the first non-zero element of A, such that the ith row is the first non-zero row and the (i,j)-entry is the first non-zero element of row i. In Figure 5, the first leading element is $(1,1)$.

The reconfiguration-set S can be constructed as follows. Initially, the leading element of A is selected; then all entries along either the row and/or column of the leading element are replaced with zero entries. This process is repeated until the diagnostic matrix is a zero matrix.

Selection of leading element is based on spare availability. If either all rows or columns have been utilized, the leading element is defaulted to either a_{0j}, or a_{i0} (the index corresponding to the unavailable spare is given by 0).

If S_R and S_C are the sets of rows and columns of the elements of S, the algorithm to construct the reconfiguration-set is as follows.

ALGORITHM 1: RECONFIGURATION-SET GENERATION.

Step 1:   Initialize, $S = \{\}$; $S_R = \{\}$; $S_C = \{\}$; $var = 0$.

Step 2:   Find the leading element of A, say a_{ij}.

Step 3:   If $R_A = 0$, $var = 1$. If $C_A = 0$, $var = 2$. If $R_A = C_A = 0$, $var = 3$.

438

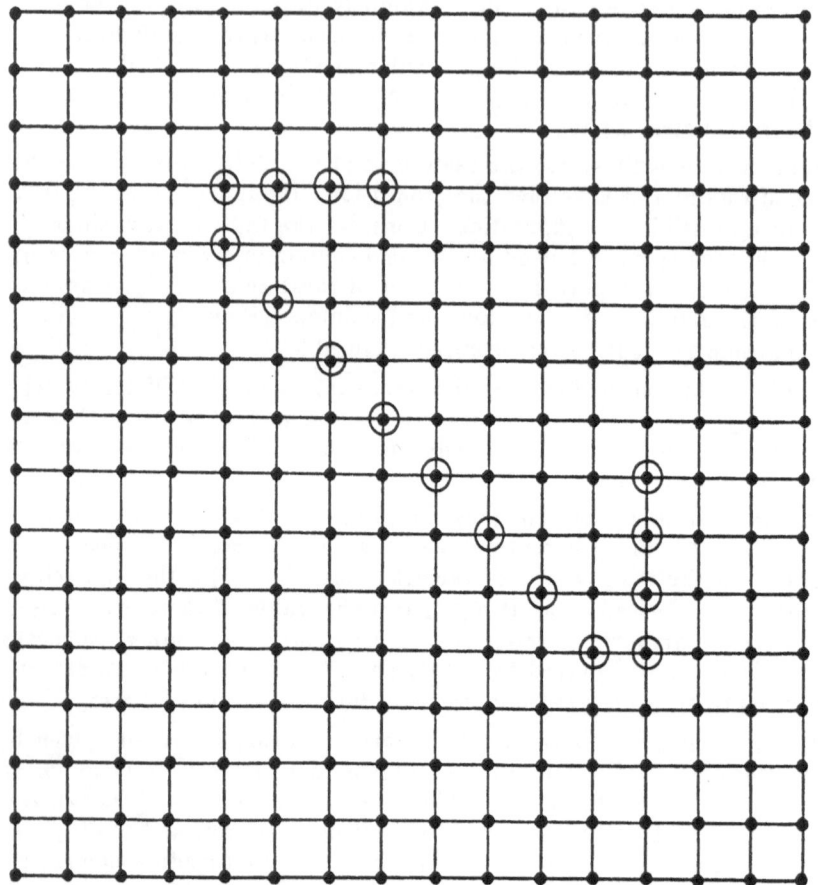

Figure 3

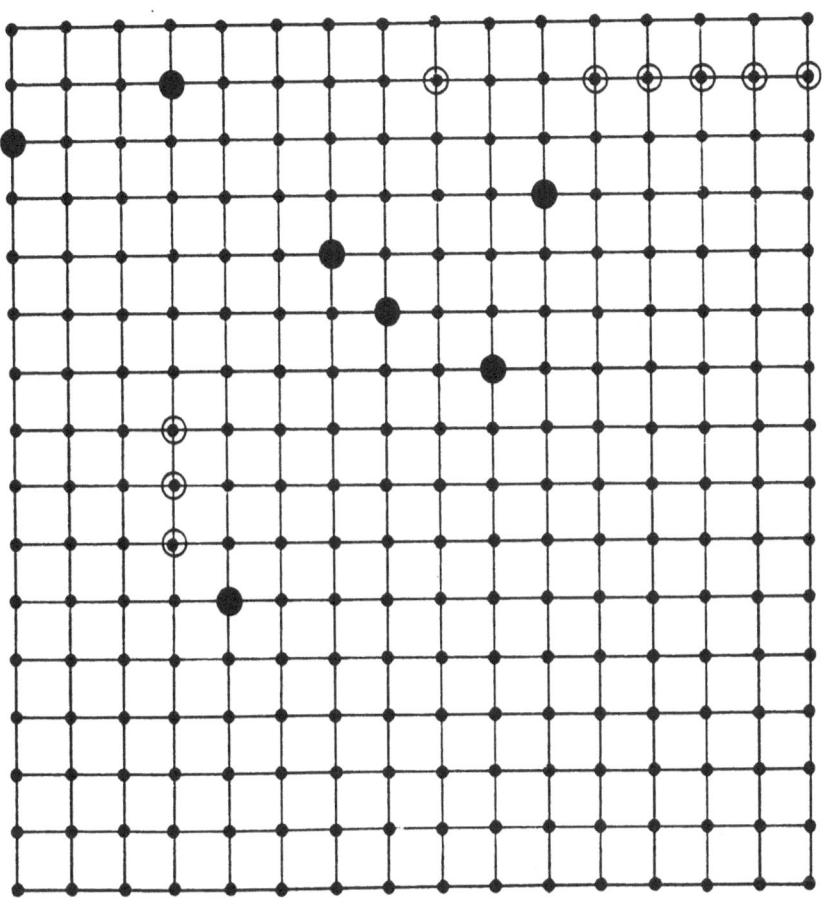

Figure 4

Step 4:   If $var = 1$, $i = 0$ and go to Step (5). Else, if $i \notin S_R$, $S_R = S_R \cup \{i\}$.

Step 5:   If $var = 2$, $j = 0$ and go to Step (6). Else, if $j \notin S_C$, $S_C = S_C \cup \{j\}$.

Step 6: $S = S \cup \{(i,j)\}$. If $var \neq 1$, reset all 1-entries along the ith row.

Step 7:   If $var \neq 2$, reset all 1-entries along the jth column.

Step 8:   If $var = 1$, $C_A = C_A - 1$. If $var = 2$ $R_A = R_A - 1$. If $var = 0$, $R_A = R_A - 1$ and $C_A = C_A - 1$. If $var = 3$, $R_A = C_A = 0$. If $i \neq M$ and $j \neq N$, $var = 0$ and go to Step (2).

Step 9: S is the reconfiguration-set.

Algorithm (1) is applicable under all availabilities of spares. For $R_A \neq C_A$, the cardinality of the reconfiguration-set reflects the difference in number of spares, i.e. $|S| = s = \max\{|S_R|, |S_C|\}$. The default value of 0 in either coordinate of an element of the reconfiguration-set is not included in either S_R or S_C. Spare utilization is considered up to spare availability: in Algorithm (1), when $R_A = C_A = 0$ ($var = 3$), coordinates of leading elements are added to S_R and S_C in a fashion similar to full availability ($R_A \neq 0$ and $C_A \neq 0$).

For the fault pattern of Figure 5, if $R_A = C_A = 3$, then $S_R = \{1,2,4\}$ and $S_C = \{1,3,6\}$. If $R_A = 2$ and $C_A = 4$, then $S_R = \{1,2\}$, $S_C = \{1,2,6,7\}$ and $S = \{(1,1),(2,3),(0,6),(0,7)\}$.

A new set denoted as D, needs to be defined for calculating the reconfiguration-solution from the reconfiguration-set. This set consists of rows and columns, that are elements of S_R and S_C; that is,

$$D = D_R \cup D_C = \{R_i | i \in S_R\} \cup \{C_j | j \in S_C\} \tag{07}$$

where $|D_R| = |S_R|$ and $|D_C| = |S_C|$.

The following Lemma is obtained for D.

LEMMA 1.

The set D covers all nonzero elements of the matrix A.

PROOF:

Omitted. It follows from the results of [21]. □

Lemma (1) can be used to prove the following Theorem that is fundamental in the reconfiguration process.

THEOREM 1.

$|S|$ is not minimal.

PROOF:

Omitted. It can be found in [21]. □

It is possible to determine R_A/C_A-reconfiguration as function of s and number of spares prior to the reconfiguration-solution. This is based on the following Theorem.

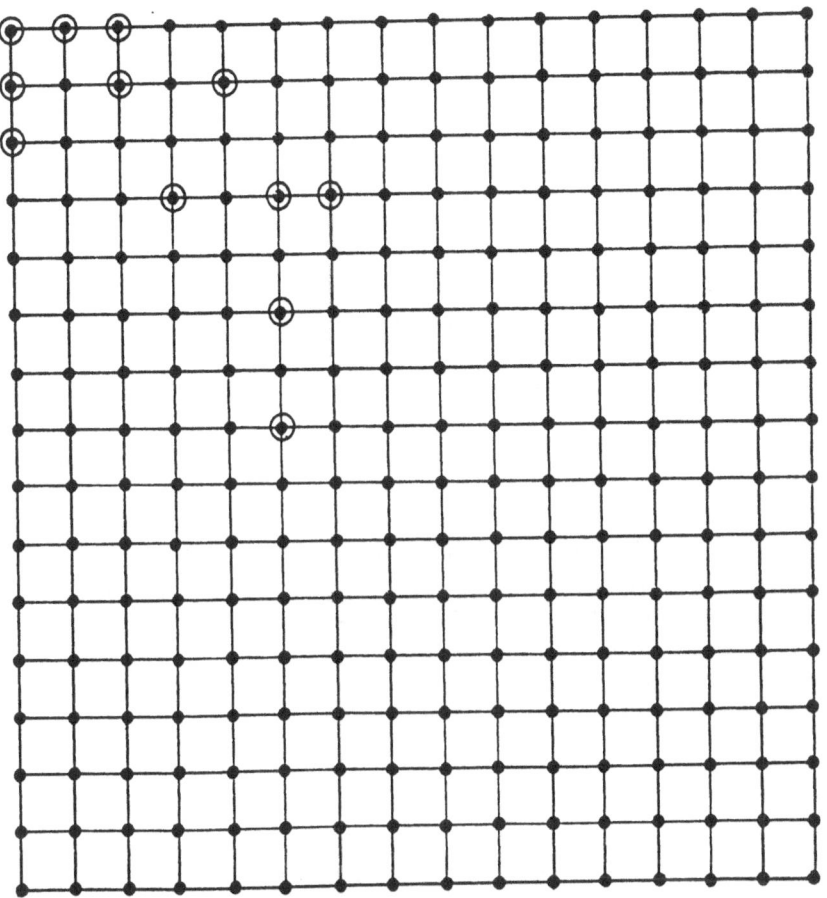

Figure 5

THEOREM 2.

If $s \le \max\{R_A, C_A\}$, the bussed array is R_A/C_A-reconfigurable.

PROOF:

Consider D; by Lemma (1), D covers all the nonzero elements of A. If $s \le \max\{R_A, C_A\}$, then $|D_R| = |S_R| \le |S| = s \le R_A$ and $|D_C| = |S_C| \le |S| = s \le C_A$. There exists a set of $|D_R|$ rows and $|D_C|$ columns, such that all nonzero elements of the diagnostic matrix A are covered and the cardinalities of D_R and D_C do not exceed R_A and C_A. The array is reconfigurable by definition. □

By Algorithm (1) and construction of the reconfiguration-set, if Theorem (2) is satisfied, then $\min\{|S_R|, |S_C|\} \le \min\{R_A, C_A\}$ is also satisfied. Theorem (2) does not establish only reconfigurability of a bussed array, but it also provides its solution. The elements of D_R and D_C are those rows and columns, that are required to be deleted.

For the fault pattern of Figure 5, if $R_A = C_A = 3$; then $|S| = s = 3$, $S = \{(1,1),(2,3),(4,6)\}$, $s = \max\{R_A, C_A\}$. If $R_A = 2$ and $C_A = 4$, then $|S| = |S_C| = \max\{R_A, C_A\} = 4$. By Theorem (2), the bussed array is reconfigurable under both these redundancies.

The reconfiguration-solution is initially given by the reconfiguration-set obtained by using Algorithm (1). This approximate solution is given by D_R and D_C, where R_i and C_i identify the ith row and column. Theorem (2) can be used to detect unreconfigurability.

THEOREM 3.

If $s > R_A + C_A$, the bussed array is R_A/C_A-unreconfigurable.

PROOF:

Suppose that the condition of s holds and the array is R_A/C_A-reconfigurable. There exists a set of R rows such that C columns in A and all nonzero elements of A are covered and $R \le R_A$ and $C \le C_A$. Consider any element (i,j) of S; a_{ij} in A is nonzero. This element is covered by D by Lemma (1). The cardinalities of S_R and S_C do not exceed R and C respectively. Furthermore,

$$s = |S| = \max\{|S_R|, |S_C|\} \le R + C \le R_A + C_A \tag{08}$$

This contradicts the original assumption, i.e. $s > R_A + C_A$. The bussed array is not R_A/C_A-reconfigurable. □

Consider Figure 4; $R_A = C_A = 3$, using Algorithm (1), $|S| = 7$. This bussed array is not 3/3-reconfigurable by Theorem (3).

The advantage of Theorems (2) and (3) is that they establish reconfigurability of a bussed array by only checking the *cardinality* of the reconfiguration-set. This condition is *independent of the distribution of the faulty cells*. The benefit of this approach is that unreconfigurability can be quickly established in parallel with the reconfiguration-set and hence the reconfiguration-solution. This is advantageous, because computational over-head is reduced [20]. If unreconfigurability is quickly established, testing and reconfiguration can be aborted. This yields a low execution-time.

Having established the regions of unreconfigurability $(s > R_A + C_A)$ and reconfigurability $(s \leq \max\{R_A, C_A\})$, a *decision problem* still remains for the *middle region* (referred as *uncertainty region*) of a bussed array i.e. $\max\{R_A, C_A\} < s \leq R_A + C_A$. New conditions in excess to those of Algorithm (1) are required for reconfiguration. Several results are derived to determine reconfiguration/unreconfigurability in the uncertainty region.

It is useful to introduce the definition of a *column complement* for an element of S_C. For a_{ij}, a column complement is denoted by S'_C, where $(i,j) \in S$. This is defined as a nonzero element, say a_{ik}, such that the kth column $C_k \notin S_C$. Similarly, a nonzero element a_{lj} is a S'_R-element (row complement) of a_{ij}, if the lth row $\notin S_R$. To illustrate the definition of complement, consider the fault pattern shown in Figure 6. If $R_A = C_A = 3$, then $S = \{(2,3),(4,1),(5,5),(6,7),(7,9)\}$, $S_R = \{2,4,5,6,7\}$ and $S_C = \{1,3,5,7,9\}$ and $s = 5$. Location (5,11) is a S'_C-element of (5,5), but (5,9) is not a S'_C-element, because $C_9 \in S_C$. By the same argument, (8,5) is a S'_R-element of (5,5); (7,13) and (9,9) are respectively a S'_C-element and S'_R-element of (7,9).

Using the above definitions, the following Lemmas are obtained.

LEMMA 2.

Consider an element, say (i,j) of S; if there exists a S'_C-element and a S'_R-element of a_{ij}, the minimum number of rows and columns, that are required for reconfiguration of a_{ij}, is 2.

PROOF:

There exists a nonzero element a_{ik} and $C_k \in S_C$; if the ith row is not selected as a reconfiguration row, two columns are required: C_j covers the nonzero element a_{ij}, while C_k covers a_{ik}. If there exists a nonzero element a_{lj} and $R_l \in S_R$, either C_j or R_l and R_i are required. If both of the above conditions hold, one of the following cases is the reconfiguration-solution:

(A) R_i and C_j
(B) C_j and C_k
(C) R_i and R_l

The minimum number of rows and columns required for reconfiguration is always two. □

Lemma (2) can be used to find the minimum set of rows and columns in the reconfiguration-solution. This is found by using the reconfiguration-set of Algorithm (1) and complementing all of its elements.

Let S' be the minimum set of rows and columns that are required for reconfiguration. The following Lemma establishes unreconfigurability.

LEMMA 3.

Let s' be the minimum number of rows and columns required for reconfiguration of a bussed array using complements $(|S'| = s')$. A bussed array is not R_A/C_A-reconfigurable if

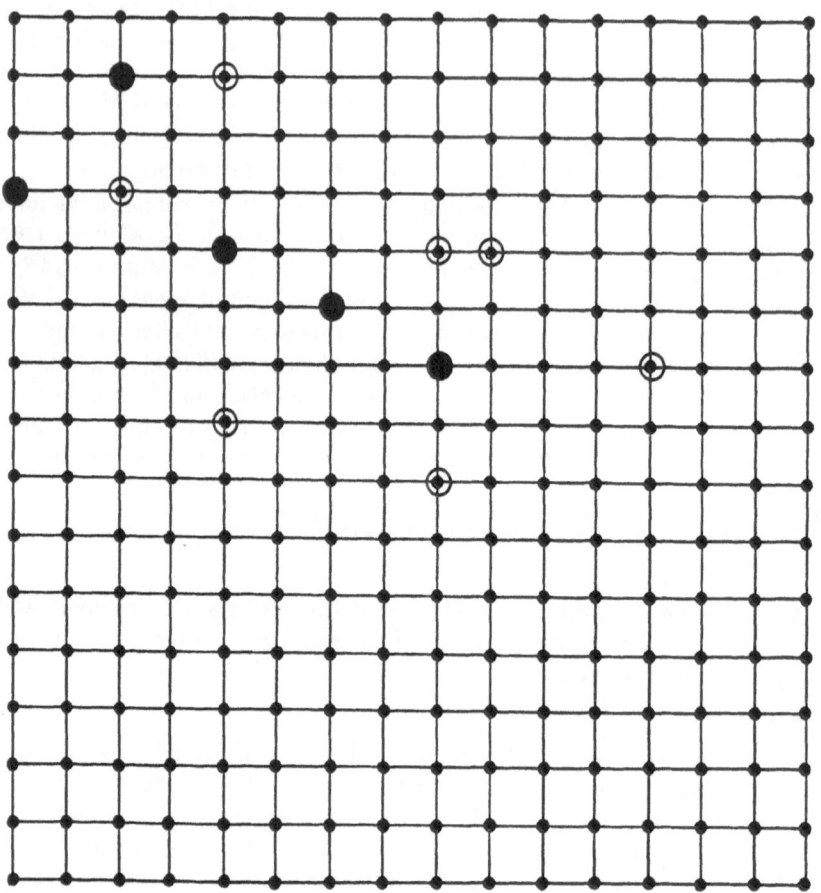

Figure 6

$$s' > R_A + C_A \qquad (09)$$

PROOF:

This proof is similar to the proof of Theorem (2). It will be derived by contradiction. Assume that (09) holds and the array is reconfigurable. By definition of a reconfigurable array, there exists R rows and C columns, such that all nonzero elements in A are covered and R and C do not exceed R_A and C_A. Hence,

$$s' \leq R + C \leq R_A + C_A \qquad (10)$$

This contradicts the initial assumption. The bussed array is not R_A/C_A-reconfigurable. □

The Theorem for unreconfigurability follows from Lemmas (2) and (3).

THEOREM 4.

A bussed array is not R_A/C_A-reconfigurable if

$$s = R_A + C_A \qquad (11)$$

and there exists an element of S, that contains row and column complements.

PROOF:

There exists an element of S that contains a S_C'-element and a S_R'-element. By Lemma (2), the minimum number of rows and columns required for reconfiguration using complements is 2. Therefore,

$$s' \geq (s-1)+2 = s+1 = R_A + C_A + 1 \qquad (12)$$

This array is not R_A/C_A-reconfigurable by Lemma (3). □

Theorem (4) can be generalized to d-elements.

THEOREM 5.

Let

$$\max\{R_A, C_A\} < s \leq R_A + C_A \qquad (13)$$

Let d elements exist in S, each of these elements contains a column complement and a row complement. Suppose that none of the column complements shares a column and none of the row complements shares a row. A bussed array is not R_A/C_A-reconfigurable if

$$s + d > R_A + C_A, \qquad (14)$$

PROOF:

Omitted. This Theorem can be proved using Lemma (3). □

For the fault pattern of Figure 6, (7,13) and (9,9) are respectively the S_C'-element and S_R'-element of (7,9), $d = 2$; by a previous example, $s = 5$, $R_A = C_A = 3$ and

$$s + d = 7 > 6 = R_A + C_A \qquad (15)$$

The bussed array, shown in Figure 6, is not 3/3-reconfigurable.

The reconfiguration-solution is denoted as S'', where S''_R and S''_C respectively denote the row and column sets in the reconfiguration-solution. The basic row/column reconfiguration Algorithm is given as follows.

ALGORITHM 2: BASIC ROW/COLUMN RECONFIGURATION.

Step 1: Execute Algorithm (1) for generation of the reconfiguration-set.

Step 2: Determine if the bussed array is unreconfigurable. If so go to Step (9). Else continue.

Step 3: Using Theorem (2), check if the bussed array is reconfigurable. If so, $var = 0$ and go to Step (5). Else continue.

Step 4: The bussed array has a fault pattern that lies in the uncertainty region; $var = 1$.

Step 5: For each element of S, determine column and row complements.

Step 6: Using Lemma (2), calculate the minimum number of rows and columns required for reconfiguration.

Step 7: If $var = 1$, establish if the array is unreconfigurable using Lemma (3). If so, go to Step (9). Else continue.

Step 8: Determine the number of elements in the reconfiguration-set that satisfy the condition of Theorem (5) and generate the reconfiguration-solution S''. Determine if the bussed array is reconfigurable. If not so, go to Step (9). Else, delete each row and column in the reconfiguration-set.

Step 9: Basic row/column reconfiguration has been completed.

If the content of the *fault counter* for the kth row (column) is denoted as $C_R(k)$ $(C_C(k))$, a total fault count can be generated and used for array reconfiguration. The sum of the contents of all fault row (column) counters is equal to the total count of faulty cells in the array. Reconfigurability using a total fault count is given by the following Theorem.

THEOREM 6.

Let τ be the total fault count and $\max\{R_A, C_A\} < s \leq R_A + C_A$; the array is reconfigurable if either

$$\tau \leq 2(R_A + C_A) - s - 1 \tag{16}$$

for

$$s \geq R_A + C_A - \max\{R_A, C_A\} \tag{17}$$

or

$$\tau \leq 2\min\{R_A, C_A\} + \max\{R_A, C_A\} \tag{18}$$

for

$$s < R_A + C_A - \max\{R_A, C_A\} \tag{19}$$

PROOF:

Consider (17). Suppose that there exist d elements that satisfy Theorem (5). The array is unreconfigurable, because

$$s = R_A + C_A - d \tag{20}$$

Suppose that each of these d elements belongs to a fault pattern in which a faulty cell has an S_R'-element, an S_C' element and an S-element. Each of the remaining $(|S| - d)$ elements contain only an S-element.

By using (20),

$$d = R_A + C_A - s \leq \max\{R_A, C_A\} \tag{21}$$

These d elements require at most d distinct rows and d distinct columns. The $(s - d)$ elements must be reconfigured by using only $(R_A - d)$ rows and $(C_A - d)$ columns. The number of faulty cells is

$$\tau = 3d + s - d = 2d + s \tag{22}$$

$$= 2(R_A + C_A - s) + s$$

$$= 2(R_A + C_A) - s$$

by using (21). If another faulty cell is added to share either a row, or a column with one of the $(s - d)$ elements, (16) is obtained by rearranging (22). (16) is the maximum fault count for which the array is reconfigurable. The proof of (18) is obtained in a similar manner and it is omitted due to lack of space. $\square$

Since

$$\tau = \sum_{k=1}^{M} C_R(k) = \sum_{k=1}^{N} C_C(k) \tag{23}$$

Theorem (6) can be integrated in Algorithm (2) to establish array reconfigurability.

Algorithm (2) relies on a simple generation of complements to reconfigure the bussed array. This requires the calculation of $|S'|$ as the reconfiguration-solution found by inspecting the reconfiguration-set. The criterion that relates S' to complement generation in Algorithm (2) is given by Lemma (2). This criterion is good in the average case. A more sophisticated criterion is required for the worst case analysis and better row/column selection. Algorithm (2) can be optimized for the following reasons.

I. Complement generation aims at covering all those faulty cells, whose either row and/or column is not in S_R and/or S_C respectively. No selection can be made for the number of elements in the reconfiguration-solution.

II. Row/column selection should abide by a differentiation between spares and their available number (i.e. R_A and C_A).

Algorithm (2) satisfies the first three benchmark criteria for a reconfiguration algorithm (given previously in Section 2). The fourth criterion requires a maximization of fault free cells in the array. This requirement must be satisfied if and only if the bussed array is reconfigurable. *Complement counters* are introduced to accomplish a better row/column selection than Algorithm (2). If the ith row (jth column) belongs to S_R (S_C), the number of columns (rows) of faulty cells that would be reconfigured using R_i (C_j) and that do not belong to S_C (S_R), is given by the content of the complement counter of R_i (C_j). This is denoted as $C(R_i)(C(C_j))$. The following Algorithm referred as the Row/Column Selection Algorithm, uses complement counters to differentiate between columns and rows and availability of spares.

ALGORITHM 3: ROW/COLUMN SELECTION.

Step 1: $var = 1$. If $|R_A| > |C_A|$, then $var = 0$ and go to Step (2); if $|R_A| < |C_A|$, then $var = 0$ and go to Step (3).

Step 2: For each R_i that belongs to S_R calculate $C(R_i)$. Sort all row complement counters in a list L_R by ascending order of magnitude. If $var = 0$, select the row corresponding to the largest entry; go to Step (6).

Step 3: For each C_j that belongs to S_C, calculate $C(C_j)$. Sort all column complement counters in a list L_C by ascending order of magnitude. If $var = 0$, select the column corresponding to the largest entry; go to Step (5).

Step 4: If the first entry of $L_R(L_C)$ is greater than the first entry of $L_C(L_R)$, then select the row (column) corresponding to the largest entry. If row (column) is selected, then $Sel = 1(0)$; go to Step (7).

Step 5: $Sel = 1$, go to Step (7).

Step 6: $Sel = 0$.

Step 7: If $Sel = 1$, then $R_A = R_A - 1$, and $|S_R| = |S_R| - 1$; else, $C_A = C_A - 1$ and $|S_C| = |S_C| - 1$.

Step 8: If $|S_R| = |S_C| = 0$ then continue; else, $Sel = 0$ and go to Step (1).

Step 9: Row/column selection is complete.

Algorithms (3) is more efficient than Algorithm (2) when the number of rows and columns required for reconfiguration by using the reconfiguration-set, is much less than redundancy, i.e. $|S_R| < R_A$ and/or $|S_C| < C_A$.

5. COMPLEXITY AND PARTITION.

A critical issue for application of Algorithm (2) is represented by its execution complexity. For the worst case analysis, the largest overhead is given by generation of complements of each element of the reconfiguration-set to satisfy Theorem (5). Complement generation consists of a sequential inspection of column and row of each faulty cell. For F_i faulty cells, F_i^2 basic operations are required (the power of 2 is due to inspection of coordinates in every faulty cell in a reconfigurable array). Hence, Algorithm (2) is $O((R_A + C_A)^2)$ by substituting the maximum value for F_t given by the available redundancy (this is due to the independence of reconfiguration on the distribution of faulty cells.)

The order of complexity of Algorithm (2) restricts its applicability to arrays with small redundancy. A different approach must be implemented if either a large number of faulty cells is present, or the reconfiguration-set is large and a long execution-time is required for finding the reconfiguration-solution. The proposed technique partitions the bussed array into equal subarrays to meet the following objectives.

I. To reconfigure each subarray as best as possible within a given execution-time.

II. A global reconfiguration-solution to the bussed array has to be found by using the reconfiguration-solution of each subarray.

The following criteria must be met to accomplish the above objectives without a significant increase in redundancy and decrease of reconfigurability.

I. Partition of the array into equal subarray does not divide the full availability of redundancy. Redundancy of each subarray is still given by the same number of spare rows and columns as in the unpartitioned array, i.e. the original redundancy is logically assigned to each subarray.

II. Deletion of a row (column) in a subarray implies that the same row (column) must be also deleted in all those subarrays that share this row (column). The global reconfiguration-solution must take into account each subarray solution.

III. A tradeoff between an earliest detection of unreconfigurability and a good reconfiguration-solution must be assessed.

This approach is referred as *reconfiguration by partial partition* due to condition (I), i.e. effectively only the operational bank is divided.

The following Theorem gives unreconfigurability for a partially partitioned bussed array.

THEOREM 7.

An array is unreconfigurable by using partial partition, if at least one subarray is unreconfigurable.

PROOf:

This Theorem will be proven by contradiction. Subarrays are numbered as in Figure 7. The subarray number is identified by a subscript. Assume that one subarray, the ith subarray, is unreconfigurable and the whole bussed array is still reconfigurable. By Theorem (5) and Lemma (3),

$$|S_i'| > R_A + C_A \qquad (24)$$

$$s_i + d > R_A + C_A \qquad (25)$$

where

$$s_i' \leq s_i \qquad (26)$$

For a reconfigurable bussed array,

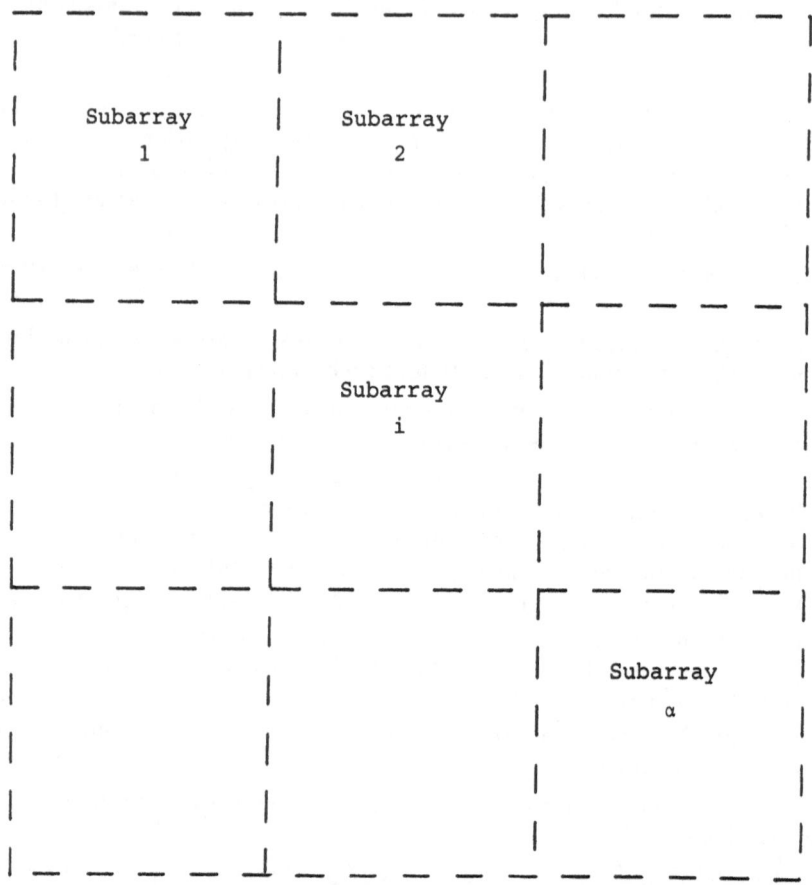

Figure 7: Array Partition

$$|S'| \le R_A + C_A \tag{27}$$

and

$$s + d \le R_A + C_A \tag{28}$$

Faulty cells in the bussed array constitute a superset of faulty cells in the ith subarray. A faulty cell (i,j) that belongs to the subarray, is covered by both S' and S_i' (by Theorem (1)). Then,

$$|S'| = s' \ge |S_i'| = s_i' \tag{29}$$

and

$$S \supseteq S_i \tag{30}$$

(24) and (25) can be equated to (27) and (28). This results in a contradiction. The Theorem is proved. □

The reconfiguration solution of each ith subarray (S_i'') is obtained by using the reconfiguration-solution to the whole bussed array found by Algorithm (2). A new reconfiguration-solution referred as the *global reconfiguration-solution*, must be established without considering the whole array. The global reconfiguration-solution can be obtained by using individual subarray solutions. The global reconfiguration-solution is denoted by $S(G)$, where

$$S(G) = S_1'' \cup S_2'' \cdots \cup S_i'' \cup S_{i+1}'' \cdots \cup S_\alpha'' \tag{31}$$

and α is the number of subarrays of the bussed array. $S(G)$ is calculated only and only if each subarray is reconfigurable where $S(G) \supseteq S$. Hence, by Theorem (2) $S(G)$ is not an optimal reconfiguration-solution. The following conclusions may be drawn by using partition.

I. Unreconfigurability of a bussed array can be established with a lower complexity by using partition and Theorem (7). If every subarray is reconfigurable, the whole bussed array is reconfigurable. Unreconfigurability is given by replacing S with $S(G)$ in Theorems (4) and (5); (31) must be calculated also.

II. (31) is $O(\alpha)$; $S(G)$ must be considered with respect to a reduced complexity of the execution-time.

III. The global reconfiguration-solution does not lead to maximization of fault free cells. This is caused because only subarray reconfiguration-solutions are considered. These solutions do not provide detailed information on faulty cells in each subarray.

The following Algorithm describes row/column reconfiguration by partial partition.

ALGORITHM 4: RECONFIGURATION BY PARTIAL PARTITION.

Step 1: Given the number of equal subarrays, execute Algorithm (2).

Step 2: If at least one subarray is unreconfigurable, then the whole bussed array is unreconfigurable; go to Step (5).

Step 3: Calculate the global reconfiguration-solution $S(G)$ using (31).

Step 4: Using $S(G)$ in Theorems (4) and (5), establish if the array is reconfigurable. If not so, continue. If the bussed array is reconfigurable, go to Step (6).

Step 5: The array is unreconfigurable by partial partition; to Step (7).

Step 6: $S(G)$ is the global reconfiguration-solution.

Step 7: Reconfiguration by partial partition is complete.

Algorithm (4) relies on a full availability of spares to each partition. (31) provides a rather effective unreconfigurability detection. This is achieved at the expense of a good reconfiguration-solution. A different approach to reconfiguration by partition is to divide the overall redundancy in equal parts between subarrays. This is referred as reconfiguration by *full partition*.

For an $M \times N$ array, that has been divided into α equal subarrays, the number of cells in each subarray is given by

$$n_\alpha = \frac{n_a}{\alpha} \tag{32}$$

R_α and C_α spare rows and columns are allocated to each subarray. Unreconfigurability/reconfigurability of a subarray can be determined by substituting these values in Theorems (2), (3) and (5) and using Algorithm (2). The following Theorems establish unreconfigurability and reconfigurability of a bussed array.

THEOREM 8.

An array is R_A/C_A unreconfigurable if at least one subarray is R_α/C_α unreconfigurable using full partition,

PROOF:

Omitted, it can be easily derived similarly to the proof of Theorem (7). □

THEOREM 9.

A bussed array is R_A/C_A reconfigurable by full partition if each subarray is R_α/C_α reconfigurable.

PROOF:

Assume that there exists a subarray that is not R_α/C_α reconfigurable in a R_A/C_A reconfigurable bussed array. This contradicts Theorem (8). The Theorem is proved. □

The following Algorithm determines reconfiguration by full partition.

ALGORITHM 5: ROW/COLUMN RECONFIGURATION BY FULL PARTITION.

Step 1: Given the number of equal subarrays (α), determine if each subarray is R_α/C_α reconfigurable by using Algorithm (2). Establish the reconfiguration-solution of each subarray.

Step 2: If at least one subarray is not R_a/C_a reconfigurable, go to Step (4).

Step 3: The reconfiguration-solution of the bussed array is given by all solutions of the subarrays using (31). Go to Step (5).

Step 4: The bussed array is not reconfigurable using full partition.

Step 5: Reconfiguration by full partition is complete.

Reconfiguration by partition (either partial or full) is still $O(F_f^2)$ for the worst case analysis (all faulty cells could be in a single subarray); however, a net improvement can be expected in the average case with respect to Algorithm (2). Prior to calculate the global reconfiguration-solution, for the worst case analysis Algorithm (4) is still $O((R_A + C_A)^2)$. (31) is $O(\alpha)$, Algorithm (4) is $O((R_A + C_A)^2 + \alpha)$. For fault independence and random distribution, faults will be equally spread in all α subarrays. The average case execution of Algorithm (4) is $O((R_A + C_A)^2/\alpha^2 + \alpha)$. For Algorithm (5) the worst case and average case executions have the same complexity, i.e. $O((R_A + C_A)^2/\alpha^2 + \alpha)$.

6. SIMULATION RESULTS.

A simulation program has been developed to show the effectiveness of the proposed approach to reconfiguration of bussed arrays. Different square arrays have been simulated. Redundancy consists of $\sqrt{N}$ spare rows and $\sqrt{M}$ spare columns. Faulty cells have been simulated using a random number generator. Choice of this distribution has been determined in accordance with the nature of faults in VLSI. A sample of 500 equally like chips has been used for simulation.

Figure 8 illustrates the status of the array (percentile of reconfigurable arrays) versus number of faulty cells. Figure 9 shows unreconfigurability and reconfigurability found using Algorithm (3). The following features of row/column reconfiguration of orthogonal bussed arrays can be deduced from the figures.

I.   Execution of the reconfigurability algorithm is proportional in the average case to the number of faulty cells.

II. No simple solution exists for front deletion. In the presence of many faulty cells, knowledge of their locations enhances probability of reconfiguration by reconfigurability detection.

The 16×16 array has been considered under the following partitions.

1) Partition A: each partition consists of an 8×16 subarray.

2) Partition B: each partition consists of an 8×8 subarray.

3) Partition C: each partition consists of a 4×4 subarray.

Figures 10 and 11 show unreconfigurability and reconfigurability for the above partitions by using reconfiguration by partial partition. Figures 12 and 13 show unreconfigurability and reconfigurability by using Algorithm (4) for reconfiguration by full partition. The following conclusions may be drawn for reconfiguration by partition.

Probability
of
Reconfiguration

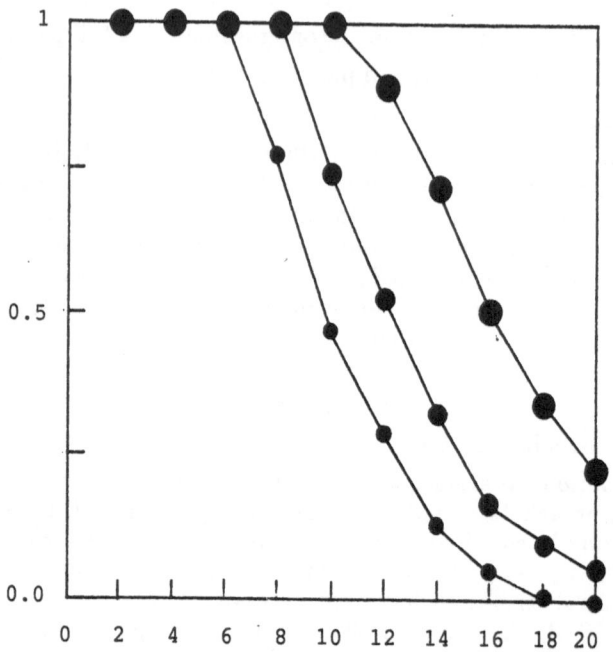

Figure 8: Array Status

Faulty Cells

●————● 9x9 Array

●————● 16x16 Array

●————● 25x25 Array

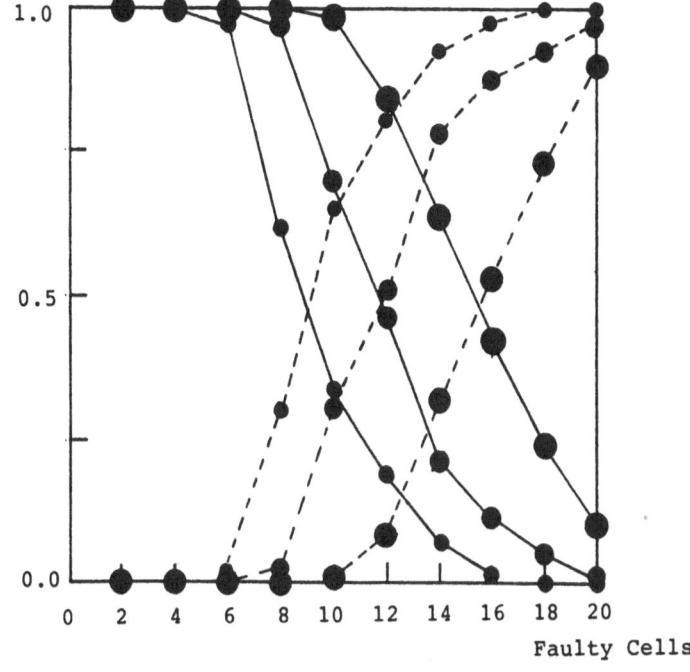

Figure 9

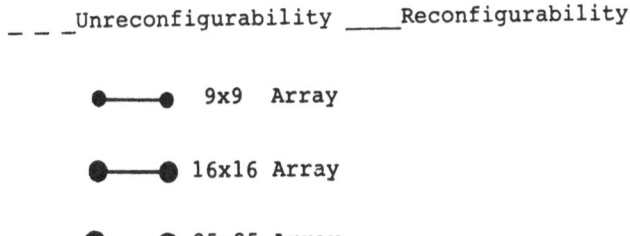

_ _ _Unreconfigurability ____Reconfigurability

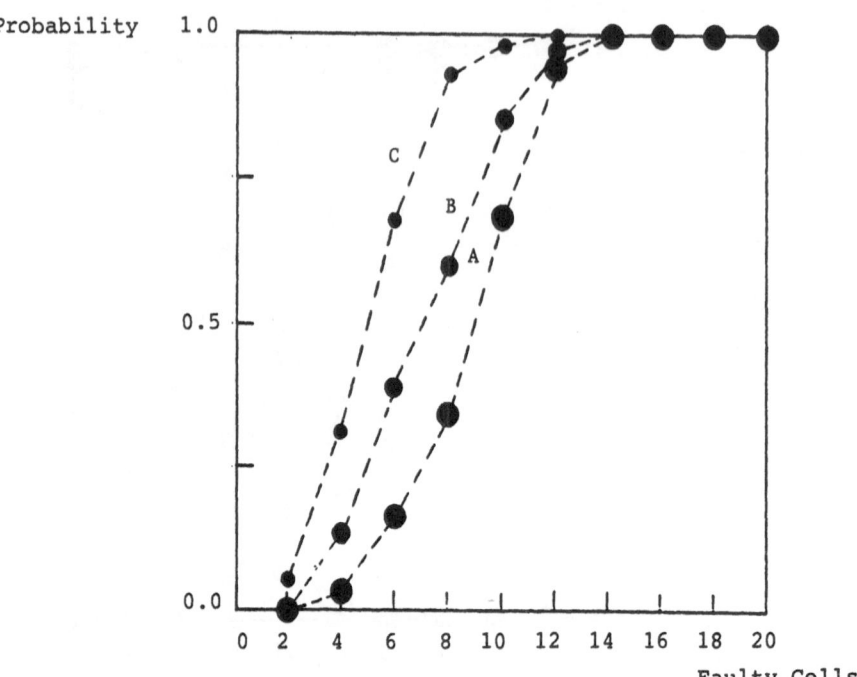

Figure 10: Unreconfigurability
for Different Partitions

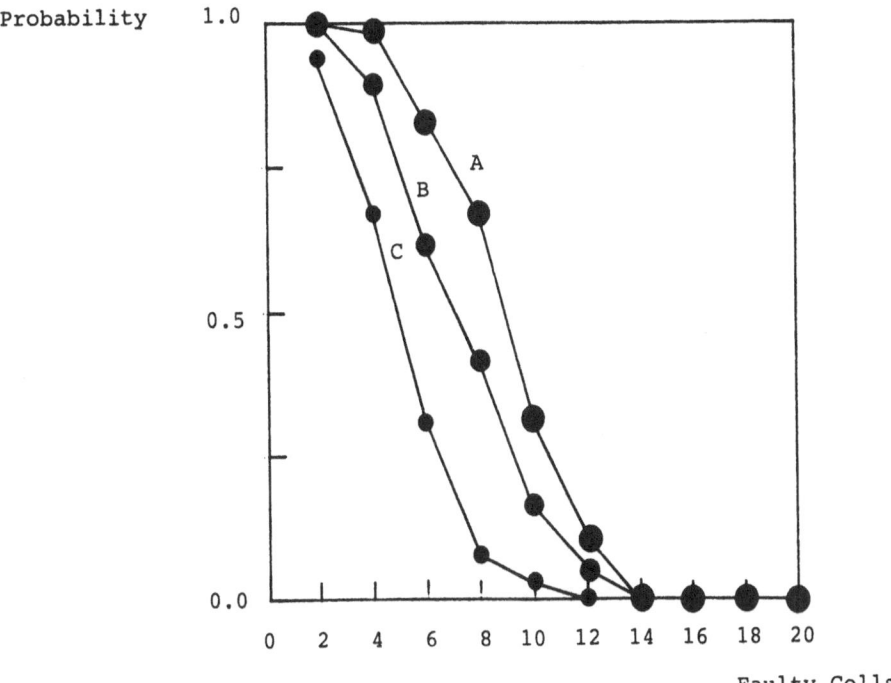

Figure 11: Reconfigurability
for Different Partitions

458

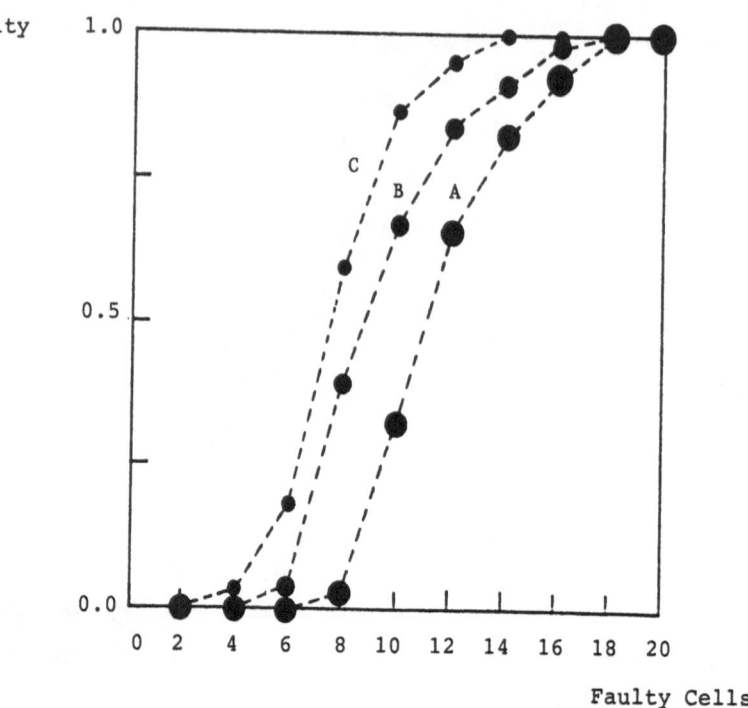

Figure 12: Unreconfigurability
for Different Partitions

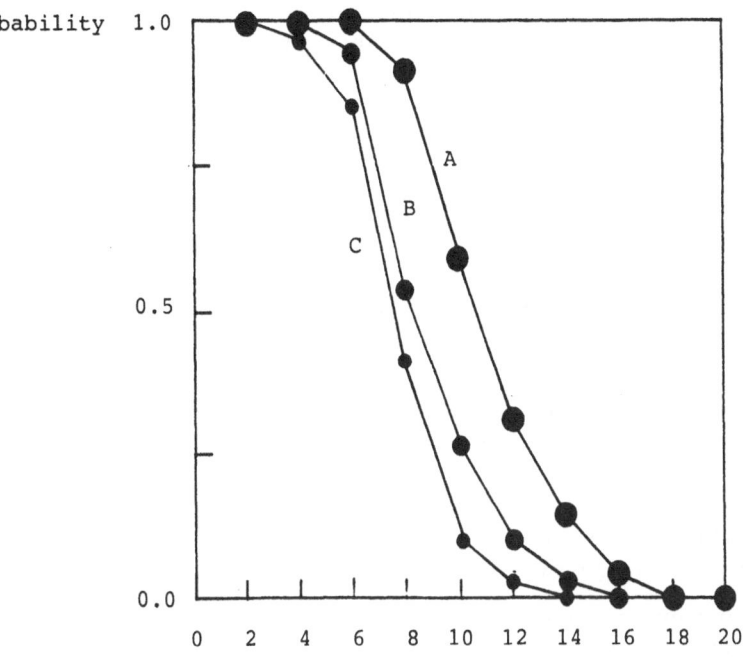

Figure 13: Reconfigurability

for Different Partitions

I. In the presence of a large number of faulty cells, reconfiguration by partial partition loses effectiveness in reconfigurability/unreconfigurability. By using (31), if the array is reconfigurable, subarray complement generation is not very efficient in finding the minimum number of rows and columns.

II. Reconfiguration by full partition is very effective at a low number of faulty cells. At higher number of faulty cells, a larger number of partitions are unnecessary, because this corresponds to a large degradation in reconfigurability/unreconfigurability.

Simulation has pointed out the following features of front deletion.

I. Reconfiguration by front deletion is probabilistically effective, provided that the number of faulty cells does not greatly exceed the number of spare rows and columns.

II. The proposed algorithms for reconfiguration by front deletion can be used in connection with the algorithmic reduction approach of [17] to provide a more complete fault tolerance of VLSI arrays.

III. Unreconfigurability/reconfigurability detection is a very effective early abort technique that can be assessed at low algorithmic complexity.

7. ARRAY DIMENSIONS AND FRONT DELETION.

In the previous sections, reconfiguration has been considered with respect to an $M \times N$ array with R_A/C_A redundancy. Analysis has been restricted to find the condition of reconfigurability and the reconfiguration-solution. In this section a different problem is analyzed. This is the inverse problem, i.e. given a number of cells in the operational bank as in (02) arranged in an $M' \times N'$ array, where $M' = M - R_A$ and $N' = N - C_A$, how can the cells in the spare bank be arranged such that the number of faulty cells that can be tolerated, is maximized?

In the following analysis it will be assumed that $N' > M'$. For orthogonal bussed arrays, front deletion is independent of selection, because faulty cells can be covered by using either kind of spares (either rows or columns). In an $M' \times N'$ array the minimum number of faulty cells that are shown in the diagnostic matrix as leading elements and require a deletion of either a new column or a row, is given by all those cells along the major diagonal, i.e.

$$s_{max} = \sqrt{M' \times N'} = \sqrt{n_0} \tag{33}$$

by using (02).

For the worst case analysis $\sqrt{n_0}$ faulty cells are assumed as upper bound to F_t. It is therefore imperative to organize the redundancy such that given a fixed number of cells in the redundancy bank, the dimensions of the target array can be maximized.

Since $N' > M'$ by assumption, spare cells can be arranged as redundant columns, i.e.

$$\frac{n_a}{N'} = C'_A + r \tag{34}$$

where C'_A is the highest number of redundant columns ($C'_A = C_{Amax}$) and r is the cell remainder. Using (34), the reconfigured array will be at least of $M' \times N'$ dimension, if

$$C'_A \geq \sqrt{n_0} \tag{35}$$

If

$$M' = K'N' \tag{36}$$

($K' < 1$, a fraction), by substituting (36) in (35)

$$C'_A \geq N'\sqrt{K'} \tag{37}$$

or optimally for $r = 0$,

$$\lfloor \frac{n_a}{N'} \rfloor = N'\sqrt{K'} \tag{38}$$

Hence,

$$n_a = \frac{n_0}{\sqrt{K'}} \tag{39}$$

which satisfies the constraint of (05). Table (1) shows some arrangements of redundancy for different values of M' and N'.

This yields the following Lemma for redundancy allocation in orthogonal bussed arrays.

Table (1).

M'	N'	n_a	C'_A	r
12	12	112	9	4
8	18	112	14	0
4	36	112	28	0

LEMMA 4.

In an $M' \times N'$ orthogonal bussed array, for $N' > M'$

$$\frac{n_a}{n_0} \geq \frac{C'_A}{N'} \tag{40}$$

PROOF:

This Lemma can be proved by substituting (39) for n_a and $N' \times M'$ for n_0. Since in general $r \neq 0$, hence the sign of (40). □

In an orthogonal bussed array redundancy utilization can be optimized by adding spares to the smallest dimension of the operational bank. In practice this technique is restricted by manufacturing limitations. In general, spares must be added as redundant rows and columns. This causes an overlap region between spare rows and columns to be present in the redundant bank. For R_A/C_A redundancy, this region consists of $(R_A C_A)$ spare cells that are shared between spare rows and columns. The overlap region has negative effects on redundancy allocation because a larger number of cells can be deleted by a single front deletion.

This characterizes the following Lemma.

LEMMA 5.

The overlap region is maximized in number of cells when $R_A = C_A(N = M$ for $N' = M')$.

PROOF:

Omitted due to lack of space. □

8. SWITCHING

Front reconfiguration requires different switching circuits than those commonly found in the literatures. In [17], a switch configuration has been proposed for row/column deletion. This configuration is more applicable to algorithmic reduction than to physical reconfiguration. The proposed switch configuration is shown in Figures 14 and 15. The basic switching element is a 2×2 Banyan switch [23]. In this switch, the two inputs can be connected to the two outputs by either straight or cross connections. This switching configuration can be made fault tolerant at low hardware cost [24]. The differences between this switching arrangement and commonly found interconnection networks [24] are represented by the sequential clocking of the D lines and the connection between adjacent Banyan switches. The D (disable) lines are used to disable all those cells that lie along elements of the reconfiguration-solution. As shown in Figures 14 and 15, the switch which is associated with each (i,j) cell, is denoted as $SW_{i,j}$. This switch consists of two parts: a row switch, denoted as $SW_R(i,j)$ and a column switch, denoted as $SW_C(i,j)$. The operation of $SW_{i,j}$ is shown in Table (2).

Table (2).

$SW_R(i,j)$			$SW_C(i,j)$		
Connection	D_{1R}	D_{2R}	Connection	D_{1C}	D_{2C}
$(i,j) \rightarrow (i,j+1)$	0	0	$(i,j) \rightarrow (i+1,j)$	0	0
$(i,j) \rightarrow$ Out	0	1	$(i,j) \rightarrow$ Out	0	1
In $\rightarrow (i,j+1)$	1	0	In $\rightarrow (i+1,j)$	1	0
In $\rightarrow$ Out	1	1	In $\rightarrow$ Out	1	1

Operation of the switch is controlled by data placed on the D lines. These lines are loaded with the reconfiguration information provided by the host. Data is loaded in each register and shifted synchronously either along the row or the column. This loads all switches with the correct information to bypass any of the cells which lie along either a column, or a row of the reconfiguration-solution.

An algorithm for reconfiguration of an orthogonal bussed array can be formulated. This algorithm is based on the generation of two switching matrices DR (for row connections) and DC (for column connections). The content of an (i,j) entry in DC is given as in Table (3).

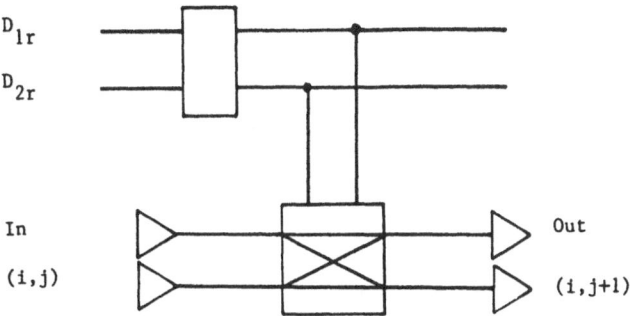

Figure 14: Row Switch

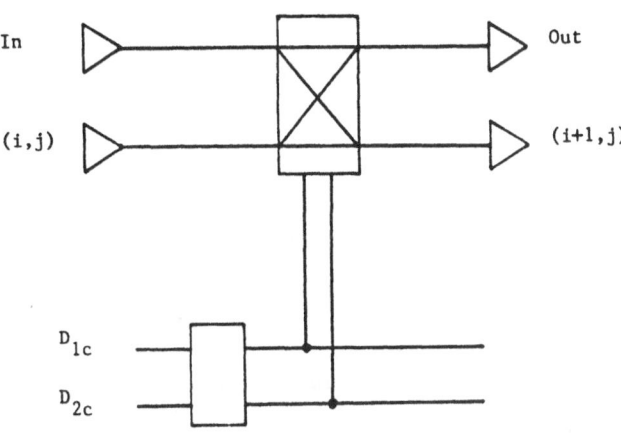

Figure 15: Column Switch

Table (3).

Content of (i,j)	Conditions
0 0	$R_i, R_{i+1} \notin S_R''$
0 1	$R_i \notin S_R'', R_{i+1} \in S_R''$
1 0	$R_i \in S_R'', R_{i+1} \notin S_R''$
1 1	$R_i, R_{i-1}, R_{i+1} \in S_R''$

Table (4) shows the content of an (i,j) entry in DR.

Table (4)

Content of (i,j)	Conditions
0 0	$C_j, C_{j+1} \notin S_C''$
0 1	$C_j \notin S_C'', C_{j+1} \in S_C''$
1 0	$C_j \in S_C'', C_{j+1} \notin S_C''$
1 1	$C_j, C_{j-1}, C_{j+1} \in C_C''$

The algorithm is given as follows.

ALGORITHM 6: SWITCHING IN ORTHOGONAL BUSSED ARRAYS.

Step 1: Using the reconfiguration-solution, generate the switching matrices DR and DC.

Step 2: Load the DR matrix from right to left through the D_R lines.

Step 3: Load the DC matrix from bottom-up through the D_C lines.

Step 4: Switching has been completed.

Figure 16 shows a reconfigured array. Large black dots identify cells that lie along the reconfiguration-solution.

9. CONCLUSIONS.

A new approach to array reconfiguration has been presented. Reconfiguration is implemented by deletion of cells along the front of computation. Various algorithms have been proposed for either row or column deletion. The algorithms compute unreconfigurability, reconfigurability and reconfiguration-solution of arrays. It has been proved that by using this approach, time overhead is small for reconfiguraton. This is accomplished by an early detection of unreconfigurability.

Different techniques have been proposed. These techniques rely on manipulating information provided by counters for each front of the array. Using these counters, simulation results have proven that very good measures for both reconfigurability and reconfiguration are possible. Switching circuits that are required for implementation of the proposed reconfiguration approach, have been presented.

10. REFERENCES.

1 IEEE Spectrum, "The one-month chip", Vol. 21, No. 9, pp. 40-46, Sept. 1984.

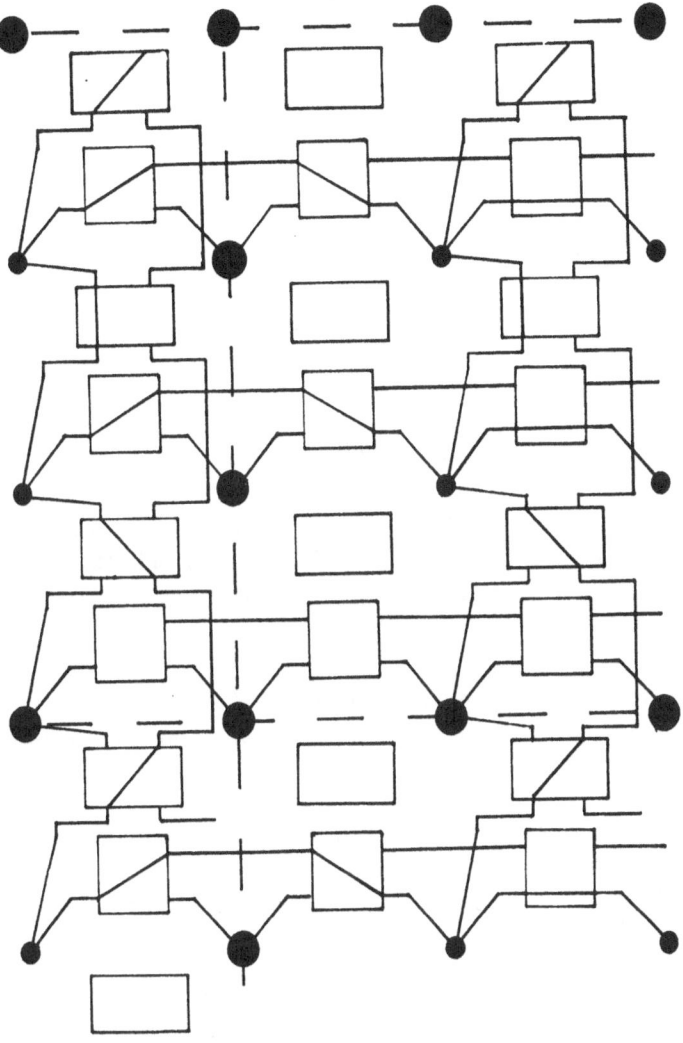

Figure 16: Switching

2   Schuster, S.E., "Multiple word/bit line redundancy for semiconductor memories", *IEEE J. Solid-State Circuits*, Vol. SC-13, pp. 698-703, 1978.

3   Heckelman, R.W. and D. K. Bhausar, "Self-testing VLSI" *Proc. IEEE Solid State Circuits Conf.*, pp. 174-175, 1981.

4   William, T.W., "VLSI Testing" *IEEE Computer*, Vol. 17, No. 10, pp. 126-136, 1984.

5   Kung, H.T. and C.E. Leiserson, "Algorithms for VLSI Processor Arrays" in *Introduction to VLSI Systems*, C. Mead and L. Conway eds., Section 8.3, Addison Wesley, Reading, Mass, 1980.

6   Uhr, L., "Computer Arrays and Networks: Algorithm-structured Parallel Architectures", Academic Press, New York, NY, 1982.

7   *IEEE ISSCC Digest of Technical Papers*, Section on Memories and Redundancy Techniques, pp. 80-87, 1981.

8   Rosenberg, A.L., "The Diogenes Approach to Testable Fault Tolerant VLSI Processor Arrays," *IEEE Trans. on Comp.*, Vol. C32, No. 10, pp. 902-910, 1983.

9   Leighton, T. and C.E. Leiserson, "Wafer-Scale Integration of Systolic Arrays," *IEEE Trans. on Comp.*, Vol. C34, No. 5, pp. 448-461, 1985.

10   Batcher, K.Z., "Architecture of a Massively Parallel Processor," *Proc. 7th Symp. Comp. Arch.*, pp. 169-173, 1980.

11   Snyder, L., "Introduction to the Configurable Highly Parallel Computer," *IEEE Computer*, Vol. 15, No. 1, pp. 47-56, 1982.

12   Kulkarni, A.V. and J. Vuillemin, "Systolic Processing and an Implementation for Signal and Image Processing," *IEEE Trans. on Comput.*, Vol. C31, No. 10, pp. 1000-1009, 1982.

13   Leiserson, C.E. and J.B. Saxe, "Optimizing Synchronous Systems," *Proc. ACM FOCS*, pp. 23-36, 1981.

14   Kung, H.T., "Let's Design Algorithms for VLSI Systems," *Proc. Caltech Conf. on VLSI*, pp. 59-90, 1979.

15   Garey, M.R., and D. S. Johnson, "Computers and Intractability," Freeman, San Francisco, CA, 1979.

16   Kung, S.Y., Arun, K.S., Gal-Ezer, R.J. and D.V.B. Rao, "Wavefront Array Processor: Language, Architecture and Applications," *IEEE Trans. on Comp.*, Vol. C31, No. 11, pp. 1054-1066, 1982.

17   Fortes, J.A.B. and C.S. Raghavendra, "Gracefully Degradable Processor Arrays," *IEEE Trans. on Comp.*, Vol. C34, No. 11, pp. 1033-1044, 1985.

18   Malek, M. and Y.H. Choi, "Real Time Diagnosis of Homogeneous Systems," *Proc. Real-Time Systems Symp.*, pp. 155-159, 1985.

19   Sami, M.G. and R. Stefanelli, "Fault Tolerance of VLSI Processing Arrays: the Time Redundancy Approach", *Proc. Real-Time Systems Symp.*, pp. 200-207, 1984.

20 Negrini, R., Stefanelli, R., and M.G. Sami, "Fault Tolerance Technique for Array Structures used in Supercomputing", *IEEE Computer*, Vol. 19, No. 2, pp. 78-87, 1986.

21 Wey, C-L and F. Lombardi, "On the Repair of Redundant RAMs," *IEEE Trans. on CAD of ICs and Systems,* Vol. CAD6, No. 2, pp. 222-231, 1987.

22 Kung, H.T., "Systolic Algorithms for the CMU Warp Processor," *Proc. 7th Int. Conf. on Pattern Recogn.*, pp. 570-577, 1984.

23 Goke, L.R. and G.J. Lipovski, "Banyan Networks for Partitioning Multiprocessor Systems", *Proc. 1st Ann. Symp. on Comp. Arch.*, pp 21-28, 1973.

24 Cherkassky, V., Opper, E. and M. Malek, "Reliability and Fault Diagnosis Analysis of Fault-Tolerant Multistage Interconnection Networks", *Proc. FTCS*, pp. 246-251, 1984.

DEVICE TESTING AND SEM TESTING TOOLS

J.P. COLLIN
IBM FRANCE
Dept 1093 - BP 27
33610 CESTAS
FRANCE

B. COURTOIS
IMAG/TIM3
46 Av. Félix Viallet
38031 GRENOBLE Cedex
FRANCE

ABSTRACT

The aim of this paper is three fold. In a first part, the basics of electron beam testing are reviewed. Next, a second part details experiences in using electron-beam testing for failure analysis of memories and microprocessors. The third part gives results on the use of an electron beam tester in a more structured methodology for failure analysis and debug of ICs.

A conclusion presents limits and alternatives to electron-beam testing.

I - INTRODUCTION

E. Beam testing and more generally internal non contact testing are considered today as powerful techniques to cut VLSI debugging time and face the challenges of VLSI failure analysis. These chalenges are not only related to the evolution of VLSI components themselves ; they are conditionned by the overall environment : time of response of the failure analysis in a semiconductor line or in vendor relationships, IC's redesign and debugging cycles time. Business and technical decisions will be taken according to the results and conclusions of such analysis ; failure analysis is moreover considered as the ultimate feedback loop to estimate the reliability and quality of the products. This explains the strategic importance of failure analysis and diagnostic testing in IC's industry. Several basic features of VLSI technology and architecture limit seriously today the standard solution of mechanical microprobing :

- small features sizes : between 1μm and 2,5 μm for the bulk production compare to 2/3 μm final diameter of a mechanical microprobe.

- speed of the circuits (4) : very large measurement bandwidths are required today in order to measure electrical performances of a gate (1 to 10 GHz in ECL technology). The voltage resolution must be also compatible with the internal electrical swings inside the chip (100 mV in Drams). To face these requirements mechanical probing does not present an infinite impedance with no capacitance relatively to the typical impedance of the net to be tested. This limitation will not only introduce perturbations for the measure itself but can also sufficiently disturb the circuit in such a way it can be no more functional in nominal conditions (bit lines in Drams).

- high density and complexity : (1, 2, 3) the ratio number of internal gates to the number of external access pins is even higher. This leads to a real bottleneck for diagnostic testing and failure analysis which generally require both direct and random access to internal nodes. It will result non ambigous reasurements and will not require multivectors testing for faults hypothesis validation.

- technology evolution : the actual technology presents often more than three metallurgy levels, the last one being often reserved for power supply and clocks distributions. Conventional mechanical probing does not permit of course a direct probing of buried layers through the passivation or the interlevel oxydes.

- test throughput and test strategy implementation : complex failures require often to test a large number of internal modes in a random and nonrepetitive way. The mechanical probing even with large probe cards does not give the sufficient flexibility and electrical contact reliability, when the test method requires hundreds of random measurements.

469

F. Lombardi and M. Sami (eds.), Testing and Diagnosis of VLSI and ULSI, 469–508.
© 1988 by Kluwer Academic Publishers.

To each and all of the problems listed above, E. Beam testing gives attractive solutions (13). Its main characteristics are the following (6) (7) :

- many modes of operations which include internal waveforms measurements, dynamic electrical and logical states imaging, selective functional imaging, logical timing analysis, etc...

- non contact testing : this includes an infinite impedance with no capacitance and non loading of the circuit.

- beam positionning accuracy well under 0,1 µm in a few microseconds for multinodes testing.

- subnanosecond temporal resolution (100 ps in commercial equipments).

- submicron spatial resolution.

All these capabilities are in fact powerful motivation for development efforts on the tool itself (in order to increase performances and its feasibility) but also on applications techniques (see 5, 9, 11, 12, 13, 19, 20, 21, 22, 24, 25, 26, 27) as they will appear in this paper, which will first introduce the basic principles.

II - BASICS OF E. BEAM TESTING (6, 34, 8)

A focused primary E. Beam (energy of a few kiloelectronvolts) products secondary electrons with the following characteristics : energy of a few electronvolts and linear dependancy of the total energy versus the electrical potential of the emission point.
The secondary electrons are collected and sensed by a detector which forms the useful signal of the E. Beam tester.
The detector is in charge of collecting the maximum of secondary electrons, of discriminating the voltage levels of the emission points (the scanned IC's surface) and of minimizing the parasitic influence effects of the neighbouring nets.

There is a positive retarding field around + 5 volts nets that tends to reduce the amount of secondary electrons compare to 0 volt net. It is responsible for the "qualitative" voltage contrast effect. A retarding grid inside the detector is generally used to make an energy high pass filter, which accept all electrons, which have an energy greater than a threshold fixed by the retarding grid. By adjusting this level, the detector scans the entire secondary electrons energy spectrum ; the detector becomes a real energy spectrometer allowing a quantitative voltage contrast effect measure. The collector efficiency and the local field parasitic effect are both optimized by an extraction grid. The voltage resolution of such a detector can be as low as a few millivolts. The linearity of the measurements is obtained with the aid of a feedback loop which adjusts the value of the filter grid voltage to obtain always the same amount of secondary electrons : the variations of the filter grid voltage are so equal to the variations of the emission point voltage. Better than 1 % linearity can be achieved with such a closed loop.
In order to obtain the best temporal resolution the primary electron beam is sampled. The internal IC's waveforms are so digitalized by the beam as a sampling oscilloscope. With this principle subnanosecond temporal resolution can be achieved. An E. Beam tester will be so an assembly of a column and a gun, which form, focuse, scan and sample the beam which impacts the surface of the chip, a chamber with external electrical connections to a tester, electronics of control, images and waveforms displays and interfaces with computers and CAD environment. The chamber and the column operate at a standard level of vacuum. Many techniques are based on the physical and electronics principle described above (see 7, 8, 9, 10, 11, 12). They are differentiated by the way they use the available basic signals of the E. Beam tester : the test vectors, the scanning synchronization signals, the sampling signal itself. We will briefly describe each of these techniques beginning by the simplest one.

- low frequency and static E. Beam Testing : this is generally the first step of the analysis ; it allows the visualization of static voltages, low frequency voltages (a few tenths of hertz in order to be followed by the human eye in real time or on a video tape) or high frequency voltages (without discrimination of frequency, phase or voltage level but with a grey value which corresponds to the duty cycle of the visualized signal). The buried layers can be visualized when their corresponding signal is not static (this phenomena called capacitive coupling corresponds to the charge and discharge of the oxyde under the polarization of the active layer and the charge of the primary E. Beam).

- window contrast or video gating (see 9) : the static and low frequency E. Beam testing does not allow the discrimination of the signal in phase and frequency. The study of dynamic signals is possible by using the principle of stroboscopy. The stroboscopy can be achieved by sampling the signal chain of the E. Beam tester (from the source of primary electrons to the last video amplifier). For obvious reasons the most efficient sampling has to be implemented before the slowest block of the E. Beam tester (the detector with its closed loop presents generally a narrow bandwidth). Video gating is a stroboscopy which is installed on the video chain of the detector. It is very cheap and easy to implement. The principle of the method is to fix a "window" of observation inside the test sequence, in order to fix the observation of the signal. Generally a boxcar average is used for a proper signal to noise ratio. The observation of a complete cyclic test sequence is obtained by moving the "window" phase.The temporal resolution is dependent on the type of detector : it is around 100 nanoseconds ; that is generally sufficient for logical timing analysis and stroboscopic imaging. One of the main advantage of this technique compare to other stroboscopic modes is its high signal to noise ratio ; that gives better pictures and low integration times.

- voltage coding (see 10) ; the technique of voltage coding allows a transposition of a temporal test sequence of a set of signals into a spatial sequence on all the points of the IC's surface which are concerned by the test sequence. This transposition is achieved by synchronizing the scanning signals of the E. Beam tester (line and frame blanking) with the test sequence. This synchronization will make appear on the image horizontal or vertical white and black stripes, the lengths of which are dependent on the various duty cycles of the signals. The voltage coding is so a "logic analyser" with a lay out background. This is visible on the first picture 10, which shows an address buffer on a static RAM : the generation of address and inverted address corresponds to out-of-phase stripes on the corresponding conductors. The next picture 10 shows the generation of two phases inside a microprocessor from the clock frequency. This division per two of the initial frequency are transposed on the image and doubles the corresponding stripes on the conductors. An interface module between the electrical tester and the scanning signals of the E. Beam tester is generally required, in order to divide or multiply, and ensure a good phase synchronization. Voltage coding is limited in frequency by the resolution of the display and the detector bandwith : a few megahertzs is the maximum.

PHASE SELECTIVE VOLTAGE CONTRAST (PSVC) (11) (9)

The information included in an E. Beam testing image is relatively rich and we can say that often this quantity of informations can be an obstacle for a quick analysis of a given function or signal inside the chip. PSVC was historically introduced (11) as a tentative of non voltage contrast effects substraction (topographical informations, surface charging effects, noise, etc...).The inherent secondary electrons signal complexity rather mades from this effort on optimization of electrical voltage contrast effects. This application is still interesting for optimal buried layers imaging (see pictures III). In fact the main advantage of PSVC is its capability to select an internal function of the component. The selective imaging of internal suboperations corresponding to an external function is a powerful tool to verify a set of faults hypothesis and make alternative choices. For example a failed memory write sequence can be due to bad cell addressing, a defective cell... (see pictures II). A second advantage is to give to the an.Valysi a quick functional subclocks identification (see pictures I). The basic electronic principle of PSC is a substraction of two samples of the secondary electrons signal. The result will be an enhancement of the electrical and logical phenomena which changes from a sampling phase to the other and a reduction of the events which do not synchronously change with the sampling clocks. So the electrical test has to be an image of the expected selective effects. It means that the test program has to be symetrical in order to present the two electrical/logical states of the internal phenomena.

```
                            A
      ___     ___    __    __     __    __
Test: ___     ____    ____    ___   _____     ____    _____
                                                                  Ā


PSVC:          __                                      __
clocks _____    _____    ____
```

RESULT : visualization of the phenomenon "A" on every concerned conductor.

	PSCV (a)	CLOCKS (b)
TEST : Load "1" ===> register A	1	0
"1" ===> register B	0	1
"0" ===> register A	1	0
"0" ===> fegister B	0	1

RESULTS : case (a) visualization of register A
case (b) visualization of register B

The module of PSVC is installed inside the video chain of the E. Beam tester. It is therefore limited in frequency to a few megahertzs but its functional interest covers most of the first steps of the failure analysis.
This module is a "push button" rack inside the E. Beam tester with a software module inside the PC microcomputer which controls the phases of the clocks of the module in reference to the electrical tester. An evolution of the hardware module is foreseen in order to replace the video signal sampling by the E. Beam sampling ; this will give better temporal resolution.

FREQUENCY TRACING (see 12)

Frequency tracing has been introduced in order to extend the frequency range of PSVC and selective voltage contrast.
This technique uses the well known beteradyn principle : the primary E. Beam is sampled at a frequency FS, the observed signal has a frequency FT, the resulted video signal will then present the resulting frequencies FS + FT and FS - FT. The last one is very interesting because it can be very low and so enter into the bandwidth of the detector. The temporal resolution of such a technique is so no more limited by the detector bandwidth but only by the capability to form a short sampling pulse of primary electron beams. The beam scans the interesting area giving a selective imaging of the conductors which are concerned by the frequency FT.

FREQUENCY MAPPING (12)

Frequency mapping is a variation of frequency tracing. One dimension of the spatial scanning is removed and replaced by a frequency scanning around a "running" frequency. The interest of such a technique is that this frequency can be unkown and revealed by the imaging. This can evidence interferences between two electrical or logical phenomenas, parasitic frequencies, etc...

LOGIC STATE TRACING (12)

This technique allows a selective visualization of conductors which are characterized by a given logical sequence, a "signature". That is a very powerful technique to "follow" a complete sequence inside the chip and to compress complex information. This technique is realized with the aim of a correlator which is installed inside the detector chain.

STROBOSCOPIC IMAGING AND PROBING

As explained for video gating the stroboscopic principle allows the static visualization of dynamic events. The implementation of a sampling module in the column of the E. Beam tester gives a subnanosecond temportal resolution for most of the equipments. The pictures XI show the two main modes : imaging and probing. In the first case the electron beam scans the chip to form an image at a given sampling phase ; the result is a frosen state image of the chip. The photos XI show four states of an address buffer of a static memory at form phases with a temporal resolution of one nanosecond.

If the beam does not scan the surface but status at an internal node, and the sampling phase scans a range inside the test sequence, the result (see picture XI) is a waveform with the temporal resolution corresponding to the pulse width of the E. Beam probe and the voltage resolution of the detector.

The pictures XII show an other example of probing inside the 8048 microcontroller RAM memory. The waveforms indicate a stuck at "1" of a cell from the "external testing point of view" (1/2) and from the "internal testing point of view" (3/4).

A mixed mode is obtained when the phase is scanned during the spatial scanning of the surface of the chip. The result (see picture XIII) is a stretched "moire" of a complete test sequence : we can see here a double accumulator complementation sequence and the corresponding bus activity of a 8085 microprocessor.

("black" is 5 volts, "white" is 0 volt)

III - EXPERIENCING ELECTRON-BEAM FOR FAILURE ANALYSIS

The objective of VLSI diagnostic testing and failure analysis is to identify the design bugs, the technological defects and the marginal defects.

The quantity and quality of the informations is one of the key parameters of the success of such analysis. In the best case the simulator, the line data base, the product life data base will point out the place of the defect, the analysis will just correlate the diagnostic and precise the failure mechanism. In the worst case (i.e. vendor product) a set of vectors and a "data book" level product knowledge will constitute the inputs of the analysis. Between these two extreme situations, one can imagine several levels of information : logical design data base, reverse engineering of the vendor product etc...

This quantity and quality of information will condition the methodology of defects search.

In the worst case a very powerful methodology will consist on the utilization of "vision tools" of the internal IC's activity like E. Beam testing ; we will then compare a golden reference with the failed chip. This comparison can be achieved in an unique chip if the failed function can be compared to an equivalent (i.e. addressing problems in a memory, bit fails in a microprocessor register etc...).

The pictures VII illustrate quite well the procedure. An external test (picture VII.1) evidences a stuck at 1 of the bit 3 at the address 56 of the RAM memory of a 8048 microcontroller. The picture VII.2 shows the corresponding read activity inside the RAM for various neighbouring cells. The defective cell (cell 16) is characterized by a lack of activity on the bit lines. The physical reveals an oxyde defect (picture VII.3).

The pictures VIII (1/2) show an "in situ" comparison inside a 8085 microprocessor : they reveal a nonoactivation of the bit 1 of the H, L, D registers compare to the other bits. After removal of the aluminum level a crack on a polysilicon land is evidenced. (Pictures IX 1.2). The last example of failure analysis concerns a 64 K DRAM. The external failed pattern was a "march" type test pattern.

The external diagnostic was a problem of addressing of an entire row (row 16). Picture IV.2 shows a pair of good row ; picture IV.3 shows a good row pair with the failed row. This last one was addressed when a certain addresses profile was achieved. This was confirmed by the waveform analysis (figure V.2) : the word line 16 presents a parasitic signal which is sufficient to addres it (the largest signal is a good reference one). An electrical analysis based on a local reverse engineering of the suspected area has guided the physical analysis, which has revealed a local diffusion scrash.

These analysis were performed without any precise information on the internal architecture of the product ; the defects were moreover tenous and for each case located at a buried layer. The methodology was based on a good external test analyses in order to well characterize the problem and make faults hypothesis based on the behaviour of the circuit.

Then the analyst will open the failed chip and if necessary a reference one. After a quick optical visual of the chip, he will put it into the E. Beam tester chamber (with the reference) and proceed to the comparisons using the E. Beam testing techniques. He will generally use at the beginning techniques like phase contrast in order to identify the physical area corresponding to the performed function.

The pictures I.1 show the activation of the AX register in a 8088 microprocessor. Pictures I.2 and I.3 show the activity of the cary flag of this microprocessor during a "clear/set carry" sequence. This step in the analysis can be essential to orientate the analysis towards the interesting area. The phase contrast technique is very much productive for this step because it visualizes selectively the functions, for an unknow structure this advantage is unique.

This technique or the frequency tracing technique can also be used to follow a given functional signal inside the entire chip (a clock or an address bit. See pictures III.1/2/3).

The analyst will then use more analytical techniques like stroboscopic imaging or probing.

IV - MORE STRUCTERED APPROACHES FOR FAILURE ANALYSIS AND DEBUG

In this section, structured approaches are addressed for failure analysis and debug of ICs.

IV-1 - FAILURE ANALYSIS OF CIRCUITS (20)

In this section, failure analysis is addressed supposing that the electrical (logical) scheme of the circuit is not known.

It is just supposed that some functional tests are known, that manifest the defect (a single one). The circuit can be either combinational or sequential. It is firstly supposed to be combinational.

The SEM observation is thus giving an image of the voltage of the points of the circuit, for some input vector (a rectangular grid for example). An observation of the same area (some grid) for a circuit without failure and for the analysed circuit allows to note all points for which there is a discrepancy. These points include the location of the defect and the locations of errors manifested by the defect (i.e. locations for which the defect has been propagated). Of course, such an input vector manifesting the defect is supposed to be known. This is illustrated by the figure 1: the points marked x on figure 1 are the possible locations of the defect. Suppose now we know another input, different of the first one, manifesting the defect. An other image similar to the figure 1-c can be obtained, for which the marked points include the defect and the errors manifested by the second input. It is clear that the defect has to be located among the marked points that appear on the first figure 1-c and on the second one.

It is understandable that an ongoing process will restrict the possible locations of the defect, until to end with a single location, according to figure 2.

The specific problem due to a sequential circuit instead of a combinational one is that errors can be memorized in this circuit, creating other errors for following inputs, but without manifesting the defect. In this case, only the first image containing discrepancies must be retained, among the series of images taken for several successive inputs.

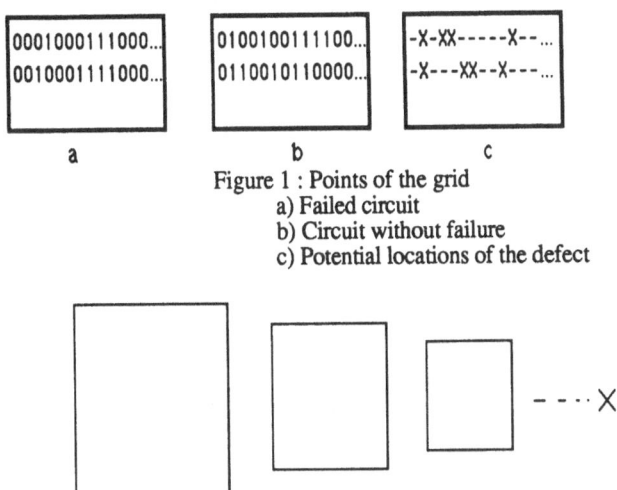

```
0001000111000...    0100100111100...    -X-XX-----X--...
0010001111000...    0110010110000...    -X---XX--X---...
```

a   b   c

Figure 1 : Points of the grid
    a) Failed circuit
    b) Circuit without failure
    c) Potential locations of the defect

Figure 2 : Possible locations of the defect

The single defect hypothesis is an ideal hypothesis if the exact location of the defect is required. But multiple defects should be localised using frequencies of discrepancies. Indeed, in case of a multiple, say a double defect, an input could manifest a first defect, but not a second one, while another input could manifest the second defect, but not the first one etc... But it may be reasonably supposed that the two locations of the defect would be the points for which discrepancies would appear the most frequently. Images comparison must be made between a good circuit picture and the failed circuit picture. To do this, both chips must be put inside the SEM chamber with the same position. A method which can be automatized, has been employed and a specific algorithm has been implemented. The basic idea is as follows : When the good circuit is in the chamber, an edge extraction is made with a classical Söbel algorithm and then a binarization is applied. Edge extraction allows to store only pertinent information (contrast points) for vertical and horizontal correlation. Binarization allows to store a reduced image (image size is divided by 8). When the second chip is analyzed the same sequence (edge extraction, binarization) is applied. With the two binarized images a third one is built to display simultaneously with different colors the two edges. Differences between the positions appear immediatly. At the present time, due to the absence of a motorized stage, a displacement is made manually after registration. In the future the differences will be communicated to a motorized stage. This operation is easy and takes about five minutes with a precision better than one micron. One drawback of our experimental tool is that the circuits displacement is not visualized in real time. This is due to the fact that a gradient is obtained by software computing. This could be avoided by using a specific real time edge extractor card.

All the procedure described above is sufficient if only one registration is necessary. It becomes unsufficient if a registration becomes necessary for each picture covering a part of the circuit. Indeed one problem is that one picture might not be sufficient to cover all the surface of the chip since in this case image resolution would become too bad. A higher electronic zoom should be used but only a part of the chip is analysed. To avoid this, a special image capture process was programmed on the image processing system : a data processing zoom (figure 3).

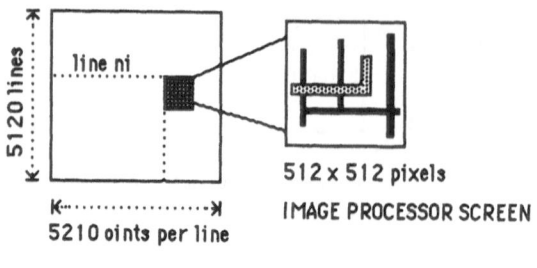

Figure 3 : Data processing zoom

With a very fine spot, it is possible to scan the whole circuit with for example 5120 lines. If each line is sampled to have the same number of pixels (5120) then, for each picture recorded by image processing, a zoom effect is obtained. This effect is not visible on the SEM video display.

For this example, the image processing system coding 512 x 512 pixels, 10 x 10 = 100 images may be recorded, each of them being processed by the image processing system. With this zoom effect, a greater surface may be analyzed without moving the stage ; registration is not necessary for each image recorded. To be applied, a trade-off between the data processing zoom and the electronic zoom is to be reached, since :

   - the more the electronic zoom is large, the more the surface of circuits becomes electrically charged,
   - the surface covered by the electronic zoom should be maximum, but the data processing zoom should be limited in order to get not noisy images,
   - each elementary connection must be represented by a minimal number of pixels.
Optimal values have been determined for our equipment. The limit of the data processing zoom is about 100 (10 x 10 image processing system images for one SEM image). Also the limit of the electronic zoom is about 100. But these values depend on the image processing system.

Experiments have been made to test the efficiency and to prove that the methodology is feasible. For that, image processing has been carried out on 6800 and 8088 microprocessors. Two kinds or technologies have been used : NMOS and CMOS. Although problems have been encountered (20), the feasibility of failure analysis has been demonstrated. The photographs 1 to 10 detail an experiment using 8088 microprocessors. The next steps are to fully automatize the process, in order to handle automatically large circuits, extensively. A point which should be mentioned is that the test method proposed in this paper may be compared to the way proposed in (5). The strategy developed in (5) is aimed at the confinement of errors occuring in a series of images taken for the series of instant times. This applies to sequential circuits, in which the origin (the first time an error occur) of a defect occurs. This strategy is exactly what was proposed in (49) in order to apply the test method proposed in this paper to sequential circuits : it is necessary to retain for comparisons only the first image for which there is a discrepancy. Thus, the two methods are fully complementary : in case of a sequential circuit, the origin of the errors should be firstly determined, and next intersections should take place.

IV-2 - DEBUG OF INTEGRATED CIRCUITS (55)

In this section, **design errors** are looked for. Two successive steps are necessary, and briefly detailed in this paper.
The first one concerns the way to acquire the logical value of a node, given by its name or pointed out on a screen. The second one concerns the way to use the basic facility provided by the first one.

IV-2.1 - DETERMINATION OF THE LOGICAL LEVEL OF A NODE (28)

A link has been achieved between the SEM and a CALMA description of a circuit. It is interesting to consider a GDS II format for the description of a circuit, since this is a very standard one. This way, the full debugging integrated system can be used by a broad class of users. The three invoked steps are presented hereafter.

IV-2.1.1 - Localization of the observation node

This point is not detailed, since it is quite easy to solve: it is necessary to determine, according to the CALMA description of the circuit, where the node of interest is located, i.e. to determine its coordinates. The problems come because experiments show that the node of interest is not located exactly where it is expected to be.
This is due to errors, aberrations of the SEM, etc... Hence another step is necessary.

IV-2.1.2 - Localization of the observation nodes within the SEM images

The chosen way is to look for corrections allowing to compensate the distortions of the SEM images, and to consider the shape of the analysed boundaries. Some programs have been developped to superimpose a mask image with a SEM one prior to the determination of the nodes practical coordinates. This process involves the following steps :

- Creation of the theoretical mask image in the image processor memory : Firstly, the designer selects the masks area to superimpose with a SEM image. During this step, a file is created: it contains the area number of boundaries, and for each boundary, its segments number, its first point coordinates, and the Δx and Δy displacements of each segment. After the transfer of this file to the image processor, a program draws the segments in the image processing system memory.

- Extraction of the practical outline of the connections : After a proper adjustment of the SEM parameters (magnification and stage position), a SEM image which covers approximately the same area, is acquired. The outline of the connections is then extracted : the value of a pixel is deduced from the grey levels of its eight neighbours. Finally, the noise is reduced by suppressing the elements having less than a given number of pixels in both x and y directions.

- Creation of an image containing the two outlines : Such an image, made up of a grey background, a black theoretical outline, and a white practical outline, allows to study the distortions of the SEM images. So a scale correction coefficient can be applied to the mask image independently in both x and y directions. We can notice that the main distortion is observed in the x direction : the dimension in pixels of an element increases along the scan line. After correction of the theoretical outline, all the elements have the same scale in both the theoretical and the practical images : the two outlines can be superimposed.

- Superimposition of the theoretical and the practical outlines : A program based on a correlation technique, determines the horizontal and vertical discards between the two outlines. A window W, which is a part of the theoretical outline, is searched within a larger search area S belonging to the practical outline.

The window is stored as a set of segments. The search area is a NxM pixels array, each pixel being either a background black pixel, or an outline white pixel. Its dimensions depend on the (K,L) window dimensions in pixels and on the dislocation between the two images. The image processor asks for an indice n allowing to compute the Δx and Δy values (which are added to both sides of the window (figure 4.1)) according to the following formulas :

$$\Delta x = K / 2^n \quad \text{and} \quad \Delta y = L / 2^n, \quad \text{with } 0 \le n \le 7$$

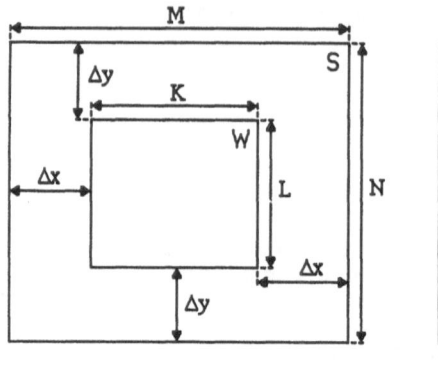

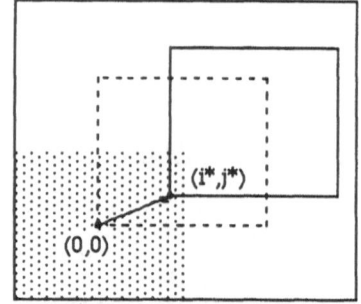

⠿ Variation space of the
subimages lower left corner

Figure 4.1 Window W and Search area S   Figure 4.2 Subimages of S

A KxL subimage of S is defined by the (i,j) coordinates of its lower left corner. As the problem comes to search the subimage which is the most similar to W, the window is compared to each of the $(2\Delta x+1) \times (2\Delta y+1)$ subimages (figure 4.2). A coefficient of correspondence is computed for each subimage : it is incremented each time a segment point of W is white in the practical subimage. The subimage giving the greatest coefficient is the one that best matches the window. Let (i^*, j^*) be the coordinates of its lower left corner. To reduce the correlation time, a test is performed after the process of each segment : the current coefficient is added to the number of remaining pixels; if the result is lower than the current maximum coefficient, the current subimage is not the best one. So the computation of its coefficient is stopped, and the next position is considered.

The theoretical outline is then shifted according to the optimum values (i^* pixels for x, j^* pixels for y), and the two outlines are displayed again.

The following images illustrate this correlation process :

 - The first one shows the SEM image acquired with the images processor. It is made up of 512x512 pixels and covers about 140x140 microns.
 - On the photograph 12, we can see the connections outline extracted from the initial SEM image.
 - The next image shows the metal mask of the same area.
 - The white SEM outline and the black mask are then simultaneously displayed : the image 14 allows to visualize the x and y discards between these two outlines, and the last one shows the result of the correlation process. For this example, the window, displayed on the photograph 16, contained 10 segments or about 820 pixels, and about 3000 coefficients were computed ; the process time was of 25 seconds.

IV-2.1.3 - Determination of the observation nodes logical level

The nodes logical level must be determined from their grey level which can be read in the image memory once the practical coordinates are known. In order to achieve a correspondence between a grey level and a logical level, we use two thresholds which allow to divide the image pixels in three classes : when the grey level is below the lowest threshold, we consider that the pixel is at a low level. We assume it is at a high level if its grey level is upper the greatest threshold, and at an intermediate state if its grey level is between the two thresholds.

Presently, the thresholds are manually given by the operator. The three grey levels image issued from these values is displayed in order to detect untrue thresholds and to modify them. In the future, we envisage to determine these thresholds using two reference grey levels which respectively correspond to a ground level and a power level. They will be read in the image memory after the designer has indicated the location of a ground line and a power line within the mask description.

IV-2.2 - DIAGNOSTICS USING EXPERT SYSTEMS TECHNIQUES (56, 57)

This way is choosen because it is the most promising for the following reasons.
It is considered, in a first step, that the expert system will not be able to do more than a conventional way, like a fault dictionary for example. But in the future, other problems will be possibly taken into account. Another point is that although a conventional approach is maybe not realistic for large circuits, it might be more efficient in the sense that the solving might be quicker. But this is not the main problem for diagnostics. It is not important to diagnose using 10 ms less, when the full process will take some hours (but it is important to diagnose within some hours using electron beam testing compared to some days using conventional tools).
The advanced e-beam testing tool should be associated to expert systems type methods for diagnostics (this would not be true for test pattern generation, to be used further for fabrication tests, since in this case 10 ms are important).

The problem solver we have been developping is an expert system (E.S.), dedicated to design validation of VLSI circuits. The choice of this test method, based on artificial intelligence techniques, can be justified by the ability of expert systems to handle a large mass of information, and by their adaptability to many kind of problems, since they are not limited to a single, predetermined, solving principle, such as those algorithmic methods are based on.

IV-2.2.1 - General architecture of the E.S.

The E.S., described using the PROLOG language, includes (figure 5) :

- A knowledge database, including every knowledge available about the circuit under test (e.g. structure, observability nodes, etc.).
- "Dynamic" data about the test currently carried out (e.g. status of the circuit, errors already detected, etc.).
- A control program containing search strategies and inference rules.

The circuit is modelized as a set of logical blocks, linked by connections, which constitute interfaces between blocks (figure 6).

The basic principle is to localize roughly in a first time the faulty block, by using logical and structural properties of VLSI circuits. Once the faulty block have been identified, the system can localize the defect inside this block, by applying more specific rules (50).

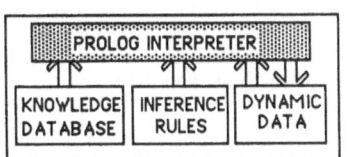

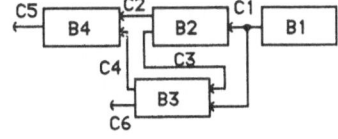

Bi : Blocks – Ci: Connections

Figure 5 General architecture of the E.S.   Figure 6 Modelization of the structure

IV-2.2.2 - Knowledge database and "dynamic" data

For the present, the knowledge database only contains data about the structure of the circuit under test and its observability nodes. These data represent the inputs and the outputs of the logical blocks, and describe how these blocks are connected.
The "dynamic" data represents the status of the connections, that is, if their logical value are erroneous or not, and may include some hypotheses about the current test, formulated by an

operator. These hypotheses may concern, for example, the type of connections (unidirectional, bidirectional, etc.).

The control program is functionally divided into three modules, which roles are :
- To verify the consistency of the databases.
- To localize the faulty block(s).
- To apply specific treatments to these blocks.

IV-2.2.3.1 Verifying the consistency of the data.

The consistency of the informations about the circuit structure have to be verified only if the extraction of the block structure is hand generated.
The different types of inconsistency are :

- A connection is declared erroneous and not erroneous.
- A connection is declared as output of several blocks. This could be an inconsistency or not, according to the fact that wired functions may be realized or not (wired-OR, wired-AND).
- A connection is declared as input and output of a block. This could be an inconsistency or not, according to the hypothesis taken (the connection is unidirectional or bidirectional).
- A bloc has no output.

An other type of inconsistency occurs if two (or more) incompatible hypotheses are formulated. Note that this type must be verified for each diagnosing session, because hypotheses are formulated by an operator.

IV-2.2.3.2 Localizing faulty block(s).

This module is divided into two parts, the first one establishing the connections between blocks, and detecting which block(s) is (are) connected to the primary inputs of the circuit under test, and the second one localizing effectively the faulty block(s).
For the present, the control program includes two search strategies, related to two hypotheses the operator may formulate. These strategies are :

- The strategy applied in the single fault hypothesis case.
- The strategy applied in the multiple fault hypothesis case.

Consequently, there are two types of inference rules, corresponding to the two strategies, but the methodology adopted for each set of rules is the same, that is, to start from the entire set of blocks constituting the circuit structure, and to reduce progressively this set in order to converge towards the expected result. This methodology is known as "start big" methodology.The inference rules are simply the expression of the classical error propagation rules. For example, the strategy adopted in the single fault case starts by gathering the erroneous connections, and then, for each of these connections, proceeds as follows (figure 7) :

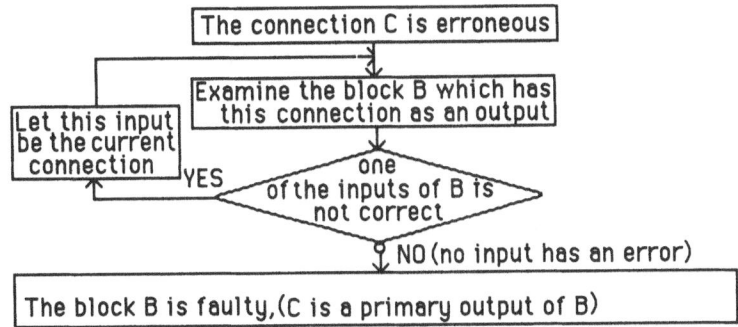

Figure 7 : Strategy adopted in single fault case

Further investigations are needed concerning the expert system for diagnostics, and the link between the basic tool giving the logical value of a node and the expert system, must be implemented (the structure can be given to the expert system by hand, or it can be extracted from CALMA description of the circuit).

V - SYSTEMS USED WITHIN IBM-CESTAS AND IMAG/TIM3

The pictures XIV and XV describe the system used within IBM CESTAS. It is based on commercial equipments (except the software and some signal processing modules). An image processor is used for image comparison, storage and treatments facilities. The entire system has been architectured with the idea in mind to do automatic images comparison.
A software is installed inside the PC in order to give to the analyst access to every subparts of the system from a level he can make choices (in electrical test, in the E. Beam testing techniques choice etc...) very easily and quickly. The software also aids to decision (25).

The system used within IMAG/TIM3 depicted in figure 8 and picture XXXII.

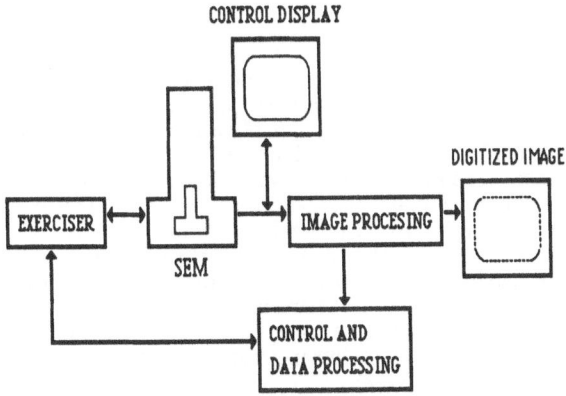

Figure 8 : IMAG/TIM3 System

The SEM is a JEOL 35. It is not a microscope especially designed for integrated circuit test : the detector is a classical THORNLEY & EVERHART detector and the lowest primary beam energy available is 2 Kev.
For the lowest range of energy only a step of 1 Kev is possible. In addition, no beam blanking is

available. Hence only static circuits have been analyzed. The stage is not motorized and this SEM is not equipped for the addressing of the beam. This equipment requires that circuit passivation layer is removed before analysis. Nevertheless, such a not sophisticated equipment allows experiments.

The exerciser allows control of the analysed circuit. It allows simulation of the environment of the circuit under test, functioning either step by step, or continously, or in a loop or with break point, or with external events. An especially designed exerciser has been developped in our laboratory for those manipulations. One advantage of this system is that an interface board is not necessary when different circuits are analysed ; an interconnection matrix allows to describe each circuit pin by placing short-circuit pin on the matrix. The circuit is thus easily controlled.

Image processing allows image capture and image processing. In our experimental tool, this machine is ROMUALD (51). Its particularity is to have a distributed architecture allowing parallelism during programs execution. Its structure is based on the distribution of the whole image between 8 processors (Z 8002) called local processors each being given 512/8 consecutive lines to process. Picture size is 512 x 512 pixels, each pixel is coded with 8 bits (allowing 256 grey levels).

The data processing systems are an IBM Series 1 computer for failure analysis, and a DEC VAX for the debug application.

VI - LIMITS AND ALTERNATIVES TO ELECTRON-BEAM

The main performances of E. Beam are described in terms of temporal resolution, voltage resolution, spatial resolution, precision and linearity.

One can precise in each of these axis what are the absolute theoritical limits but this is not sufficient because these performances of different nature are in fact related to each other in a well know formula (a), which represents the overall signal to noise of the E. Beam tester.

$$\text{(a) } S/N = \frac{\Delta V}{C} \frac{(IPE)^{1/2}}{(\Delta F)^{1/2}}$$

ΔF = bandwidth
IPE = mean current value
ΔV = voltage resolution
C = constant

This fact can be illustrated by a simple example : if we want to test a signal of 2 MHz of frequency and obtain a resolution of 10 millivolts and 2 nanoseconds on a 2 microns conductor lines it will require 500 seconds for a 1024 samples waveform. If we decrease the temporal resolution to 200 picoseconds, we will obtain 5000 seconds of testing time !

The absolute performances of a mean commercial system today are the followings : a few millivolts of voltage resolution, 100 picoseconds of temporal resolution and a few hundred of agström of spatial resolution. A mean value for C factor in the formule (a) is about 10 - 9. For the implementation, of methodologies which require massive measurements, for high sped integrated components, for new technologies these limitations of E. Beam testing is the test of buried layers : this problem, although it has received some beginning of responses, is always pending. It is today difficult to test the multilayers circuit, the passivated circuits and the dirty surfaces. A consequence of this problem is the difficulty to do absolute measurements wtith E. Beam testing with a good precision and reliability (15) (16) (17) (18).

For multilayers chip without special test modes these are nevertheless sample preparation techniques like microtechnique or Focused Ion Beam assisted metal deposition, which create conductive vias for the access of the buried layers. Some other noncontact internal techniques can compete with E. Beam.

The utilization of Laser Beam in place on Electron Beam brings more signal (no space change limited current), shorter pulses for sampling measurements of very high speed components. Either for the technique of Photosensitive samples testing or for the technique of Electro-optic samples testing, subpicosecond pulses and submillivolts voltages resolution has been achieved in experimental equipments (see 46, 47, 48). The photoexitation technique (42), which through the detection of photocurrents generated by a laser beam allows the analysis of the logical levels, presents less temporal resolution than E. Beam (50 ns), no analogical measurements capabilities, no imaging, but has the advantage of being no sensitive to the passivation layer and working in the atmosphere. We have not spoken in this paper on E. Beam as an influence probe. For the strict problem of testing it is obvious that this phenomena has to be limited for non perturbation (see 29). For other applications this perturbation can be usefull (37, 38, 40, 41, 44) and searched as a method to reveal marginal defects by a local modification of the temperature, of the leakage current to the substrate etc... or by a triggering of the latch-up phenomena in CMOS circuit (30). This method of influence can also use the laser beam in place of the electron beam with some advantages concerning the modulation of the wavelength of the laser (for example in the infrared) and the test at the atmosphere. This method is now available in commercial equipments (35, 36, 37).

The use of liquid crystals for the visualization of the internal activity of the chip is an interesting technique although it can not compete with E. Beam concerning its analytical performances and its capability to probe waveforms (see 43, 45).

As conclusion we can say that E.Beam is certainly the most complete internal non contact testing technique available today for diagnostic testing and failure analysis (31) (32) (33). Numerous examples prove it : this technique has permitted to face some challenges. A market, specialists, specialized conferences have appeared. The next step now for this tool is its integration in the CAD environment ; this evolution is in fact not dependant on the nature of the probe : laser or E.Beam but rather of the methodological choices in diagnostic testing and failure analysis of every company (19) (21) (26) (27) (28).

REFERENCES

1. "Design for testability : A survey" : T. Williams, K. Parker
   IEEE Transaction on computers. Vol C31 n°1 / 01.82
2. "Managing VLSI complexity : an outlook" : C. Sequin
   Proceedings of the IEEE. Vol 71 n°1 / 01.83
3. "On the possible limits of external testing" : R.M. Sedmark
   Proceedings of the International test conference 1986
4. "Requirements and trends for high seep testing" : G. Chiu and J.M. Halbout
   Proceeding on the International test conference 1986
5. "Dynamic fault imaging of VLSI random logic devices" : T.C. May, G.L. Scott,
   E.S. Meiearn, P. Winer, V.R. Rao
   Proceedings of IEEE/IRPS 1984
6. "Fundamentals of electron beam testing of integrated circuits" : E. Menzel, E. Kubalek
   Scanning Vol. 5 103.122 (1983)
7. "Electron beam testing in electronics" : K. Ura
   Proc. Int. Cong. on Electron. microscopy Kyoto 1986
8. "Various methods of observing working statesd of LSI circuits and other electronics devices"
   E. Wolfgang
   Proc. Int. Cong. on Electron. microscopy Kyoto 1986
9. "Electron beam testing : Image and signal processing for the failure analysis of VLSI
   components" : J.P. Collin
   Proceedings of ISTFA 85
10. "Voltage coding : temporal versus spatial frequencies" : G.V. Lukianoff, T.R. Touw
    SEM / 1975 pp 465-71
11. "Phase dependent voltage contrast : an inexpensive SEM addition for LSI failure analysis" :
    D. Younkin
    Proceedings of the IEEE / IRPS 1981
12. "Frequency tracing and mapping in theory and practice" : H.D. Brust
    Microelectronic Engineering Vol. 2 n° 4 1985
13. "A practical E.Beam system for high speed continuity testing of conductor networks" :
    H.C Pfeiffer, S.D. Golladay, F.J. Hohn
    Proc. Int. Cong. on Electron. microscopy Kyoto 1986
14. "Dynamic testing of a passivated device with electron beam tester" : M. Ekuni, Y. Komoto,
    T. Gobara, Y. Hadara
    Proc. Int. Cong. on Electron. microscopy Kyoto 1986
15. "Contrast mechanism of SEM images over the passivation layer of microelectronic devices"
    N. Sugiyama, S. Ikeda, Y. Uchikawa
    Proc. Int. Cong. on Electron. microscopy Kyoto 1986
16. "Sampling and real time electron beam probing on passivated and non passivated IC's"
    E. Menzel, R. Buchanan
    Proc. Int. Cong. on Electron. microscopy Kyoto 1986
17. "A multi-sampling waveform measurement method in the E.Beam tester" : T. Sano,
    M. Miyoshi, S. Asami, K. Okumura
    Proc. Int. Cong. on Electron. microscopy Kyoto 1986
18. "Quantitative voltage waveform measurement technique for an IC internal electrode with a
    passivation film" : K. Ookubo, Y. Goto, Y. Furukawwa, T. Inagaki
    Proc. Int. Cong. on Electron. microscopy Kyoto 1986
19. "Fully automated electron beam tester combined with CAD database" : T. Tamama, N. Kuji
    Proc. Int. Cong. on Electron. microscopy Kyoto 1986
20. "Towards automatic failure analysis of complex ICs through E.Beam testing" : L. Bergher,
    B. Courtois, J. Laurent, J.P. Collin
    Proc. of the international test conference 1986

21. "An automated E.Beam tester with CAD interface "Finder" : a powerfull tool for fault diagnosis" : N. Kuji, T. Tamama
Proc. of the international test conference 1986
22. "Automatic positioning for electron beam probing" : M. Battu, G. Bestente, P. Cremonese, A. DiJanni, P. Garino
Proc. Int. Cong. on Electron. microscopy Kyoto 1986
23. "Automating detector system for LSI failure during operation" : T. Shinkawa, K. Uchida, T. Takagi, S. Takashima
Proc. Int. Cong. on Electron. microscopy Kyoto 1986
24. "Computer control of electron beam testing for design validation of VLSI circuits" : D.W. Ranasinghe, G. Proctor, M. Cocito, G. Bestente
Proc. Int. Cong. on Electron. microscopy Kyoto 1986
25. "A fully computer controlled scanning electron microscope for electron beam testing" : H. Fujioka, K. Nakamae, H. Tottori, K. Ura, S. Takashima, Y. Harada, N. Date, K. Harasawa
Proc. Int. Cong. on Electron. microscopy Kyoto 1986
26. "Electron beam tester with a CAD pattern data" : F. Lomatsu, M. Miyoshi, T. Sano, K. Sekiwa, K. Okumura
Proc. Int. Cong. on Electron. microscopy Kyoto 1986
27. "A SEM based workstation for design validation" : Y.J. Vernay, R. Mignone, P. Rivoire
ESSIRC'86 Delft
28. "Electron beam observability and controlability for the debugging of integrated circuits" I. Guiguet, D. Micollet, J. Laurent, B. Courtois
ESSIRC'86 Delft
29. "Electron beam effects on VLSI MOS : conditions for testing and reconfiguration" : P. Girard, F.M. Roche, B. Pistoulet
Proceedings of the workshops on "Wafer scale integration" Grenoble, mars 1986
30. "An SEM Based system for a complete characterization of Latch up in CMOS integrated circuits" : C. Canali, F. Fantini, M. Giannini, A. Senin, M. Vanzi, E. Zanoni
Scanning Vol. 8 1986
31. "Resistive contrast imaging a new SEM mode for failure analysis" : C. Smith, C.R. Bagnell, E.I. Cole, F.A. Dibianca, D.G. Jonhson, W.V. Oxford, R.H. Propst
IEEE Transactions on Electron Devices Vol. ED-33 N°2 Feb. 1986
32. "Electrical junction delineation by scanning electron beam technique" : G.V. Lukianoff
Solid state technology - March 1971
33. "Time resolved EBIC : a non destructive method for an accurate determination of P-N junction depth' : A. Georges, J.M. Fournier, D. Bois
SEM / 1982 / I p. 147-156
34. "A non contact voltage measurement technique using Auger spectroscopy" : J. Patterson, M. Smith
Proc. of the IEEE / IRPS 1983
35. "Optical scanning microscopy : the laser scan microscope" : V. Wilke
Scanning Vol 7/1985
36. "Scanning optical microscopy" : T. Wilson
Scanning Vol 7/1985
37. "Scanning laser microscopy" : H. Hinkelmann
92 / Semiconducteur International Feb. 1985
38. "Localisation de défauts de tension marginale de circuits intégrés par simulation photonique ou electronique" : G. Auvert, J.M. Fournier
4ème Colloque international sur la fiabilité et la maintenance 1984
39. "The application of marginal voltage measurements to detect and locate defects in digital micocircuits" : D. Ager, G. Cornwell, W. Stanley
Microelectronics and reliability Vol. 22, 1982
40. "Laser die probing for complex CMOS" : M. Pronobis, D. Burns
Proceedings ISTFA 1982

41. "An investigation of flaws in complex CMOS devices by a scanning photoexitation technique" : D. Miguel, E. Levy
Proceedings IRPS 1977

42. "Test of hermetically sealed LSI/VLSI devices by laser photoexitation logic analysis" : F. Henley
Proc. of the international test conference 1986

43. "Les cristaux liquides : une nouvelle méthode de visualisation du fonctionnement des circuits VLSI" : T. Viacroze, C. Baumaun
5ème Colloque de fiabilité et de maintenabilité Biarritz 1986

44. "Localisation par irradation au faisceau laser des zones sensibles au Latch-up des circuits intégrés en technologie CMOS " : G. Auvert
5ème Colloque de fiabilité et de maintenabilité Biarritz 1986

45. "Etude expérimentale de l'électrohydrodynamique des cristaux liquides : application à la visualisation du fonctionnement des circuits intégrés complexes" : C. Bauman
Thèse de l'université de Bordeaux, 1986

46. "Energy and time resolved photoemission in a promising new approach for contactless integrated circuit testing" : H.K. Seitz, A. Blacha, R. Clauberg, H. Beha
Microelectronic engineering 5 (1986) 547-533

47. "Two-dimensional E-field mapping with subpicosecond tempora resolution" : K.E. Meyer, G. Mourou
Electronics letters 20/06/19845 Vol. 21 N°13

48. "Direct Electro-optic sampling of GaAs integrated circuits" : K.J. Weingarten, M.J.W. Rodwell, H.K. Heinrich, B.H Kolner, D. Bloom
Electronics letters 15/06/1985 Vol. 21 N°17

49. " Analyse de défaillances de VLSI par microscopie électronique" : G. Baille, B. Courtois, J. Laurent
Congrès AFCET, Lille Novembre 1982

50. " E.Beam testing strategies for VLSI" : G. Berger-Sabbatel, B.Courtois
European Conference on Circuit theory and Design. Stuttgart, September 5-9 1983 FRG

51. " Architecture of a multiprocessor for pictures capture and Processing" : B. Bretagnolle, C. Rubat du Merac
International Micro and Mini Conference. Houston November 1979

52. "Software integration in a workstation-based E-beam tester" : S. Concina, G.Liu, L. Lattanzi, S. Reyfam, N. Richardson
International Test Conference, Washington 1986

53. " Electron beam testing for microprocessors" : G. Crichton, P. Fazekas, E.Wolfgang
1980 Digest of papers IEEE Test Conference, Philadelphia USA

54. " E-Beam testing : methods and applications" : H. P. Feurbaum
1982 October Microcircuit Engineering : microlithography Grenoble France

55. "Electron-beam testing : failure analysis and debug of integrated circuits" : L. Bergher, I.Guiguet, M. Marzouki, B.Courtois
New directions for IC testing. Winnipeg, Canada, April 8-10 1987

56. "An integrated debugging system based on e-beam test" : I. Guiguet, M. Marzouki, J. Laurent, B. Courtois
IMAG/TIM3 Internal Report

57. "PESTICIDE (a Prolog-written Expert System as a Tool for Integrated CIrcuits DEbugging" M. Marzouki, B. Courtois
IMAG/TIM3 Internal Report

PICTURES I: 80C88 μP REVERSE ENGINEERING

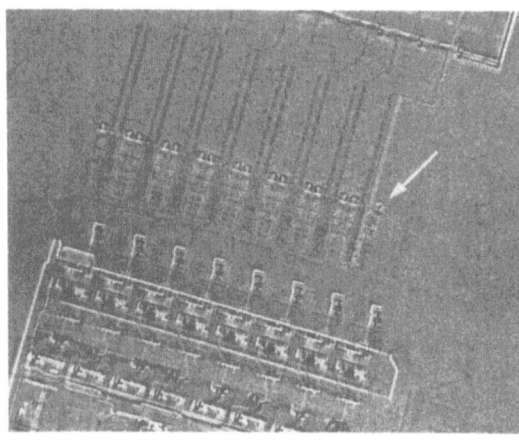

PICTURE I.1

Visualization of the bit 0 of the AX register. (see arrow.)

MOV sequence

FFFE -> AX
0001 -> AX

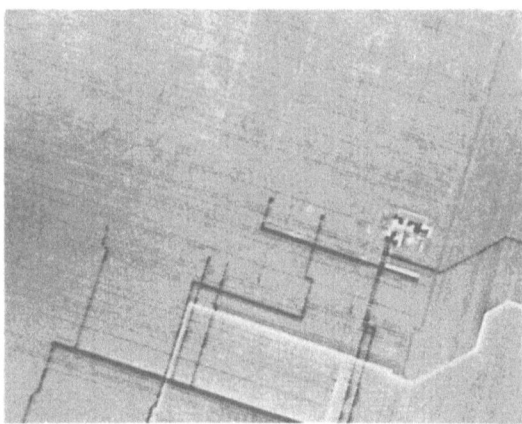

PICTURE I.2

activation of the carry flag by complementary instructions.

CLC: clear carry
STC: set carry

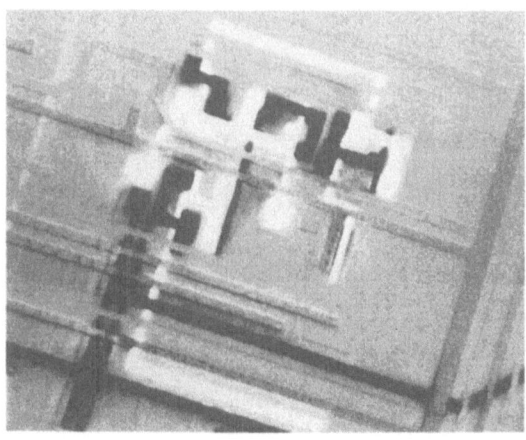

PICTURE I.3

Enlarged view of the carry flag with the same test sequence.
CLC/STC

PICTURES II: ADDRESSING VISUALIZATION ON 64 K DRAM

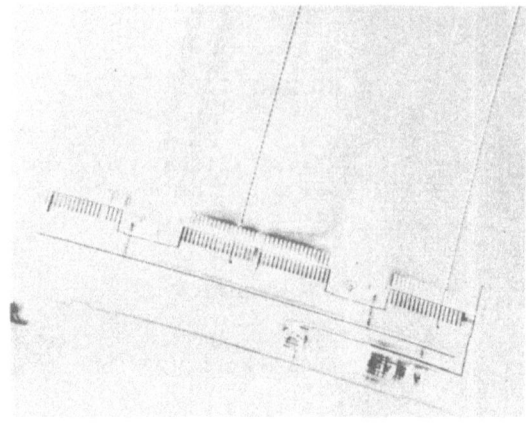

PICTURE II.1

64 K DRAM row decoding activity

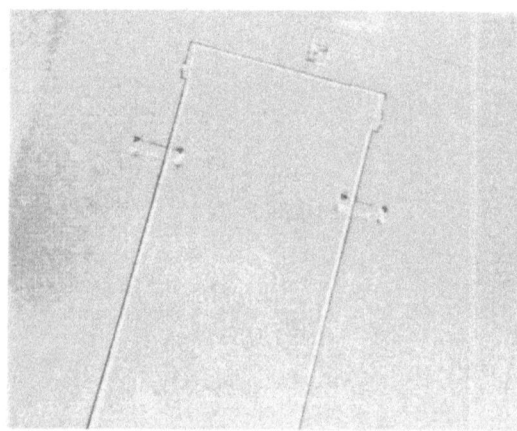

PICTURE II.2

Column decoding activity

PICTURE II.3

Row/column decoding activity

489

PICTURES III: BURIED LAYERS VISUALIZATION

PICTURE III.1

Chip select input

signal visualization

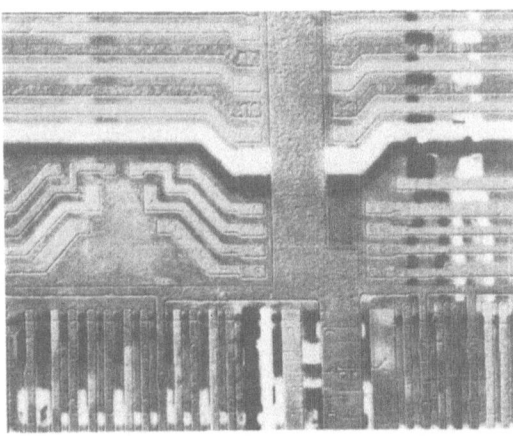

PICTURE III.2

static ram decoder ac-
tivity

PICTURE III.3

Details on the Chip Se-
lect circuit.

490

PICTURES IV: FAILURE ANALYSIS ON 64 K DRAM

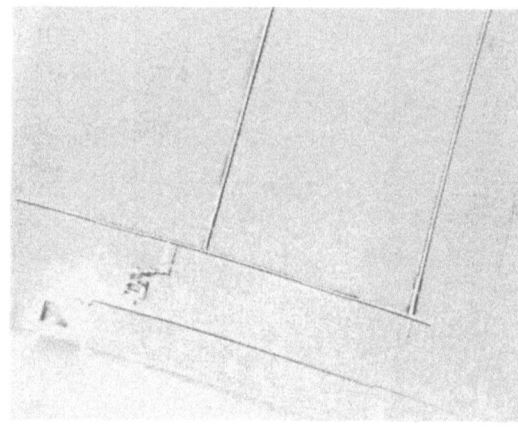

PICTURE IV.1

Comparative rows
decoding activity

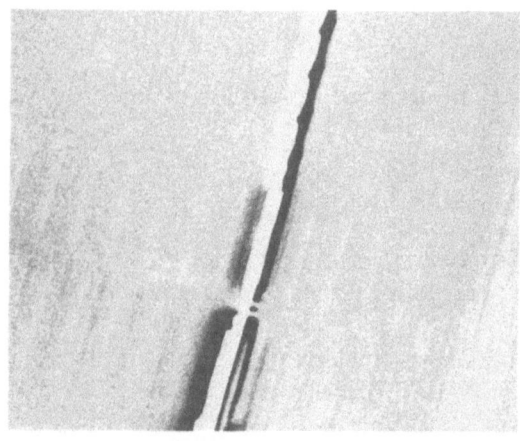

PICTURE IV.2

Good row pair

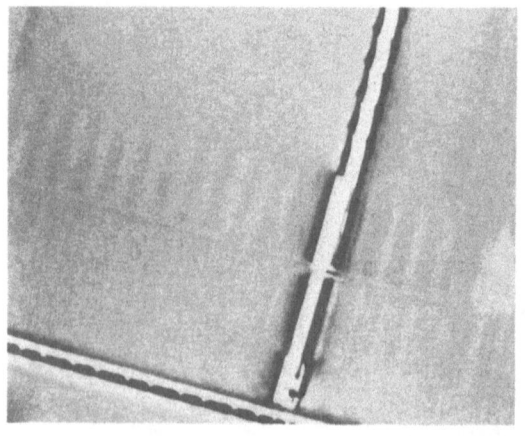

PICTURE IV.3

Goog row pair and the
failed one.

FIGURES V: FAILURE ANALYSIS ON 64 K DRAM

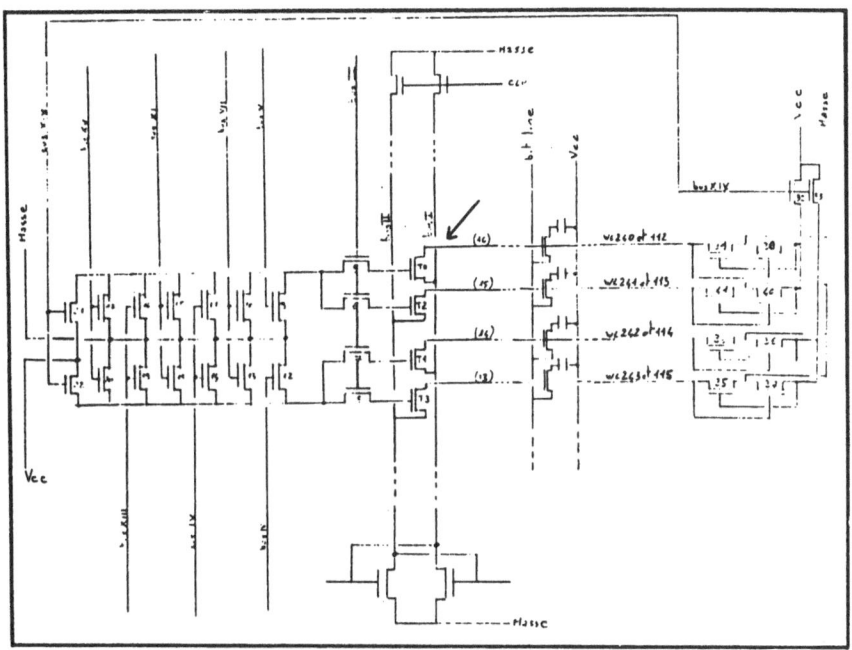

FIGURE V.1

Reverse engineering of the defective area.

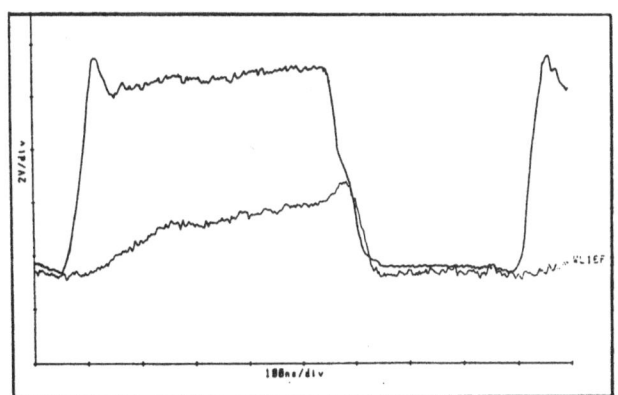

FIGURE V.2

Comparative waveforms on a good row and the failed one.

PICTURES VI: FAILURE ANALYSIS ON 64 K DRAM

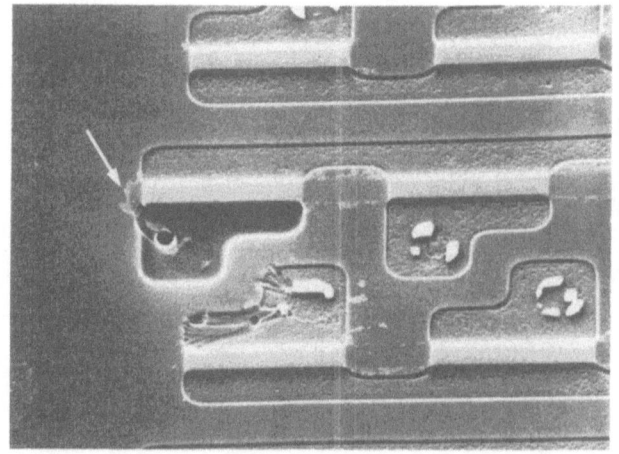

PICTURE VI.1

Defect view of the tx T0
(see picture V.1)

PICTURES VII: FAILURE ANALYSIS OF THE 8048 μP

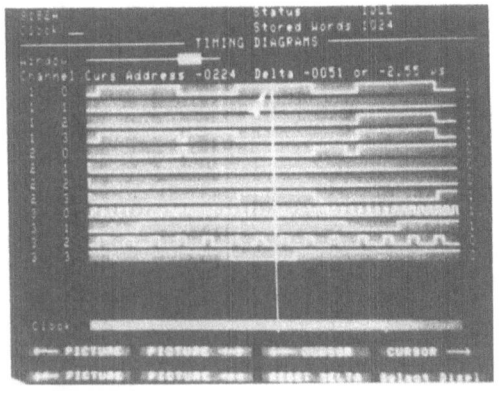

PICTURE VII.1

Timing of a read se-
quence of the address
56.
Bit 3 stuck at "1".

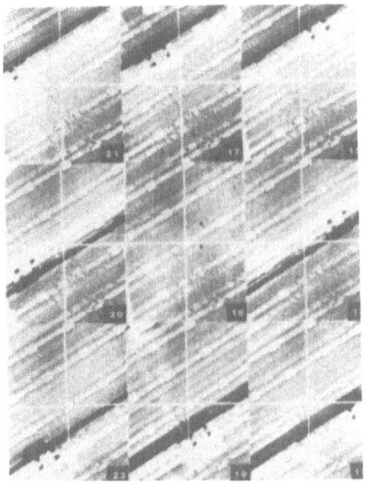

PICTURE VII.2

PSVC view: activity
around the defective
cell of the memory
area.

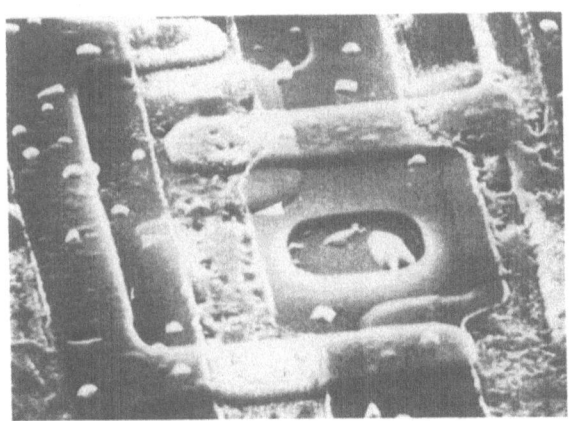

PICTURE VII.3

SEM VIEW: oxide aspect
after passivation and
aluminum etching.

PICTURES VIII: FAILURE ANALYSIS OF THE 8085 μP

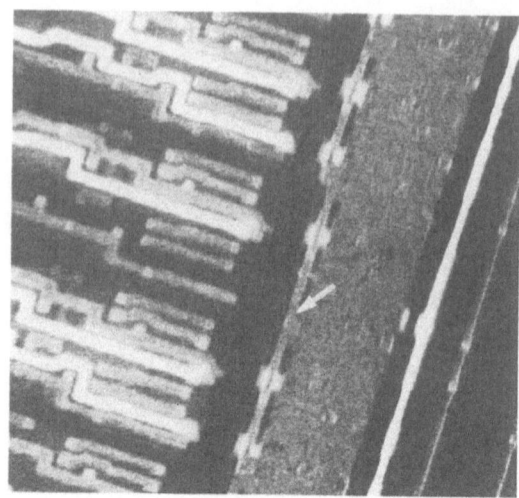

PICTURE VIII.1/2

The signal is stopped
under the large alumi-
num land . the line of
the bit 1 of the H,L,D
registers is not acti-
vated.

The two pictures show
the same test sequence,
but   in   phase   and
out-of-phase with   the
PSVC clock.

PICTURES IX: FAILURE ANALYSIS OF THE 8085 µP

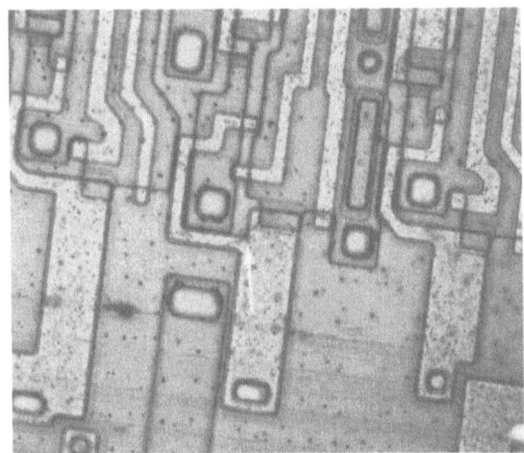

PICTURE IX.1

OPTICAL VIEW: a crack
in the polysilicon can
be observed through the
oxide after the
passivation and oxide
etching.

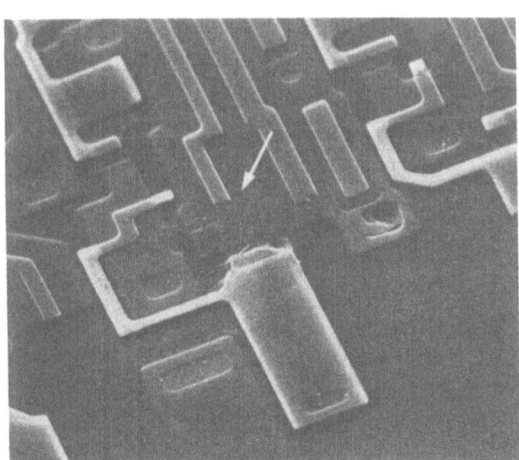

PICTURE IX.2

SEM VIEW: aspect of the
polysilicon after oxide
removal.

496

PICTURE X

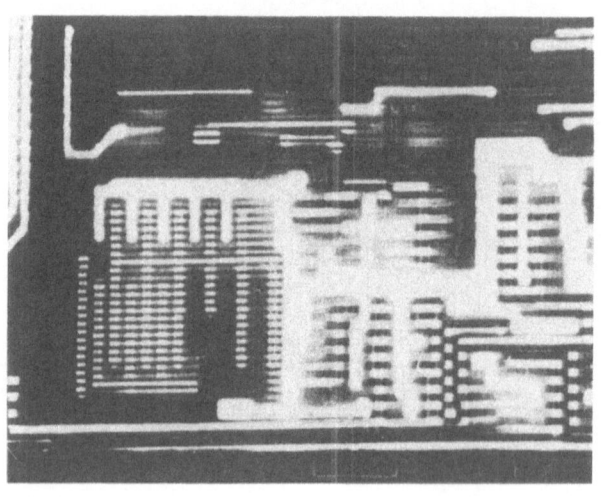

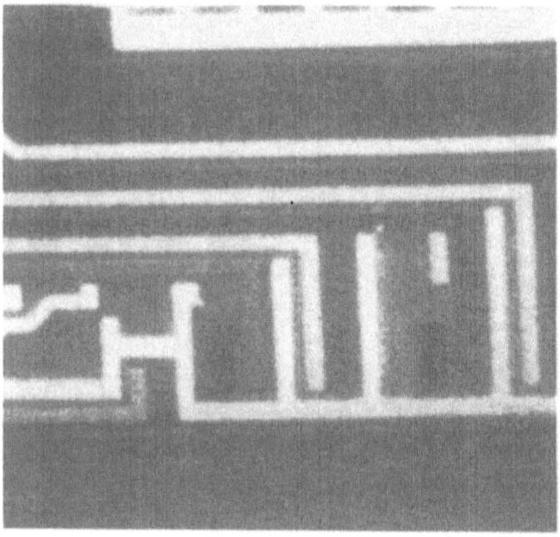

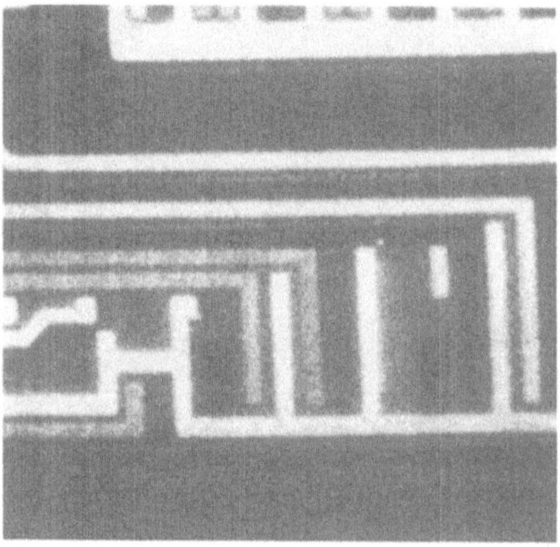

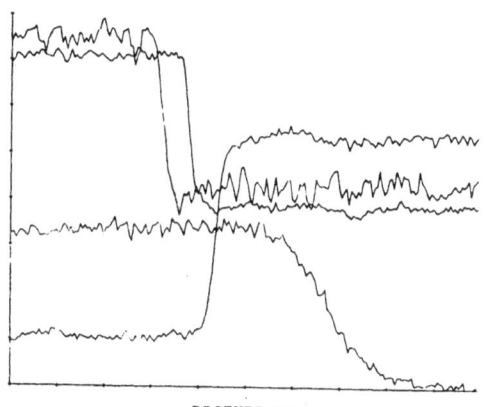

PICTURE XI

PICTURE XII

EXTERNAL ELECTRICAL TEST

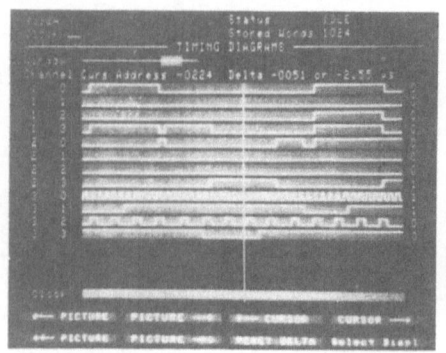 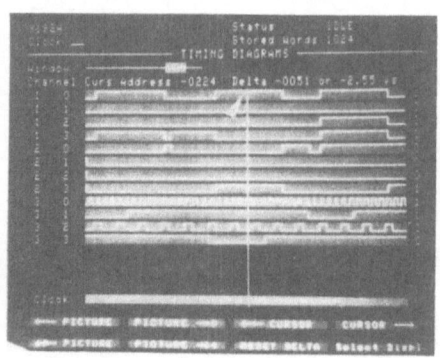

1/ Good memory cell.

2/ Failed memory cell:
bit 8 stuck   at "1"
(see arrow.)

INTERNAL ELECTRICAL TEST

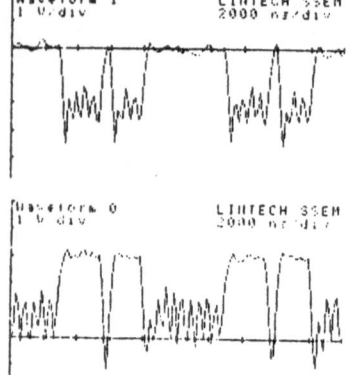 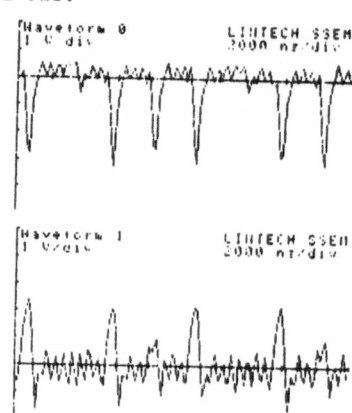

3/ Good memory cell.

4/ Failed memory cell.
(bit lines stuck at "0","1"
during a read/write sequence

Internal electrical test with STROBOSCOPIC
WAVEFORM MODE.(Ebeam tester.)
(internal nodes:A,B see sheet = 3.)

PICTURE XIII

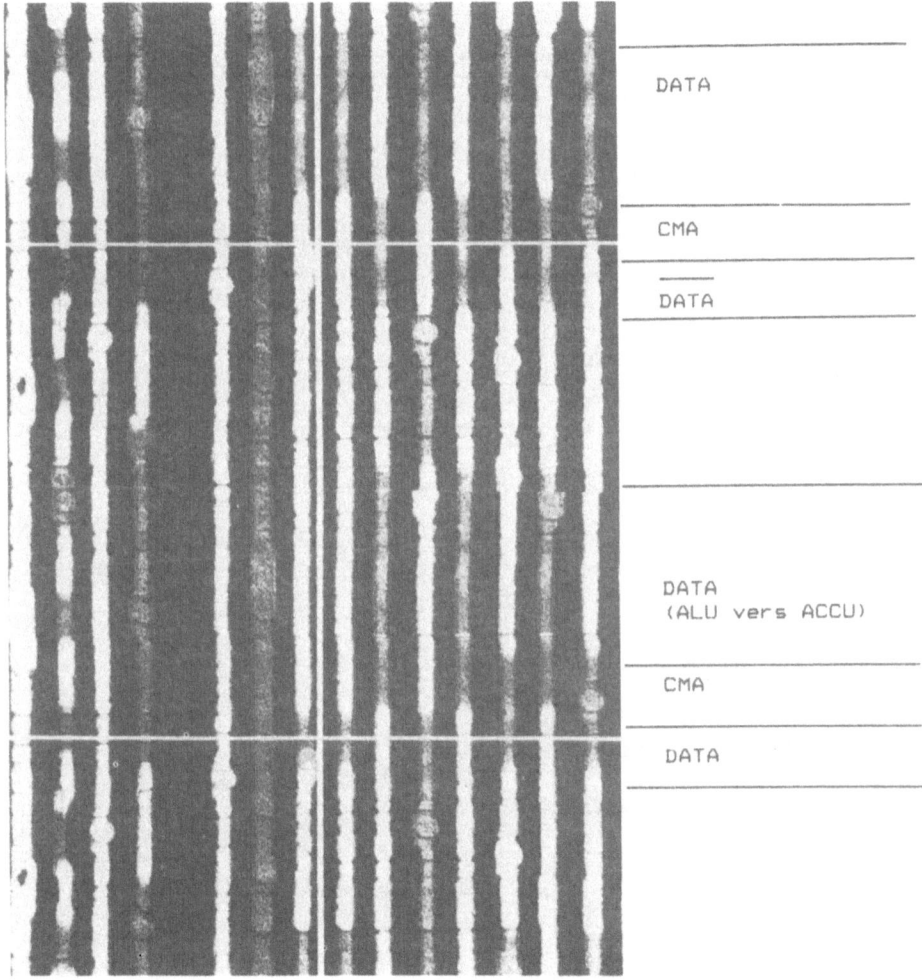

DATA

CMA

DATA

DATA
(ALU vers ACCU)

CMA

DATA

500

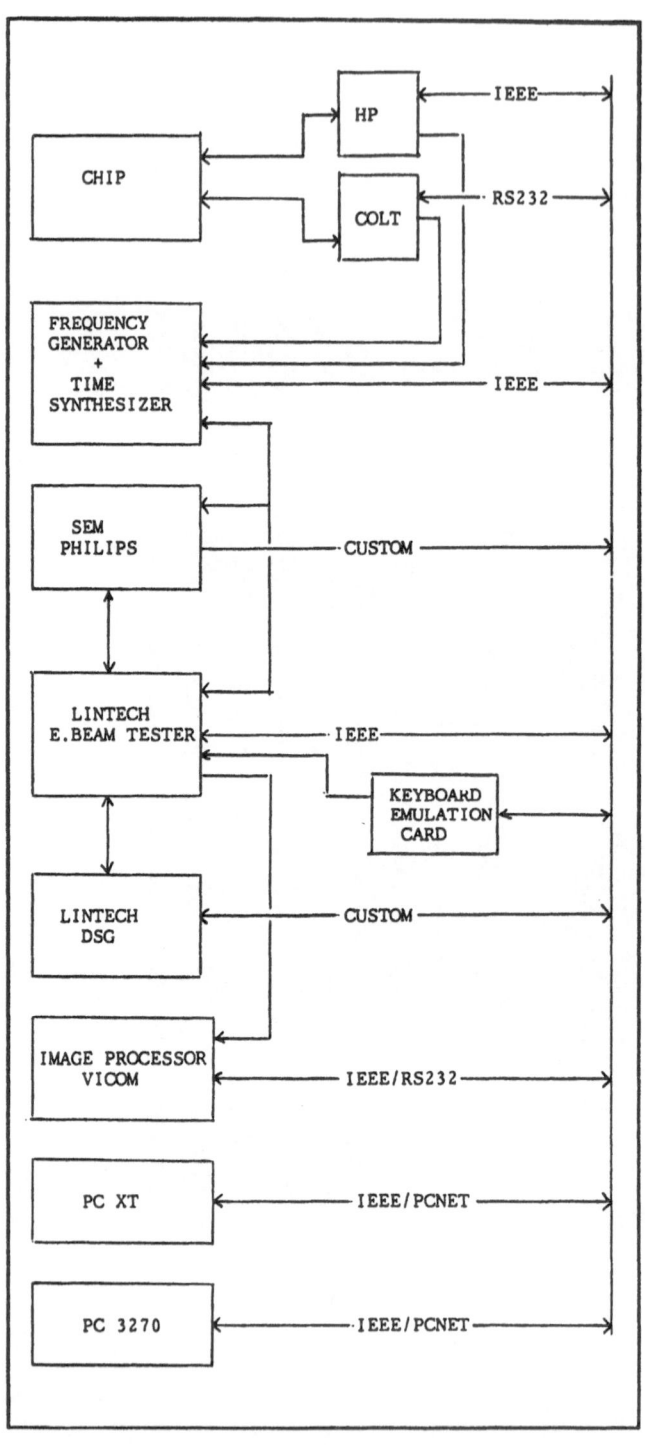

PICTURE X V

WORKING STATION

1) PHILIPS SEM
2) LINTECH E_BEAM CONTROL SYSTEM WITH DIGITAL
   SCAN GENERATOR
3) LINTECH CONTROL MONITOR
4) IBM PC AT/2 KEYBOARD
5) PC/AT MONITOR.MOUSE ORIENTED
   MANAGEMENT MENU
5')MICROSOFT MOUSE FOR SYSTEM DRIVING
6) ENGINEERING TESTER FOR MICROPROCESSORS
   MICROCONTROLLERS AND GATE ARRAYS.
7) MEMORY TESTER (COLT).
8) INTERACTIVE INPUT DEVICES FOR IMAGING
   MANAGEMENT.
9) TREATED IMAGE.

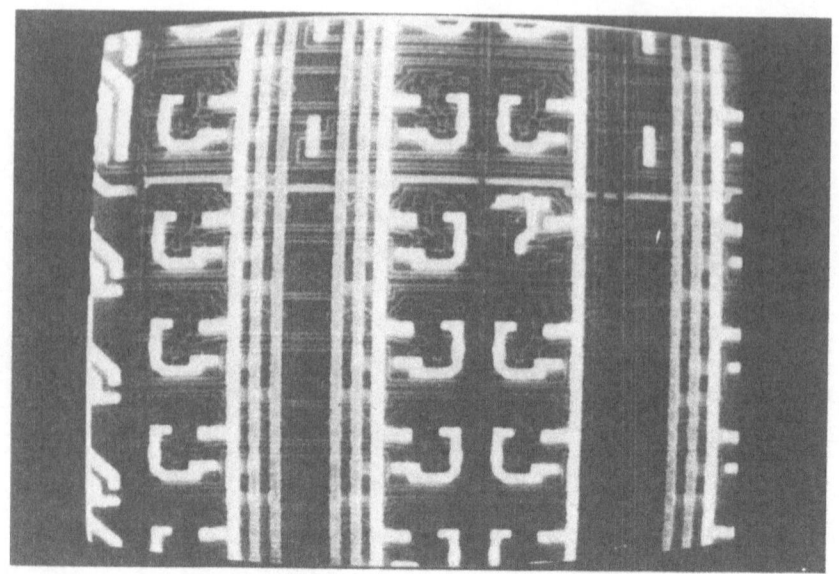

PICTURE XVI: Circuit n° 1 Input n° 1

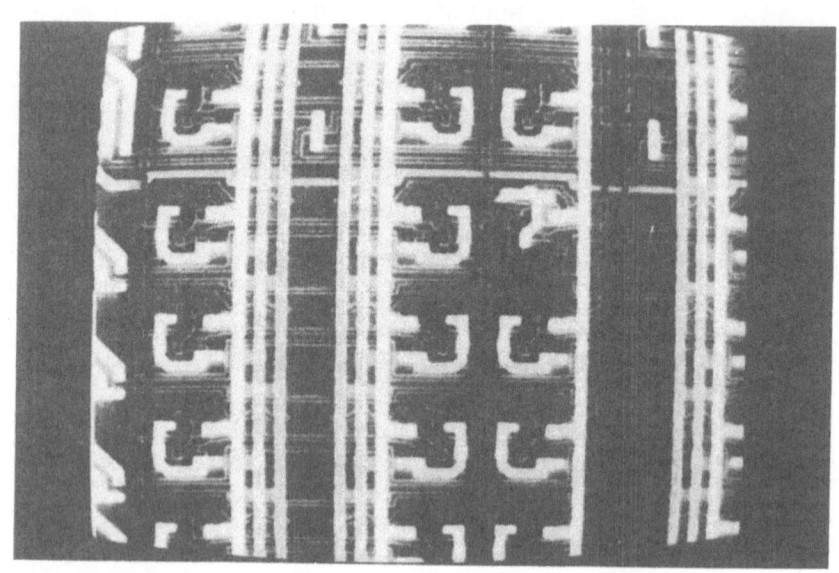

PICTURE XVII: Equalization

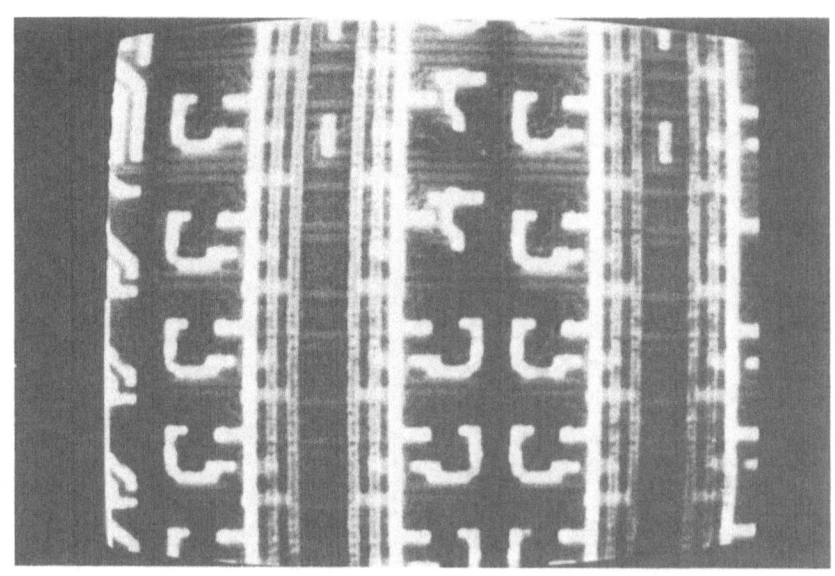

PICTURE XVIII: Circuit n° 2 Input n° 2

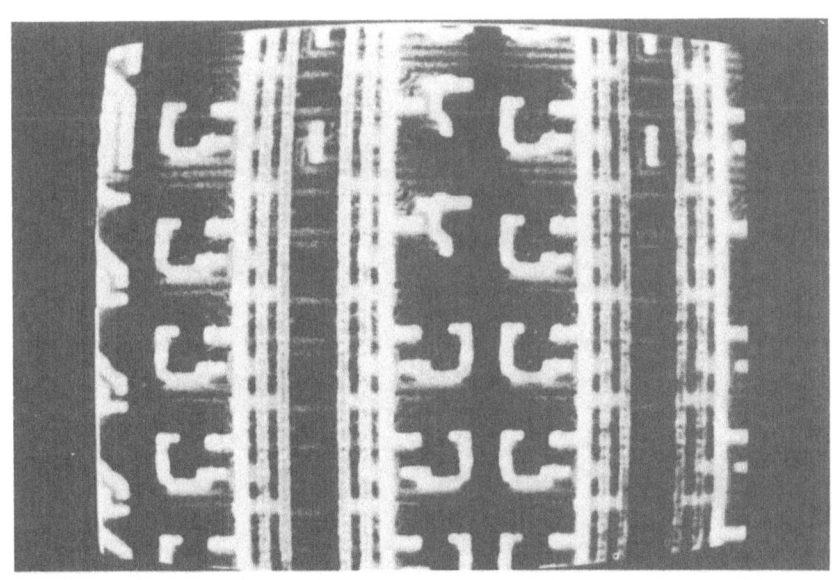

PICTURE XIX: Equalization

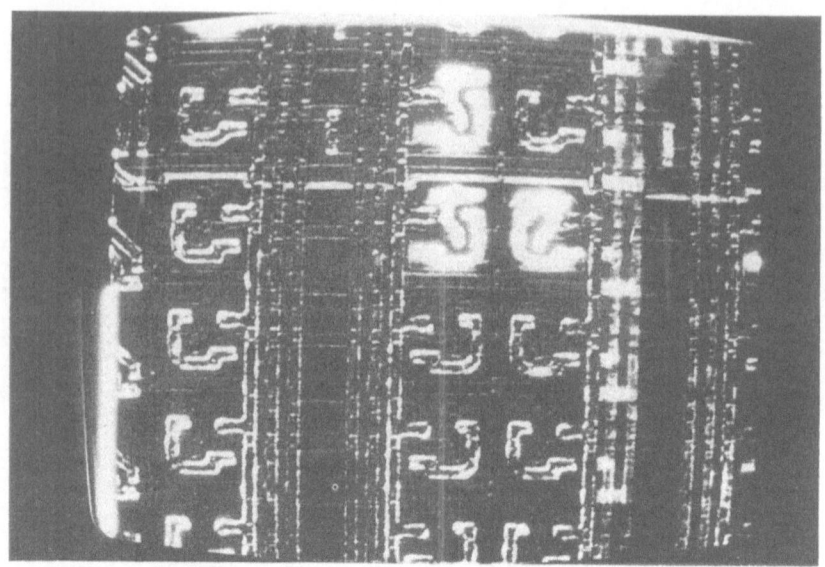

PICTURE XX: Comparison of trinarized images

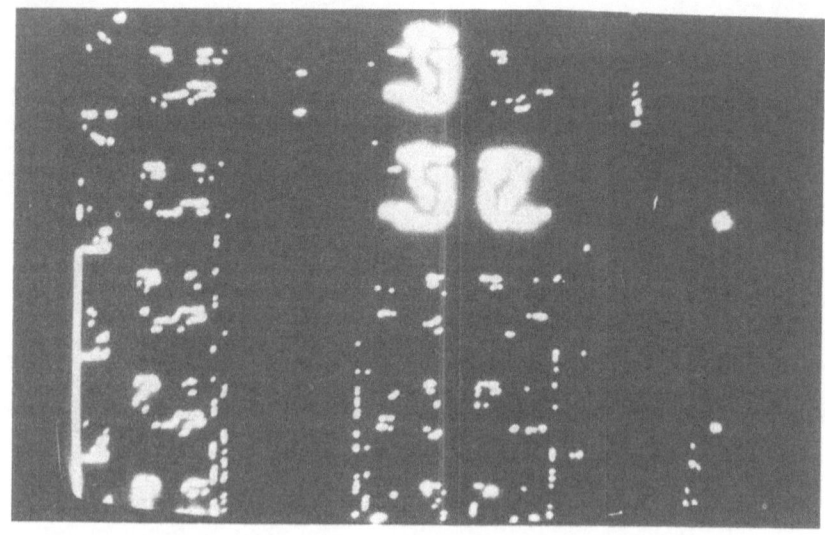

PICTURE XXI: Binarization of the comparison image

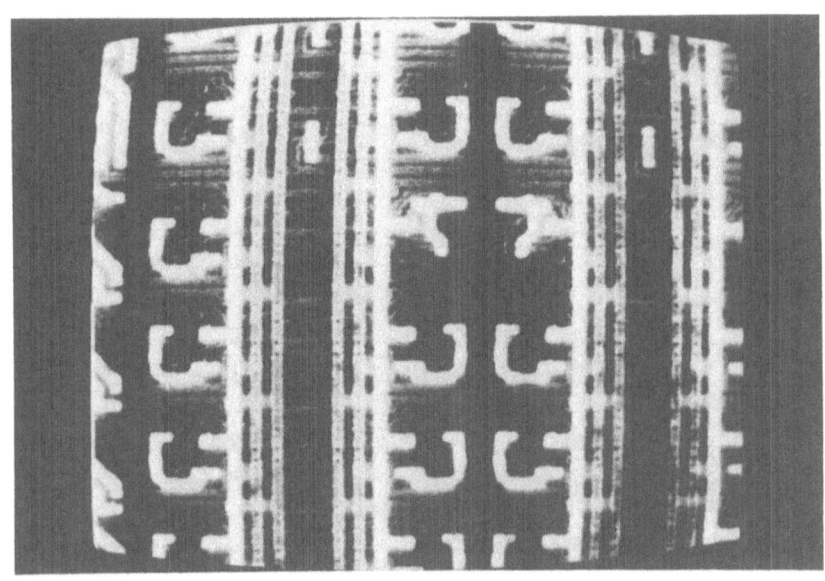

PICTURE XXII: Circuit n° 2 Input vector n°3

PICTURE XXIII: Comparison of equalized images
            (circuit n° 1, input vector n° 1; circuit n° 2, input vector n° 3)

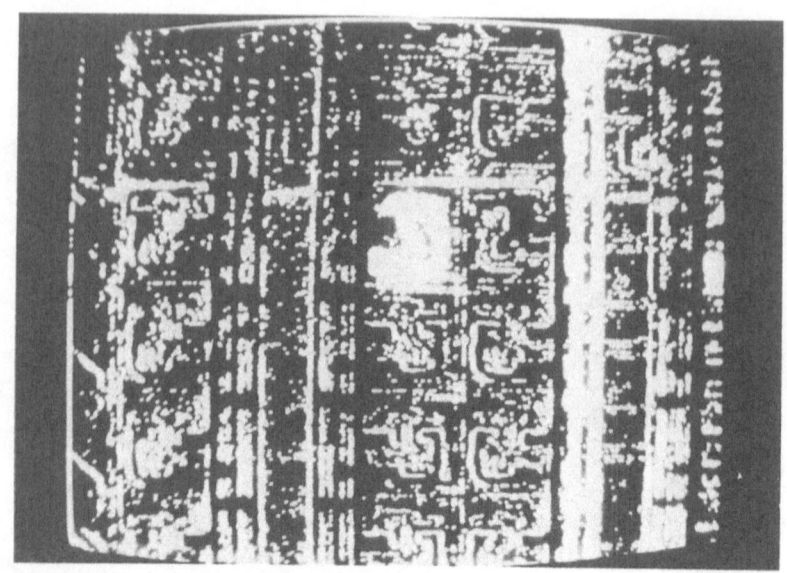

PICTURE XXIV: Binarization of image n° 8

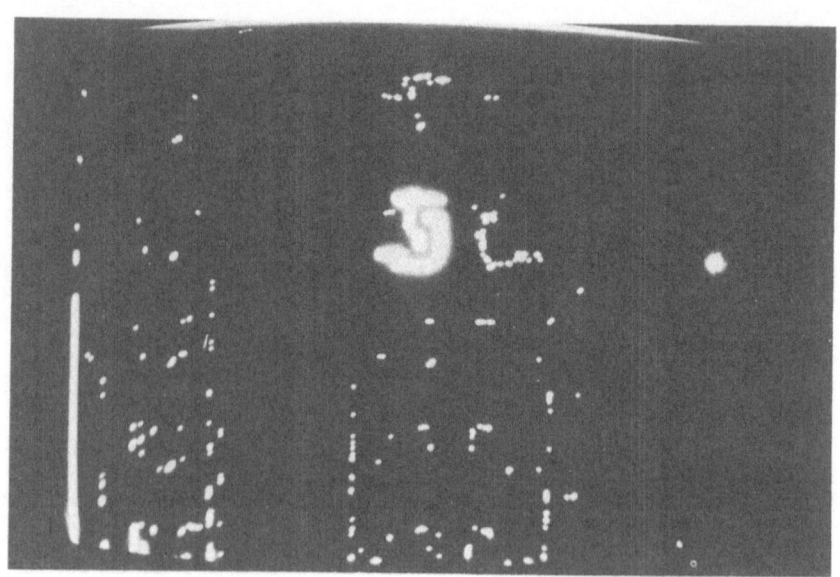

PICTURE XXV: Intersection of binarized images

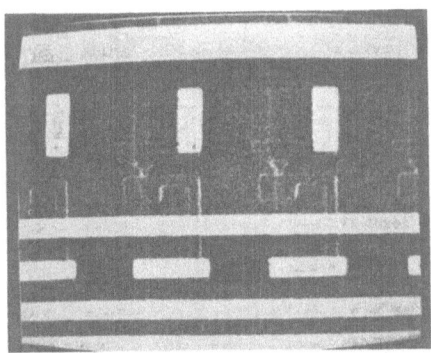

PICTURE XXVI:
Connections outline

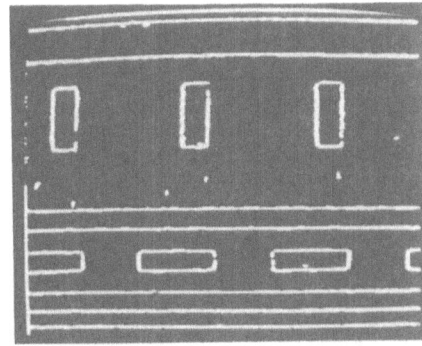

PICTURE XXVII:
Grey levels image

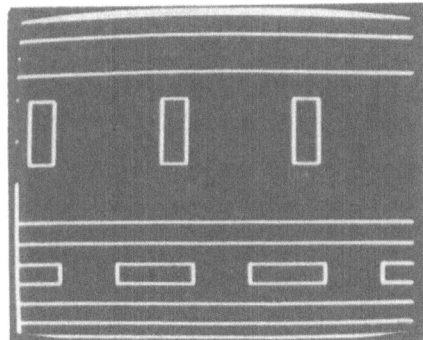

PICTURE XXVIII:
Metal masks

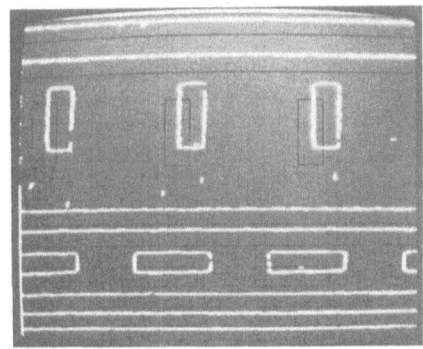

PICTURE XXIX:
X and Y discard

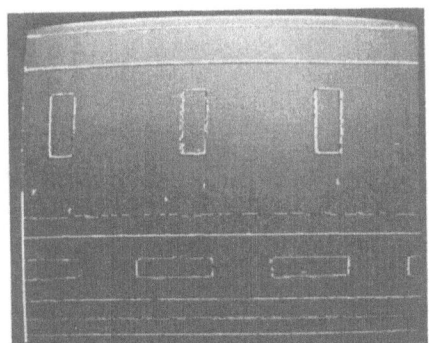

PICTURE XXX:
Correlation result

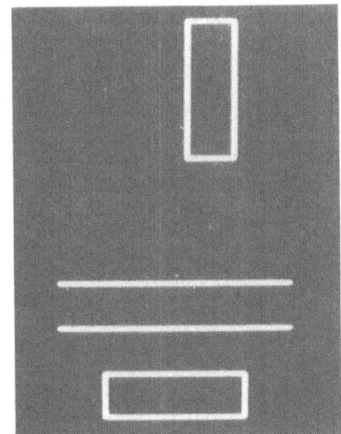

PICTURE XXXI
Correlation window

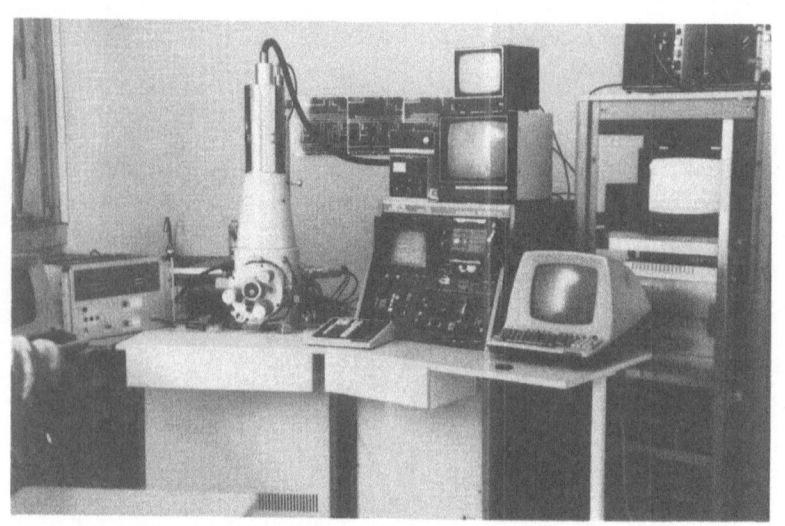

PICTURE XXXII IMAG/TIM3 SYSTEM

D.W.Ranasinghe
British Telecom Res.Lab. plc
Martlesham Heath
Ipswich.IP5 7RE.
Suffolk
U.K.

INTRODUCTION:

The steerable electron beam probe has been an ideal tool for the testing of internal nodes in VLSI circuits for many years(1,2). However, in the past as shown in Figure 1,

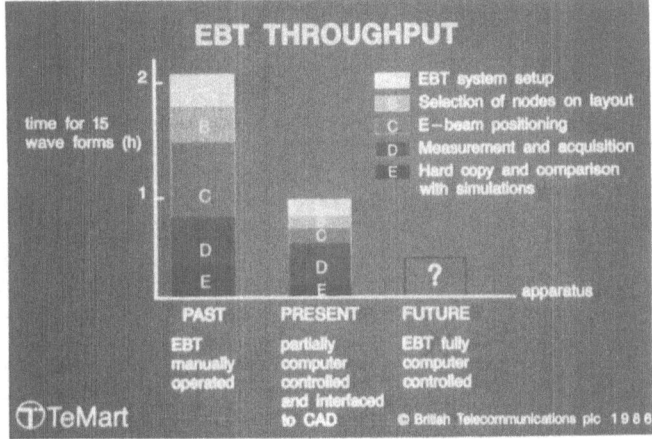

FIGURE 1.

the time taken to extract waveforms from the internal nodes in the circuit could take around two hours to acquire fifteen waveforms. But with the arrival of commercial e-beam test systems, a list of which is shown in Figure 2,

COMMERCIAL E-BEAM TEST SYSTEMS

FIGURE 2.

* AKASHI (Japan) Ltd

* ABT GmbH
* AMRAY-ICT (USA)

* CAMBRIDGE INSTRUMENTS Ltd

* ICT GmbH
* JEOL (Japan) Ltd

* JEOL-ICT (Europe) Ltd

* PHILIPS-ICT (Europe & USA)

* LINTECH INSTRUMENTS Ltd

* SENTRY-SCHLUMBERGER (USA)

F. Lombardi and M. Sami (eds.), Testing and Diagnosis of VLSI and ULSI, 509–526.
© 1988 by Kluwer Academic Publishers.

a number of improvements have been introduced eg.e-beam system setup, measurement and acquisition of waveforms from the circuit.

These systems are partially computer controlled to achieve the time saving. However the selection of nodes from the layout, e-beam positioning and comparison of the acquired waveforms with the simulation results are still dependent on the circuit designer. If a large number of nodal voltages are to be acquired, node selection and e-beam positioning could take several hours. In addition, manual comparison of the acquired results with simulations is likely to be unreliable as well as time consuming.

As the circuit complexity increases, the number of nodes which need to be accessed can be exceedingly large(3). Hence, the design validation times will become prohibitive unless more efficient methods are adopted. Five key areas in which time saving has yet to be made as seen in Figure 1 are (A)"EBT system setup",(B)"node selection",(C)"beam positioning",(D)"measurement and acquisition"and (E)"comparison". Waveform display in commercial e-beam test systems is an output to a plotter and the time required to extract and compare waveforms with simulations is very long.

Hence considerable improvements are required in the waveform measurement and display capabilities in order to meet the needs of the test engineer who would normally use an oscilloscope or a logic analyser. Therefore, logic analysis to include synchronous acquisition of waveforms is essential. Since, all commercial e-beam systems use boxcar signal averagers or digital storage oscilloscopes which are asynchronous and are calibrated in time rather than by external clock pulses(3).

In VLSI testing the number of test vectors required increases with its complexity. As a result the data repetition rate decreases proportionally and it becomes important to acquire as much data as possible from the circuit per repetition of the signal in order to keep the test time down. This is especially important in e-beam testing, since only one probe is available for testing the whole circuit.

Figure 3 shows how a reduction in measurement time can be achieved by comparing two techniques used for signal averaging in commercial e-beam systems, with a dedicated high speed signal averager(3).

time in seconds for 256 samples, 1000 time points				
boxcar	7D20	averager	time/div	rep. rate
25,600	2,240	# 2,560	100 ns	100 ms
25,600	130	256	1.0 μs	100 ms
25,600	25.6	25.6	10.0 μs	100 ms
25,600	25.6	25.6	100.0 μs	100 ms
256	54	25.6	100.0 ns	1 ms
256	28	2.56	1.0 μs	1 ms
** 256	20	0.256	10.0 μs	1 ms
256	20	0.256	100.0 μs	1 ms
2.56	40	0.256	100.0 ns	10 μs
2.56	28	0.0256	1.0 μs	10 μs
2.56	20	0.00256	10.0 μs	10 μs
5	2	10	computer access time (secs)	
boxcar = rep. rate • no. of samples • no. of points 7D20 = rep. rate • no. of samples • processing time dedicated fast averager = rep. rate • no. of samples # greater than 10 MHz sampling rate ** typical times for external clock				

FIGURE 3.

Signal averagers "boxcar"or Tektronics 7D20 digitising oscilloscope have
long averaging time when the repetition rate is low eg.100mS repetition
rate would require 25,600 seconds in a boxcar averager and 2,240 seconds in
a 7D20 oscilloscope. For typical external clock rates of VLSI circuits
eg.1mS the two commercial signal averagers have long acquisition time,
Figure 3 in comparison to the dedicated fast signal averager.

BACKGROUND:
For some time it has been possible to test complex circuitry on printed
wiring boards using a computer-guided probe technique. Access to internal
circuitry is by a probe which is positioned manually under the direction of
a computer probing algorithm(4)as seen in Figure 4.

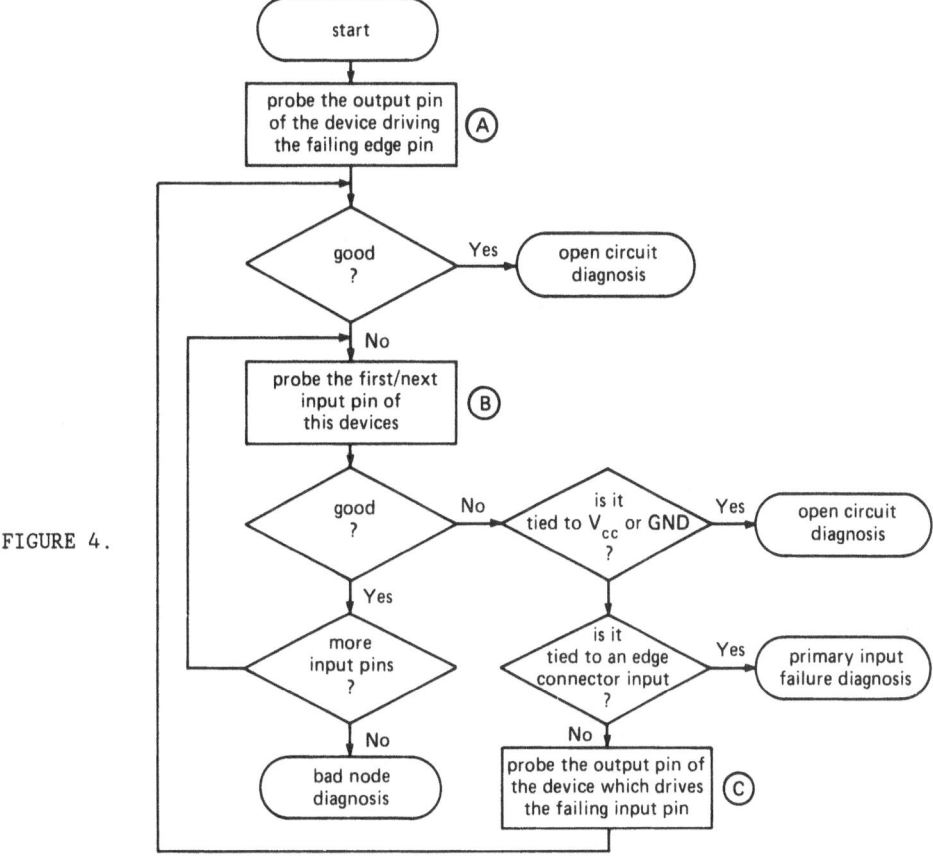

FIGURE 4.

The actual points selected by the algorithm for probing are keyed back to
the flowchart. In order for the probe to automatically select the points,
to be probed by the operator, it must have a knowledge of the layout of the
circuit board. This circuit topology database describes the interconnection
of all devices in the circuit and is usually a by-product of circuit
simulation stored for efficiency in a compact form.

A diagram of a simplified e-beam test system (5,6) is shown in Figure 5.

automatic E – beam test system

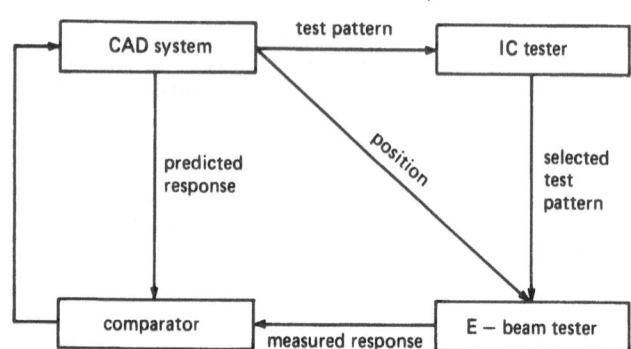

FIGURE 5.

The design of most modern VLSI circuits is computer assisted, with circuit models available for logic simulation and automatic test pattern generation. Linking a circuit model to a file of test point coordinates forms the basis of an automatic system for design checking(7,8,9). Comparison of simulated nodal responses to an input stimuli of the real circuit and that of the logic pattern generator stimuli, the position of first occurrence of faults can be determined. Since the dimensions of an e-beam is much smaller than a mechanical probe, it is more suitable for data extraction with the added advantage of being able to direct it automatically.

Although design for testability using selftest or scan path techniques are steps towards automatic testing. Design verification of VLSI circuits containing thousands of transistors requires access to internal circuit nodes to determine signal levels and timing information. Figure 6 is a typical example of eight analogue waveforms measured using the e-beam probe from a custom designed circuit (vertical axis: node name, voltage and horizontal axis: time in uS).

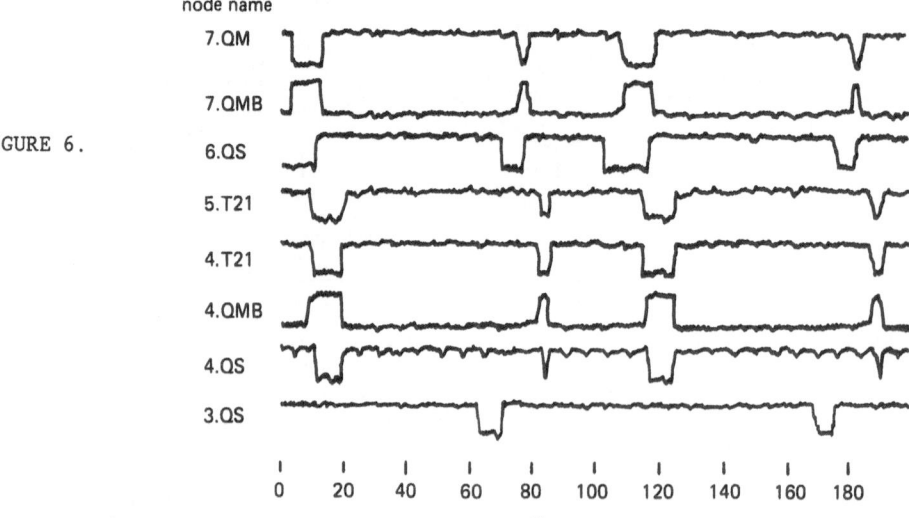

FIGURE 6.

By comparison with a simulation of the same nodal voltages it is possible to deduce faults within the circuit. However, this task is extremely demanding on the time spent by a circuit designer and an SEM operator to search for the required nodes with a layout map and the subsequent comparisons.

AUTOMATIC DESIGN VALIDATION OF INTEGRATED CIRCUITS USING ELECTRON BEAM

"ADVICE"

The ESPRIT-ADVICE is a collaborative project between CSELT (Italy), BTRL (U.K), CNET (France), IMAG (France) and TCDU (Ireland) under the sponsership of the EEC and is aimed at addressing the problems in relation to e-beam testing of integrated circuits. The project objective is to develop an automatic facility for VLSI circuit debug and failure analysis based on the electron beam tester connected to the CAD system. A task list of the project is shown below(10,11,12).

1. Computer control of the electron beam system

2. Interfacing to CAD software

3. Identification of circuit elements

4. Methodology for design error diagnosis

5. Test pattern generation for electron beam debugging

6. Design for electron beam testability

7. Electron beam equipment development

These tasks are tackled by the expertise available from each partner in collaboration. Figure 7 shows the block diagram of a computer controlled electron beam test system (1) and the computer system electronics is shown in Figure 8.

FIGURE 7

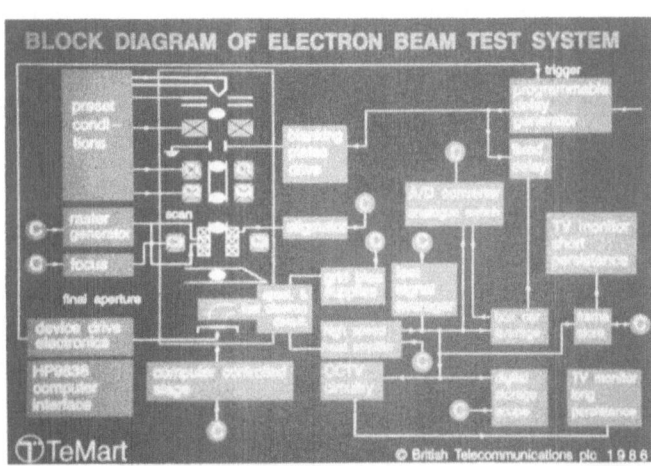

```
┌─────────────────────────────┐
│ COMPUTER CONTROLLED         │
│ SYSTEM ELECTRONICS          │
│   3 MByte RAM               │
│  70 MByte Winchester        │
│  Menu Driven System         │
│  15 Function Keys           │
│ SOFTWARE PACKAGE            │
│ Signal Processing           │
│ Measurement Control         │
│ Waveform Storage            │
│ Waveform Processing         │
└─────────────────────────────┘
```

FIGURE 8.

Typical electron optical parameters eg. focus, stigmator, rotation, dynamic focus maintains an adequate beam integrity for waveform extraction thus reducing the block A Figure 1. The motor driven stage for coarse positioning in conjunction with the pattern recognition technique for precise e-beam placement is aimed at saving considerable time and effort in block C Figure 1. The computer controlled functions also include signal processing Figure 9 (box-car, Tektronics 7D20 and fast signal averager) and waveform presentation Figure 10 (logic analyser display) aimed at reducing block D Figure 1.

COMPUTER CONTROL	COMPUTER CONTROL
Electron Optics	Waveform Presentation
* X & Y movement of TV raster	* Waveform scrolling
* Zoom control of TV raster	* Waveform expansion
* Rotation of TV raster	* Waveform rearrangement
* Focus adjustment of TV raster	* Waveform state display
* Dynamic focus control	* Scrolling of waveform "state
* Test point location	display"
Signal Processing	* Jump to address
* Waveform acquisition using Box-Car signal averager	* Boolean simulation
* Waveform acquisition using Tektronix 7D20 oscilloscope	Video Frame Store
* Waveform acquisition using fast signal averager	* Frame acquisition with low-filter
	* Frame acquisition with high-filter
	* Frame comparison
FIGURE 9.	FIGURE 10.

Application of pattern recognition(13)to achieve precise e-beam positioning is essentially a task of recognising shapes in the SEM images and correlating them with CAD data taken from the design environment. This requires image transformations as shown in Figure 11.

Image Transformations

* Shear
* Thresholding
* Filtering

FIGURE 11.

      1. Average

      2. Median

Pattern Recognition

*Template Matching
* Edge Extraction

The poor quality images from the SEM are captured by noise filtering into a frame store. This image is then subjected to shear corrections and is thresholded by choosing sixteen grey levels between peak white and peak black and is sbsequently filtered before it is used in template matching. Template matching is one of the techniques used for image matching. The second technique is edge extraction. This technique requires less computing time and has been chosen as the most viable of the two techniques.

Image registration using extracted lines is chosen because the image contains a high proportion of paraxial edges. This simplifies the line detection process and avoids the extra computation associated with the handling of angled lines. The edge extraction algorithm has been found eminently suited to processing of VLSI images as they rarely contain many non-paraxial shapes.

Figure 12 shows two key areas in which the CAD tools are addressed in the project(14).

FIGURE 12.

CAD Connections and Tools

* Interfacing to CAD system

* Automatic Test Pattern Generation

(ATPG)

516

For complex integrated circuits designed on CAD systems eg. ASTRA at British Telecom Research Laboratories (BTRL), is a full custom design system engineered to meet the needs of VLSI designers. In this system hierarchical floor plans and symbolic cell layouts capture the design data, which is entered and edited via interactive graphical editors.

The design verification of custom and semi-custom circuits, at BTRL is performed using yet another tool eg. HITEST. The information stored in ASTRA is not relevant to the exploration of the key features of HITEST eg. accurate simulation of MOS circuits, Waveform display in a logic-analyser format, highlevel waveform language to enable compact specification of lengthy test sequences.

It is the aim of this task to bring together ASTRA and HITEST to perform two tasks, one is to guide the e-beam to the required cells using the design data from ASTRA and the second is to generate test patterns automatically with HITEST for electron beam testing. This would eliminate the extremely time consuming task of searching through layout maps and execute tight programme loops which would otherwise be required in e-beam testing and has the consequences of excessive local heating and irradiation effects on the circuit under test(15).

Figure 13 highlights one aspect of the project developed during its first two years.

FIGURE 13.

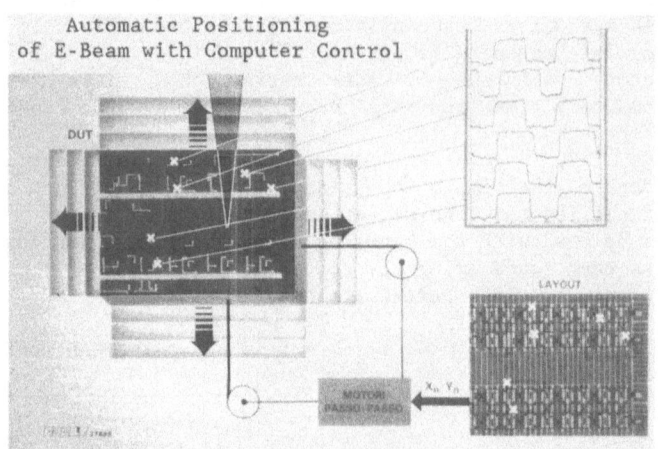

The circuit layout is presented on the bottom right window, the nodes are marked manually with a cross. X,Y coordinates extracted from the CAD data are presented to the stepper motors for coarse positioning the e-beam on the selected nodes of the circuit. The nodal waveforms are shown on the right top window of the workstation.

Figure 14 shows the key areas selected for the methodology of electron beam debugging(16).

FIGURE 14.

> Methodology for Assisted Electron Beam Debugging
>
> * strategy fo VLSI design validation
>
> * modelling design error
>
> * interfacing to automatic test equipment (ATE)
>
> * design for electron beam testing

The progress of tasks 1, 2, 3 and 7 of the task list outlined before has provided practical tools for real implementation of an industrially usable automatic electron beam test system. To complete the requirements of full automation of the e-beam test system several strategic points has been investigated. Research activity into an approach based on an expert system and one using a fault dictionary in combination with a probing algorithm has selected the latter approach as the most viable technique in the short term. Modelling design errors based on stuck-at faults may be implemented on commercial software tools and an approach similar to PCB testers will be adopted in the task. A probing algorithm which acts as the central core of the automated e-beam system, intends to provide the user to find faults in VLSI circuits without resorting to guess work. This is estimated to be the most cost effective approach to debugging VLSI circuits.

Integrated circuit design engineers have not as yet been fully aware of the test engineers requirements Although probe pads are sometimes provided to access the internal logic cells during a debugging session on a mechanical probe station. The local electric fields produced by the circuit voltages as shown in Figure 15

FIGURE 15.

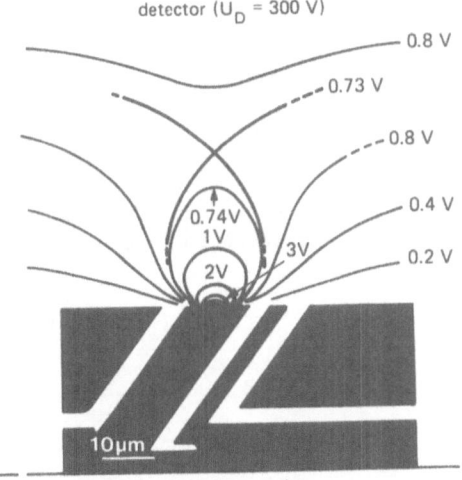

electrostatic fields at the IC — surface

may occasionally introduce cross-talk effects (or local field effects) on the acquired waveforms with an e-beam system as shown in Figure 16.

FIGURE 16.

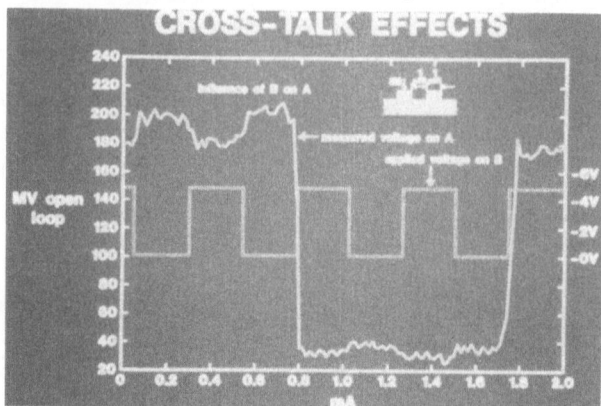

This may inevitably introduce unwanted glitches into a timing diagram which is obtained by thresholding the waveforms. The task of circuit design for e-beam testing is essential to obtain waveform integrity. Many simple solutions to this problem have been researched and the results proved adequate for this task eg. introduction of a field plate during circuit design under an address bus (connected to earth potential) has the effect of reducing cross-talk effect on the acquired waveforms(17).

Providing the stimuli to the device under test (DUT) with automatic test equipment (ATE) and collecting the responses at primary outputs is to be initiated from the workstation which controls the e-beam test system. The implementation of a comparator for measured device response with circuit simulations gives a significant reduction in time for fault detection process, thus reducing the size of block E in Figure 1. A flow chart(18,19) illustrating the operating steps of the automatic fault diagnostic procedure, is shown in Figure 17.

FIGURE 17.

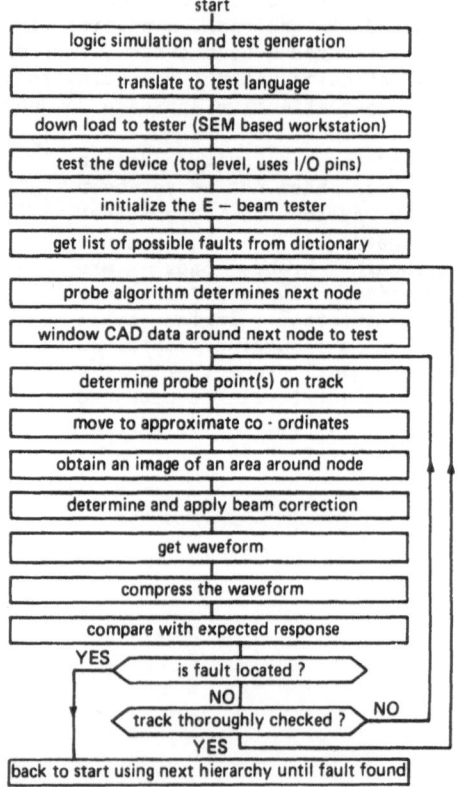

The block schematic for the complete "ADVICE" VLSI test system(10)is shown
in Figure 18.

ADVICE

automatic E − beam test system

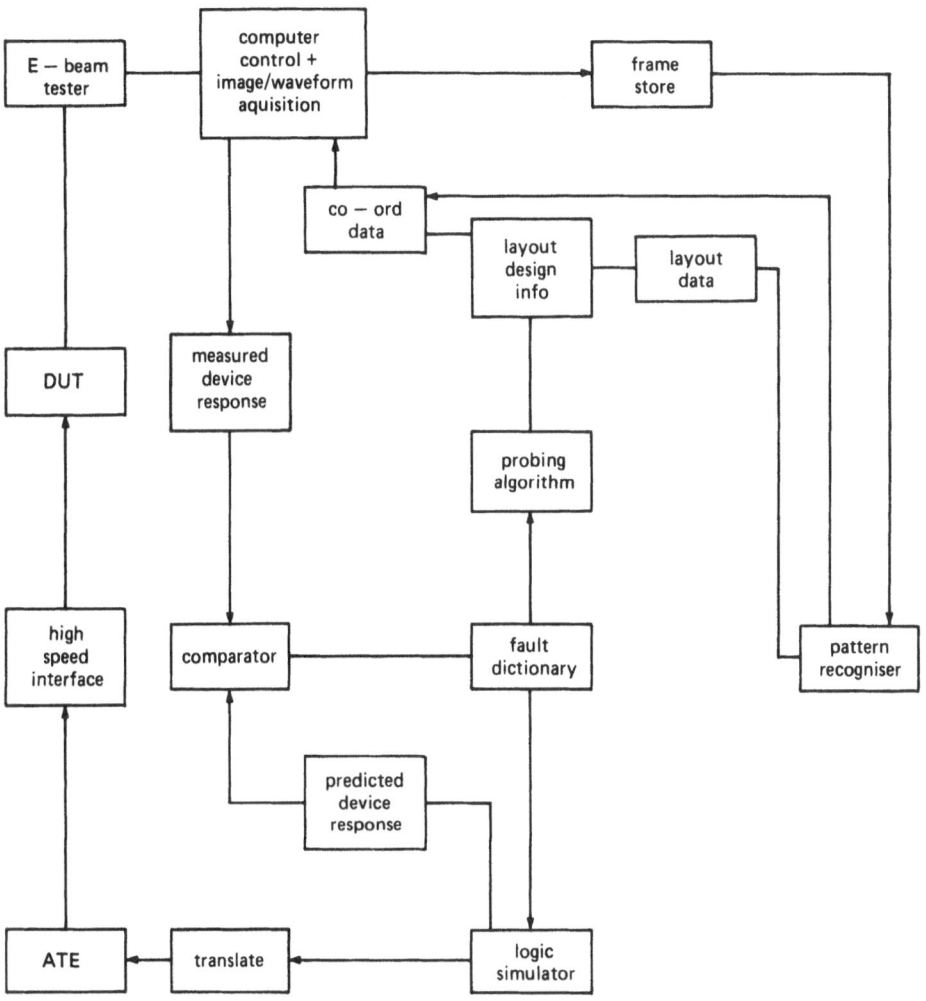

FIGURE 18.

The blocks labelled "co-ord data","probing algorithm","layout data", "fault
dictionary" and "comparator" in Figure 18 are software tools resident in
the workstation. A hardware based pattern recogniser is to be implemented
with its control algorithm executed from the workstation. The device under
test (DUT) receiving the stimuli from the automatic test equipment (ATE),
which is local to the e-beam test system and is controlled from the
workstation for fast signal acquisition and comparison.

The target hardware system for "ADVICE" (10) is shown in Figure 19.

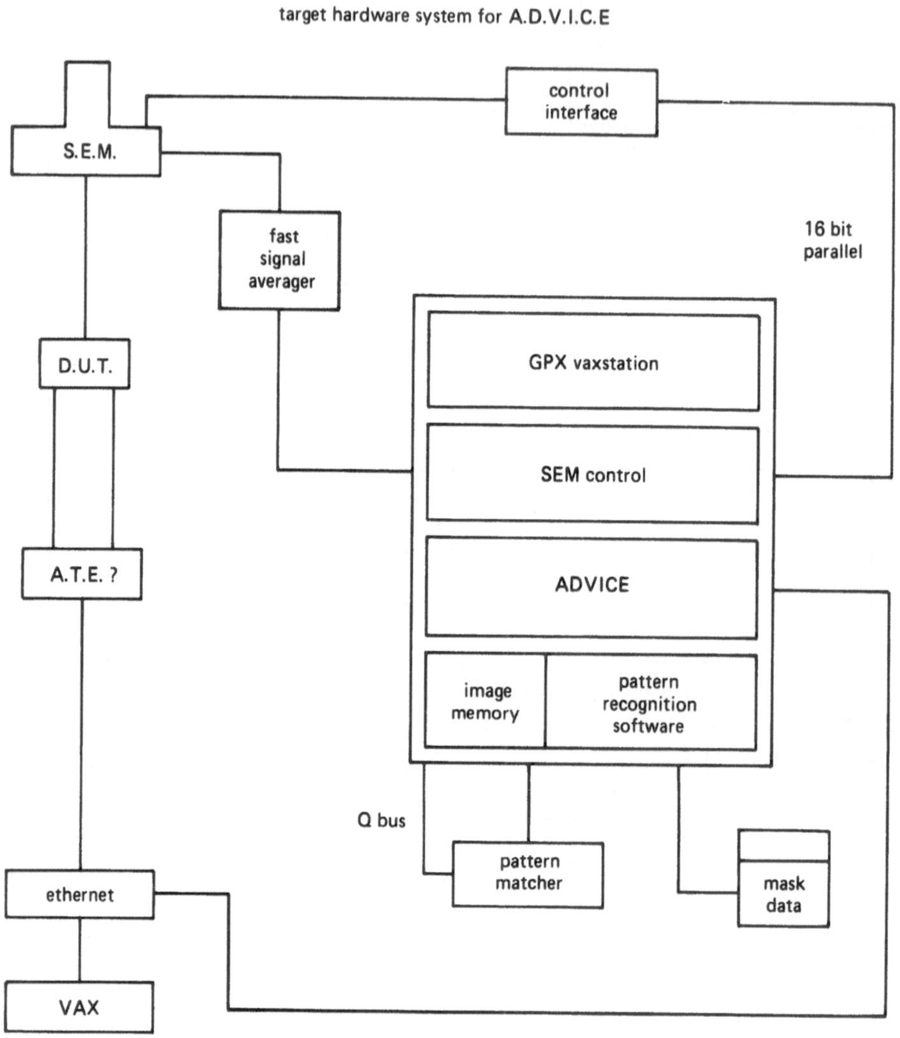

target hardware system for A.D.V.I.C.E

FIGURE 19.

A GPX VAX station containing the "ADVICE" shell (a stack of menus) and the control programs are connected to a host computer via the ethernet. The host computer holds the circuit design and layout data. A hardware based, fast signal averager and an edge extractor for pattern matcher are linked to the "ADVICE" system. A 16 bit parallel bus connects the e-beam tester hardware via a control interface to the VAX station. Figure 20 shows the system "setup" menus eg. (a) operator selectable parameters of system, (b) electron optics control and (c) waveform acquisition files.

521

(a)

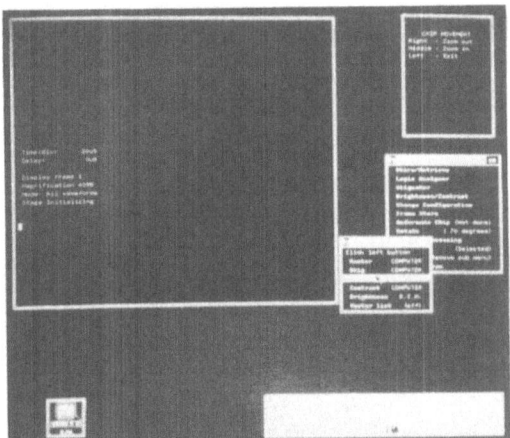

FIGURE 20.

(b)

(c)

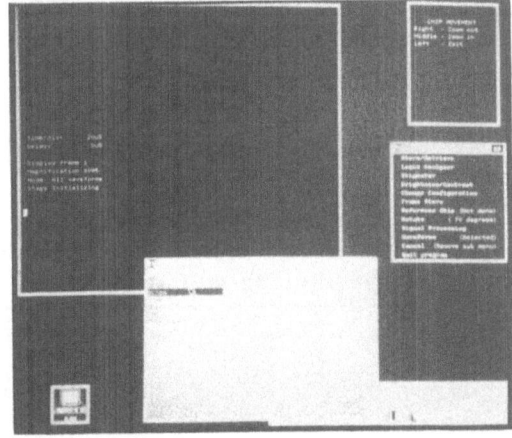

522

Figure 21 shows the waveform display acquired from several nodes on a circuit eg. (a) real time waveform display from a number of nodes, (b) timing diagram and (c) logic analyser display.

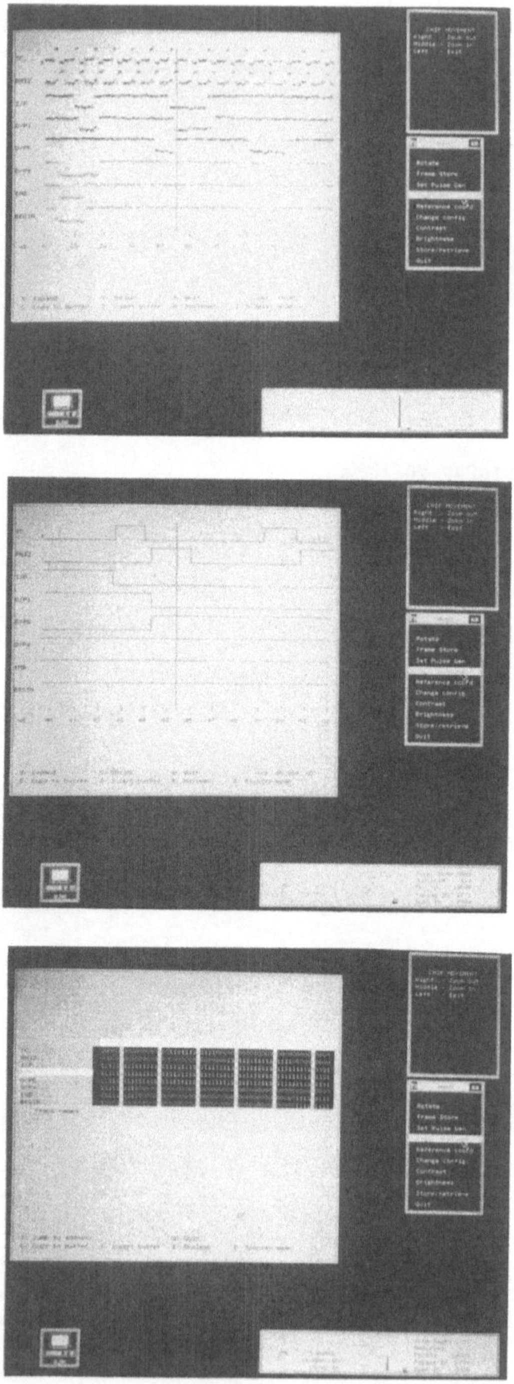

(a)

(b)

FIGURE 21.

(c)

Figure 22 shows (a) asynchronous (synchronised to an external pulse) and (b) synchronous (synchronised to an external clock) acquisition of waveforms from a circuit.

(a)

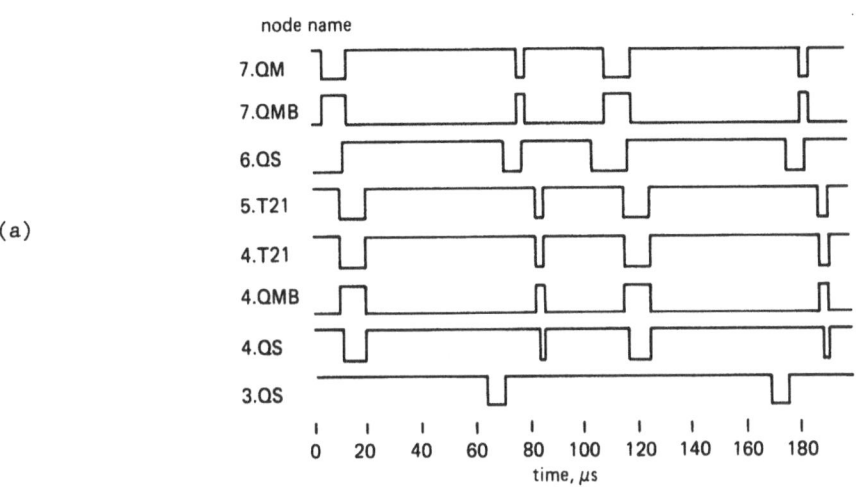

FIGURE 22.

(b)

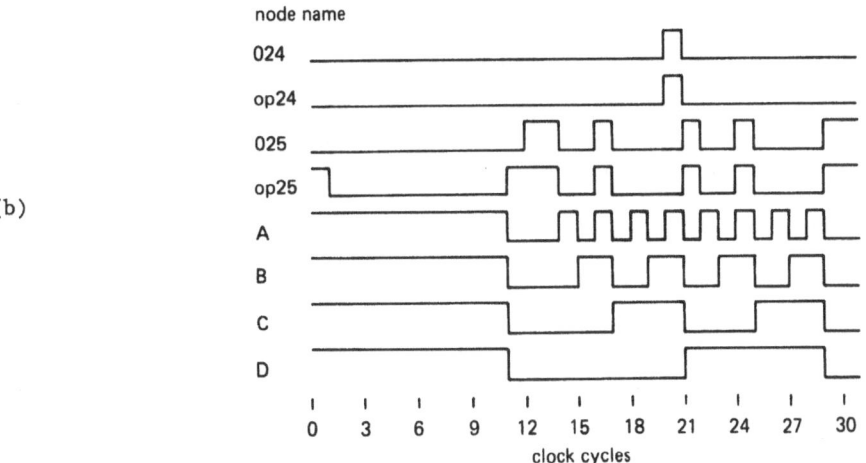

SUMMARY OF "ADVICE" PROJECT

A 50 manyear project costing 4 MECU (Million European Currency Units) has completed its 2.5 year period. The project has achieved its targets well in advance of time. A list of achievements during the 2.5 year period is shown below.

* Fully Computer Controlled E-Beam Tester (Electron Optics)

* A High Speed Wide-Band Signal Detector (Real Time Signal Acquisition)

* Waveform Acquisition with a High Speed Signal Averager (synchronous and asynchronous)

* Image Processing with Frame Store (Filtering, Comparison and Subtraction)

* Capturing Layout Data from CAD

* Pattern Recognition Algorithms

* Node Name to X,Y Translation (Coordinate Extraction from Design CAD)

* Translation of Logic Simulations to Automatic Test Equipment(ATE)

* Coarse E-Beam Placement with CAD Data under Stepper Motor Control

* A Target system Hardware Configuration and its Implementation

* Automatic Comparator for Measured and Predicted Device Response

* Evaluation of Commercial E-Beam Systems (suitability for integration into "ADVICE" system and its application to sub-micron technologies)

The work in progress for the next 2.5 year period is shown below.

* Implementation of the Comparator

* Generation of a Probing Algorithm

* Generation of a Fault Dictionary

* Fully Automatic E-Beam Placement with CAD

* System Integration

The system integration is due for completion in December 1989. The "ADVICE" diagnostic system for VLSI design validation is expected to be a commercially viable tool.

REFERENCES

1. Computer Control of Electron Beam Testing for Design Validation of
   VLSI Circuits. D.W.Ranasinghe, G.Proctor, M.Cocito and G.Bestente
   Proc. 11 th Int.Cong. on Electron Microscopy, Kyoto, Japan, 1986

2. Dynamic Testing of a Passivated Device with Electron Beam Tester.
   M.Ekuni, Y.Komoto, T.Gobara and Y.Harada. Proc. 11 th Int.Cong.
   on Electron Microscopy, Kyoto, Japan, 1986

3. A High Speed Signal Averager for Electron Beam Test Systems.
   D.J.Machin, D.W.Ranasinghe and G.Proctor, Proc. of 1st European
   Conf. on Electron and Optical Beam Testing of Integrated Circuits,
   Grenoble, France, 1987 (to be published)

4. Computer-Guided Probing Techniques. S.Kochan, N.Landis and
   D.Monson. 1981 IEEE Test Conference.

5. Software Integration in a Workstation-Based E-Beam Tester.
   S.Concina, G.Lieu, L.Lattanzi, S.Reyfman and N.Richardson.
   Proc.Int.Test Conf. 1986

6. A New Tool Dramatically Cuts VLSI Debugging Time.
   Electronics/April 30, 1987

7. "Finder" A CAD System-Based Electron Beam Tester for Fault
   Diagnosis of VLSI Circuits. N.Kuji, T.Tamama and M.Nagatani.
   IEEE, CAD-5, 1986

8. Integrating an Electron-Beam System into VLSI Fault Diagnosis.
   T.Tamama and N.Kuji. IEEE Design & Test Conf. 1986

9. Integration of CAD, CAT and Electron Beam Testing for IC-Internal
   Logic Verification. S.Gorlich, H.Harbeck, P.Kessler, E.Wolfgang
   and K.Zibert. Corporate R & D, Siemans AG, Munich, FDR.

10. European Effort ADVICE. M.Cocito, Proc. of 1st European
    Conf. on Electron and Optical Beam Testing of Integrated Circuits,
    Grenoble, France, 1987 (to be published)

11. A SEM-Based Workstation for Design Validation.
    Y.J.Vernay, R.Mignone and P.Rivoire. Proc. ESSCIRC 1986

12. Fully Automatic VLSI Diagnosis in a CAD-Linked E-Beam Probing
    System. M.Melgara, M.Battu, P.Garino, J.Dowe and M.Marzouki,
    Proc. of 1st European Conf. on Electron and Optical Beam
    Testing of Integrated Circuits, Grenoble, France, 1987 (to be
    published)

13. Automatic Registration of Scanning Electron Microscope Images.
    F.W.M.Stentiford and T.J.Twell, Proc. of 1st European Conf. on
    Electron and Optical Beam Testing of Integrated Circuits,
    Grenoble, France, 1987 (to be published)

14. Electron Beam Tester Linked with a CAD Pattern Data.
    F.Komatsu, M.Miyoshi, T.Sano, K.Sekiwa and K.Okumura
    Proc. 11 th Int.Cong. on Electron Microscopy, Kyoto, Japan, 1986

15. Electron Beam Irradiation Effects on MOS-Transistors and its Significance to E-Beam Testing. D.W.Ranasinghe, D.J.Machin and G.Proctor, Proc. of 1st European Conf. on Electron and Optical Beam Testing of Integrated Circuits, Grenoble, France, 1987 (to be published)

16. Methodology for the use of an Electron Beam Tester for Integrated Circuits. B.Courtois. Proc.Int.Symp.on Circuits and Systems. 1984

17. Design for E-Beam Testability a Demand for the Testing of Future Device Generations ? K.D.Hermann and E.Kubalek, Proc. of 1st European Conf. on Electron and Optical Beam Testing of Integrated Circuits, Grenoble, France, 1987 (to be published)

18. Automatic Failure Analysis of VLSI Circuits using E-Beam. D.Savart and B.Courtois, Proc. of 1st European Conf. on Electron and Optical Beam Testing of Integrated Circuits, Grenoble, France, 1987 (to be published)

19. An Integrated Debugging System Based on E-Beam Test. I.Guiguet, M.Marzouki and B.Courtois, Proc. of 1st European Conf. on Electron and Optical Beam Testing of Integrated Circuits, Grenoble, France, 1987 (to be published)

Subject Index

530